2011
Senior
Biology 1
Student Workbook

🦎 BIOZONE

Senior *Biology 1* 2011
Student Workbook

Previous annual editions 2002-2009
Ninth Edition 2011

THIRD PRINTING with corrections

ISBN 978-1-877462-59-7

Copyright © **2010** Richard Allan
Published by **BIOZONE International Ltd**

Printed by REPLIKA PRESS PVT LTD using paper
produced from renewable and waste materials

About the Writing Team

Tracey Greenwood joined the staff of Biozone at the beginning of 1993. She has a Ph.D in biology, specializing in lake ecology, and taught undergraduate and graduate biology at the University of Waikato for four years.

Kent Pryor has a BSc from Massey University majoring in zoology and ecology. He was a secondary school teacher in biology and chemistry for 9 years before joining Biozone as an author in 2009.

Richard Allan has had 11 years experience teaching senior biology at Hillcrest High School in Hamilton, New Zealand. He attained a Masters degree in biology at Waikato University, New Zealand.

Purchases of this workbook may be made direct from the publisher:

 **BIO**ZONE

www.thebiozone.com

NORTH & SOUTH AMERICA, AFRICA:

BIOZONE International Ltd.
P.O. Box 13-034, Hamilton 3251, **New Zealand**
Telephone: +64 7-856-8104
FREE Fax: 1-800-717-8751 (USA-Canada)
FAX: +64 7-856-9243
E-mail: sales@biozone.co.nz

UNITED KINGDOM:

BIOZONE Learning Media (UK) Ltd.
Bretby Business Park, Ashby Road, Bretby,
Burton upon Trent, DE15 0YZ, **UK**
Telephone: +44 1283-553-257
FAX: +44 1283-553-258
E-mail: sales@biozone.co.uk

ASIA & AUSTRALIA:

BIOZONE Learning Media Australia
P.O. Box 2841, Burleigh BC,
QLD 4220, **Australia**
Telephone: +61 7-5535-4896
FAX: +61 7-5508-2432
E-mail: sales@biozone.com.au

Preface to the 2011 Edition

This is the ninth edition of Biozone's **Student Workbook** for students in biology programs at grades 11 and 12 or equivalent. This title, and its companion volume, are particularly well suited to students taking International Baccalaureate (IB) Biology, Advanced Placement (AP) Biology, or Honors Biology. Biozone's aim with this release is to build on the successful features of previous editions, while specifically focusing on scientific literacy and learning within relevant contexts. The workbook is a substantial revision and marks an important shift in several respects from earlier versions of this product.

▶ Content reorganization. In this edition we have brought together the content for molecular genetics, inheritance, and evolution into one volume, and shifted classification and ecology into Senior Biology 2. This reorganization of content follows the schedule of the AP biology program and provides a more cohesive coverage of material in these areas. Extension material is provided on the Teacher Resource CD-ROM (for separate purchase).

▶ International Baccalaureate students and teachers will find that, although core content has shifted between volumes (Senior Biology 1 and 2), it should be easier to locate material. Activities suitable for HL-only are indicated in the 'Contents', as is material that is not required under the IB scheme (there is a limited amount of this). **IB Options (A-H) are provided as complete units on the IB Options CD-ROM (for separate purchase). Options C-E are also adequately covered within the workbooks for those making those option choices.**

▶ A contextual approach. We encourage students to become thinkers through the application of their knowledge in appropriate contexts. Many chapters are prefaced with an account examining a 'biological story' related to the theme of the chapter. This approach provides a context for the material to follow and an opportunity to focus on comprehension and the synthesis of ideas.

▶ Concept maps introduce each main part of the workbook, integrating the content across chapters to encourage linking of ideas.

▶ A lead by example approach to teaching techniques and applications in biotechnology. Gene technology can be a difficult area for students so, this year, each application is treated using the same systematic, explanatory approach. This should help students to identify techniques and place them within the appropriate context.

▶ An easy-to-use chapter introduction comprising brief learning objectives, a list of key terms, and a short summary of key concepts.

▶ An emphasis on acquiring skills in scientific literacy. Each chapter includes a comprehension and/or literacy activity, and the appendix (a new feature) includes references for works cited throughout the text.

▶ Web links and Related Activities support the material provided on each activity page.

A Note to the Teacher

This workbook is a student-centered resource, and benefits students by facilitating independent learning and critical thinking. This workbook is just that; a place for your answers notes, asides, and corrections. It is **not a textbook** and annual revisions are our commitment to providing a current, flexible, and engaging resource. The low price is a reflection of this commitment. Please **do not photocopy** the activities. If you think it is worth using, then we recommend that the students themselves own this resource and keep it for their own use. I thank you for your support.
Richard Allan

Acknowledgements

We would like to thank those who have contributed to this edition:
• Ben Lowe, University of Minnesota for advice and input on *Ensatina* subspecies and *Canis* distribution • Stacey Farmer and Greg Baillie, Waikato DNA Sequencing Facility, University of Waikato, for their assistance with material on PCR, DNA sequencing, and genetic profiling • Raewyn Poole, University of Waikato, for information provided in her MSc thesis: Culture and transformation of *Acacia* • Dr. John Stencil for his data on the albino gray squirrel population • Mary McDougall, Sue FitzGerald and Gwen Gilbert for their efficient handling of the office • TechPool Studios, for their clipart collection of human anatomy: Copyright ©1994, TechPool Studios Corp. USA (some of these images were modified by R. Allan and T. Greenwood) • Totem Graphics, for their clipart collection • Corel Corporation, for vector clipart from the Corel MEGAGALLERY collection • 3D artwork created using Poser IV, Curious Labs and Bryce.

Photo Credits

Royalty free images, purchased by Biozone International Ltd, are used throughout this workbook and have been obtained from the following sources: Corel Corporation from various titles in their Professional Photos CD-ROM collection; IMSI (International Microcomputer Software Inc.) images from IMSI's MasterClips® and MasterPhotosTM Collection, 1895 Francisco Blvd. East, San Rafael, CA 94901-5506, USA; ©1996 **Digital Stock**, Medicine and Health Care collection; ©**Hemera** Technologies Inc, 1997-2001; © 2005 JupiterImages Corporation www.clipart.com; ©1994., ©**Digital Vision**; Gazelle Technologies Inc.; ©**istockphotos** (www.istockphoto.com); **PhotoDisc®**, inc. USA, www.photodisc.com

The writing team would like to thank the following individuals and institutions who kindly provided photographs: • Charles W. Brown for the photographs of the 7 subspecies of *Ensatina* • Dena Borchardt at HGSI for photos of large scale DNA sequencing • Campus Photography at the Uni. of Waikato (NZ) for photographs of monitoring equipment • Dept. of Natural Resources, Illinois, for the photograph of the threatened prairie chicken • Dartmouth College for TEMs of cell structures • Genesis Research & Development Corp. Auckland (NZ), for the photo used on the HGP activity • Wadsworth Centre (NYSDH) for the photo of the cell undergoing cytokinesis,• Missouri Botanical Gardens for their photograph of egg mimicry in *Passiflora* • Alex Wild for his photograph of swollen thorn *Acacia* • The late Ron Lind for his photograph of stromatolites • Marc King for photographs of comb types in poultry • Pharmacia (Aust) Ltd. for providing the photographs of DNA gel sequencing • The Roslin Institute, for their photographs of Dolly • Dr. Nita Scobie, Cytogenetics Department, Waikato Hospital (NZ) for chromosome photographs • Dr. David Wells, AgResearch, NZ, for his photos on livestock cloning, • Alan Sheldon Sheldon's Nature Photography, Wisconsin for the photo of the lizard without its tail • Jeremy Kemp for the tick photos • Ed Uthman for the image of the nine week human embryo • Seotaro for photo of topminnow • Leo Sanchez and Burkhard Budel for use of their photographs in the activities on Antarctic springtails • Greenpeace for photos used for the ethics of gene technology • Adam Luckenbach and the North Carolina State University for use of the poster image on sex determination in flounder • The three-spined stickleback image was originally prepared by Ellen Edmonson as part of the 1927-1940 New York Biological Survey. Permission for use granted by the New York State Department of Environmental Conservation • ms.donna for the photo of the albino child • California Academy of Sciences for the photo of the ground finch • Rita Willaert, Flickr, for the photograph of the Nuba woman • Aptychus, Flickr for use of the photograph of the Tamil girl

We also acknowledge the photographers that have made their images available through **Wikimedia Commons** under Creative Commons Licences 2.5. or 3.0: • Indian Nomad • Cereal Research Centre, AAFC • Georgetown University Hospital • Jacoplane • Velela • UtahCamera • Alan & Elaine Wilson • Onno Zweers • Bruce Marlin • Lorax • Dirk Bayer • AKA • Ian Beatty • Velela • Karl Magnacca • Graham Colm • Andreas Trepte

Contributors identified by coded credits are as follows: **BF**: Brian Finerran (Uni. of Canterbury), **BH**: Brendan Hicks (Uni. of Waikato), **BOB**: Barry O'Brien (Uni. of Waikato), **CDC**: Centers for Disease Control and Prevention, Atlanta, USA, **COD**: Colin O'Donnell, **DNRI**: Dept of Natural Resources, Illinois, **EII**: Education Interactive Imaging, **GW**: Graham Walker, **HGSI**: Human Genome Sciences Inc., **IF**: I. Flux (DoC), **JB-BU**: Jason Biggerstaff, Brandeis University, **JDG**: John Green (Uni. of Waikato), **MPI**: Max Planck Institute for Developmental Biology, Germany; **NASA**: National Aeronautics and Space Administration, **NOAA**: National Oceanic and Atmospheric Administration www.photolib. noaa.gov **RA**: Richard Allan, **RCN**: Ralph Cocklin, **RL**: Ron Lind, **TG**: Tracey Greenwood, **USDA**: US Dept of Agriculture, **WMU**: Waikato Microscope Unit.

Special thanks to all the partners of the Biozone team for their support.

Cover Photographs

Main photograph: The red eyed tree frog (*Agalychnis callidryas*) is a slender, delicate frog found in the neotropical forests of central America. This species is nocturnal and completely arboreal (tree-dwelling). Bright markings along the sides of the body and on the limbs startle and distract predators with a bright flash of color as the frog moves away.
PHOTO: ©PhotoLibrary

Background photograph: Autumn leaves, Image ©2005 JupiterImages Corporation www.clipart.com

Contents

CODES: △   Upgraded   ☆   New activity   * Not for IB-SL   ** Not for IB   † IB Option

CONTENTS (continued)

CODES: Δ   Upgraded   ☆   New activity   * Not for IB-SL   ** Not for IB   † IB Option

CONTENTS (continued)

CODES: △ Upgraded   ☆ New activity   * Not for IB-SL   ** Not for IB   † IB Option

Getting The Most From This Resource

This workbook is designed as a resource to increase your understanding and enjoyment of biology. While this workbook meets the needs of most general biology courses, it also provides specific keyed objectives for the **International Baccalaureate** (IB) and **Advanced Placement** (AP) courses. Consult the Syllabus Guides on pages 6-8 of this workbook to establish where material for your syllabus is covered. It is hoped that this workbook will reinforce and extend the ideas developed by your teacher. It must be emphasized that this workbook is **not a textbook**. It is designed to complement the biology textbooks provided for your course. Each topic in the workbook includes the following useful features:

Features of the Concept Map

Each major section of the workbook has a central theme:
Part 1: Asking questions, finding answers
Part 2: Molecules and cells
Part 3: Heredity and molecular genetics
Part 4: Evolutionary biology
The themes in Senior Biology 1 and 2 also encapsulate the recurring themes (I-VIII) of the AP scheme.

Each section of the workbook emphasizes skills and knowledge to be gained.

Encouraging Key Competencies

Thinking - bringing ideas together
Relating to others - communicating
Using language, symbols, and text
Managing self - independence
Participating and contributing

Chapter panels identify and summarize the material covered within each chapter.

A summary of why this material is important and where it fits into your understanding of your course content.

Features of the Chapter Topic Page

The part of the AP or IB (SL/HL) scheme to which this chapter applies. For other courses, objectives can be assigned at the teacher's discretion.

The important key ideas in this chapter. You should have a thorough understanding of the concepts summarized here.

The page numbers for the activities covering the material in this subsection of objectives.

The objectives provide a point by point summary of what you should have achieved by the end of the chapter. An equivalent set of objectives, for teacher's-only reference, is provided in the Teacher's Handbook. These provide extra explanatory detail and examples.

A list of key terms used in the chapter. These terms appear in the chapter's vocab activity and can be used to create a glossary for revision purposes. The list represents the minimum literacy requirement for the chapter.

Periodicals of interest are identified by title on a tab on the activity page to which they are relevant. The full citation appears in the **Appendix** on the page indicated.

You can use the check boxes to mark objectives to be completed (a **dot** to be done; a **tick** when completed).

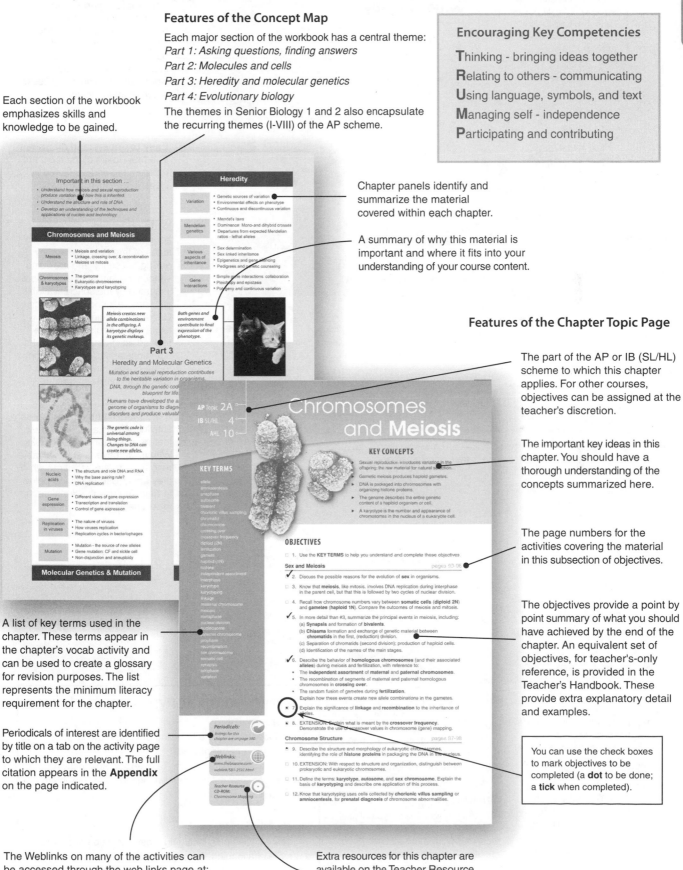

The Weblinks on many of the activities can be accessed through the web links page at: *www.thebiozone.com/weblink/SB1-2597.html* See page 5 for more details.

Extra resources for this chapter are available on the Teacher Resource CD-ROM (for separate purchase).

Using the Activities

The activities make up most of the content of this book. Your teacher may use the activity pages to introduce a topic for the first time, or you may use them to revise ideas already covered by other means. They are excellent for use in the classroom, as homework exercises and topic revision, and for self-directed study and personal reference.

Perforations allow easy removal so that pages can be submitted for grading or kept in a separate folder of related work.

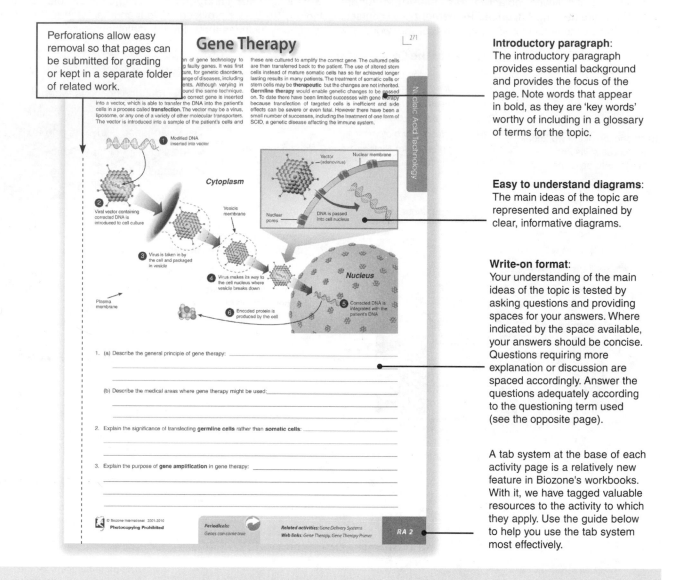

Introductory paragraph:
The introductory paragraph provides essential background and provides the focus of the page. Note words that appear in bold, as they are 'key words' worthy of including in a glossary of terms for the topic.

Easy to understand diagrams:
The main ideas of the topic are represented and explained by clear, informative diagrams.

Write-on format:
Your understanding of the main ideas of the topic is tested by asking questions and providing spaces for your answers. Where indicated by the space available, your answers should be concise. Questions requiring more explanation or discussion are spaced accordingly. Answer the questions adequately according to the questioning term used (see the opposite page).

A tab system at the base of each activity page is a relatively new feature in Biozone's workbooks. With it, we have tagged valuable resources to the activity to which they apply. Use the guide below to help you use the tab system most effectively.

Using page tabs more effectively

Periodicals:
Genes can come true

Related activities: Gene Delivery Systems
Web links: Gene Therapy, Gene Therapy Primer

RA 2

Students (and teachers) who would like to know more about this topic area are encouraged to locate the periodical cited on the Periodicals tab.
Articles of interest directly relevant to the topic content are cited. The full citation appears in the Appendix as indicated at the beginning of the topic chapter.

Related activities
Other activities in the workbook cover related topics or may help answer the questions on the page. In most cases, extra information for activities that are coded R can be found on the pages indicated here.

Web links
This citation indicates a valuable video clip or animation that can be accessed from the web links page specifically for this workbook. www.thebiozone.com/weblink/SB1-2597.html

INTERPRETING THE ACTIVITY CODING SYSTEM
Type of Activity
D = includes some data handling or interpretation
P = includes a paper practical
R = *may* require extra reading (e.g. text or other activity)
A = includes application of knowledge to solve a problem
E = extension material

Level of Activity
1 = generally simpler, including mostly describe questions
2 = more challenging, including explain questions
3 = challenging content and/or questions, including discuss

Introduction

Explanation of Terms

Questions come in a variety of forms. Whether you are studying for an exam, or writing an essay, it is important to understand exactly what the question is asking. A question has two parts to it: one part of the question will provide you with information, the second part of the question will provide you with instructions as to how to answer the question. Following these instructions is most important.

Commonly used Terms in Biology

The following terms are frequently used when asking questions in examinations and assessments. Most of these are listed in the IB syllabus document as action verbs indicating the depth of treatment required for a given statement. Students should have a clear understanding of each of the following terms and use this understanding to answer questions appropriately.

Account for: Provide a satisfactory explanation or reason for an observation.

Analyze: Interpret data to reach stated conclusions.

Annotate: Add **brief** notes to a diagram, drawing or graph.

Apply: Use an idea, equation, principle, theory, or law in a new situation.

Appreciate: To understand the meaning or relevance of a particular situation.

Calculate: Find an answer using mathematical methods. Show the working unless instructed not to.

Compare: Give an account of similarities and differences between two or more items, referring to both (or all) of them throughout. Comparisons can be given using a table. Comparisons generally ask for similarities more than differences (see contrast).

Construct: Represent or develop in graphical form.

Contrast: Show differences. Set in opposition.

Deduce: Reach a conclusion from information given.

Define: Give the precise meaning of a word or phrase as concisely as possible.

Derive: Manipulate a mathematical equation to give a new equation or result.

Describe: Give an account, including all the relevant information.

Design: Produce a plan, object, simulation or model.

Determine: Find the only possible answer.

Discuss: Give an account including, where possible, a range of arguments, assessments of the relative importance of various factors, or comparison of alternative hypotheses.

Distinguish: Give the difference(s) between two or more different items.

Often students in examinations know the material but fail to follow instructions and, as a consequence, do not answer the question appropriately. Examiners often use certain key words to introduce questions. Look out for them and be absolutely clear as to what they mean. Below is a list of commonly used terms that you will come across and a brief explanation of each.

Draw: Represent by means of pencil lines. Add labels unless told not to do so.

Estimate: Find an approximate value for an unknown quantity, based on the information provided and application of scientific knowledge.

Evaluate: Assess the implications and limitations.

Explain: Give a clear account including causes, reasons, or mechanisms.

Identify: Find an answer from a number of possibilities.

Illustrate: Give concrete examples. Explain clearly by using comparisons or examples.

Interpret: Comment upon, give examples, describe relationships. Describe, then evaluate.

List: Give a sequence of names or other brief answers with no elaboration. Each one should be clearly distinguishable from the others.

Measure: Find a value for a quantity.

Outline: Give a brief account or summary. Include essential information only.

Predict: Give an expected result.

Solve: Obtain an answer using algebraic and/or numerical methods.

State: Give a specific name, value, or other answer. No supporting argument or calculation is necessary.

Suggest: Propose a hypothesis or other possible explanation.

Summarize: Give a brief, condensed account. Include conclusions and avoid unnecessary details.

In Conclusion

Students should familiarize themselves with this list of terms and, where necessary throughout the course, they should refer back to them when answering questions. The list of terms mentioned above is not exhaustive and students should compare this list with past examination papers and essays etc. and add any new terms (and their meaning) to the list above. The aim is to become familiar with interpreting the question and answering it appropriately.

Resources Information

Your set textbook should be a starting point for information about the content of your course. There are also many other resources available, including journals, magazines, supplementary texts, dictionaries, computer software, and the internet. Your teacher will have some prescribed resources for your use, but a few of the readily available periodicals are listed here for quick reference. The titles of relevant articles are listed with the activity to which they relate and are cited in the appendix. Please note that listing any product in this workbook does not, in any way, denote Biozone's endorsement of that product and Biozone does not have any business affiliation with the publishers listed herein.

Supplementary Texts

Supplementary texts are those that cover a specific topic or range of topics, rather than an entire course. All titles are available in North America unless indicated by (§).
For further details or to make purchases, link to the publisher via Biozone's resources hub: **www.thebiozone.com > Resources > Supplementary > International**

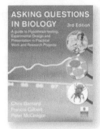

Barnard, C., F. Gilbert, & P. McGregor, 2007
Asking Questions in Biology: Key Skills for Practical Assessments & Project Work, 256 pp.
Publisher: Benjamin Cummings
ISBN: 978-0132224352
Comments: *Covers many aspects of design, analysis and presentation of practical work in senior level biology.*

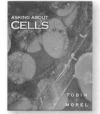

Barnum, S.R., 2nd edn 2005
Biotechnology: An Introduction, 336 pp.
Publisher: Thomson Brooks/Cole
ISBN: 978-0495112051
Comments: *A broad view of biotechnology, integrating historical and modern topics. Processes and methods are described, and numerous examples describe applications.*

Tobin, A.J. and R.E Morel, 1997
Asking About Cells, 698 pp (paperback)
Publisher: Thomson Brooks/Cole
ISBN: 0-030-98018-6
Comments: *An introduction to cell biology, cellular processes and specialization, DNA and gene expression, and inheritance. The focus is on presenting material through inquiry.*

Periodicals, Magazines and Journals

Details of the periodicals referenced in this workbook are listed below. For enquiries and further details regarding subscriptions, link to the relevant publisher via Biozone's resources hub or by going to: **www.thebiozone.com > Resources > Journals**

Biological Sciences Review (Biol. Sci. Rev.)
An excellent quarterly publication for teachers and students of biology. The content is current and the language is accessible. Subscriptions available from Philip Allan Publishers, Market Place, Deddington, Oxfordshire OX 15 OSE.
Tel. 01869 338652
Fax: 01869 338803
E-mail: sales@philipallan.co.uk

New Scientist: *Published weekly and found in many libraries. It often summarizes the findings published in other journals. Articles range from news releases to features.*
Subscription enquiries:
Tel. (UK and international): +44 (0)1444 475636. (US & Canada) 1 888 822 3242.
E-mail: ns.subs@qss-uk.com

Scientific American: *A monthly magazine containing mostly specialist feature articles. Articles range in level of reading difficulty and assumed knowledge.*
Subscription enquiries:
Tel. (US & Canada) 800-333-1199.
Tel. (outside North America): 515-247-7631
Web: www.sciam.com

The American Biology Teacher: *The official, peer-reviewed journal of the National Association of Biology Teachers. Published nine times a year and containing information and activities relevant to the teaching of biology in the US and elsewhere.* Enquiries: NABT, 12030 Sunrise Valley Drive, #110, Reston, VA 20191-3409
Web: www.nabt.org

Biology Dictionaries

Access to a good biology dictionary is of great value when dealing with the technical terms used in biology. Below are some biology dictionaries that you may wish to locate or purchase. They can usually be obtained directly from the publisher or they are all available (at the time of printing) from www.amazon.com. For further details of text content, or to make purchases, link to the relevant publisher via Biozone's resources hub or by typing: **www.thebiozone.com > Resources > Dictionaries**

Hale, W.G. **Collins: Dictionary of Biology** 4 ed. 2005, 528 pp. Collins.
ISBN: 0-00-720734-4.
Updated to take in the latest developments in biology and now internet-linked. (§ This latest edition is currently available only in the UK. The earlier edition, ISBN: 0-00-714709-0, is available though amazon.com in North America).

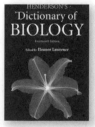

Henderson, E. Lawrence. **Henderson's Dictionary of Biological Terms**, 2008, 776 pp. Benjamin Cummings. **ISBN:** 978-0321505798
This edition has been updated, rewritten for clarity, and reorganised for ease of use. An essential reference and the dictionary of choice for many.

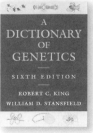

King, R.C. and W.D. Stansfield
A Dictionary of Genetics, 6 ed., 2002, 544 pp. Oxford University Press.
ISBN: 0195143256
A good source for the specialized terminology associated with genetics and related disciplines. Genera and species important to genetics are included, cross linked to an appendix

Making www.thebiozone.com Work For You

The current internet address (URL) for the web site is displayed here. You can type a new address directly into this space.

Use Google to search for web sites of interest. The more precise your search words are, the better the list of results. EXAMPLE: If you type in "biotechnology", your search will return an overwhelmingly large number of sites, many of which will not be useful to you. Be more specific, e.g. "biotechnology medicine DNA uses".

Find out about our superb **Presentation Media**. These slide shows are designed to provide in-depth, highly accessible illustrative material and notes on specific areas of biology.

Podcasts: Access the latest news as audio files (mp3) that may be downloaded or played directly off your computer.

News: Find out about product announcements, shipping dates, and workshops and trade displays by Biozone at teachers' conferences around the world.

RSS Newsfeeds: See breaking news and major new discoveries in biology directly from our web site.

Access the **BioLinks** database of web sites related to each major area of biology. It's a great way to quickly find out more on topics of interest.

Weblinks: www.thebiozone.com/weblink/SB1-2597.html

BOOKMARK WEBLINKS BY TYPING IN THE ADDRESS: IT IS NOT ACCESSIBLE DIRECTLY FROM BIOZONE'S WEBSITE

Throughout this workbook, some pages make reference to web links and periodicals that are particularly relevant to the activity on which they are cited. They provide great support to aid understanding of basic concepts:

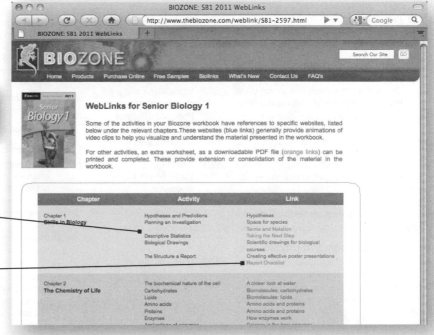

Periodicals: Full citations are provided in the Appendix for those that wish to read further on a topic.

Web Link: Provides a link to an **external web site** with supporting information for the activity.

There are also color coded links to downloadable **Acrobat (PDF) files** which may provide an additional activity or the same activity from a different teaching perspective.

International Baccalaureate Course

The International Baccalaureate (IB) biology course is divided into three sections: core, additional higher level material, and option material. All **IB candidates** must complete the **core** topics. Higher level students are also required to undertake Additional Higher Level **(AHL)** material as part of the core. Options fall into three categories (see the following page): those specific to standard level students **(OPT-SL)**, one only specific to higher level students **(OPT-HL)** and those offered to both **(OPT-SL/HL)**. All candidates are required to study two options. All candidates must also carry out **practical work** and must participate in **the group 4 project**. In the guide below, we have indicated where the relevant material can be found: SB1 for Senior Biology 1 and SB2 for Senior Biology 2.

Topic		See workbook
CORE:	*(All students)*	
1	**Statistical analysis**	
1.1	Mean and SD, t-test, correlation.	SB1 — Skills in Biology
	● *For this CORE topic also see the TRC: Spreadsheets and Statistics*	
2	**Cells**	
2.1	Cell theory. Cell and organelle sizes. Surface area to volume ratio. Emergent properties. Cell specialization and differentiation. Stem cells.	SB1 — Cell Structure, Processes in Cells
2.2	Prokaryotic cells: ultrastructure & function.	SB1 — Cell Structure
2.3	Eukaryotic cells: ultrastructure & function. Prokaryotic vs eukaryotic cells. Plant vs animal cells. Extracellular components.	SB1 — Cell Structure, Processes in Cells
2.4	Membrane structure. Active and passive transport. Diffusion and osmosis.	SB1 — Processes in Cells
2.5	Cell division and the origins of cancer.	SB1 — Processes in Cells
	● *For extension on this topic also see the TRC: The Cell Theory*	
3	**The chemistry of life**	
3.1	Elements of life. The properties and importance of water.	SB1 — The Chemistry of Life
3.2	Structure and function of carbohydrates, lipids, and proteins.	SB1 — The Chemistry of Life
3.3	Nucleotides and the structure of DNA.	SB1 — The Chemistry of Life
3.4	Semi-conservative DNA replication.	SB1 — Molecular Genetics
3.5	RNA and DNA structure. The genetic code. Transcription. Translation.	SB1 — Molecular Genetics
3.6	Enzyme structure and function.	SB1 — The Chemistry of Life
3.7	Cellular respiration and ATP production.	SB1 — Cellular Energetics
3.8	Biochemistry of photosynthesis. Factors affecting photosynthetic rates.	SB1 — Cellular Energetics
4	**Genetics**	
4.1	Eukaryote chromosomes. Genomes. Gene mutations and consequences.	SB1 — Chromosomes & Meiosis
4.2	Meiosis and non-disjunction. Karyotyping and pre-natal diagnosis.	SB1 — Chromosomes & Meiosis
4.3	Theoretical genetics: alleles and single gene inheritance, sex linkage, pedigrees.	SB1 — Heredity
4.4	Genetic engineering and biotechnology: PCR, gel electrophoresis, DNA profiling. HGP. Transformation. GMOs. Cloning.	SB1 — Nucleic Acid Technology
	● *For extension on this topic also see the TRC: Engineering Solutions*	
5	**Ecology and evolution**	
5.1	Ecosystems. Food chains and webs. Trophic levels. Ecological pyramids. The role of decomposers in recycling nutrients.	SB2 — Habitat & Distribution, Community Ecology
5.2	The greenhouse effect. The carbon cycle. Precautionary principle. Global warming.	SB2 — Ecosystem Ecology & Human Impact
5.3	Factors influencing population size. Population growth.	SB2 — Population Ecology
5.4	Genetic variation. Sexual reproduction as a source of variation in species.	SB1 — Chromosomes & Meiosis,
	Evidence for evolution: natural selection. Evolution in response to environmental change.	SB1 — The Origin & Evolution of Life, Speciation
5.5	Classification. Binomial nomenclature. Features of plant & animal phyla. Keys.	SB2 — Classification
	● *For extension on this topic also see the TRC: Classification of Life*	
6	**Human health and physiology**	
6.1	Role of enzymes in digestion. Structure and function of the digestive system.	SB2 — Eating to Live
6.2	Structure and function of the heart. The control of heart activity. Blood & vessels.	SB2 — Life Blood

Topic		See workbook
6.3	Pathogens and their transmission. Antibiotics. Role of skin as a barrier to infection. Role of phagocytic leukocytes. Antigens & antibody production. HIV/AIDS.	SB2 — Defending Against Disease
6.4	Gas exchange. Ventilation systems. Control of breathing.	SB2 — Breath of Life
6.5	Principles of homeostasis. Control of body temperature and blood glucose. Diabetes. Role of the nervous and endocrine systems in homeostasis.	SB2 — Keeping in Balance, Responding to the Environment
6.6	Human reproduction and the role of hormones. Reproductive technologies and ethical issues.	SB2 — The Next Generation
COMPULSORY: AHL Topics	*(HL students only)*	
7	**Nucleic acids and proteins**	
7.1	DNA structure, exons & introns (junk DNA)	SB1 — Molecular Genetics
7.2	DNA replication, including the role of enzymes and Okazaki fragments.	SB1 — Molecular Genetics
7.3	DNA alignment, transcription. The removal of introns to form mature mRNA.	SB1 — Molecular Genetics
7.4	The structure of tRNA and ribosomes. The process of translation. Peptide bonds.	SB1 — Molecular Genetics, The Chemistry of Life
7.5	Protein structure and function.	SB1 — The Chemistry of Life
7.6	Enzymes: induced fit model. Inhibition. Allostery in the control of metabolism.	SB1 — The Chemistry of Life
	● *For extension on this topic see the TRC: The Meselson-Stahl Experiment*	
8	**Cell respiration and photosynthesis**	
8.1	Structure and function of mitochondria. Biochemistry of cellular respiration.	SB1 — Cellular Energetics
8.2	Chloroplasts, the biochemistry and control of photosynthesis, chemiosmosis.	SB1 — Cellular Energetics
9	**Plant science**	
9.1	Structure and growth of a dicot plant. Function and distribution of tissues in leaves. Dicots vs monocots. Plant modifications. Auxins.	SB2 — Plant Structure & Growth Responses
9.2	Support in terrestrial plants. Transport in angiosperms: ion movement through soil, active ion uptake by roots, transpiration, translocation. Abscisic acid. Xerophytes.	SB2 — Plant Support & Transport
9.3	Dicot flowers. Pollination and fertilization. Seeds: structure, germination, dispersal. Flowering and phytochrome.	SB2 — Plant Reproduction
10	**Genetics**	
10.1	Meiosis, and the process of crossing over. Mendel's law of independent assortment.	SB1 — Chromosomes & Meiosis
10.2	Dihybrid crosses. Types of chromosomes.	SB1 — Heredity
10.3	Polygenic inheritance.	SB1 — Heredity
	● *For extension on this topic also see the TRC: Chromosome Mapping*	
11	**Human health and physiology**	
11.1	Blood clotting. Clonal selection. Acquired immunity. Antibodies and monoclonal antibodies. Vaccination.	SB2 — Defending Against Disease
11.2	Nerves, muscles, bones and movement. Joints. Skeletal muscle and contraction.	SB2 — Muscles & Movement
11.3	Excretion. Structure and function of the human kidney. Urine production. Diabetes.	SB2 — Regulating Fluids and Removing Wastes
11.4	Testis and ovarian structure. Spermatogenesis and oogenesis. Fertilization and embryonic development. The placenta. Birth. Role of hormones.	SB2 — The Next Generation

International Baccalaureate Course *continued*

Topic	See workbook	Topic	See workbook

OPTIONS: **OPT - SL** *(SL students only)*

A   **Human nutrition and health**

A.1   Diet and malnutrition. Deficiency & supplements. PKU.

A.2   Energy content of food types. BMI. Obesity and anorexia. Appetite control.

A.3   Special diet issues; breastfeeding vs bottle-feeding, type II diabetes, cholesterol.

     ● *Provided as a separate complete unit on the IB OPTIONS CD-ROM*

B   **Physiology of exercise**

B.1   Locomotion in animals. Roles of nerves, muscles, and bones in movement. Joints. Skeletal muscle and contraction.

B.2   Training and the pulmonary system.

B.3   Training and the cardiovascular system.

B.4   Respiration and exercise intensity. Roles of myoglobin and adrenaline. Oxygen debt and lactate in muscle fatigue.

B.5   Exercise induced injuries and treatment.

     ● *Provided as a separate complete unit on the IB OPTIONS CD-ROM*

C   **Cells and energy**

C.1   Protein structure and function. Fibrous and globular proteins.   SB1   The Chemistry of Life

C.2   Enzymes: induced fit model. Inhibition. Allostery in the control of metabolism.   SB1   The Chemistry of Life

C.3   Biochemistry of cellular respiration.   SB1   Cellular Energetics

C.4   The biochemistry of photosynthesis including chemiosmosis. Action and absorption spectra. Limiting factors.   SB1   Cellular Energetics

     ● *Provided as a separate complete unit on the IB OPTIONS CD-ROM*

OPTIONS: **OPT - SL/HL** *(SL and HL students)*

D   **Evolution**

D.1   Prebiotic experiments. Comets. Protobionts and prokaryotes. Endosymbiotic theory.   SB1   The Origin & Evolution of Life

D.2   Species, gene pools, speciation. Types and pace of evolution. Transient vs balanced polymorphism.   SB1   Speciation, Patterns of Evolution

D.3   Fossil dating. Primate features. Hominid features. Diet and brain size correlation. Genetic and cultural evolution.   ●   The Evolution of Humans (**TRC**)

D.4-D.5 is extension for HL only

D.4   The Hardy-Weinberg principle.   SB1   Speciation

D.5   Biochemical evidence for evolution. Biochemical variations indicating phylogenetic relationships.   SB1   The Origin and Evolution of Life

     Classification. Cladistics and cladograms   SB2   Classification

     ● *Provided as a separate complete unit on the IB OPTIONS CD-ROM*

E   **Neurobiology and behavior**

E.1   Stimuli, responses and reflexes in the context of animal behavior. Animal responses and natural selection.   SB2   Nerves, Muscles & Movement, Animal Behavior

E.2   Sensory receptors. Structure and function of the human eye and ear.   SB2   Nerves, Muscles & Movement

E.3   Innate vs learned behavior and its role in survival. Learned behavior and birdsong.   SB2   Animal Behavior

E.4   Presynaptic neurons at synapses. Examples of excitatory and inhibitory psychoactive drugs. Effects of drugs on synaptic transmission. Causes of addiction.   SB2   Aspects covered in Nerves, Muscles & Movement

E.5-E.6 is extension for HL only

E.5   Structure and function of the human brain. ANS control. Pupil reflex and its use in testing for death. Hormones as painkillers.   SB2   Aspects covered in Nerves, Muscles & Movement

E.6   Social behavior and organization. The role of altruism in sociality. Foraging behavior. Mate selection. Rhythmical behavior.   SB2   Animal Behavior

     ● *Provided as a separate complete unit on the IB OPTIONS CD-ROM*

F   **Microbes and Biotechnology**

F.1   Classification. Diversity of Archaea and Eubacteria. Diversity of viruses. Diversity of microscopic eukaryotes.

F.2   Roles of microbes in ecosystems. Details of the nitrogen cycle including the role of bacteria. Sewage treatment. Biofuels.

F.3   Reverse transcription. Somatic vs germline, gene therapy. Viral vectors.

F.4   Microbes involved in food production of beer, wine, bread, and soy sauce. Food preservation. Food poisoning.

F.5-F.6 is extension for HL only

F.5   Metabolism of microbes. Modes of nutrition. Cyanobacterium. Bioremediation.

F.6   Pathogens and disease: influenza virus, malaria, bacterial infections. Controlling microbes. Epidemiology. Prion hypothesis.

     ● *Provided as a separate complete unit on the IB OPTIONS CD-ROM*

G   **Ecology and conservation**

G.1   Factors affecting plant and animal distribution. Sampling. Ecological niche and the competitive exclusion principle. Species interactions. Measuring biomass.

G.2   Trophic levels. Ecological pyramids. Primary vs secondary succession. Biome vs biosphere. Plant productivity (includes calculating gross and net production, and biomass).

G.3   Conservation of biodiversity. Diversity index. Human impact on ecosystems: alien species. Biological control. Effect of CFCs on ozone layer. UV radiation absorption.

G.4-G.5 is extension for HL only

G.4   Monitoring environmental change. Biodiversity. Endangered species. Conservation Strategies. Extinction.

G.5   *r*-strategies and K-strategies.Mark-and-recapture sampling. Fisheries conservation.

     ● *Provided as a separate complete unit on the IB OPTIONS CD-ROM*

OPTION: **OPT - HL** *(HL students only)*

H   **Further human physiology**

H.1   Hormones and their modes of action. Hypothalamus and pituitary gland. Control of ADH secretion.

H.2   Digestion and digestive juices. Stomach ulcers and stomach cancers. Role of bile.

H.3   Structure of villi. Absorption of nutrients and transport of digested food.

H.4   The structure and function of the liver (including role in nutrient processing and detoxification). Liver damage from alcohol.

H.5   The cardiac cycle and control of heart rhythm. Atherosclerosis, coronary thrombosis and coronary heart disease.

H.6   Gas exchange: oxygen dissociation curves and the Bohr shift. Ventilation rate and exercise. Breathing at high altitude. Causes and effects of asthma.

     ● *Provided as a separate complete unit on the IB OPTIONS CD-ROM*

Practical Work *(All students)*

Practical work consists of short and long term investigations, and an interdisciplinary project (The Group 4 project). Also see the "Guide to Practical Work" on the last page of this introductory section.

Short and long term investigations

Investigations should reflect the breadth and depth of the subjects taught at each level, and include a spread of content material from the core, options, and AHL material, where relevant.

The Group 4 project

All candidates must participate in the group 4 project. In this project it is intended that students analyze a topic or problem suitable for investigation in each of the science disciplines offered by the school (not just in biology). This project emphasizes the processes involved in scientific investigations rather than the products of an investigation.

Advanced Placement Course

The Advanced Placement (AP) biology course is designed to be equivalent to a college introductory biology course. It is to be taken by students after successful completion of first courses in high school biology and chemistry. In the guide below, we have indicated where the relevant material can be found: SB1 for Senior Biology 1 and SB2 for Senior Biology 2. Because of the general nature of the AP curriculum document, the detail given here is based on Biozone's interpretation of the scheme.

Topic		See workbook

Topic I: Molecules and Cells

A   Chemistry of life

1. The chemical & physical properties of water. The importance of water to life. — SB1 The Chemistry of Life

2. The role of carbon. Structure and function of carbohydrates, lipids, nucleic acids, and proteins. The synthesis and breakdown of macromolecules. — SB1 The Chemistry of Life, Molecular Genetics, Processes in Cells

3. The laws of thermodynamics and their relationship to biochemical processes. Free energy changes. — SB1 The Chemistry of Life

4. The action of enzymes and their role in the regulation of metabolism. Enzyme specificity. Factors affecting enzyme activity. Applications of enzymes. — SB1 The Chemistry of Life

● *For extension on this topic also see the TRC: Industrial Microbiology*

B   Cells

1. Comparison of prokaryotic and eukaryotic cells, including evolutionary relationships. — SB1 Cell Structure, SB1 The Origin & Evolution of Life

● *For extension on this topic also see the TRC: The Cell Theory*

2. Membrane structure: fluid mosaic model. Active and passive transport. — SB1 Processes in Cells

3. Structure and function of organelles. Comparison of plant and animal cells. Cell size and surface area: volume ratio. Organization of cell function. — SB1 Cell Structure, Processes in Cells

4. Mitosis and the cell cycle. Mechanisms of cytokinesis. Cancer (tumour formation) as the result of uncontrolled cell division. — SB1 Processes in Cells

C   Cellular energetics

1. Nature and role of ATP. Anabolic and catabolic processes. Chemiosmosis. — SB1 Cellular Energetics

2. Structure and function of mitochondria. Biochemistry of cellular respiration, including the role of oxygen in energy yielding pathways. Anaerobic systems. — SB1 Cellular Energetics

3. Structure and function of chloroplasts. Biochemistry of photosynthesis. Adaptations for photosynthesis in different environments. — SB1 Cellular Energetics

● *For extension on this topic also see the TRC: Events in Biochemistry*

Topic II: Heredity and Evolution

A   Heredity

1. The importance of meiosis in heredity. Gametogenesis. Similarities and differences between gametogenesis in animals and plants. — SB1 Chromosomes & Meiosis, SB1 Processes in Cells

2. Structure of eukaryotic chromosomes. Heredity of genetic information. — SB1 Chromosomes & Meiosis

3. Mendel's laws. Inheritance patterns. — SB1 Heredity

● *For extension on this topic also see the TRC: Chromsome Mapping*

B   Molecular genetics

1. RNA and DNA structure and function. Eukaryotic and prokaryotic genomes. — SB1 Molecular Genetics

   For extension on this topic see the TRC: The Meselsohn-Stahl Experiment

2. Gene expression in prokaryotes and eukaryotes. The *Lac* operon model. — SB1 Molecular Genetics

3. Causes of mutations. Gene mutations (e.g. sickle cell disease). Chromosomal mutations (e.g. Down syndrome). — SB1 Mutation

4. Viral structure and replication. — SB1 Molecular Genetics

5. Nucleic acid technology and applications. legal and ethical issues. — SB1 Nucleic Acid Technology

● *For extension on this topic also see the TRC: Engineering Solutions*

C   Evolutionary biology

1. The origins of life on Earth. Prebiotic experiments. Origins of prokaryotic cells. Endosymbiotic theory. — SB1 The Origin & Evolution of Life

2. Evidence for evolution. Dating of fossils. — SB1 The Origin & Evolution of Life

● *For extension on this topic also see the TRC: Dating the Past*

3. The species concept. Mechanisms of evolution: natural selection, speciation, macroevolution. — SB1 Speciation, Patterns of Evolution

● *For extension on this topic also see the TRC: A Case Study in Evolution*

Topic III: Organisms and Populations

A   Diversity of organisms

1. Evolutionary patterns: major body plans of plants and animals. — SB2 Classification

2. Diversity of life: representative members from the five kingdoms Monera (=Prokaryotae), Fungi, Protista (=Protoctista), Animalia and Plantae. — SB2 Classification

3. Phylogenetic classification. Binomial nomenclature. Five kingdom classification. Use of dichotomous keys. — SB2 Classification

4. Evolutionary relationships: genetic and morphological characters. Phylogenies. — SB2 Classification

● *For extension on this topic also see the TRC: Practical Classification*

B   Structure and function of plants and animals

1. Plant and animal reproduction and development (includes humans). Adaptive significance of reproductive features and their regulation. — SB2 The Next Generation, Plant Reproduction

● *For extension material see the TRC: Mammalian Patterns of Reproduction*

2. Organization of cells, tissues & organs.

   The structure and function of animal and plant organ systems. Adaptive features that have contributed to the success of plants and animals in occupying particular terrestrial niches. — SB2 PART 2 (Animals), chapters as required. PART 3 (Plants), chapters as required

3. Plant and animal responses to environmental cues. The role of hormones in these responses. — SB2 Respondng to the Environment, Plant Structure and Growth Responses

● *For extension material see the TRC: Migratory Navigation in Birds*

C   Ecology

1. Factors influencing population size. Population growth curves. — SB2 Population Ecology

2. Abiotic and biotic factors: effects on community structure and ecosystem function. Trophic levels: energy flows through ecosystems and relationship to trophic structure. Nutrient cycles. — SB2 Habitat and Distribution, Community Ecology, Ecosystem Ecology and Human Impact

● *For extension material see the TRC: Production and Trophic Efficiency*

3. Human influence on biogeochemical cycles: (e.g. use of fertilizers). — SB2 Ecosystem Ecology and Human Impact

● *For extension material see the TRC: Sustainable Futures*

Practical Work

Integrated practicals as appropriate
"Guide to Practical Work" in this introductory section.
Senior Biology 1: Skills in Biology (for reference throughout the course).
● TRC: *Spreadsheets and Statistics*

Guide to Practical Work

A practical or laboratory component is an essential part of any biology course, especially at senior level. It is through your practical sessions that you are challenged to carry out experiments drawn from many areas within modern biology. Both AP and IB courses have a strong practical component, aimed at providing a framework for your laboratory experience. Well executed laboratory and field sessions will help you to understand problems, observe accurately, make hypotheses, design and implement controlled experiments, collect and analyze data, think analytically, and communicate your findings in an appropriate way using tables and graphs. The outline below provides some guidelines for AP and IB students undertaking their practical work. Be sure to follow required safety procedures at all times during practical work.

International Baccalaureate Practical Work

The practical work carried out by IB biology students should reflect the depth and breadth of the subject syllabus, although there may not be an investigation for every syllabus topic. All candidates must participate in the group 4 project, and the internal assessment (IA) requirements should be met via a spread of content from the core, options and, where relevant, AHL material. A wide range of IA investigations is possible: short laboratory practicals and longer term practicals or projects, computer simulations, data gathering and analysis exercises, and general laboratory and field work.

Suitable material, or background preparation, for this component can be found in this workbook and its companion title, Senior Biology 2.

College Board's AP® Biology Lab Topics

Each of the 12 set laboratory sessions in the AP course is designed to complement a particular topic area within the course. The basic structure of the lab course is outlined below:

LAB 1:   Diffusion and osmosis
Overview: To investigate diffusion and osmosis in dialysis tubing. To investigate the effect of solute concentration on water potential (ψ) in plant tissues.

Aims: An understanding of passive transport mechanisms in cells, and an understanding of the concept of water potential, solute potential, and pressure potential, and how these are measured.

LAB 2:   Enzyme catalysis
Overview: To investigate the conversion of hydrogen peroxide to water and oxygen gas by catalase.

Aims: An understanding of the effects of environmental factors on the rate of enzyme catalyzed reactions.

LAB 3:   Mitosis and meiosis
Overview: To use prepared slides of onion root tips to study plant mitosis. To simulate the phases of meiosis by using chromosome models.

Aims: Recognition of stages in mitosis in plant cells and calculation of relative duration of cell cycle stages. An understanding of chromosome activity during meiosis and an ability to calculate map distances for genes.

LAB 4:   Plant pigments and photosynthesis
Overview: To separate plant pigments using chromatography. To measure photosynthetic rate in chloroplasts.

Aims: An understanding of Rf values. An understanding of the techniques used to determine photosynthetic rates. An ability to explain variations in photosynthetic rate under different environmental conditions.

LAB 5:   Cell(ular) respiration
Overview: To investigate oxygen consumption during germination (including the effect of temperature).

Aims: An understanding of how cell respiration rates can be calculated from experimental data. An understanding of the relationship between gas production and respiration rate, and the effect of temperature on this.

LAB 6:   Molecular biology
Overview: To investigate the basic principles of molecular biology through the transformation of E.coli cells. To investigate the use of restriction digestion and gel electrophoresis.

Aims: An understanding of the role of plasmids as vectors, and the use of gel electrophoresis to separate DNA fragments of varying size. An ability to design appropriate experimental procedures and use multiple experimental controls.

LAB 7:   Genetics of organisms
Overview: Use Drosophila to perform genetic crosses. To collect and analyze the data from these crosses.

Aims: An understanding of the independent assortment of two genes and an ability to determine if genes are autosomal or sex linked from the analysis of the results of multigeneration genetic crosses.

LAB 8:   Population genetics and evolution
Overview: To learn about the Hardy-Weinberg law of genetic equilibrium and study the relationship between evolution and changes in allele frequency.

Aims: An ability to calculate allele and genotype frequencies using the Hardy-Weinberg formula. An understanding of natural selection and other causes of microevolution.

LAB 9:   Transpiration
Overview: To investigate transpiration in plants under controlled conditions. To examine the organization of plant stems and leaves as they relate to this.

Aims: An understanding of the effects of environmental variables on transpiration rates. An understanding of the relationship between the structure and function of the tissues involved.

LAB 10:   Physiology of the circulatory system
Overview: To measure (human) blood pressure and pulse rate under different conditions. To analyze these variables and relate them to an index of fitness. To investigate the effect of temperature on heart rate in Daphnia.

Aims: An understanding of blood pressure and pulse rate, and their measurement and significance with respect to fitness. An understanding of the relationship between heart rate and temperature in a poikilotherm.

LAB 11:   Animal behavior
Overview: To investigate responses in pillbugs (woodlice). To investigate mating behavior in fruit flies.

Aims: To understand and describe aspects of animal behavior. To understand the adaptiveness of appropriate behaviors.

LAB 12:   Dissolved oxygen & aquatic primary productivity
Overview: To measure & analyze dissolved oxygen concentration in water samples. To measure and analyze the primary productivity of natural waters or lab cultures.

Aims: An understanding of primary productivity and its measurement. To use a controlled experiment to investigate the effect of changing light intensity on primary productivity.

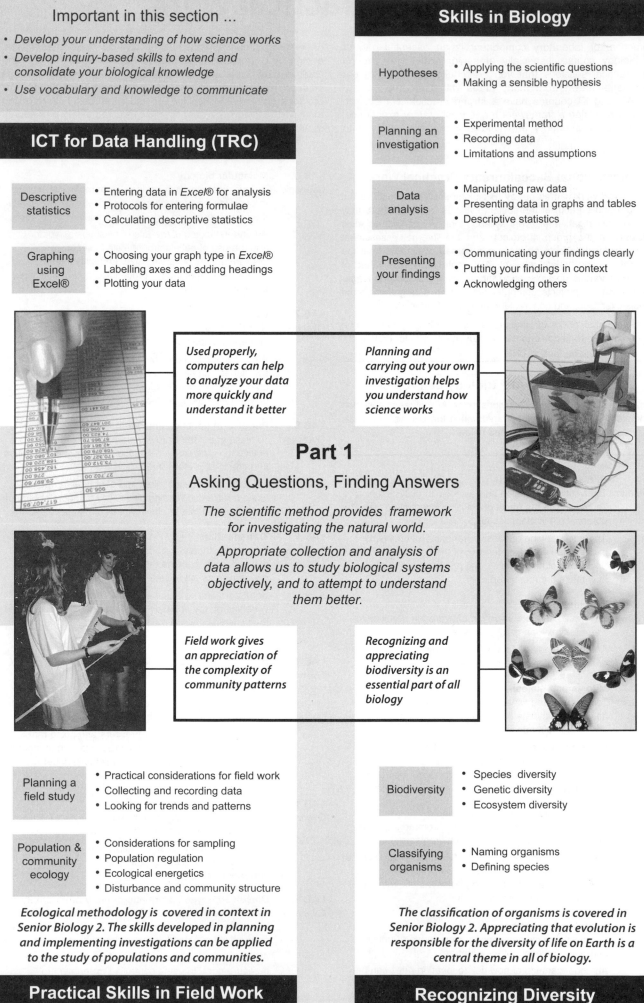

Important in this section ...

- *Develop your understanding of how science works*
- *Develop inquiry-based skills to extend and consolidate your biological knowledge*
- *Use vocabulary and knowledge to communicate*

Skills in Biology

| Hypotheses | • Applying the scientific questions
• Making a sensible hypothesis |

| Planning an investigation | • Experimental method
• Recording data
• Limitations and assumptions |

| Data analysis | • Manipulating raw data
• Presenting data in graphs and tables
• Descriptive statistics |

| Presenting your findings | • Communicating your findings clearly
• Putting your findings in context
• Acknowledging others |

ICT for Data Handling (TRC)

| Descriptive statistics | • Entering data in *Excel®* for analysis
• Protocols for entering formulae
• Calculating descriptive statistics |

| Graphing using Excel® | • Choosing your graph type in *Excel®*
• Labelling axes and adding headings
• Plotting your data |

Used properly, computers can help to analyze your data more quickly and understand it better

Planning and carrying out your own investigation helps you understand how science works

Part 1

Asking Questions, Finding Answers

The scientific method provides framework for investigating the natural world.

Appropriate collection and analysis of data allows us to study biological systems objectively, and to attempt to understand them better.

Field work gives an appreciation of the complexity of community patterns

Recognizing and appreciating biodiversity is an essential part of all biology

| Planning a field study | • Practical considerations for field work
• Collecting and recording data
• Looking for trends and patterns |

| Population & community ecology | • Considerations for sampling
• Population regulation
• Ecological energetics
• Disturbance and community structure |

Ecological methodology is covered in context in Senior Biology 2. The skills developed in planning and implementing investigations can be applied to the study of populations and communities.

| Biodiversity | • Species diversity
• Genetic diversity
• Ecosystem diversity |

| Classifying organisms | • Naming organisms
• Defining species |

The classification of organisms is covered in Senior Biology 2. Appreciating that evolution is responsible for the diversity of life on Earth is a central theme in all of biology.

Practical Skills in Field Work

Recognizing Diversity

Skills in **Biology**

KEY CONCEPTS

▶ The basis of all science is observation, hypothesis, and investigation.

▶ Scientists collect and analyze data to test their hypotheses.

▶ Data can be analyzed and presented in various ways, including in graphs and tables.

▶ A scientific report summarizes the results of a scientific investigation and makes the findings accessible.

KEY TERMS

accuracy
bibliography
biological drawing
citation
control
controlled variable
data
datalogger
dependent variable
graph
histogram
hypothesis
independent variable
mean
measurement
median
mode
observation
precision
qualitative data
quantitative data
random sampling
raw data
report
sample
scientific method
standard deviation
statistic
table
transformation (of data)
trend (of data)
variable
X axis
Y axis

OBJECTIVES

☐ 1. Use the **KEY TERMS** to help you understand and complete these objectives.

☐ 2. Describe and explain the basic principles of the scientific method.

Making Investigations   pages 12-20

☐ 3. Produce an outline of your practical biological investigation, including your aim and hypothesis, and all information relevant to the study design.

☐ 4. Identify your dependent and independent variables, their range, and how you will measure them. Identify controlled variables and their significance. Evaluate any sources of error.

☐ 5. Explain the difference between qualitative and quantitative data and give examples of their appropriate use.

☐ 6. Demonstrate an ability to **systematically record** data. Evaluate the accuracy and precision of any recording or measurements you make.

☐ 7. Demonstrate an ability or make accurate biological drawings.

Data Analysis   pages 21-38

☐ 8. Demonstrate an ability to process raw data. Calculate percentages, rates, and frequencies for raw data and explain the reason for these manipulations.

☐ 9. Describe the benefits of tabulating data and present different types of data appropriately in a table, including any calculated values.

☐ 10. Describe the benefits of graphing data and present different types of data appropriately in both graphs.

Presenting Your Report   pages 39-44

☐ 11. Present the findings of your investigation in a well organized scientific report. The report may be a written document, a seminar, poster, web page, or multimedia presentation

☐ 12. Identify and explain the important features of a scientific report including
   - materials and methods
   - recorded and processed data, including figures
   - conclusions based on analysis of the data and the experimental aims
   - discussion of the biological concepts involved
   - evaluation of the investigation

Periodicals:
listings for this chapter are on page 379

Weblinks:
www.thebiozone.com/
weblink/SB1-2597.html

Teacher Resource CD-ROM:
Spreadsheets and Statistics

Investigations in Field Science

Investigating the habitat preferences of an organism is frequently complicated by the fact that organisms are often found in a range of quite variable habitats. What is required is a method by which the habitat can be described quantitatively, so that numerical data can be used to predict occurrence and distribution outside the study area. The information and data below relate to catches of black mudfish in a rural area of New Zealand and are part of a habitat preference study for mudfish, which occupy swamps, drains and wetlands, including those that periodically dry out requiring the fish to estivate (go into a torpid state) in the mud.

Describing the mudfish environment

The vegetation at each of 80 capture sites was used to rank sites on a 5 point scale. The vegetation reflected the influence of invasive exotic plants and the extent of human disturbance. Five plant species, including willow, rushes, and grasses, were used to indicate the extent of disturbance at the 80 sites. Direct alteration was defined as a visible change to the substrate or vegetation resulting from activities such as drain digging, dam construction, or tree felling and was given the number 5 on the **disturbance scale rating** (DSR). Indirect alteration was defined as change from the natural state caused by disturbances elsewhere, such as degradation of water quality as a result of human activity, and was given a rating of 3. Nil or very little disturbance to the environment was given a rating of 1. Types of cover potentially important for mudfish were recorded for each site. Cover types were recorded as being present only if they occupied more than 20% of the available habitat. This precaution was to prevent one or two scattered plants being classified as significant fish cover. For example, overhanging vegetation was classified as any vegetation that overhung more than 20% of water surface over the area trapped.

Data provided by Assoc. Prof. Brendan Hicks and Rhys Barrier, (University of Waikato).

3. Describe how the researchers ensured there was no bias in the way the mudfish were trapped:

_____

_____

_____

_____

_____

4. Describe which aspects of the mudfish were measured:

_____

_____

_____

Analyzing the data

The catch rate of fish per trap per night was plotted against the DSR. Preference for the DSR of 1 was assumed to be the same as for the DSR of 2, as the small number of sites with DSR of 1 and 2 made their separate preferences unreliable.

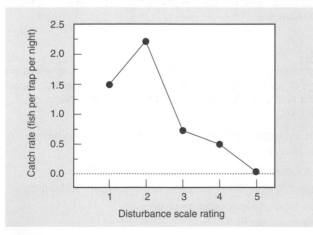

1. Describe how the qualitative description of the sample site habitat was transformed into quantitative data:

_____

_____

_____

_____

_____

2. Describe how the researchers ensured that the cover described was relevant to the overall habitat description:

_____

_____

Making the catch

Fish were trapped between 20 May and 21 October, and wet-season data were collected then. There was no trapping over the dry summer months because many sites had only damp mud with no surface water at that time. To ensure the complete range of likely mudfish habitats was sampled, an approximately equal number of sites predicted to be with and without mudfish were chosen in shallow water. Care was taken to fully submerge the trap entrance while leaving an air space at the top of the trap, otherwise in low oxygen conditions, where mudfish have to surface to breathe air, trapping could kill them. Traps were spaced uniformly over each site covering and area of approximately 25 m² wherever possible. Traps were set between 10am and 2pm, left overnight, and picked up the same time on the following day. The mudfish caught were weighed and their length was measured, whereupon they were returned to their point of capture.

Nick Ling

5. Describe the environment preferred by the mudfish:

_____

_____

_____

_____

6. Explain why a preference for DSR of 1 and 2 were assumed to be the same:

_____

_____

_____

Hypotheses and Predictions

Scientific knowledge grows through a process called the **scientific method**. This process involves observation and measurement, hypothesizing and predicting, and planning and executing investigations designed to test formulated **hypotheses**. A scientific hypothesis is a tentative explanation for an observation, which is capable of being tested by experimentation. Hypotheses lead to **predictions** about the system involved and they are accepted or rejected on the basis of findings arising from the investigation. Rejection of the hypothesis may lead to new, alternative explanations (hypotheses) for the observations. Acceptance of the hypothesis as a valid explanation is not necessarily permanent: explanations may be rejected at a later date in light of new findings. This process eventually leads to new knowledge (theory, laws, or models).

Making Observations

These may involve the observation of certain behaviors in wild populations, physiological measurements made during previous experiments, or 'accidental' results obtained when seeking answers to completely unrelated questions.

Testing predictions may lead to new observations

Asking Questions

The observations lead to the formation of questions about the system being studied.

Testing the Predictions

The predictions are tested out in the practical part of an investigation.

Accept or reject the hypothesis

Forming a Hypothesis

Features of a sound hypothesis:

- It is based on observations and prior knowledge of the system.
- It offers an explanation for an observation.
- It refers to only one independent variable.
- It is written as a definite statement and not as a question.
- It is testable by experimentation.
- It leads to predictions about the system.

Designing an Investigation

Investigations are planned so that the predictions about the system made in the hypothesis can be tested. Investigations may be laboratory or field based.

Generating a Null Hypothesis

A hypothesis based on observations is used to generate the **null hypothesis (H_0)**; the hypothesis of no difference or no effect. Hypotheses are expressed in the null form for the purposes of statistical testing. H_0 may be rejected in favor of accepting the alternative hypothesis, H_A.

Making Predictions

Based on a hypothesis, **predictions** (expected, repeatable outcomes) can be generated about the behavior of the system. Predictions may be made on any aspect of the material of interest, e.g. how different variables (factors) relate to each other.

Periodicals:
The truth is out there

Related activities: Experimental Method
Web links: Hypotheses, Terms and Notation

A 2

Useful Types of Hypotheses

A hypothesis offers a tentative explanation to questions generated by observations. Some examples are described below. Hypotheses are often constructed in a form that allows them to be tested statistically. For every hypothesis, there is a corresponding **null hypothesis**; a hypothesis against the prediction. Predictions are tested with laboratory and field experiments and carefully focused observations. For a hypothesis to be accepted it should be possible for anyone to test the predictions with the same methods and get a similar result each time.

Hypothesis involving manipulation
Used when the effect of manipulating a variable on a biological entity is being investigated. **Example**: The composition of applied fertilizer influences the rate of growth of plant A.

Hypothesis of choice
Used when species preference, e.g. for a particular habitat type or microclimate, is being investigated. **Example**: Woodpeckers (species A) show a preference for tree type when nesting.

Hypothesis involving observation
Used when organisms are being studied in their natural environment and conditions cannot be changed. **Example**: Fern abundance is influenced by the degree to which the canopy is established.

1. Generate a prediction for the hypothesis: *"Moisture level of the microhabitat influences woodlouse distribution"*:

   _____

2. During the course of any investigation, new information may arise as a result of observations unrelated to the original hypothesis. This can lead to the generation of further hypotheses about the system. For each of the incidental observations described below, formulate a prediction, and an outline of an investigation to test it. *The observation described in each case was not related to the hypothesis the experiment was designed to test:*

   (a) **Bacterial cultures**

   Prediction: _____

   _____

   Outline of the investigation: _____

   _____

   _____

   _____

   _____

   _____

Bacterial Cultures

Observation: During an experiment on bacterial growth, these girls noticed that the cultures grew at different rates when the dishes were left overnight in different parts of the laboratory.

   (b) **Plant cloning**

   Prediction: _____

   _____

   _____

   Outline of the investigation: _____

   _____

   _____

   _____

   _____

   _____

   _____

Plant Cloning

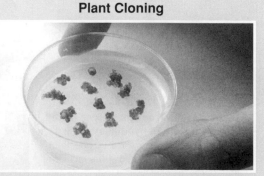

Observation: During an experiment on plant cloning, a scientist noticed that the root length of plant clones varied depending on the concentration of a hormone added to the agar.

Planning an Investigation

Investigations involve written stages (planning and reporting), at the start and end. The middle stage is the practical work when the data are collected. Practical work may be laboratory or field based. Typical lab based studies involve investigating how a biological response is affected by manipulating a particular **variable**, e.g. temperature. Field work often involves investigating features of a population or community. These may be interrelationships, such as competition, or patterns, such as zonation. Where quantitative information must be gathered from the population or community, particular techniques (such as quadrat sampling) and protocols (e.g. random placement of sampling units) apply. These aspects of practical work are covered in *Senior Biology 2*. Investigations in the field are usually more complex than those in the laboratory because natural systems have many more variables that cannot easily be controlled or accounted for.

Planning	Execution	Analysis and Reporting

- Formulate your hypothesis from an observation.
- Use a checklist (see the next activity) or a template (above) to construct a plan.

- Spend time (as appropriate to your study) collecting the data.
- Record the data in a systematic format (e.g. a table or spreadsheet).

- Analyze the data using graphs, tables, or statistics to look for trends or patterns.
- Write up your report including all the necessary sections.

Identifying Variables

A variable is any characteristic or property able to take any one of a range of values. Investigations often look at the effect of changing one variable on another. It is important to identify all variables in an investigation: independent, dependent, and controlled, although there may be nuisance factors of which you are unaware. In all fair tests, only one variable is changed by the investigator.

Dependent variable
- Measured during the investigation.
- Recorded on the y axis of the graph.

Controlled variables
- Factors that are kept the same or controlled.
- List these in the method, as appropriate to your own investigation.

Independent variable
- Set by the person carrying out the investigation.
- Recorded on the x axis of the graph.

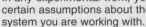

Assumptions

In any experimental work, you will make certain assumptions about the biological system you are working with.

Assumptions are features of the system (and your experiment) that you assume to be true but do not (or cannot) test.

Examples of Investigations

Aim		Variables	
Investigate the effect of varying ...	on the following ...	Independent variable	Dependent variable
Temperature	Leaf width	Temperature	Leaf width
Light intensity	Activity of woodlice	Light intensity	Woodlice activity
Soil pH	Plant height at age 6 months	pH	Plant height

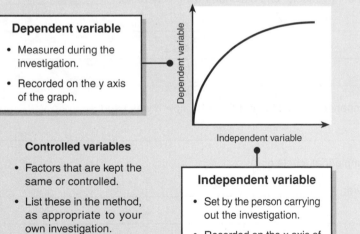

Related activities: Variables and Data
Web links: Space for Species, Terms and Notation

DA 2

In order to write a sound method for your investigation, you need to determine how the independent, dependent, and controlled variables will be set and measured (or monitored). A good understanding of your methodology is crucial to a successful investigation. You must be clear about how much data, and what type of data, you will collect. You should also have a good idea about how you plan to analyze the data. Use the example below to practise your skills in identifying this type of information.

Case Study: Catalase Activity

Catalase is an enzyme that converts hydrogen peroxide (H_2O_2) to oxygen and water. An experiment investigated the effect of temperature on the rate of the catalase reaction. Small ($10 \ cm^3$) test tubes were used for the reactions, each containing $0.5 \ cm^3$ of enzyme and 4 cm^3 of hydrogen peroxide. Reaction rates were assessed at four temperatures ($10°C$, $20°C$, $30°C$, and $60°C$). For each temperature, there were two reaction tubes (e.g. tubes 1 and 2 were both kept at $10°C$). The height of oxygen bubbles present after one minute of reaction was used as a measure of the reaction rate; a faster reaction rate produced more bubbles. The entire experiment, involving eight tubes, was repeated on two separate days.

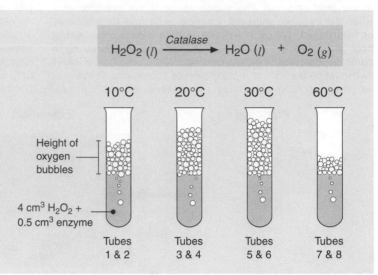

1. Write a suitable aim for this experiment: _____
   _____

2. Write a suitable hypothesis for this experiment: _____
   _____

3. (a) Name the **independent variable**: _____

   (b) State the range of values for the independent variable: _____

   (c) Name the unit for the independent variable: _____

   (d) List the equipment needed to set the independent variable, and describe how it was used: _____
   _____

4. (a) Name the **dependent variable**: _____

   (b) Name the unit for the dependent variable: _____

   (c) List the equipment needed to measure the dependent variable, and describe how it was used: _____
   _____

5. (a) Each temperature represents a treatment/sample/trial (circle one):

   (b) State the number of tubes at each temperature: _____

   (c) State the sample size for each treatment: _____

   (d) State how many times the whole investigation was repeated: _____

6. Explain why it would have been desirable to have included an extra tube containing no enzyme: _____
   _____

7. Identify three variables that might have been controlled in this experiment, and how they could have been monitored:

   (a) _____

   (b) _____

   (c) _____

8. Explain why controlled variables should be monitored carefully: _____
   _____

Experimental Method

An aim, hypothesis, and method for an experiment are described below. Explanations of the types of variables for which data are collected, and methods of recording these, are provided in the next two activities. The method described below includes numbered steps and incorporates other features identified in the previous activity. The method can be thought of as a 'statement of intent' for the practical work, and it may need slight changes during execution. The investigation described below was based on the observation that plant species 'A' was found growing in soil with a low pH (pH 4-5). The investigators wondered whether plant species 'A' was adapted to grow more vigorously under acid conditions than under alkaline or neutral conditions.

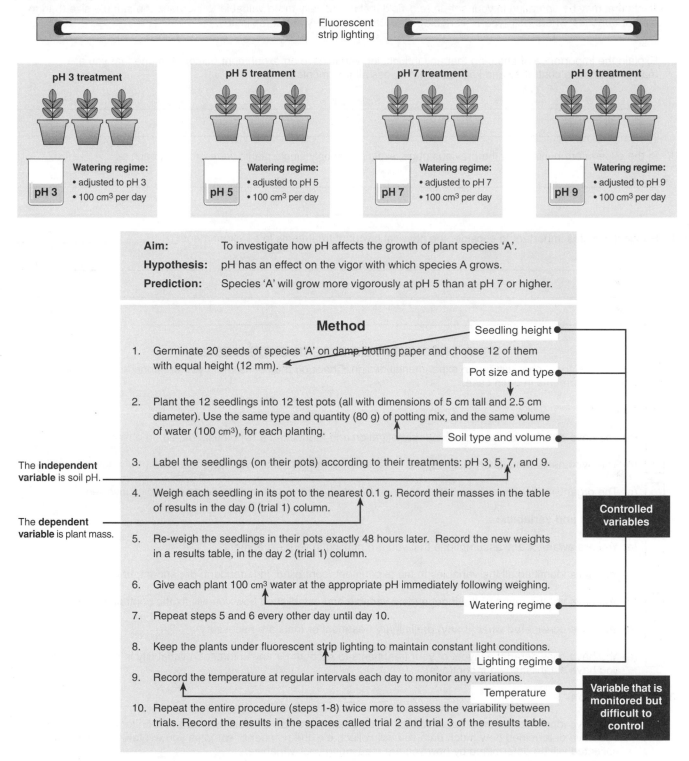

Fluorescent strip lighting

pH 3 treatment

Watering regime:
• adjusted to pH 3
• 100 cm³ per day

pH 3

pH 5 treatment

Watering regime:
• adjusted to pH 5
• 100 cm³ per day

pH 5

pH 7 treatment

Watering regime:
• adjusted to pH 7
• 100 cm³ per day

pH 7

pH 9 treatment

Watering regime:
• adjusted to pH 9
• 100 cm³ per day

pH 9

Skills in Biology

Aim: To investigate how pH affects the growth of plant species 'A'.

Hypothesis: pH has an effect on the vigor with which species A grows.

Prediction: Species 'A' will grow more vigorously at pH 5 than at pH 7 or higher.

Method

Seedling height

1. Germinate 20 seeds of species 'A' on damp blotting paper and choose 12 of them with equal height (12 mm).

Pot size and type

2. Plant the 12 seedlings into 12 test pots (all with dimensions of 5 cm tall and 2.5 cm diameter). Use the same type and quantity (80 g) of potting mix, and the same volume of water (100 cm³), for each planting.

Soil type and volume

The **independent variable** is soil pH.

3. Label the seedlings (on their pots) according to their treatments: pH 3, 5, 7, and 9.

4. Weigh each seedling in its pot to the nearest 0.1 g. Record their masses in the table of results in the day 0 (trial 1) column.

Controlled variables

The **dependent variable** is plant mass.

5. Re-weigh the seedlings in their pots exactly 48 hours later. Record the new weights in a results table, in the day 2 (trial 1) column.

6. Give each plant 100 cm³ water at the appropriate pH immediately following weighing.

Watering regime

7. Repeat steps 5 and 6 every other day until day 10.

8. Keep the plants under fluorescent strip lighting to maintain constant light conditions.

Lighting regime

9. Record the temperature at regular intervals each day to monitor any variations.

Temperature

Variable that is monitored but difficult to control

10. Repeat the entire procedure (steps 1-8) twice more to assess the variability between trials. Record the results in the spaces called trial 2 and trial 3 of the results table.

1. Explain the best way to take account of natural variability between individuals when designing an experiment:

_____

_____

_____

_____

Replication in Experiments

Replication refers to the number of times you repeat your entire experimental design (including controls). True replication is not the same as increasing the sample size (*n*) although it is often used to mean the same thing. Replication accounts for any unusual and unforeseen effects that may be operating in your set-up (e.g. field trials of plant varieties where soil type is variable). Replication is necessary when you expect that the response of treatments will vary because of factors outside your control. It is a feature of higher level experimental designs, and complex statistics are needed to separate differences between replicate treatments. For simple experiments, it is usually more valuable to increase the sample size than to worry about replicates.

2. Explain the importance of ensuring that any influencing variables in an experiment (except the one that you are manipulating) are controlled and kept constant across all treatments:

_____

_____

3. In the experiment outlined on the previous page, explain why only single plants were grown in each pot:

_____

_____

4. Suggest why it is important to consider the physical layout of treatments in an experiment: _____

_____

_____

_____

YOUR CHECKLIST FOR EXPERIMENTAL DESIGN

The following provides a checklist for an experimental design. Check off the points when you are confident that you have satisfied the requirements in each case:

1. **Preliminary:**

   ☐ (a) You have determined the aim of your investigation and formulated a hypothesis based on observation(s).

   ☐ (b) The hypothesis (and its predictions) are testable using the resources you have available (the study is feasible).

   ☐ (c) The organism you have chosen is suitable for the study and you have considered the ethics involved.

2. **Assumptions and variables:**

   ☐ (a) You are aware of any assumptions that you are making in your experiment.

   ☐ (b) You have identified all the variables in the experiment (controlled, independent, dependent, uncontrollable).

   ☐ (c) You have set the range of the independent variable and established how you will fix the controlled variables.

   ☐ (d) You have considered what (if any) preliminary treatment or trials are necessary.

   ☐ (e) You have considered the layout of your treatments to account for any unforeseen variability in your set-up and you have established your control(s).

3. **Data collection:**

   ☐ (a) You have identified the units for all variables and determined how you will measure or monitor each variable. You have determined how much data you will collect, e.g. the number of samples you will take. The type of data collected will be determined by how you are measuring your variables.

   ☐ (b) You have considered how you will analyze the data you collect and made sure that your experimental design allows you to answer the questions you have asked.

   ☐ (c) You have designed a method for systematically recording your results and had this checked with a teacher. The format of your results table or spreadsheet accommodates all your raw results, any transformations you intend to make, and all trials and treatments.

   ☐ (d) You have recorded data from any preliminary trials and any necessary changes to your methodology.

Recording Results

Designing a table to record your results is part of planning your investigation. Once you have collected all your data, you will need to analyze and present it. To do this, it may be necessary to transform your data first, by calculating a mean or a rate. An example of a table for recording results is presented below. This example relates to the investigation described in the previous activity, but it represents a relatively standardized layout. The labels on the columns and rows are chosen to represent the design features of the investigation. The first column contains the entire range chosen for the independent variable. There are spaces for multiple sampling units, repeats (trials), and averages. A version of this table should be presented in your final report.

Dependent variable and its units

Space for repeats of the experimental design (in this case, three trials).

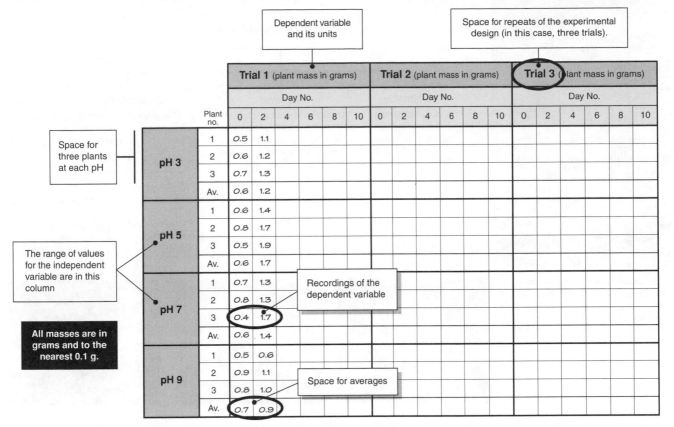

	Plant no.	Trial 1 (plant mass in grams)						Trial 2 (plant mass in grams)						Trial 3 (plant mass in grams)					
		Day No.						Day No.						Day No.					
		0	2	4	6	8	10	0	2	4	6	8	10	0	2	4	6	8	10
pH 3	1	0.5	1.1																
	2	0.6	1.2																
	3	0.7	1.3																
	Av.	0.6	1.2																
pH 5	1	0.6	1.4																
	2	0.8	1.7																
	3	0.5	1.9																
	Av.	0.6	1.7																
pH 7	1	0.7	1.3																
	2	0.8	1.3																
	3	0.4	1.7																
	Av.	0.6	1.4																
pH 9	1	0.5	0.6																
	2	0.9	1.1																
	3	0.8	1.0																
	Av.	0.7	0.9																

Space for three plants at each pH

The range of values for the independent variable are in this column

All masses are in grams and to the nearest 0.1 g.

Recordings of the dependent variable

Space for averages

1. In the space (below) design a table to collect data from the case study below. Include space for individual results and averages from the three set ups (use the table above as a guide).

Case Study
Carbon dioxide levels in a respiration chamber

A datalogger was used to monitor the concentrations of carbon dioxide (CO_2) in respiration chambers containing five green leaves from one plant species. The entire study was performed in conditions of full light (quantified) and involved three identical set-ups. The CO_2 concentrations were measured every minute, over a period of ten minutes, using a CO_2 sensor. A mean CO_2 concentration (for the three set-ups) was calculated. The study was carried out two more times, two days apart.

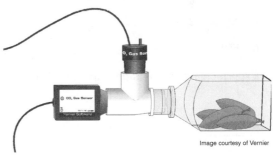

Image courtesy of Vernier

2. Next, the effect of various light intensities (low light, half-light, and full light) on CO_2 concentration was investigated. Describe how the results table for this investigation would differ from the one you have drawn above (for full light only):

_____

_____

Variables and Data

When planning any kind of biological investigation, it is important to consider the type of data that will be collected. It is best, whenever possible, to collect quantitative or numerical data, as these data lend themselves well to analysis and statistical testing. Recording data in a systematic way as you collect it, e.g. using a table or spreadsheet, is important, especially if data manipulation and transformation are required. It is also useful to calculate summary, descriptive statistics (e.g. mean, median) as you proceed. These will help you to recognize important trends and features in your data as they become apparent.

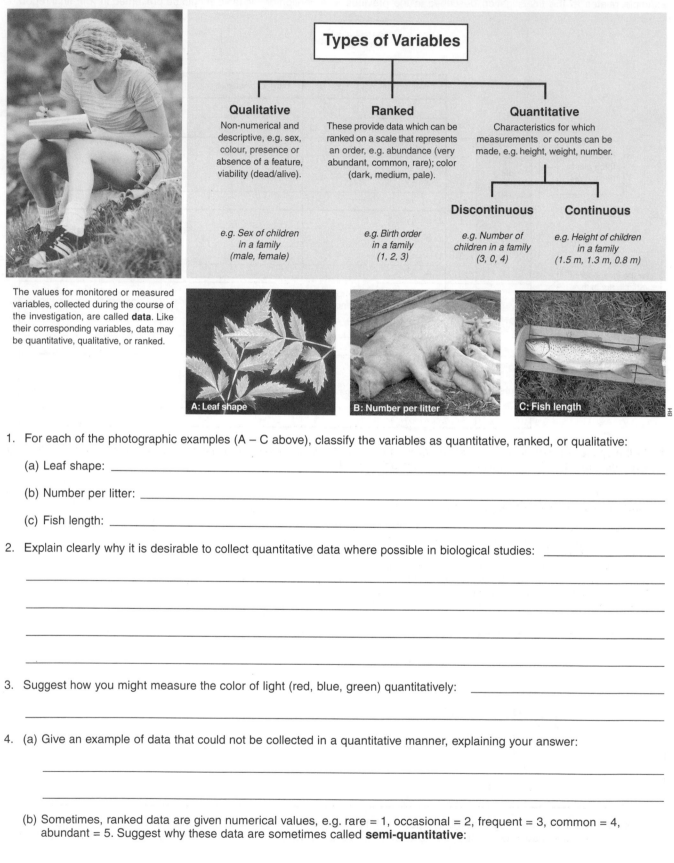

Types of Variables

Qualitative
Non-numerical and descriptive, e.g. sex, colour, presence or absence of a feature, viability (dead/alive).

Ranked
These provide data which can be ranked on a scale that represents an order, e.g. abundance (very abundant, common, rare); color (dark, medium, pale).

Quantitative
Characteristics for which measurements or counts can be made, e.g. height, weight, number.

Discontinuous

Continuous

e.g. Sex of children in a family (male, female)

e.g. Birth order in a family (1, 2, 3)

e.g. Number of children in a family (3, 0, 4)

e.g. Height of children in a family (1.5 m, 1.3 m, 0.8 m)

The values for monitored or measured variables, collected during the course of the investigation, are called **data**. Like their corresponding variables, data may be quantitative, qualitative, or ranked.

A: Leaf shape

B: Number per litter

C: Fish length

1. For each of the photographic examples (A – C above), classify the variables as quantitative, ranked, or qualitative:

   (a) Leaf shape: _____

   (b) Number per litter: _____

   (c) Fish length: _____

2. Explain clearly why it is desirable to collect quantitative data where possible in biological studies: _____

   _____

   _____

   _____

   _____

3. Suggest how you might measure the color of light (red, blue, green) quantitatively: _____

   _____

4. (a) Give an example of data that could not be collected in a quantitative manner, explaining your answer:

   _____

   _____

   (b) Sometimes, ranked data are given numerical values, e.g. rare = 1, occasional = 2, frequent = 3, common = 4, abundant = 5. Suggest why these data are sometimes called **semi-quantitative**:

   _____

   _____

Related activities: Descriptive Statistics

Periodicals:
Descriptive statistics

Manipulating Raw Data

The data collected by measuring or counting in the field or laboratory are called **raw data**. They often need to be changed (**transformed**) into a form that makes it easier to identify important features of the data (e.g. trends). Some basic calculations, such as totals (the sum of all data values for a variable), are made as a matter of course to compare replicates or as a prelude to other transformations. The calculation of **rate** (amount per unit time) is another example of a commonly performed calculation, and is appropriate for many biological situations (e.g. measuring growth or weight loss or gain). For a line graph, with time as the independent variable plotted against the values of the biological response, the slope of the line is a measure of the rate. Biological investigations often compare the rates of events in different situations (e.g. the rate of photosynthesis in the light and in the dark). Other typical transformations include frequencies (number of times a value occurs) and percentages (fraction of 100).

Skills in Biology

Tally Chart

Records the number of times a value occurs in a data set

HEIGHT (cm)	TALLY	TOTAL
0 - 0.99	III	3
1 - 1.99	HHH I	6
2 - 2.99	HHH HHH	10
3 - 3.99	HHH HHH II	12
4 - 4.99	III	3
5 - 5.99	II	2

- A useful first step in analysis; a neatly constructed tally chart doubles as a simple histogram.

- Cross out each value on the list as you tally it to prevent double entries. Check all values are crossed out at the end and that totals agree.

Example: Height of 6d old seedlings

Percentages

Expressed as a fraction of 100

Women	Body mass (kg)	Lean body mass (kg)	% lean body mass
Athlete	50	38	76.0
Lean	56	41	73.2
Normal weight	65	46	70.8
Overweight	80	48	60.0
Obese	95	52	54.7

- Percentages provide a clear expression of what proportion of data fall into any particular category, e.g. for pie graphs.

- Allows meaningful comparision between different samples.

- Useful to monitor change (e.g. % increase from one year to the next).

Example: Percentage of lean body mass in women

Rates

Expressed as a measure per unit time

Time (minutes)	Cumulative sweat loss (mL)	Rate of sweat loss (mL min^{-1})
0	0	0
10	50	5
20	130	8
30	220	9
60	560	11.3

- Rates show how a variable changes over a standard time period (e.g. one second, one minute, or one hour).

- Rates allow meaningful comparison of data that may have been recorded over different time periods.

Example: Rate of sweat loss in exercise

---

1. Explain why you might perform basic data transformations: _____

_____

2. (a) Describe a transformation for data relating to the relative abundance of plant species in different habitats:

_____

(b) Explain your answer: _____

_____

3. Complete the transformations on the table (right). The first value is given for you.

Table: *Incidence of cyanogenic clover in different areas*

Working: 120 ÷ 158 = 0.76 = 76%

This is the number of cyanogenic clover out of the total.

Incidence of cyanogenic clover in different areas

Clover plant type	Frost free area		Frost prone area		Totals
	Number	%	Number	%	
Cyanogenic	120	76	22		
Acyanogenic	38		120		
Total	158				

Periodicals:

Periodicals:

Related activities: Variables and Data

DA 2

Constructing Tables

Tables provide a convenient way to systematically record and condense a large amount of information for later presentation and analysis. The protocol for creating tables for recording data during the course of an investigation is provided elsewhere, but tables can also provide a useful summary in the results section of a finished report. They provide an accurate record of numerical values and allow you to organize your data in a way that allows you to clarify the relationships and trends that are apparent. Columns can be provided to display the results of any data transformations such as rates. Some basic descriptive statistics (such as mean or standard deviation) may also be included prior to the data being plotted. For complex data sets, graphs tend to be used in preference to tables, although the latter may be provided as an appendix.

Presenting Data in Tables

Tables should have an accurate, descriptive title. Number tables consecutively through the report.

Independent variable in the left column.

Heading and subheadings identify each set of data and show units of measurement.

Table 1: Length and growth of the third internode of bean plants receiving three different hormone treatments (data are given ± standard deviation).

Treatment	Sample size	Mean rate of internode growth (mm day $^{-1}$)	Mean internode length (mm)	Mean mass of tissue added (g day $^{-1}$)
Control	50	0.60 ± 0.04	32.3 ± 3.4	0.36 ± 0.025
Hormone 1	46	1.52 ± 0.08	41.6 ± 3.1	0.51 ± 0.030
Hormone 2	98	0.82 ± 0.05	38.4 ± 2.9	0.56 ± 0.028
Hormone 3	85	2.06 ± 0.19	50.2 ± 1.8	0.68 ± 0.020

Control values (if present) should be placed at the beginning of the table.

Each row should show a different experimental treatment, organism, sampling site etc.

Columns for comparison should be placed alongside each other. Show values only to the level of significance allowable by your measuring technique.

Organise the columns so that each category of like numbers or attributes is listed vertically.

Tables can be used to show a calculated measure of spread of the values about the mean.

1. Describe two advantages of using a table format for data presentation:

   (a) _____

   _____

   (b) _____

   _____

2. Explain why you might tabulate data before you presented it in a graph format: _____

   _____

   _____

3. (a) Explain the value of tabulating basic descriptive statistics rather than the raw data: _____

   _____

   _____

   (b) Explain the value of including a measure of spread (dispersion) for a calculated statistic in a table:

   _____

   _____

4. Explain why control values should be placed at the beginning of a table: _____

   _____

   _____

Related activities: Variables and Data, Manipulating Raw Data, Constructing Graphs

Periodicals: Descriptive statistics

Constructing Graphs

Presenting results in a graph format provides a visual image of trends in data in a minimum of space. The choice between graphing or tabulation depends on the type and complexity of the data and the information that you are wanting to convey. Presenting graphs properly requires attention to a few basic details, including correct orientation and labeling of the axes, and accurate plotting of points. Common graphs include scatter plots and line graphs (for continuous data), and bar charts and histograms (for categorical data). Where there is an implied trend, a line of best fit can be drawn through the data points, as indicated in the figure below. Further guidelines for drawing graphs are provided on the following pages.

Presenting Data in Graph Format

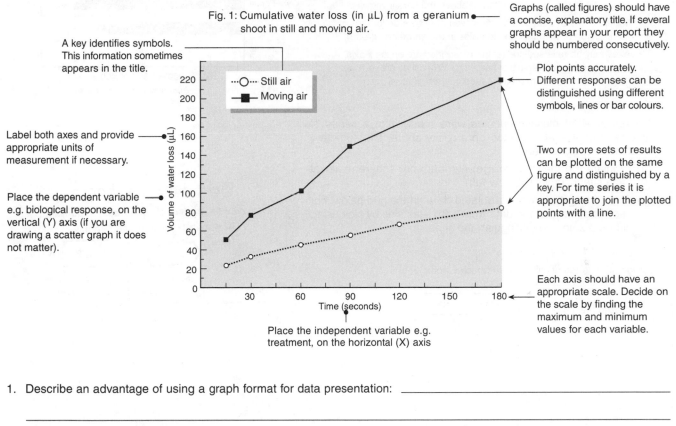

Fig. 1: Cumulative water loss (in μL) from a geranium shoot in still and moving air.

A key identifies symbols. This information sometimes appears in the title.

Label both axes and provide appropriate units of measurement if necessary.

Place the dependent variable e.g. biological response, on the vertical (Y) axis (if you are drawing a scatter graph it does not matter).

Place the independent variable e.g. treatment, on the horizontal (X) axis

Graphs (called figures) should have a concise, explanatory title. If several graphs appear in your report they should be numbered consecutively.

Plot points accurately. Different responses can be distinguished using different symbols, lines or bar colours.

Two or more sets of results can be plotted on the same figure and distinguished by a key. For time series it is appropriate to join the plotted points with a line.

Each axis should have an appropriate scale. Decide on the scale by finding the maximum and minimum values for each variable.

Skills in Biology

1. Describe an advantage of using a graph format for data presentation: _____

_____

2. (a) Explain the importance of using an appropriate scale on a graph: _____

_____

   (b) Scales on X and Y axes may sometimes be "floating" (not meeting in the lower left corner), or they may be broken using a double slash and recontinued. Explain the purpose of these techniques:

_____

_____

3. (a) Explain what is wrong with the graph plotted to the right:

   _____

   _____

   _____

   _____

   (b) Describe the graph's appearance if it were plotted correctly:

   _____

   _____

   _____

   _____

   _____

Fig. 1: Yeast growth against time

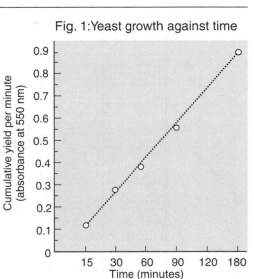

Periodicals:
It's a plot!
Dealing with data

Related activities: Descriptive Statistics

DA 2

Drawing Bar Graphs

Guidelines for Bar Graphs

Bar graphs are appropriate for data that are non-numerical and **discrete** for at least one variable, i.e. they are grouped into separate categories. There are no dependent or independent variables. Important features of this type of graph include:

- Data are collected for discontinuous, non-numerical categories (e.g. place, color, and species), so the bars do not touch.

- Data values may be entered on or above the bars if you wish.

- Multiple sets of data can be displayed side by side for direct comparison (e.g. males and females in the same age group).

- Axes may be reversed so that the categories are on the x axis, i.e. the bars can be vertical or horizontal. When they are vertical, these graphs are sometimes called column graphs.

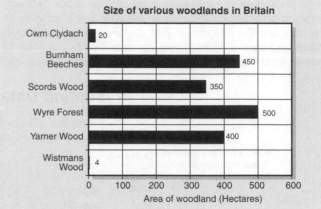

Size of various woodlands in Britain

1. Counts of eight mollusc species were made from a series of quadrat samples at two sites on a rocky shore. The summary data are presented here.

    (a) Tabulate the mean (**average**) numbers per square meter at each site in Table 1 (below left).

    (b) Plot a **bar graph** of the tabulated data on the grid below. For each species, plot the data from both sites side by side using different colors to distinguish the sites.

Average abundance of 8 molluscan species from two sites along a rocky shore.

Species	Mean (no. m^{-2})	
	Site 1	Site 2

Field data notebook

Total counts at site 1 (11 quadrats) and site 2 (10 quadrats). Quadrats 1 sq m.

Species	Site 1 Total	Site 1 Mean	Site 2 Total	Site 2 Mean
	No m^{-2}		No m^{-2}	
Ornate limpet	232	21	299	30
Radiate limpet	68	6	344	34
Limpet sp. A	420	38	0	0
Cats-eye	68	6	16	2
Top shell	16	2	43	4
Limpet sp. B	628	57	389	39
Limpet sp. C	0	0	22	2
Chiton	12	1	30	3

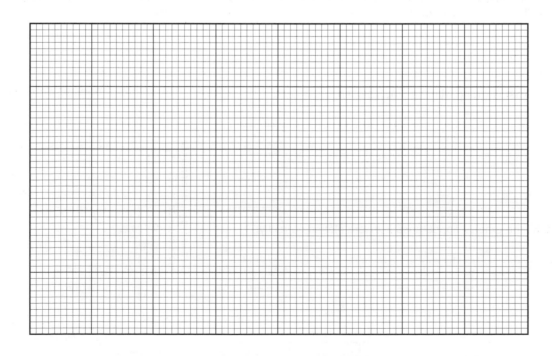

Related activities: Constructing Tables, Descriptive Statistics

Periodicals: Drawing graphs

Drawing Histograms

Guidelines for Histograms

Histograms are plots of **continuous** data and are often used to represent frequency distributions, where the y-axis shows the number of times a particular measurement or value was obtained. For this reason, they are often called frequency histograms. Important features of this type of graph include:

- The data are numerical and continuous (e.g. height or weight), so the bars touch.

- The x-axis usually records the class interval. The y-axis usually records the number of individuals in each class interval (frequency).

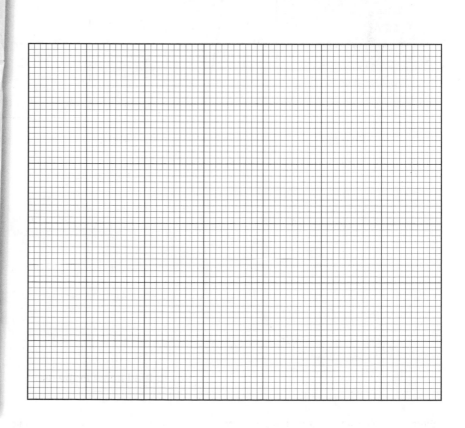

Frequency of different mass classes of animals in a population.

1. The weight data provided below were recorded from 95 individuals (male and female), older than 17 years.

   (a) Create a tally chart (frequency table) in the frame provided, organizing the weight data into a form suitable for plotting. An example of the tally for the weight grouping 55-59.9 kg has been completed for you as an example. Note that the raw data values, once they are recorded as counts on the tally chart, are crossed off the data set in the notebook. It is important to do this in order to prevent data entry errors.

   (b) Plot a **frequency histogram** of the tallied data on the grid provided below.

Weight (kg)	Tally	Total
45-49.9		
50-54.9		
55-59.9	⊥⊥⊤ //	7
60-64.9		
65-69.9		
70-74.9		
75-79.9		
80-84.9		
85-89.9		
90-94.9		
95-99.9		
100-104.9		
105-109.9		

Lab notebook

Weight (in kg) of 95 individuals

63.4	81.2	65
56.5	83.3	75.6
84	95	76.8
81.5	105.5	67.8
73.4	82	68.3
56	73.5	63.5
60.4	75.2	58
83.5	63	58.5
82	70.4	50
61	82.2	92
55.2	87.8	91.5
48	86.5	88.3
53.5	85.5	81
63.8	87	72
69	98	66.5
82.8	71	61.5
68.5	76	66
67.2	72.5	65.5
82.5	61	67.4
83	60.5	73
78.4	67	67
76.5	86	71
83.4	85	70.5
77.5	93.5	65.5
77	62	68
87	62.5	90
89	63	83.5
93.4	60	73
83	71.5	66
80	73.8	57.5
76	77.5	76
56	74	

Periodicals:
Drawing graphs

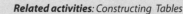

Related activities: Constructing Tables

DA 2

Skills in Biology

Drawing Pie Graphs

Guidelines for Pie Graphs

Pie graphs can be used instead of bar graphs, generally in cases where there are six or fewer categories involved. A pie graph provides strong visual impact of the relative proportions in each category, particularly where one of the categories is very dominant. Features of pie graphs include:

- The data for one variable are discontinuous (non-numerical or categories).

- The data for the dependent variable are usually in the form of counts, proportions, or percentages.

- Pie graphs are good for visual impact and showing relative proportions.

- They are not suitable for data sets with a large number of categories.

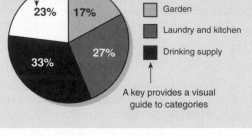

Average residential water use

Values may be shown

Key
- Bath, shower, toilet
- Garden
- Laundry and kitchen
- Drinking supply

A key provides a visual guide to categories

1. The data provided below are from a study of the diets of three vertebrates.

   (a) Tabulate the data from the notebook in the frame provided. Calculate the angle for each percentage, given that each percentage point is equal to 3.6° (the first example is provided: 23.6 x 3.6 = 85).

   (b) Plot a pie graph for each animal in the circles provided. The circles have been marked at 5° intervals to enable you to do this exercise without a protractor. For the purposes of this exercise, begin your pie graphs at the 0° (= 360°) mark and work in a clockwise direction from the largest to the smallest percentage. Use one key for all three pie graphs.

Field data notebook

% of different food items in the diet

Food item	Ferrets	Rats	Cats
Birds	23.6	1.4	6.9
Crickets	15.3	23.6	0
Other insects (not crickets)	15.3	20.8	1.9
Voles	9.2	0	19.4
Rabbits	8.3	0	18.1
Rats	6.1	0	43.1
Mice	13.9	0	10.6
Fruits and seeds	0	40.3	0
Green leaves	0	13.9	0
Unidentified	8.3	0	0

Percentage occurrence of different foods in the diet of ferrets, rats, and cats. Graph angle representing the % is shown to assist plotting.

Food item in diet	Ferrets		Rats		Cats	
	% in diet	Angle (°)	% in diet	Angle (°)	% in diet	Angle (°)
Birds	23.6	85				

Ferrets
0°

Rats
0°

Cats
0°

Key to food items in the diet

| Birds | Crickets | Other insects | Voles | Rabbits | Rats | Mice | Green leaves | Fruits & seeds | Unidentified |

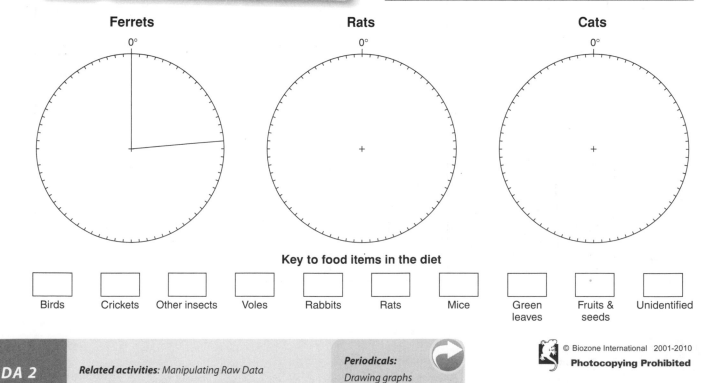

Related activities: Manipulating Raw Data

Periodicals:
Drawing graphs

Drawing Kite Graphs

Guidelines for Kite Graphs

Kite graphs are ideal for representing distributional data, e.g. abundance along an environmental gradient. They are elongated figures drawn along a baseline. Important features of kite graphs include:

- Each kite represents changes in species abundance across a landscape. The abundance can be calculated from the kite width.
- They often involve plots for more than one species; this makes them good for highlighting probable differences in habitat preferences between species.
- A thin line on a kite graph represents species absence.
- The axes can be reversed depending on preference.
- Kite graphs may also be used to show changes in distribution with time, for example, with daily or seasonal cycles of movement.

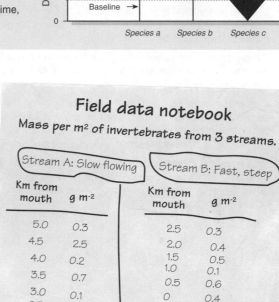

Species abundance along a rocky shoreline

Skills in Biology

1. The following data were collected from three streams of different lengths and flow rates. Invertebrates were collected at 0.5 km intervals from the headwaters (0 km) to the stream mouth. Their wet weight was measured and recorded (per m^2).

   (a) Tabulate the data below for plotting.

   (b) Plot a **kite graph** of the data from all three streams on the grid provided below. Do not forget to include a scale so that the weight at each point on the kite can be calculated.

Wet mass of invertebrates along three different streams

Distance from mouth (km)	Wet weight (g m^{-2})		
	Stream A	Stream B	Stream C

Field data notebook
Mass per m^2 of invertebrates from 3 streams.

Stream A: Slow flowing

Km from mouth	g m^{-2}
5.0	0.3
4.5	2.5
4.0	0.2
3.5	0.7
3.0	0.1
2.5	0.6
2.0	0.3
1.5	0.3
1.0	0.4
0.5	0.5
0	0.4

Stream B: Fast, steep

Km from mouth	g m^{-2}
2.5	0.3
2.0	0.4
1.5	0.5
1.0	0.1
0.5	0.6
0	0.4

Stream C: Steep torrent

Km from mouth	g m^{-2}
1.5	0.2
1.0	0
0.5	0.5
0	0

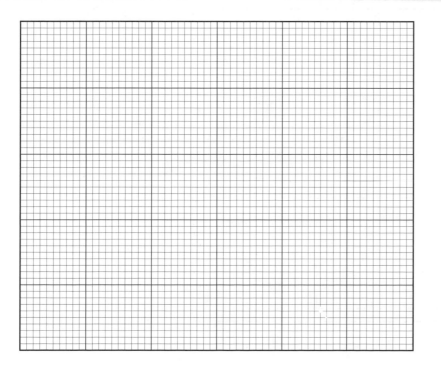

Periodicals:
Drawing graphs

Related activities: Transect Sampling

DA 2

Drawing Line Graphs

Guidelines for Line Graphs

Line graphs are used when one variable (the independent variable) affects another, the dependent variable. Line graphs can be drawn without a measure of spread (top figure, right) or with some calculated measure of data variability (bottom figure, right). Important features of line graphs include:

- The data must be continuous for both variables.

- The dependent variable is usually the biological response.

- The independent variable is often time or the experimental treatment.

- In cases where there is an implied trend (e.g. one variable increases with the other), a line of best fit is usually plotted through the data points to show the relationship.

- If fluctuations in the data are likely to be important (e.g. with climate and other environmental data) the data points are usually connected directly (point to point).

- Line graphs may be drawn with measure of error. The data are presented as points (the calculated means), with bars above and below, indicating a measure of variability or spread in the data (e.g. standard error, standard deviation, or 95% confidence intervals).

- Where no error value has been calculated, the scatter can be shown by plotting the individual data points vertically above and below the mean. By convention, bars are not used to indicate the range of raw values in a data set.

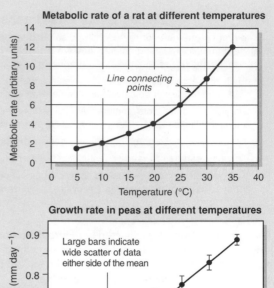

Metabolic rate of a rat at different temperatures

Line connecting points

Growth rate in peas at different temperatures

Large bars indicate wide scatter of data either side of the mean

1. The results (shown right) were collected in a study investigating the effect of temperature on the activity of an enzyme.

   (a) Using the results provided in the table (right), plot a line graph on the grid below:

   (b) Estimate the rate of reaction at 15°C: _____

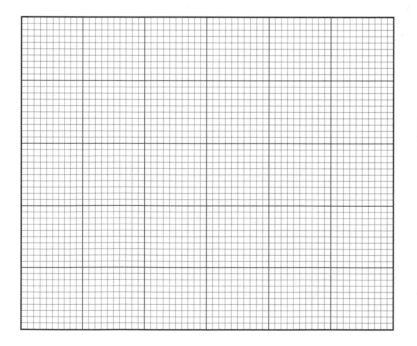

Lab Notebook

An enzyme's activity at different temperatures

Temperature (°C)	Rate of reaction (mg of product formed per minute)
10	1.0
20	2.1
30	3.2
35	3.7
40	4.1
45	3.7
50	2.7
60	0

Related activities: Manipulating Raw Data, Constructing Graphs, Interpreting Line Graphs

Periodicals: *Drawing graphs*

Plotting Multiple Data Sets

A single figure can be used to show two or more data sets, i.e. more than one curve can be plotted per set of axes. This type of presentation is useful when you want to visually compare the trends for two or more treatments, or the response of one species against the response of another. Important points regarding this format are:

- If the two data sets use the same measurement units and a similar range of values for the independent variable, one scale on the y axis is used.

- If the two data sets use different units and/or have a very different range of values for the independent variable, two scales for the y axis are used (see example provided). The scales can be adjusted if necessary to avoid overlapping plots.

- The two curves must be distinguished with a key.

Transpiration and root uptake rates in peas at different relative humidity

2. A census of a deer population on an island indicated a population of 2000 animals in 1960. In 1961, ten wolves (natural predators of deer) were brought to the island in an attempt to control deer numbers. Over the next nine years, the numbers of deer and wolves were monitored. The results of these population surveys are presented in the table, right.

(a) Plot a line graph (joining the data points) for the tabulated results. Use one scale (on the left) for numbers of deer and another scale (on the right) for the number of wolves. Use different symbols or colors to distinguish the lines and include a key.

Field data notebook
Results of a population survey on an island

Time (yr)	Wolf numbers	Deer numbers
1961	10	2000
1962	12	2300
1963	16	2500
1964	22	2360
1965	28	2244
1966	24	2094
1967	21	1968
1968	18	1916
1969	19	1952

(b) Study the line graph that you plotted for the wolf and deer census on the previous page. Provide a plausible explanation for the pattern in the data, stating the evidence available to support your reasoning:

_____

_____

_____

_____

3. In a sampling program, the number of perch and trout in a hydro-electric reservoir were monitored over a period of time. A colony of black shag was also present. Shags take large numbers of perch and (to a lesser extent) trout. In 1960-61, 424 shags were removed from the lake during the nesting season and nest counts were made every spring in subsequent years. In 1971, 60 shags were removed from the lake, and all existing nests dismantled. The results of the population survey are tabulated below (for reasons of space, the entire table format has been repeated to the right for 1970-1978).

(a) Plot a line graph (joining the data points) for the survey results. Use one scale (on the left) for numbers of perch and trout and another scale for the number of shag nests. Use different symbols to distinguish the lines and include a key.

(b) Use a vertical arrow to indicate the point at which shags and their nests were removed.

Results of population survey at a reservoir

Time (yr)	Fish number (average per haul)		Shag nest numbers	Time (yr) continued	Fish number (average per haul)		Shag nest numbers
	Trout	Perch			Trout	Perch	
1960	–	–	16	1970	1.5	6	35
1961	–	–	4	1971	0.5	0.7	42
1962	1.5	11	5	1972	1	0.8	0
1963	0.8	9	10	1973	0.2	4	0
1964	0	5	22	1974	0.5	6.5	0
1965	1	1	25	1975	0.6	7.6	2
1966	1	2.9	35	1976	1	1.2	10
1967	2	5	40	1977	1.2	1.5	32
1968	1.5	4.6	26	1978	0.7	2	28
1969	1.5	6	32				

Source: Data adapted from 1987 Bursary Examination

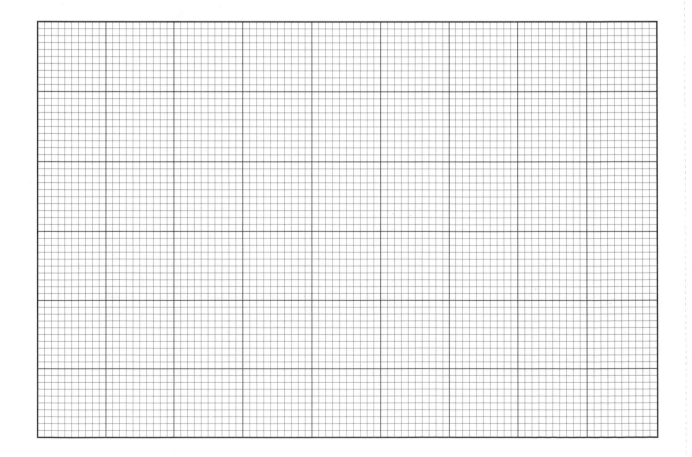

Interpreting Line Graphs

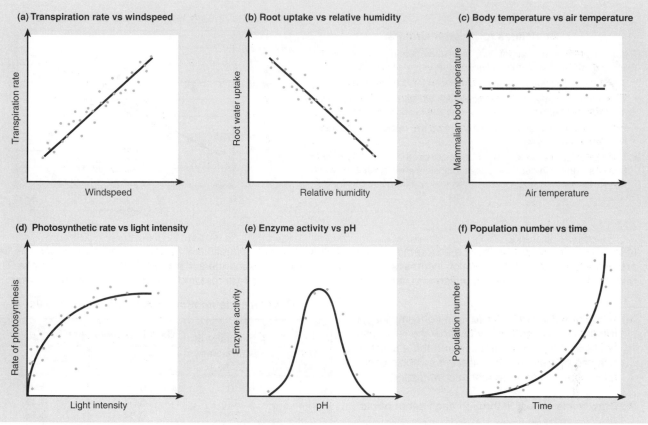

(a) Transpiration rate vs windspeed — Transpiration rate vs Windspeed

(b) Root uptake vs relative humidity — Root water uptake vs Relative humidity

(c) Body temperature vs air temperature — Mammalian body temperature vs Air temperature

(d) Photosynthetic rate vs light intensity — Rate of photosynthesis vs Light intensity

(e) Enzyme activity vs pH — Enzyme activity vs pH

(f) Population number vs time — Population number vs Time

1. For each of the graphs (b-f) above, give a description of the slope and an interpretation of how one variable changes with respect to the other. For the purposes of your description, call the independent variable (horizontal or x-axis) in each example "variable X" and the dependent variable (vertical or y-axis) "variable Y". Be aware that the existence of a relationship between two variables does not necessarily mean that the relationship is causative (although it may be).

(a) Slope: _Positive linear relationship, with constantly rising slope_

   Interpretation: _Variable Y (transpiration) increases regularly with increase in variable X (windspeed)_

(b) Slope: _____

   Interpretation: _____

(c) Slope: _____

   Interpretation: _____

(d) Slope: _____

   Interpretation: _____

(e) Slope: _____

   Interpretation: _____

(f) Slope: _____

   Interpretation: _____

2. Study the line graph of trout, perch and shag numbers that you plotted on the previous page:

   (a) Describe the evidence suggesting that the shag population is exercising some control over perch numbers:

   _____

   (b) Describe evidence that the fluctuations in shag numbers are related to fluctuations in trout numbers: _____

   _____

   _____

Periodicals:
Dealing with data

Related activities: Drawing Line Graphs

RA 2

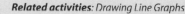

Drawing Scatter Plots

Guidelines for Scatter Graphs

A scatter graph is a common way to display continuous data where there is a relationship between two interdependent variables.

- The data for this graph must be continuous for both variables.

- There is no independent (manipulated) variable, but the variables are often correlated, i.e. they vary together in some predictable way.

- Scatter graphs are useful for determining the relationship between two variables.

- The points on the graph need not be connected, but a line of best fit is often drawn through the points to show the relationship between the variables (this may be drawn be eye or computer generated).

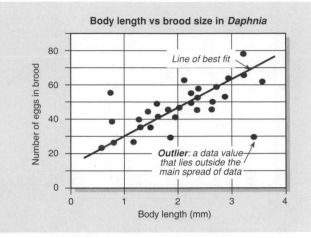

Body length vs brood size in *Daphnia*

1. In the example below, metabolic measurements were taken from seven Antarctic fish *Pagothenia borchgrevinski*. The fish are affected by a gill disease, which increases the thickness of the gas exchange surfaces and affects oxygen uptake. The results of oxygen consumption of fish with varying amounts of affected gill (at rest and swimming) are tabulated below.

   (a) Using **one** scale only for oxygen consumption, plot the data on the grid below to show the relationship between oxygen consumption and the amount of gill affected by disease. Use different symbols or colors for each set of data (at rest and swimming).

   (b) Draw a line of best fit through each set of points.

2. Describe the relationship between the amount of gill affected and oxygen consumption in the fish:

   (a) For the **at rest** data set:

   _____

   _____

   _____

   _____

   _____

   _____

   _____

   (b) For the **swimming** data set:

   _____

   _____

   _____

   _____

   _____

   _____

   _____

Oxygen consumption of fish with affected gills

Fish number	Percentage of gill affected	Oxygen consumption $(cm^3\ g^{-1}\ h^{-1})$	
		At rest	**Swimming**
1	0	0.05	0.29
2	95	0.04	0.11
3	60	0.04	0.14
4	30	0.05	0.22
5	90	0.05	0.08
6	65	0.04	0.18
7	45	0.04	0.20

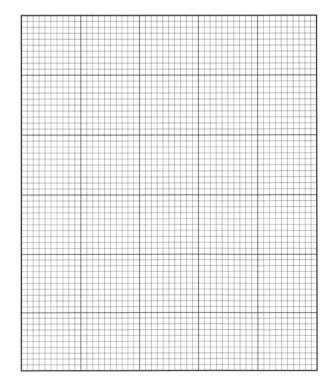

3. Describe how the gill disease affects oxygen uptake in resting fish:

   _____

   _____

Related activities: Interpreting Line Graphs

Biological Drawings

Microscopes are a powerful tool for examining cells and cell structures. In order to make a permanent record of what is seen when examining a specimen, it is useful to make a drawing. It is important to draw **what is actually seen**. This will depend on the **resolution** of the microscope being used. Resolution refers to the ability of a microscope to separate small objects that are very close together. Making drawings from mounted specimens is a skill. Drawing forces you to observe closely and accurately. While photographs are limited to representing appearance at a single moment in time, drawings can be composites of the observer's cumulative experience, with many different specimens of the same material. The total picture of an object thus represented can often communicate information much more effectively than a photograph. Your attention to the outline of suggestions below will help you to make more effective drawings. If you are careful to follow the suggestions at the beginning, the techniques will soon become habitual.

1. **Drawing materials**: All drawings should be done with a clear pencil line on good quality paper. A sharp HB pencil is recommended. A soft eraser of good quality is essential. Diagrams in ballpoint or fountain pen are unacceptable because they cannot be corrected.

2. **Positioning**: Center your diagram on the page. Do not draw it in a corner. This will leave plenty of room for the addition of labels once the diagram is completed.

3. **Size**: A drawing should be large enough to easily represent all the details you see without crowding. Rarely, if ever, are drawings too large, but they are often too small. Show only as much as is necessary for an understanding of the structure; a small section shown in detail will often suffice. It is time consuming and unnecessary, for example, to reproduce accurately the entire contents of a microscope field.

4. **Accuracy**: Your drawing should be a complete, accurate representation of the material you have observed, and should communicate your understanding of the material to anyone who looks at it. Avoid making "idealized" drawings; your drawing should be a picture of what you actually see, not what you imagine should be there. Proportions should be accurate. If necessary, measure the lengths of various

parts with a ruler. If viewing through a microscope, estimate them as a proportion of the field of view, then translate these proportions onto the page. When drawing shapes that indicate an outline, make sure the line is complete. Where two ends of a line do not meet (as in drawing a cell outline) then this would indicate that it has a hole in it.

5. **Technique**: Use only simple, narrow lines. Represent depth by stippling (dots close together). Indicate depth only when it is essential to your drawing (usually it is not). Do not use shading. Look at the specimen while you are drawing it.

6. **Labels**: Leave a good margin for labels. All parts of your diagram must be labeled accurately. Labeling lines should be drawn with a ruler and should not cross. Where possible, keep label lines vertical or horizontal. Label the drawing with:
   - A title, which should identify the material (organism, tissues or cells).
   - Magnification under which it was observed, or a scale to indicate the size of the object.
   - Names of structures.
   - In living materials, any movements you have seen.

Remember that drawings are intended as records for you, and as a means of encouraging close observation; artistic ability is not necessary. Before you turn in a drawing, ask yourself if you know what every line represents. If you do not, look more closely at the material. *Take into account the rules for biological drawings and draw what you see, not what you think you see!*

Examples of acceptable biological drawings: The diagrams below show two examples of biological drawings that are acceptable. The example on the left is of a whole organism and its size is indicated by a scale. The example on the right is of plant tissue: a group of cells that are essentially identical in the structure. It is not necessary to show many cells even though your view through the microscope may show them. As few as 2-4 will suffice to show their structure and how they are arranged. Scale is indicated by stating how many times larger it has been drawn. Do not confuse this with what magnification it was viewed at under the microscope. The abbreviation **T.S.** indicates that the specimen was a *cross* or *transverse section*.

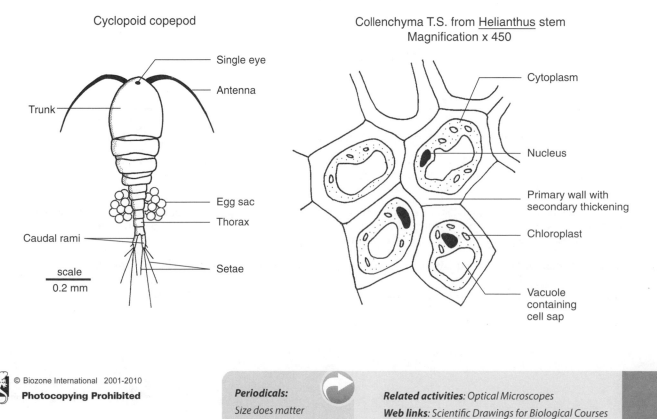

Cyclopoid copepod

Single eye
Antenna
Trunk
Egg sac
Thorax
Caudal rami
Setae
scale
0.2 mm

Collenchyma T.S. from <u>Helianthus</u> stem
Magnification x 450

Cytoplasm
Nucleus
Primary wall with secondary thickening
Chloroplast
Vacuole containing cell sap

Periodicals:
Size does matter

Related activities: Optical Microscopes
Web links: Scientific Drawings for Biological Courses

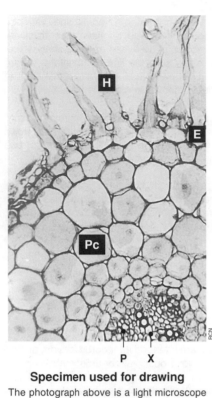

P   X

Specimen used for drawing

The photograph above is a light microscope view of a stained transverse section (cross section) of a root from a *Ranunculus* (buttercup) plant. It shows the arrangement of the different tissues in the root. The vascular bundle is at the center of the root, with the larger, central xylem vessels (**X**) and smaller phloem vessels (**P**) grouped around them. The root hair cells (**H**) are arranged on the external surface and form part of the epidermal layer (**E**). Parenchyma cells (**Pc**) make up the bulk of the root's mass. The distance from point **X** to point **E** on the photograph (above) is about 0.15 mm (150 μm).

An Unacceptable Biological Drawing

The diagram below is an example of how *not* to produce a biological drawing; it is based on the photograph to the left. There are many aspects of the drawing that are unacceptable. The exercise below asks you to identify the errors in this student's attempt.

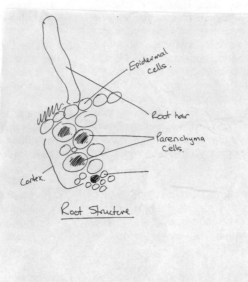

1. Identify and describe eight unacceptable features of the student's biological diagram above:

    (a) _____

    (b) _____

    (c) _____

    (d) _____

    (e) _____

    (f) _____

    (g) _____

    (h) _____

2. In the remaining space next to the 'poor example' (above) or on a blank piece of refill paper, attempt your own version of a biological drawing for the same material, based on the photograph above. Make a point of correcting all of the errors that you have identified in the sample student's attempt.

3. Explain why accurate biological drawings are more valuable to a scientific investigation than an 'artistic' approach:

    _____

    _____

    _____

Descriptive Statistics

For most investigations, measures of the biological response are made from more than one sampling unit. The sample size (the number of sampling units) will vary depending on the resources available. In lab based investigations, the sample size may be as small as two or three (e.g. two test-tubes in each treatment). In field studies, each individual may be a sampling unit, and the sample size can be very large (e.g. 100 individuals). It is useful to summarize the data collected using **descriptive statistics**.

Descriptive statistics, such as mean, median, and mode, can help to highlight patterns in data sets. Each of these statistics is appropriate to certain types of data or distributions, e.g. a mean is not appropriate for data with a skewed distribution (see below). Frequency graphs are useful for indicating the distribution of data. Standard deviation and standard error are statistics used to quantify the amount of spread in the data and evaluate the reliability of estimates of the true (population) mean.

Skills in Biology

Variation in Data

Whether they are obtained from observation or experiments, most biological data show variability. In a set of data values, it is useful to know the value about which most of the data are grouped; the center value. This value can be the mean, median, or mode depending on the type of variable involved (see schematic below). The main purpose of these statistics is to summarize important trends in your data and to provide the basis for statistical analyses.

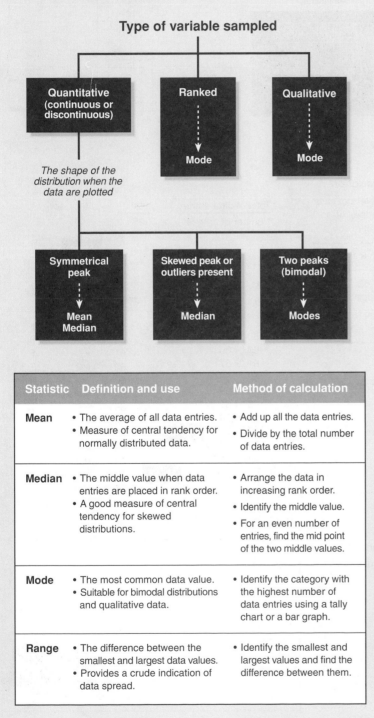

Distribution of Data

Variability in continuous data is often displayed as a **frequency distribution**. A frequency plot will indicate whether the data have a normal distribution (A), with a symmetrical spread of data about the mean, or whether the distribution is skewed (B), or bimodal (C). The shape of the distribution will determine which statistic (mean, median, or mode) best describes the central tendency of the sample data.

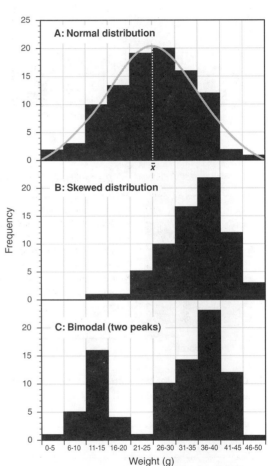

Statistic	Definition and use	Method of calculation
Mean	• The average of all data entries. • Measure of central tendency for normally distributed data.	• Add up all the data entries. • Divide by the total number of data entries.
Median	• The middle value when data entries are placed in rank order. • A good measure of central tendency for skewed distributions.	• Arrange the data in increasing rank order. • Identify the middle value. • For an even number of entries, find the mid point of the two middle values.
Mode	• The most common data value. • Suitable for bimodal distributions and qualitative data.	• Identify the category with the highest number of data entries using a tally chart or a bar graph.
Range	• The difference between the smallest and largest data values. • Provides a crude indication of data spread.	• Identify the smallest and largest values and find the difference between them.

When NOT to calculate a mean:

In certain situations, calculation of a simple arithmetic mean is inappropriate.

Remember:

• *DO NOT* calculate a mean from values that are already means (averages) themselves.

• *DO NOT* calculate a mean of ratios (e.g. percentages) for several groups of different sizes; go back to the raw values and recalculate.

• *DO NOT* calculate a mean when the measurement scale is not linear, e.g. pH units are not measured on a linear scale.

Periodicals:
Describing the normal distribution

Related activities: Variables and Data
Web links: Taking the Next Step

DA 2

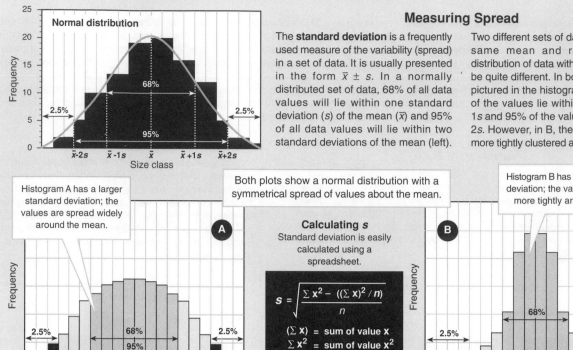

Measuring Spread

The **standard deviation** is a frequently used measure of the variability (spread) in a set of data. It is usually presented in the form $\bar{x} \pm s$. In a normally distributed set of data, 68% of all data values will lie within one standard deviation (s) of the mean ($\bar{x}$) and 95% of all data values will lie within two standard deviations of the mean (left).

Two different sets of data can have the same mean and range, yet the distribution of data within the range can be quite different. In both the data sets pictured in the histograms below, 68% of the values lie within the range $\bar{x} \pm 1s$ and 95% of the values lie within $\bar{x} \pm 2s$. However, in B, the data values are more tightly clustered around the mean.

Normal distribution

Histogram A has a larger standard deviation; the values are spread widely around the mean.

Both plots show a normal distribution with a symmetrical spread of values about the mean.

Histogram B has a smaller standard deviation; the values are clustered more tightly around the mean.

Calculating s
Standard deviation is easily calculated using a spreadsheet.

$$s = \sqrt{\frac{\sum x^2 - ((\sum x)^2 / n)}{n}}$$

$(\sum x)$ = sum of value x
$\sum x^2$ = sum of value x^2
n = sample size

Case Study: Fern Reproduction

Raw data (below) and descriptive statistics (right) from a survey of the number of sori found on the fronds of a fern plant.

Fern spores

Raw data: Number of sori per frond

64	60	64	62	68	66	63
69	70	63	70	70	63	62
71	69	59	70	66	61	70
67	64	63	64			

$$\frac{\text{Total of data entries}}{\text{Number of entries}} = \frac{1641}{25} = 66 \text{ sori}$$

Mean

Number of sori per frond (in rank order)	
59	66
60	66
61	67
62	68
62	69
63	69
63	70
63	70
63	70
64	70
64	70
64	71
64	

Median

Mode

Sori per frond	Tally	Total
59	✔	1
60	✔	1
61	✔	1
62	✔✔	2
63	✔✔✔✔	4
64	✔✔✔✔	4
65		0
66	✔✔	2
67	✔	1
68	✔	1
69	✔✔	2
70	✔✔✔✔✔	5
71	✔	1

1. Give a reason for the differences between the mean, median, and mode of the fern sori data:

_____

_____

_____

_____

2. Calculate the mean, median, and mode of the data on beetle masses below. Draw up a tally chart and show all calculations:

Beetle masses (g)		
2.2	2.1	2.6
2.5	2.4	2.8
2.5	2.7	2.5
2.6	2.6	2.5
2.2	2.8	2.4

_____

_____

_____

_____

Interpreting Sample Variability

Measures of central tendency, such as mean, attempt to identify the most representative value in a set of data, but the description of a data set also requires that we know something about how far the data values are spread around that central measure. As we have seen in the previous activity, the **standard deviation** (s) gives a simple measure of the spread or **dispersion** in data. The **variance** (s^2) is also a measure of dispersion, but the standard deviation is usually preferred because it is expressed in the original units. Two data sets could have exactly the same mean values, but very different values of dispersion. If we were simply to use the central tendency to compare these data sets, the results would (incorrectly) suggest that they were alike. The assumptions we make about a population will be affected by what the sample data tell us. This is why it is important that sample data are unbiased (e.g. collected by **random sampling**) and that the sample set is as large as practicable. This exercise will help to illustrate how our assumptions about a population are influenced by the information provided by the sample data.

Skills in Biology

Random Sampling, Sample Size, and Dispersion in Data

Sample size and sampling bias can both affect the information we obtain when we sample a population. In this exercise you will calculate some descriptive statistics for some sample data.

The complete set of sample data we are working with comprises 689 length measurements of year zero (young of the year) perch (column left). Basic descriptive statstics for the data have bee calculated for you below and the frequency histogram has also been plotted.

Look at this data set and then complete the exercise to calculate the same statistics from each of two smaller data sets (tabulated right) drawn from the same population. This excercise shows how random sampling, large sample size, and sampling bias affect our statistical assessment of variation in a population.

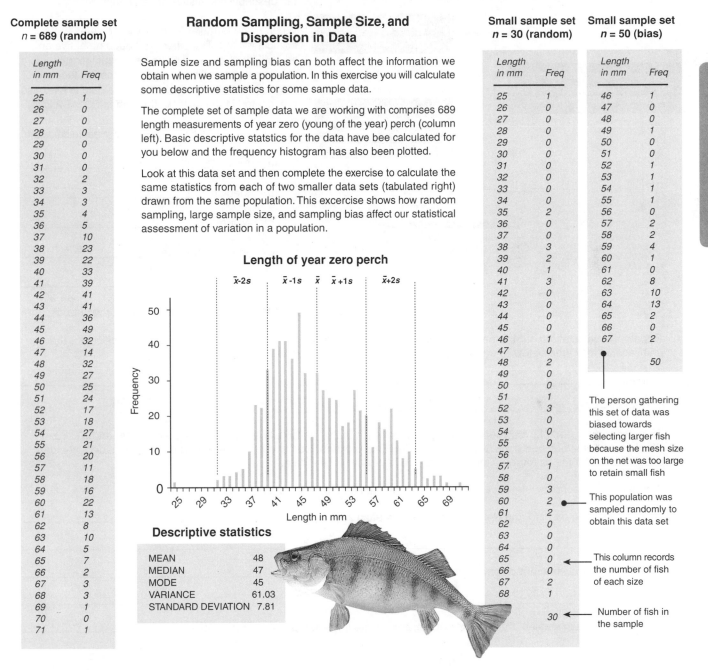

Length of year zero perch

Descriptive statistics

MEAN	48
MEDIAN	47
MODE	45
VARIANCE	61.03
STANDARD DEVIATION	7.81

Complete sample set n = 689 (random)

Length in mm	Freq
25	1
26	0
27	0
28	0
29	0
30	0
31	0
32	2
33	3
34	3
35	4
36	5
37	10
38	23
39	22
40	33
41	39
42	41
43	41
44	36
45	49
46	32
47	14
48	32
49	27
50	25
51	24
52	17
53	18
54	27
55	21
56	20
57	11
58	18
59	16
60	22
61	13
62	8
63	10
64	5
65	7
66	2
67	3
68	3
69	1
70	0
71	1

Small sample set n = 30 (random)

Length in mm	Freq
25	1
26	0
27	0
28	0
29	0
30	0
31	0
32	0
33	0
34	0
35	2
36	0
37	0
38	3
39	2
40	1
41	3
42	0
43	0
44	0
45	0
46	1
47	0
48	2
49	0
50	0
51	1
52	3
53	0
54	0
55	0
56	0
57	1
58	0
59	3
60	2
61	2
62	0
63	0
64	0
65	0
66	0
67	2
68	1
	30

Small sample set n = 50 (bias)

Length in mm	Freq
46	1
47	0
48	0
49	1
50	0
51	0
52	1
53	1
54	1
55	1
56	0
57	2
58	2
59	4
60	1
61	0
62	8
63	10
64	13
65	2
66	0
67	2
	50

The person gathering this set of data was biased towards selecting larger fish because the mesh size on the net was too large to retain small fish

This population was sampled randomly to obtain this data set

This column records the number of fish of each size

Number of fish in the sample

---

1. For the complete data set (n = 689) calculate the percentage of data falling within:

   (a) ± one standard deviation of the mean: _____

   (b) ± two standard deviations of the mean: _____

   (c) Explain what this information tells you about the distribution of year zero perch from this site: _____

   _____

2. Give another reason why you might reach the same conclusion about the distribution: _____

   _____

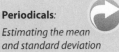 **Periodicals:** *Estimating the mean and standard deviation*

Related activities: Descriptive Statistics, Drawing Histograms

DA 3

Karori age zero perch 12-15 F

	A	B	C	D
		LENGTH	WEIGHT	
		25	0.15	
		35	0.44	
		35	0.44	
		38	0.57	
		38	0.57	
		38	0.57	
		39	0.61	
		39	0.61	
10		40	0.67	
11		41	0.72	
12		41	0.72	
13		41	0.72	
14		46	1.03	
15		48	1.18	
16		48	1.18	
17		51	1.43	
18		52	1.52	
			1.52	
			1.52	
			2.04	
			2.27	
			2.27	
24		59	2.27	
25		60	2.39	
26		60	2.39	
27		61	2.52	
28		61	2.52	
29		67	3.39	
30		67	3.39	
31		68	3.56	
32				
33				
34	N			=COUNT(B2:B31)
35	MEAN			**=AVERAGE(B2:B31)**
36	MEDIAN			=MEDIAN(B2:B31)
37	MODE			=MODE(B2:B31)
38	VARIANCE			=VAR(B2:B31)
39	STANDARD DEVIATION			=STDEV(B2:B31)
40				
41				

The variables being measured. Both length and weight were measured, but here we are working with only the length data.

Enter the data values in separate cells under an appropriate descriptor

Ignore this WEIGHT column. Sometimes the data we are interested in is part of larger data set.

The cells for the calculations below are B2 to B31

Type in the name of the statistic Excel will calculate. This gives you a reference for the row of values.

Type the formula into cell beside its label. When you press return, the cell will contain the calculated value.

Calculating Descriptive Statistics Using *Excel*

You can use *Microsoft Excel* or another similar spreadsheet program to easily calculate descriptive statistics for sample data.

In this first example, the smaller data set ($n = 30$) is shown as it would appear on an *Excel* spreadsheet, ready for the calculations to be made. Use this guide to enter your data into a spreadsheet and calculate the descriptive statistics as described.

When using formulae in *Excel*, = indicates that a formula follows. The cursor will become active and you will be able to select the cells containing the data you are interested in, or you can type the location of the data using the format shown. The data in this case are located in the cells B2 through to B31 (B2:B31).

3. For this set of data, use a spreadsheet to calculate:

   (a) Mean: _____

   (b) Median _____

   (c) Mode: _____

   (d) Sample variance: _____

   (e) Standard deviation: _____

   Staple the spreadsheet into your workbook.

4. Repeat the calculations for the second small set of sample data (n = 50) on the previous page. Again, calculate the statistics as indicated below and staple the spreadsheet into your workbook:

   (a) Mean: _____ (b) Median: _____ (c) Mode: _____

   (d) Variance: _____ (e) Standard deviation: _____

5. On a separate sheet, plot **frequency histograms** for each of the two small data sets. Label them *n* = 30 and *n* = 50. Staple them into your workbook. If you are proficient in *Excel* and you have the "Data Analysis" plug in loaded, you can use *Excel* to plot the histograms for you once you have entered the data.

6. Compare the descriptive statistics you calculated for each data set with reference to the following:

   (a) How close the median and mean to each other in each sample set: _____

   _____

   (b) The size of the standard deviation in each case: _____

   _____

   (c) How close each small of the sample sets resembles the large sample set of 689 values: _____

   _____

   _____

7. (a) Compare the two frequency histograms you have plotted for the two smaller sample data sets: _____

   _____

   _____

   (b) Explain why you think two histograms look so different: _____

   _____

   _____

The Structure of a Report

Once you have collected and analyzed your data, you can write your report. You may wish to present your findings as a written report, a poster presentation, or an oral presentation. The structure of a scientific report is described below using a poster presentation (which is necessarily very concise) as an example. When writing your report, it is useful to write the methods or the results first, followed by the discussion and conclusion. Although you should do some reading in preparation, the introduction should be one of the last sections that you write. Writing the other sections first gives you a better understanding of your investigation within the context of other work in the same area.

To view this and other examples of posters, see the excellent NC State University web site listed below

1. Title (and author)
Provides a clear and concise description of the project.

2. Introduction
Includes the aim, hypothesis, and background to the study

3. Materials and Methods
A description of the materials and procedures used.

4. Results
An account of results including tables and graphs. This section should not discuss the result, just present them.

5. Discussion
An discussion of the findings in light of the biological concepts involved. It should include comments on any limitations of the study.

6. Conclusion
A clear statement of whether tor not the findings support the hypothesis. In abbreviated poster presentations, these sections may be combined.

7. References & acknowledgements
An organised list of all sources of information. Entries should be consistent within your report. Your teacher will advise you as to the preferred format.

Flounder Exhibit Temperature-Dependent Sex Determination

J. Adam Luckenbach*, John Godwin and Russell Borski
Department of Zoology, Box 7617, North Carolina State University, Raleigh, NC 27695

Skills in Biology

Image courtesy: Adam Luckenbach, NC State University

Introduction

Southern flounder (*Paralichthys lethostigma*) support valuable fisheries and show great promise for aquaculture. Female flounder are known to grow faster and reach larger adult... Therefore, information on sex dete... might increase the ratio of female... important for aquaculture.

Objective

This study was conducted to determine whether southern flounder exhibit temperature-dependent sex determination (TSD), and if growth is affected by rearing temperature.

Methods

- Southern flounder broodstock were strip spawned to collect eggs and sperm for *in vitro* fertilization.
- Hatched larvae were weaned from a natural diet (rotifers/*Artemia*) to high protein pelleted feed and fed until satiation at least twice daily.
- Upon reaching a mean total length of 40 mm, the juvenile flounder were stocked at equal densities into one of three temperatures 18, 23, or 28°C for 245 days.
- Gonads were preserved and later sectioned at 2-6 microns.
- Sex-distinguishing markers were used to distinguish males (spermatogenesis) from females (oogenesis).

Histological Analysis

Male Differentiation   **Female Differen...**

Temperature Affects Sex Determination

Growth Does Not Differ by Sex

Results

- Sex was discernible in most fish greater than 120 mm long.
- High (28°C) temperature produced 4% females.
- Low (18°C) temperature produced 22% females.
- Mid-range (23°C) temperature produced 44% females.
- Fish raised at high or low temperatures showed reduced growth compared to those at the mid-range temperature.
- Up to 245 days, no differences in growth existed between sexes.

Conclusions

- These findings indicate that sex determination in southern flounder is temperature-sensitive and temperature has a profound effect on growth.
- A mid-range rearing temperature (23°C) appears to maximize the number of females and promote better growth in young southern flounder.
- Although adult females are known to grow larger than males, no difference in growth between sexes occurred in age-0 (< 1 year) southern flounder.

Acknowledgements

...e authors acknowledge the Salstonstall-Kennedy Program of the National Marine Fisheries Service and the University of North Carolina Sea Grant College Program for funding this research. Special thanks to Lea Ware and Beth Shimps for help with the work.

---

1. Explain the purpose of each of the following sections of a report. The first has one been completed for you:

   (a) Introduction: *Provides the reader with the background to the topic and the rationale for the study*

   (b) Methods: _____

   _____

   (c) Results: _____

   _____

   (d) Discussion: _____

   _____

   (e) References and acknowledgements: _____

   _____

2. Posters are a highly visual method of presenting the findings of a study. Describe the positive features of this format:

   _____

   _____

Related activities: *Hypotheses and Predictions*
Web links: *Report Checklist, NC State University: Creating Effective Poster Presentations*

RA 2

Writing the Methods

The materials and methods section of your report should be brief but informative. All essential details should be included but those not necessary for the repetition of the study should be omitted. The following diagram illustrates some of the important details that should be included in a methods section. Obviously, a complete list of all possible equipment and procedures is not possible because each experiment or study is different. However, the sort of information that is required for both lab and field based studies is provided.

Field Studies	Laboratory Based Studies

Study site & organisms
- Site location and features
- Why that site was chosen
- Species involved

Specialized equipment
- pH and oxygen meters
- Thermometers
- Nets and traps

Data collection
- Number and timing of observations/collections
- Time of day or year
- Sample sizes and size of the sampling unit
- Methods of preservation
- Temperature at time of sampling
- Weather conditions on the day(s) of sampling
- Methods of measurement/sampling
- Methods of recording

Data collection
- Pre-treatment of material before experiments
- Details of treatments and controls
- Duration and timing of experimental observations
- Temperature
- Sample sizes and details of replication
- Methods of measurement or sampling
- Methods of recording

Experimental organisms
- Species or strain
- Age and sex
- Number of individuals used

Specialized equipment
- pH meters
- Water baths & incubators
- Spectrophotometers
- Centrifuges
- Aquaria & choice chambers
- Microscopes and videos

Special preparations
- Techniques for the preparation of material (staining, grinding)
- Indicators, salt solutions, buffers, special dilutions

General guidelines for writing a methods section

- Choose a suitable level of detail. *Too little detail and the study could not be repeated. Too much detail obscures important features.*
- Do NOT include the details of standard procedures (e.g. how to use a balance) or standard equipment (e.g. beakers and flasks).
- Include details of any statistical analyses and data transformations.
- Outline the reasons why procedures were done in a certain way or in a certain order, if this is not self-evident.
- If your methodology involves complicated preparations (e.g. culture media) then it is acceptable to refer just to the original information source (e.g. lab manual) or include the information as an appendix.

1. The following text is part of the methods section from a report. Using the information above and on the checklist on the previous page, describe four errors (there are many) and suggest how to correct them in the methods. The errors are concerned with a lack of explanation or detail that would be necessary to repeat the experiment (they are not typographical, nor are they associated with the use of the active voice, which is now considered preferable to the passive):

"We wanted to carry out an investigation on the distribution of lawn species in the sun and the shade. We found an area on the field that gets half sun and half shade, and we pegged out a transect line. We took turns to step out 100 paces along the line, recording the name of the biggest plant under the string at each step. We ignored bare ground and unidentifiable plants. We repeated our experiment by moving the string to a similar area. When we got back to the lab we calculated the % abundance for each plant species present."

(a) _____

_____

(b) _____

_____

(c) _____

_____

(d) _____

_____

Related activities: The Structure of a Report

Writing Your Results

The results section is arguably the most important part of any research report; it is the place where you can bring together and present your findings. When properly constructed, this section will present your results clearly and in a way that shows you have organized your data and carefully considered the appropriate analysis. A portion of the results section from a scientific paper on the habitat preference of black mudfish is presented below (Hicks, B. and Barrier, R. (1996), NZJMFR. 30, 135-151). It highlights some important features of the results section and shows you how you can present information concisely, even if your results are relatively lengthy. Use it as a guide for content when you write up this section.

Results

A total of 222 black mudfish were caught in the 400 traps set be[...] Mean total length (TL) was 67 mm (range 26-139 mm, $n = 214$)[...] had black mudfish. Mudfis[...]nly amo[...] independence, $P < 0.001$: [...]at 8 out[...] at 20 out of 30 wetland sit[...]d at only[...] none of the 6 lake margin or 4 pond, dam, and lagoon sites. Categorical variables that distinguished [...]χ^2 tests of independence, $P < 0.05$: Table 4) were: [...]rate disturbance scale rating; presence of emergent [...]sed or peat bog substrate types; absence of fish [...]orphus cotidianus) and inanga (Galaxias maculatus);

Keep your statement of important findings brief.

Graphs (figures) illustrate trends in the data. Be sure to choose the correct type of graph and allocate enough space to it in the report.

Label figures and tables clearly and in sequence so that they can be referred to easily in the text.

Scientific names are included if they are known.

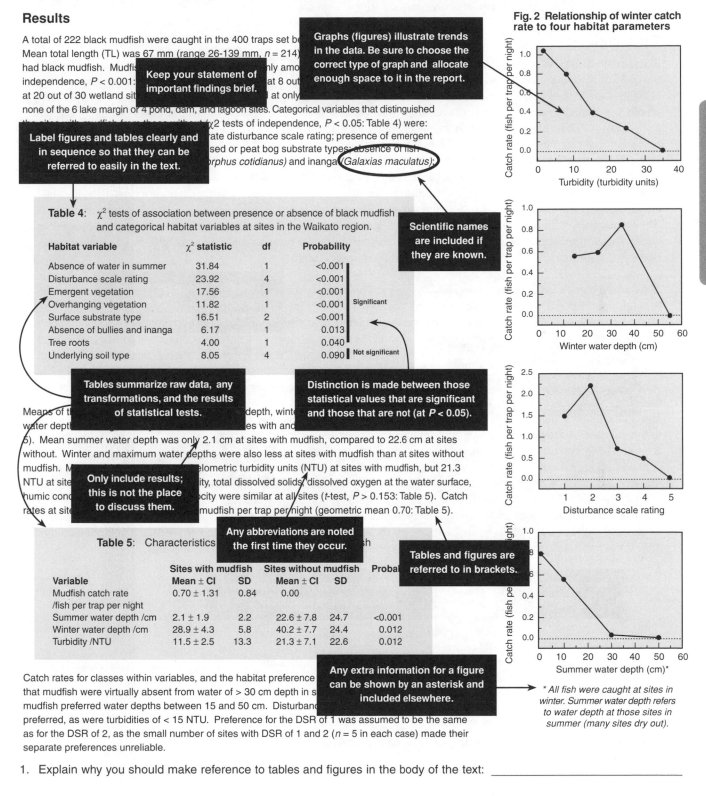

Table 4: χ^2 tests of association between presence or absence of black mudfish and categorical habitat variables at sites in the Waikato region.

Habitat variable	χ^2 statistic	df	Probability	
Absence of water in summer	31.84	1	<0.001	
Disturbance scale rating	23.92	4	<0.001	
Emergent vegetation	17.56	1	<0.001	
Overhanging vegetation	11.82	1	<0.001	Significant
Surface substrate type	16.51	2	<0.001	
Absence of bullies and inanga	6.17	1	0.013	
Tree roots	4.00	1	0.040	
Underlying soil type	8.05	4	0.090	Not significant

Tables summarize raw data, any transformations, and the results of statistical tests.

Distinction is made between those statistical values that are significant and those that are not (at $P < 0.05$).

Means of t[...] depth, winte[...]es with and[...] water depth[...]
5). Mean summer water depth was only 2.1 cm at sites with mudfish, compared to 22.6 cm at sites without. Winter and maximum water depths were also less at sites with mudfish than at sites without mudfish. M[...]elometric turbidity units (NTU) at sites with mudfish, but 21.3 NTU at sites[...]ity, total dissolved solids, dissolved oxygen at the water surface, humic conc[...]ocity were similar at all sites (t-test, $P > 0.153$: Table 5). Catch rates at site[...]mudfish per trap per night (geometric mean 0.70: Table 5).

Only include results; this is not the place to discuss them.

Any abbreviations are noted the first time they occur.

Tables and figures are referred to in brackets.

Table 5: Characteristics [...]sh

Variable	Sites with mudfish Mean ± CI	SD	Sites without mudfish Mean ± CI	SD	Probab[...]
Mudfish catch rate /fish per trap per night	0.70 ± 1.31	0.84	0.00		
Summer water depth /cm	2.1 ± 1.9	2.2	22.6 ± 7.8	24.7	<0.001
Winter water depth /cm	28.9 ± 4.3	5.8	40.2 ± 7.7	24.4	0.012
Turbidity /NTU	11.5 ± 2.5	13.3	21.3 ± 7.1	22.6	0.012

Any extra information for a figure can be shown by an asterisk and included elsewhere.

Catch rates for classes within variables, and the habitat preference [...] that mudfish were virtually absent from water of > 30 cm depth in s[...] mudfish preferred water depths between 15 and 50 cm. Disturbanc[...] preferred, as were turbidities of < 15 NTU. Preference for the DSR of 1 was assumed to be the same as for the DSR of 2, as the small number of sites with DSR of 1 and 2 ($n = 5$ in each case) made their separate preferences unreliable.

Fig. 2 Relationship of winter catch rate to four habitat parameters

* All fish were caught at sites in winter. Summer water depth refers to water depth at those sites in summer (many sites dry out).

1. Explain why you should make reference to tables and figures in the body of the text: _____

_____

2. Explain why you might present the same data in a table and as a figure: _____

_____

_____

Writing Your Discussion

In the discussion section of your report, you must interpret your results in the context of the specific questions you set out to answer in the investigation. You should also place your findings in the context of any broader relevant issues. If your results coincide exactly with what you expected, then your discussion will be relatively brief. However, be prepared to discuss any unexpected or conflicting results and critically evaluate any problems with your study design. The Discussion section may (and should) refer to the findings in the Results section, but it is not the place to introduce new results. Try to work towards a point in your discussion where the reader is lead naturally to the conclusion. The conclusion may be presented within the discussion or it may be included separately after the discussion as a separate section.

Discussion:

Black mudfish habitat in the Waikato region can be ad[**Support your statements with reference to Tables and Figures from the Results section.**]ses by four variables that are easy to measure: summer water depth, winter wa[...]cated by vegetation), and turbidity. Catch rates of black mudfish can be extreme[...]es ranged from 0.2 to 8.4 mudfish per trap per night (mean 0.70) between May and October 1992, and were similar to those of Dean (1995) in September 1993 and October 1994 in the Whangamarino Wetland complex (0.0-2.0 mudfish per trap per night). The highest mean catch rate in our study, 8.4 mudfish per trap per night, was at Site 24 (Table 1, Figure 1). The second highest (6.4 mudfish per trap per night) was at Site 32, in a drain about 4 km east of Hamilton. Black mudfish in the Waikato region were most commonly found at sites in wetlands with absence of water in summer, moderate depth of water in winter, limited modification of the vegetation (low DSR), and low turbidity (Fig. 2). There are similarities between the habitat requirements of black mudfish and those of brown mudfish and the common river galaxias *(Galaxias vulgaris)*. Brown mudfish inhabited shallow water, sometimes at the edges of deeper water bodies, but were usually absent from water deeper than about 30-50 cm (Eldon 1978). The common river galaxias also has a preference for shallow water, occupying river margins < 20 cm deep (Jowett and Richardson 1995).

Sites where black mudfish were found were not just shallow or dry in sum[**The discussion describes the relevance of the results of the investigation.**]al variation in water depth. A weakness of this study is the fact that sites were trap[...]ere spread relatively widely at each site to maximise the chance of catching any fish[...]nt for black mudfish in the form of [**State any limitations of your approach in carrying out the investigation and what further studies might be appropriate.**]erhanging vegetation, or tree roots. The significance of cover in determining the pres[...]s predictable, considering the shallow nature of their habitats. Mudfish, though noc[...]require cover during the to protect them from avian predators, such as bitterns *(Bo*[...]fishers *(Halcyon sancta vagans)*. Predation of black mudfish by a swamp bittern has [...]1). Cover is also important for brown mudfish (Eldon 1978). Black mudfish were found at sites with the predatory mosquitofish and juvenile eels, and the seasonal drying of their habitats may be a key to the successful coexistence of mudfish with their predators. Mosquitofish are known predators of mudfish fry (Barrier & Hicks 1994), and eels would presumably also prey on black mudfish, as t[**Reference is made to the work of others.**]h (Eldon 1979b). If, however, black mudfish are relatively uncompetitive and vulnerable to pr[...]s as to how they manage to coexist with juvenile eels and mosquitofish. The habitat varia[...] can be used to classify the suitability of sites for black mudfish in future. The adaptability of black mudfish allows them to survive in some altered habitats, such as farm or roadside drains. From this study, we can conclude that the continued existence of suitable habitats appears to be more important to black mudfish than the presence of predators and competitors. This study has also improved methods of identifying suitable mudfish habitats in the Waikato region.

[**Further research is suggested**]

[**A clear conclusion is made towards the end of the discussion.**]

1. Explain why it is important to discuss any weaknesses in your study design: _____

_____

_____

2. Explain why you should **critically evaluate** your results in the discussion: _____

_____

_____

_____

3. Describe the purpose of the conclusion: _____

_____

Related activities: The Structure of a Report

Citing and Listing References

Proper referencing of sources of information is an important aspect of report writing. It shows that you have explored the topic and recognize and respect the work of others. There are two aspects to consider: **citing sources** within the text (making reference to other work to support a statement or compare results) and **compiling a reference list** at the end of the report. A **bibliography** lists all sources of information, but these may not necessarily appear as citations in the report. In contrast, a reference list should contain only those texts cited in the report.

Citations in the main body of the report should include only the authors' surnames, publication date, and page numbers (or internet site), and the citation should be relevant to the statement it claims to support. Accepted methods for referencing vary, but your reference list should provide all the information necessary to locate the source material, it should be consistently presented, and it should contain only the references that you have read yourself (not those cited by others). A suggested format is described below.

Skills in Biology

Preparing a Reference List

When teachers ask students to write in "APA style", they are referring to the editorial style established by the **American Psychological Association** (APA). These guidelines for citing **electronic (online) resources** differ only slightly from the **print sources**.

For the Internet

Where you use information from the internet, you must provide the following:
- The website address (URL), the person or organization who is in charge of the web site and the date you accessed the web page.

This is written in the form: URL (person or organization's name, day, month, and year retrieved)

This goes together as follows:

> http://www.scientificamerican.com (Scientific American, 17.12.03)

For Periodicals (or Journals)

This is written in the form: author(s), date of publication, article title, periodical title, and publication information.
Example: Author's family name, A. A. (author's initials only), Author, B. B., & Author, C. C. (xxxx = year of publication in brackets). Title of article. Title of Periodical, volume number, page numbers (Note, only use "pp." before the page numbers in newspapers and magazines).

This goes together as follows:

> Bamshad M. J., & Olson S. E. (2003). Does Race Exist? Scientific American, 289(6), 50-57.

For Online Periodicals based on a Print Source

At present, the majority of periodicals retrieved from online publications are exact duplicates of those in their print versions and although they are unlikely to have additional analyses and data attached to them, this is likely to change in the future.

- If the article that is to be referenced has been viewed only in electronic form and not in print form, then you must add in brackets, "Electronic version", after the title.
  This goes together as follows:

  > Bamshad M. J., & Olson S. E. (2003). Does Race Exist? (Electronic version). Scientific American, 289(6), 50-57.

- If you have reason to believe the article has changed in its electronic form, then you will need to add the date you retrieved the document and the URL.
  This goes together as follows:

  > Bamshad M. J., & Olson S. E. (2003). Does Race Exist? (Electronic version). Scientific American, 289(6), 50-57. Retrieved December 17, 2003, from http://www.scientificamerican.com

For Books

This is written in the form: author(s), date of publication, title, and publication information.
Example: Author, A. A., Author, B. B., & Author, C. C. (xxxx). Title (any additional information to enable identification is given in brackets). City of publication: publishers name.
This goes together as follows:

> Martin, R.A. (2004). Missing Links Evolutionary Concepts & Transitions Through Time. Sudbury, MA: Jones and Bartlett

For Citation in the Text of References

This is written in the form: authors' surname(s), date of publication, page number(s) (abbreviated p.), chapter (abbreviated chap.), figure, table, equation, or internet site, in brackets at the appropriate point in text.
This goes together as follows:

> (Bamshad & Olson, 2003, p. 51) or (Bamshad & Olson, 2003, http://www.scientificamerican.com)

This can also be done in the form of footnotes. This involves the use of a superscripted number in the text next to your quoted material and the relevant information listed at the bottom of the page.
This goes together as follows:

> .... Bamshad & Olson reported that ....[1]

---

[1]Bamshad & Olson, 2003, p. 51

Related activities: The Structure of a Report

A 2

Example of a Reference List

Lab notes can be listed according to title if the author is unknown.

→ Advanced biology laboratory manual (2000). Cell membranes. pp. 16-18. Sunhigh College.

References are listed alphabetically according to the author's surname.

Cooper, G.M. (1997). *The cell: A molecular approach* (2nd ed.). Washington D.C.: ASM Press

Book title in italics (or underlined)   Place of publication: Publisher

Davis, P. (1996) Cellular factories. *New Scientist* 2057: Inside science supplement.

Publication date   Journal title in italics   A supplement may not need page references

If a single author appears more than once, then list the publications from oldest to most recent.

Indge, B. (2001). Diarrhea, digestion and dehydration. *Biological Sciences Review*, 14(1), 7-9.

Indge, B. (2002). Experiments. *Biological Sciences Review*, 14(3), 11-13.

Article title follows date

Kingsland, J. (2000). Border control. *New Scientist* 2247: Inside science supplement.

Spell out only the last name of authors. Use initials for first and middle names.

Laver, H. (1995). Osmosis and water retention in plants. *Biological Sciences Review* 7(3), 14-18

Volume (Issue number), Pages

Steward, M. (1996). Water channels in the cell membrane. *Biological Sciences Review*, 9(2), 18-22.

Internet sites change often so the date accessed is included. The person or organization in charge of the site is also included.

→ http://www.cbc.umn.edu/~mwd/cell_intro.html (Dalton, M. "Introduction to cell biology" 12.02.03)

1. Distinguish between a **reference list** and a **bibliography**: _____

_____

_____

_____

2. Explain why internet articles based on a print source are likely to have additional analyses and data attached in the future, and why this point should be noted in a reference list:

_____

_____

3. Following are the details of references and source material used by a student in preparing a report on enzymes and their uses in biotechnology. He provided his reference list in prose. From it, compile a correctly formatted reference list:

Pages 18-23 in the sixth edition of the textbook "Biology" by Neil Campbell. Published by Benjamin/Cummings in California (2002). New Scientist article by Peter Moore called "Fuelled for life" (January 1996, volume 2012, supplement). "Food biotechnology" published in the journal Biological Sciences Review, page 25, volume 8 (number 3) 1996, by Liam and Katherine O'Hare. An article called "Living factories" by Philip Ball in New Scientist, volume 2015 1996, pages 28-31. Pages 75-85 in the book "The cell: a molecular approach" by Geoffrey Cooper, published in 1997 by ASM Press, Washington D.C. An article called "Development of a procedure for purification of a recombinant therapeutic protein" in the journal "Australasian Biotechnology", by I Roberts and S. Taylor, pages 93-99 in volume 6, number 2, 1996.

REFERENCE LIST

_____

_____

_____

_____

_____

_____

_____

KEY TERMS Mix and Match

INSTRUCTIONS: Test your vocab by matching each term to its correct definition, as identified by its preceding letter code.

ACCURACY

BIBLIOGRAPHY

BIOLOGICAL DRAWING

CITATION

CONTROL

CONTROLLED VARIABLE

DATA

DATALOGGER

DEPENDENT VARIABLE

GRAPH

HISTOGRAM

HYPOTHESIS

INDEPENDENT VARIABLE

MEAN

MEASUREMENT

MEDIAN

MODE

OBSERVATION

PARAMETER

PRECISION

QUALITATIVE DATA

QUANTITATIVE DATA

RAW DATA

REPORT

SAMPLE

SCIENTIFIC METHOD

TREND (OF DATA)

VARIABLE

A Note normally appearing directly after a new fact or data that states the author of the information and the date it was published.

B A variable whose values are set, or systematically altered, by the investigator.

C A diagram drawn to accurately show what has been seen by the observer.

D Facts collected for analysis.

E The value that occurs most often in a data set.

F A pattern observed in processed data showing that data values may be linked.

G A standard (reference) treatment that helps to ensure that the responses to the other treatments can be reliably interpreted.

H The sum of the data divided by the number of data entries (n).

I A variable whose values are determined by another variable.

J Data able to be expressed in numbers. Numerical values derived from counts or measurements.

K Variable that is fixed at a specific amount as part of the design of experiment.

L A type of column graph used to display frequency distributions.

M A tentative explanation of an observation, capable of being tested by experimentation.

N The number that occurs in the middle of a set of sorted numbers. It divides the upper half of the number data set from the lower half.

O The repeatability of a measurement or (in statistics) how close a statistic is to the value of the parameter being estimated.

P The sampling of an object or substance to record numerical data that describes some aspect of the it, e.g. length or temperature.

Q Data that have not been processed or manipulated in any way.

R A diagram which often displays numerical information in a way that can be used to identify trends in the data.

S The act of seeing and noting an occurrence in the object or substance being studied.

T The use of an ordered, repeatable method to investigate, manipulate, gather, and record data.

U Data described in descriptors or terms rather than by numbers.

V A sub-set of a whole used to estimate the values that might have been obtained if every individual or response was measured.

W Device that is able to record data as it changes, in real time.

X The completed study including methods, results and discussion of the data obtained.

Y A list displaying the titles and publication information of resources used in the gathering of information.

Z The degree of closeness of a measured value to its true amount.

AA A factor in an experiment that is subject to change.

BB A quantity that defines a characteristic of a system.

RA 2

Cell Structure

Types of living things
- Characteristics of cells
- Prokaryotic vs eukaryotic cells
- The sizes of cells, SA:V ratios
- Cell structures and organelles

Microscopy
- Practical microscopy
- Electron microscopy: TEM/SEM
- Magnification and resolution

The Chemistry of Life

Water and inorganic ions
- The structure and properties of water
- Water as the universal solvent
- Inorganic ions essential for life

Lipids, carbohydrates, nucleic acids
- Structure and roles of carbohydrates
- Structure and roles of lipids
- Structure and role of nucleic acids
- Condensation and hydrolysis

Amino acids and proteins
- The structure of amino acids
- Protein structure: 1°, 2°, 3°, 4°
- Roles of proteins in the body
- Denaturation

Enzymes
- Enzymes as biological catalysts
- Enzymes reaction rates
- Cofactors and inhibitors
- Applications of enzymes

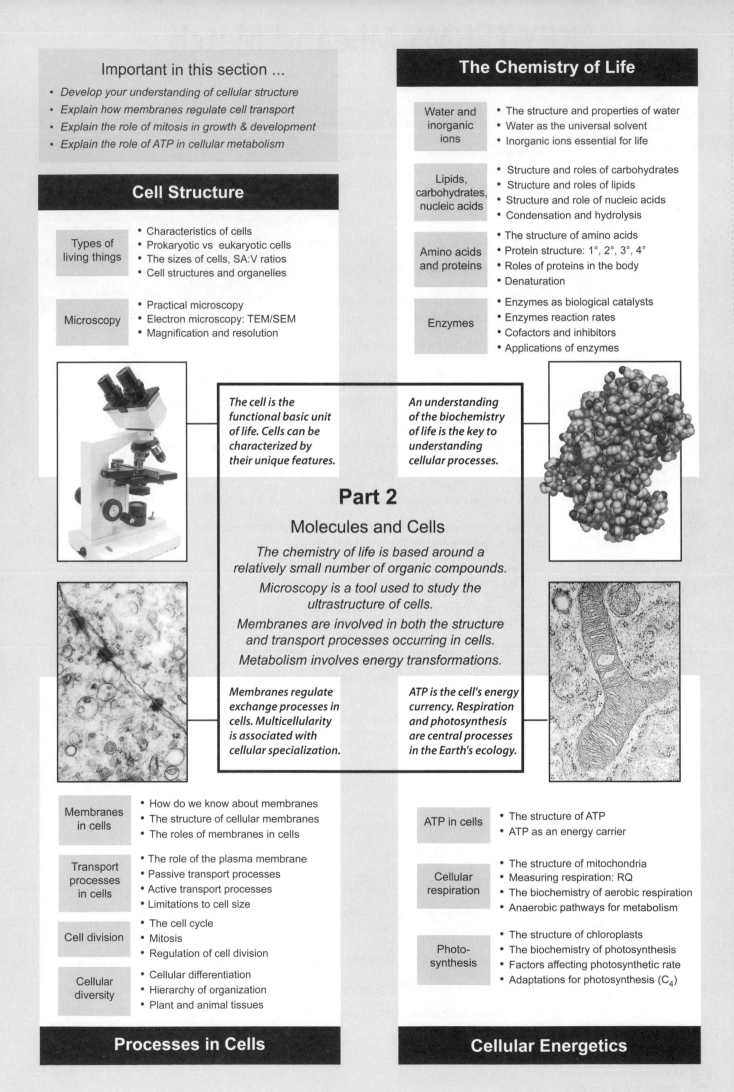

The cell is the functional basic unit of life. Cells can be characterized by their unique features.

An understanding of the biochemistry of life is the key to understanding cellular processes.

Part 2

Molecules and Cells

The chemistry of life is based around a relatively small number of organic compounds.

Microscopy is a tool used to study the ultrastructure of cells.

Membranes are involved in both the structure and transport processes occurring in cells.

Metabolism involves energy transformations.

Membranes regulate exchange processes in cells. Multicellularity is associated with cellular specialization.

ATP is the cell's energy currency. Respiration and photosynthesis are central processes in the Earth's ecology.

Membranes in cells
- How do we know about membranes
- The structure of cellular membranes
- The roles of membranes in cells

Transport processes in cells
- The role of the plasma membrane
- Passive transport processes
- Active transport processes
- Limitations to cell size

Cell division
- The cell cycle
- Mitosis
- Regulation of cell division

Cellular diversity
- Cellular differentiation
- Hierarchy of organization
- Plant and animal tissues

ATP in cells
- The structure of ATP
- ATP as an energy carrier

Cellular respiration
- The structure of mitochondria
- Measuring respiration: RQ
- The biochemistry of aerobic respiration
- Anaerobic pathways for metabolism

Photosynthesis
- The structure of chloroplasts
- The biochemistry of photosynthesis
- Factors affecting photosynthetic rate
- Adaptations for photosynthesis (C_4)

Processes in Cells

Cellular Energetics

The Chemistry of Life

KEY CONCEPTS

▶ Organic molecules are carbon-containing molecules and are central to living systems.

▶ Water's properties make it essential to life.

▶ Biological molecules have specific roles in cells.

▶ Enzymes are biological catalysts and regulate the metabolism of cells.

▶ Enzymes have many practical applications.

KEY TERMS

activation energy
active site
amino acid
anabolism
Benedict's test
biuret test
carbohydrate
catabolism
catalyst
chromatography
coenzyme
cofactor
denaturation
disaccharide
emulsion test
endergonic reaction
enzyme
enzyme inhibition
exergonic reaction
fibrous protein
globular protein
inorganic ion
I2/KI test
lipid
macromolecule
metabolic pathway
monomer
monosaccharide
nucleic acid
optimum
organic molecule
polar molecule
polymer
polysaccharide
protein
Rf value

Periodicals:
listings for this
chapter are on page 379

Weblinks:
www.thebiozone.com/
weblink/SB1-2597.html

**Teacher Resource
CD-ROM:**
Industrial Microbiology

OBJECTIVES

☐ 1. Use the **KEY TERMS** to help you understand and complete these objectives.

The Structure of Biological Molecules
pages 48-60, 67

☐ 2. Identify the common elements found in organisms and give examples of where these elements occur in cells. Describe the importance of **organic molecules** and **inorganic ions** in biological systems.

☐ 3. Distinguish between **monomers** and **polymers**. Describe the range of **macromolecules** produced by cells.

☐ 4. Describe the structure of water, including reference to its **polar** nature and the physical and chemical properties that are important in biological systems.

☐ 5. Using examples, describe the basic structure and roles of **carbohydrates**, **amino acids**, **proteins**, **lipids**, and **nucleic acids**.

☐ 6. Describe the synthesis of macromolecules by **condensation** and their breakdown by **hydrolysis**. Identify the bonds formed or broken in each case.

☐ 7. Describe where (in the cell) proteins are made. Describe the structure and functional diversity of proteins including their primary, secondary, and tertiary structure and their classification based on structure (e.g. **globular** or **fibrous**) or **function** (e.g. **catalytic**).

☐ 8. Explain what is meant by **denaturation** and explain why it destroys the activity of proteins. Describe factors that result in denaturation.

☐ 9. Recognize **chromatography** as a technique for separating and identifying biological molecules. Describe the calculation and use of **Rf values**.

☐ 10. Demonstrate an understanding of simple tests for organic compounds, including the **Benedict's test** for reducing sugars, the **I2/KI test** for starch, the **emulsion test** for lipids, and the **biuret test** for proteins.

Enzymes
pages 61-66

☐ 11. Describe the properties and mode of action of **enzymes**, including the role of the **active site**, **specificity**, and **activation energy**. Explain models of enzyme function and outline the role of enzymes in **anabolism** and **catabolism**.

☐ 12. Describe the effect of substrate concentration, enzyme concentration, pH, and temperature on enzyme activity. Explain the term **optimum** with respect to enzyme activity. Recognize that enzymes (as proteins) can be denatured.

☐ 13. Using examples, explain the role of **cofactors** in enzyme activity.

☐ 14. Distinguish between reversible and irreversible **inhibition**, and competitive and non-competitive inhibition. Explain how inhibition affects enzyme activity.

☐ 15. Describe some of the commercial applications of microbial enzymes, e.g. pectinases and rennin in the food industry and proteases in detergents.

The Biochemical Nature of the Cell

The molecules that make up living things can be grouped into five classes: water, carbohydrates, lipids, proteins, and nucleic acids. Water is the main component of organisms and provides an environment in which metabolic reactions can occur. Water molecules attract each other, forming large numbers of hydrogen bonds. It is this feature that gives water many of its unique properties, including its low viscosity and its chemical behavior as a **universal solvent**. Apart from water, most other substances in cells are compounds of carbon, hydrogen, oxygen, and nitrogen. The combination of carbon atoms with the atoms of other elements provides a huge variety of molecular structures. These are described on the following pages.

Important Properties of Water

Water is a liquid at room temperature and many substances dissolve in it. It is a medium inside cells and for aquatic life.

A lot of energy is required before water will change state so aquatic environments are thermally stable and sweating and transpiration cause rapid cooling.

Carbohydrates form the structural components of cells, they are important in energy storage, and they are involved in cellular recognition.

Proteins may be structural (e.g. collagen), catalytic (enzymes), or they may be involved in movement, message signaling, internal defense and transport, or storage.

Nucleotides and nucleic acids Nucleic acids encode information for the construction and functioning of an organism. The nucleotide, ATP, is the energy currency of the cell.

Lipids provide insulation and a concentrated source of energy. Phospholipids are a major component of cellular membranes.

Water is a major component of cells: many substances dissolve in it, metabolic reactions occur in it, and it provides support and turgor.

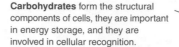

Ice is less dense than water. Consequently ice floats, insulating the underlying water and providing valuable habitat.

Water has a high surface tension and low viscosity. It forms droplets on surfaces and can flow freely through narrow vessels.

Water is colorless, with a high transmission of visible light, so light penetrates tissue and aquatic environments.

1. Explain the biological significance of each of the following physical properties of water:

   (a) Low viscosity: _____

   (b) Colorless and transparent: _____

   (c) Universal solvent: _____

   (d) Ice is less dense than water: _____

2. Identify the biologically important role of each of the following molecules:

   (a) Lipids: _____

   (b) Carbohydrates: _____

   (c) Proteins: _____

   (d) Nucleic acids: _____

Related activities: Organic Molecules, Water and Inorganic Ions
Web links: A Closer Look at Water

Periodicals:
Water, life and H bonding

Organic Molecules

Organic molecules are those chemical compounds containing carbon, and all biology are carbon-based. Specific groups of atoms, called **functional groups**, attach to a carbon-hydrogen core and confer specific chemical properties on the molecule. Some organic molecules in organisms are small and simple, containing only one or a few **functional groups**, while others are large complex assemblies called **macromolecules**. The macromolecules that make up living things can be grouped into four classes: carbohydrates, lipids, proteins, and nucleic acids. An understanding of the structure and function of these

molecules is necessary to many branches of biology, especially biochemistry, physiology, and molecular genetics. The diagram below illustrates some of the common ways in which biological molecules are portrayed. Note that the **molecular formula** expresses the number of atoms in a molecule, but does not convey its structure; this is indicated by the **structural formula**. Molecules can also be represented as **models**. A ball and stick model shows the arrangement and type of bonds while a space filling model gives a more realistic appearance of a molecule, showing how close the atoms really are.

Portraying Organic Molecules

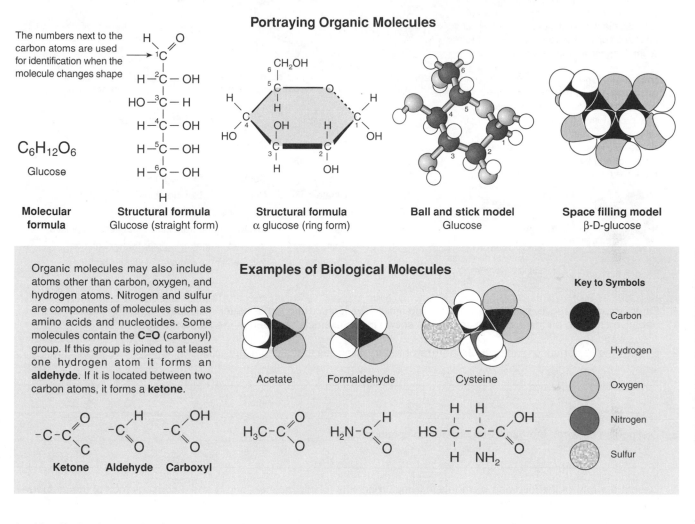

The numbers next to the carbon atoms are used for identification when the molecule changes shape

$C_6H_{12}O_6$
Glucose

Molecular formula

Structural formula
Glucose (straight form)

Structural formula
α glucose (ring form)

Ball and stick model
Glucose

Space filling model
β-D-glucose

Organic molecules may also include atoms other than carbon, oxygen, and hydrogen atoms. Nitrogen and sulfur are components of molecules such as amino acids and nucleotides. Some molecules contain the **C=O** (carbonyl) group. If this group is joined to at least one hydrogen atom it forms an **aldehyde**. If it is located between two carbon atoms, it forms a **ketone**.

Ketone   **Aldehyde**   **Carboxyl**

Examples of Biological Molecules

Acetate   Formaldehyde   Cysteine

Key to Symbols

- Carbon
- Hydrogen
- Oxygen
- Nitrogen
- Sulfur

The Chemistry of Life

1. Identify the three main elements comprising the structure of organic molecules: _____

2. Name two other elements that are also frequently part of organic molecules: _____

3. State how many covalent bonds a carbon atom can form with neighboring atoms: _____

4. Distinguish between molecular and structural formulae for a given molecule: _____

   _____

   _____

5. Describe what is meant by a functional group: _____

   _____

6. Classify formaldehyde according to the position of the C=O group: _____

7. Identify a functional group always present in amino acids: _____

8. Identify the significance of cysteine in its formation of disulfide bonds: _____

   _____

Related activities: The Biochemical Nature of the Cell, Amino Acids, Proteins

RA 2

Water and Inorganic Ions

The Earth's crust contains approximately 100 elements but only 16 are essential for life (see the table of inorganic ions below). Of the smaller molecules making up living things water is the most abundant typically making up about two-thirds of any organism's body. Water has a simple molecular structure and the molecule is very polar, with ends that exhibit partial positive and negative charges. Water molecules have a weak attraction for each other and inorganic ions, forming weak hydrogen bonds.

Water and Inorganic Ions

Water provides an environment in which metabolic reactions can happen. Water takes part in, and is a common product of, many reactions. The most important feature of the chemical behavior of water is its **dipole** nature. It has a small positive charge on each of the two hydrogens and a small negative charge on the oxygen.

Inorganic ions are important for the structure and metabolism of all living organisms. An ion is simply an atom (or group of atoms) that has gained or lost one or more electrons. Many of these ions are soluble in water. Some of the inorganic ions required by organisms are listed in the table (right).

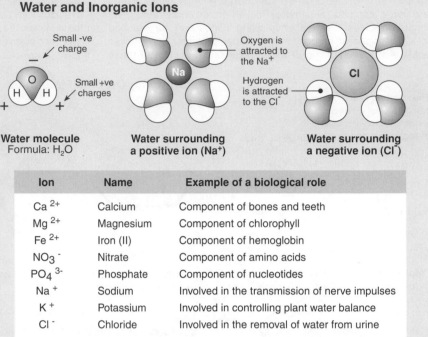

Water molecule
Formula: H_2O

Water surrounding a positive ion (Na^+)

Water surrounding a negative ion (Cl^-)

Oxygen is attracted to the Na^+

Hydrogen is attracted to the Cl^-

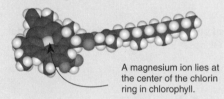

A magnesium ion lies at the center of the chlorin ring in chlorophyll.

Ion	Name	Example of a biological role
Ca^{2+}	Calcium	Component of bones and teeth
Mg^{2+}	Magnesium	Component of chlorophyll
Fe^{2+}	Iron (II)	Component of hemoglobin
NO_3^-	Nitrate	Component of amino acids
PO_4^{3-}	Phosphate	Component of nucleotides
Na^+	Sodium	Involved in the transmission of nerve impulses
K^+	Potassium	Involved in controlling plant water balance
Cl^-	Chloride	Involved in the removal of water from urine

1. On the diagram above, showing a positive and a negative ion surrounded by water molecules, draw the positive and negative charges on the water molecules (as shown in the example provided in the same panel).

2. Explain the importance of the **dipole nature** of water molecules to the chemistry of life: _____

   _____

   _____

3. Distinguish between inorganic and organic compounds: _____

   _____

4. Describe a role of the following elements in living organisms (plants, animals and) and a consequence of the element being deficient in an organism:

   (a) Calcium: _____

   _____

   (b) Iron: _____

   _____

   (c) Phosphorus: _____

   _____

   (d) Sodium: _____

   _____

   (e) Sulfur: _____

   _____

   (f) Nitrogen: _____

   _____

Related activities: The Biochemical Nature of the Cell, Organic Molecules

Periodicals:
Water, life and H bonding

Carbohydrates

Carbohydrates are a family of organic molecules made up of carbon, hydrogen, and oxygen atoms with the general formula $(CH_2O)_x$. The most common arrangements found in sugars are hexose (6 sided) or pentose (5 sided) rings. Simple sugars, or monosaccharides, may join together to form compound sugars (disaccharides and polysaccharides), releasing water in the process (**condensation**). Compound sugars can be broken down into their constituent monosaccharides by the opposite reaction (**hydrolysis**). Sugars play a central role in cells, providing energy and, in some cells, contributing to support. They are the major component of most plants (60-90% of the dry weight) and are used by humans as a cheap food source, and a source of fuel, housing, and clothing. In all carbohydrates, the structure is closely related to their functional properties (below).

Monosaccharides

Monosaccharides are used as a primary energy source for fueling cell metabolism. They are **single-sugar** molecules and include glucose (grape sugar and blood sugar) and fructose (honey and fruit juices). The commonly occurring monosaccharides contain between three and seven carbon atoms in their carbon chains and, of these, the 6C hexose sugars occur most frequently. All monosaccharides are classified as **reducing** sugars (i.e. they can participate in reduction reactions).

Single sugars (monosaccharides)

Triose

C
|
C
|
C

e.g. glyceraldehyde

Pentose

e.g. ribose, deoxyribose

Hexose

e.g. glucose, fructose, galactose

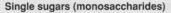

Disaccharides

Disaccharides are **double-sugar** molecules and are used as energy sources and as building blocks for larger molecules. The type of disaccharide formed depends on the monomers involved and whether they are in their α- or β- form. Only a few disaccharides (e.g. lactose) are classified as reducing sugars.

Sucrose = α-glucose + β-fructose (simple sugar found in plant sap)
Maltose = α-glucose + α-glucose (a product of starch hydrolysis)
Lactose = β-glucose + β-galactose (milk sugar)
Cellobiose = β-glucose + β-glucose (from cellulose hydrolysis)

Double sugars (disaccharides)

Examples sucrose, lactose, maltose, cellobiose

Polysaccharides

Cellulose: Cellulose is a structural material in plants and is made up of unbranched chains of β-**glucose** molecules held together by **1, 4 glycosidic links**. As many as 10 000 glucose molecules may be linked together to form a straight chain. Parallel chains become cross-linked with hydrogen bonds and form bundles of 60-70 molecules called microfibrils. Cellulose microfibrils are very strong and are a major component of the structural components of plants, such as the cell wall (photo, right).

Starch: Starch is also a polymer of glucose, but it is made up of long chains of α-**glucose** molecules linked together. It contains a mixture of 25-30% **amylose** (unbranched chains linked by α-1, 4 glycosidic bonds) and 70-75% **amylopectin** (branched chains with α-1, 6 glycosidic bonds every 24-30 glucose units). Starch is an energy storage molecule in plants and is found concentrated in insoluble **starch granules** within plant cells (see photo, right). Starch can be easily hydrolyzed by enzymes to soluble sugars when required.

Glycogen: Glycogen, like starch, is a branched polysaccharide. It is chemically similar to amylopectin, being composed of α-**glucose** molecules, but there are more α-1,6 glycosidic links mixed with α-1,4 links. This makes it more highly branched and water-soluble than starch. Glycogen is a storage compound in animal tissues and is found mainly in **liver** and **muscle** cells (photo, right). It is readily hydrolyzed by enzymes to form glucose.

Chitin: Chitin is a tough modified polysaccharide made up of chains of β-**glucose** molecules. It is chemically similar to cellulose but each glucose has an amine group (–NH2) attached. After cellulose, chitin is the second most abundant carbohydrate. It is found in the cell walls of fungi and is the main component of the **exoskeleton** of insects (right) and other arthropods.

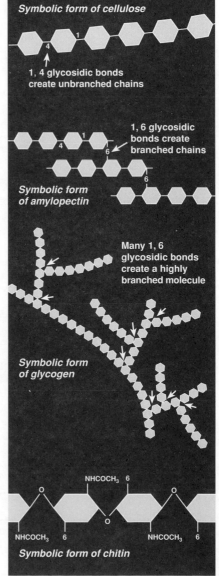

Cellulose

Starch granules

Starch granules in a plant cell

Skeletal muscle tissue

Chitinous insect exoskeleton

Symbolic form of cellulose

1, 4 glycosidic bonds create unbranched chains

1, 6 glycosidic bonds create branched chains

Symbolic form of amylopectin

Many 1, 6 glycosidic bonds create a highly branched molecule

Symbolic form of glycogen

NHCOCH₃

Symbolic form of chitin

The Chemistry of Life

Periodicals:
Glucose,
Designer starches

Related activities: Organic Molecules
Web links: Biomolecules: Carbohydrates

A 2

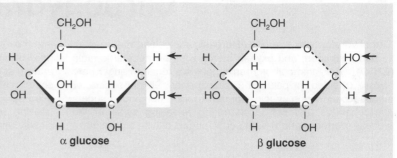

Isomerism

Compounds with the same chemical formula (same types and numbers of atoms) may differ in the arrangement of their atoms. Such variations in the arrangement of atoms in molecules are called **isomers**. In **structural isomers** (such as fructose and glucose, and the α and β glucose, right), the atoms are linked in different sequences. **Optical isomers** are identical in every way but are mirror images of each other.

Condensation and Hydrolysis Reactions

Monosaccharides can combine to form compound sugars in what is called a **condensation** reaction. Compound sugars can be broken down by **hydrolysis** to simple monosaccharides.

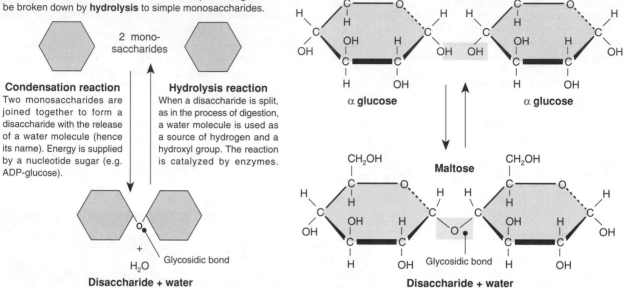

2 mono-saccharides

Condensation reaction
Two monosaccharides are joined together to form a disaccharide with the release of a water molecule (hence its name). Energy is supplied by a nucleotide sugar (e.g. ADP-glucose).

Hydrolysis reaction
When a disaccharide is split, as in the process of digestion, a water molecule is used as a source of hydrogen and a hydroxyl group. The reaction is catalyzed by enzymes.

+
H_2O Glycosidic bond

Disaccharide + water

Disaccharide + water

1. Distinguish between structural and optical isomers in carbohydrates, describing examples of each:

_____

_____

_____

2. Explain how the isomeric structure of a carbohydrate may affect its chemical behavior: _____

_____

_____

3. Explain briefly how compound sugars are formed and broken down: _____

_____

_____

4. Discuss the structural differences between the polysaccharides cellulose, starch, and glycogen, explaining how the differences in structure contribute to the functional properties of the molecule:

_____

_____

_____

_____

_____

_____

_____

Lipids

Lipids are a group of organic compounds with an oily, greasy, or waxy consistency. They are relatively insoluble in water and tend to be water-repelling (e.g. cuticle on leaf surfaces). Lipids are important biological fuels, some are hormones, and some serve as structural components in plasma membranes. Proteins and carbohydrates may be converted into fats by enzymes and stored within cells of adipose tissue. During times of plenty, this store is increased, to be used during times of food shortage.

Neutral Fats and Oils

The most abundant lipids in living things are **neutral fats**. They make up the fats and oils found in plants and animals. Fats are an economical way to store fuel reserves, since they yield more than twice as much energy as the same quantity of carbohydrate. Neutral fats are composed of a glycerol molecule attached to one (monoglyceride), two (diglyceride) or three (triglyceride) fatty acids. The fatty acid chains may be saturated or unsaturated (see below). **Waxes** are similar in structure to fats and oils, but they are formed with a complex alcohol instead of glycerol.

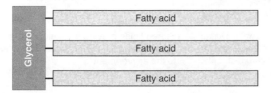

Triglyceride: an example of a neutral fat

Condensation

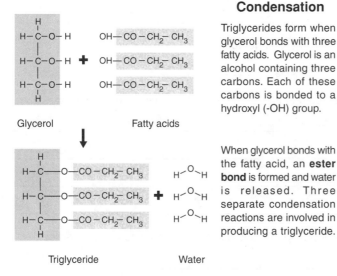

Glycerol   Fatty acids

Triglyceride   Water

Triglycerides form when glycerol bonds with three fatty acids. Glycerol is an alcohol containing three carbons. Each of these carbons is bonded to a hydroxyl (-OH) group.

When glycerol bonds with the fatty acid, an **ester bond** is formed and water is released. Three separate condensation reactions are involved in producing a triglyceride.

Saturated and Unsaturated Fatty Acids

Fatty acids are a major component of neutral fats and phospholipids. About 30 different kinds are found in animal lipids. **Saturated fatty acids** contain the maximum number of hydrogen atoms. **Unsaturated fatty acids** contain some carbon atoms that are double-bonded with each other and are not fully saturated with hydrogens. Lipids containing a high proportion of saturated fatty acids tend to be solids at room temperature (e.g. butter). Lipids with a high proportion of unsaturated fatty acids are oils and tend to be liquid at room temperature. This is because the unsaturation causes kinks in the straight chains so that the fatty acids do not pack closely together. Regardless of their degree of saturation, fatty acids yield a large amount of energy when oxidized.

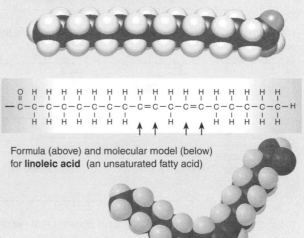

Formula (above) and molecular model (below) for **palmitic acid** (a saturated fatty acid)

Formula (above) and molecular model (below) for **linoleic acid** (an unsaturated fatty acid)

Phospholipids

Phospholipids are the main component of cellular membranes. They consist of a glycerol attached to two fatty acid chains and a phosphate (PO_4^{3-}) group. The phosphate end of the molecule is attracted to water (it is hydrophilic) while the fatty acid end is repelled (hydrophobic). The hydrophobic ends turn inwards in the membrane to form a **phospholipid bilayer**.

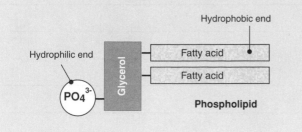

Phospholipid

Steroids

Although steroids are classified as lipids, their structure is quite different from that of other lipids. Steroids have a basic structure of three rings made of 6 carbon atoms each and a fourth ring containing 5 carbon atoms. Examples of steroids include the male and female sex hormones (testosterone and estrogen), and the hormones cortisol and aldosterone. Cholesterol, while not a steroid itself, is a sterol lipid and is a precursor to several steroid hormones.

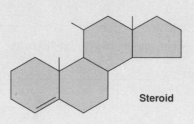

Steroid

The Chemistry of Life

Related activities: The Structure of Membranes
Web links: Biomolecules: Lipids

A 2

Important Biological Functions of Lipids

Lipids are concentrated sources of energy and provide fuel for aerobic respiration.

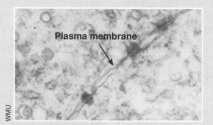

Plasma membrane

Phospholipids form the structural framework of cellular membranes.

Waxes and oils secreted on to surfaces provide waterproofing in plants and animals.

Fat absorbs shocks. Organs that are prone to bumps and shocks (e.g. kidneys) are cushioned with a relatively thick layer of fat.

Lipids are a source of metabolic water. During respiration stored lipids are metabolized for energy, producing water and carbon dioxide.

Stored lipids provide insulation. Increased body fat levels in winter reduce heat losses to the environment.

1. Outline the key **chemical** difference between a phospholipid and a triglyceride: _____

   _____

   _____

2. Name the type of fatty acids found in lipids that form the following at room temperature:

   (a) Solid fats: _____ (b) Oils: _____

3. Relate the structure of phospholipids to their chemical properties and their functional role in cellular membranes:

   _____

   _____

4. (a) Distinguish between saturated and unsaturated fatty acids: _____

   _____

   (b) Explain how the type of fatty acid present in a neutral fat or phospholipid is related to that molecule's properties:

   _____

   _____

   (c) Suggest how the cell membrane structure of an Arctic fish might differ from that of tropical fish species:

   _____

   _____

5. Describe two examples of steroids. For each example, describe its physiological function:

   (a) _____

   (b) _____

6. Explain how fats can provide an animal with:

   (a) Energy: _____

   (b) Water: _____

   (c) Insulation: _____

Nucleic Acids

Nucleic acids are a special group of chemicals in cells concerned with the transmission of inherited information. They have the capacity to store the information that controls cellular activity. The central nucleic acid is called **deoxyribonucleic acid** (DNA). DNA is a major component of chromosomes and is found primarily in the nucleus, although a small amount is found in mitochondria and chloroplasts. Other **ribonucleic acids** (RNA) are involved in the 'reading' of the DNA information. All nucleic acids are made up of simple repeating units called **nucleotides**, linked together to form chains or strands, often of great length (see the activity *DNA Molecules*). The strands vary in the sequence of the bases found on each nucleotide. It is this sequence which provides the 'genetic code' for the cell. In addition to nucleic acids, certain nucleotides and their derivatives are also important as suppliers of energy (**ATP**) or as hydrogen ion and electron carriers in respiration and photosynthesis (NAD, NADP, and FAD).

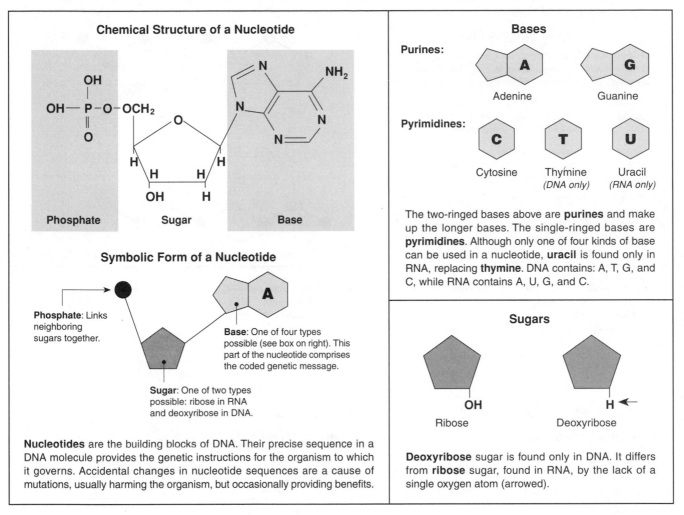

Chemical Structure of a Nucleotide

Phosphate   Sugar   Base

Symbolic Form of a Nucleotide

Phosphate: Links neighboring sugars together.

Base: One of four types possible (see box on right). This part of the nucleotide comprises the coded genetic message.

Sugar: One of two types possible: ribose in RNA and deoxyribose in DNA.

Nucleotides are the building blocks of DNA. Their precise sequence in a DNA molecule provides the genetic instructions for the organism to which it governs. Accidental changes in nucleotide sequences are a cause of mutations, usually harming the organism, but occasionally providing benefits.

Bases

Purines:

A Adenine   G Guanine

Pyrimidines:

C Cytosine   T Thymine *(DNA only)*   U Uracil *(RNA only)*

The two-ringed bases above are **purines** and make up the longer bases. The single-ringed bases are **pyrimidines**. Although only one of four kinds of base can be used in a nucleotide, **uracil** is found only in RNA, replacing **thymine**. DNA contains: A, T, G, and C, while RNA contains A, U, G, and C.

Sugars

OH Ribose   H Deoxyribose

Deoxyribose sugar is found only in DNA. It differs from **ribose** sugar, found in RNA, by the lack of a single oxygen atom (arrowed).

The Chemistry of Life

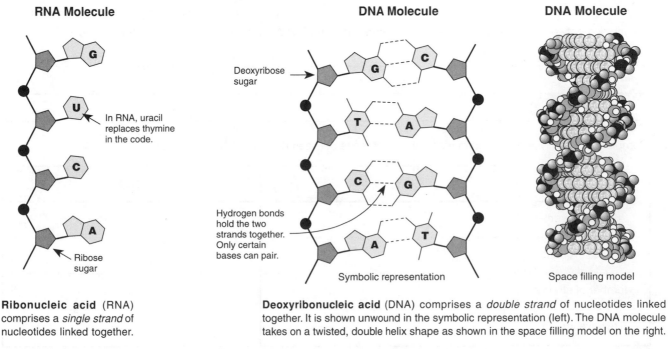

RNA Molecule

In RNA, uracil replaces thymine in the code.

Ribose sugar

DNA Molecule

Deoxyribose sugar

Hydrogen bonds hold the two strands together. Only certain bases can pair.

Symbolic representation

DNA Molecule

Space filling model

Ribonucleic acid (RNA) comprises a *single strand* of nucleotides linked together.

Deoxyribonucleic acid (DNA) comprises a *double strand* of nucleotides linked together. It is shown unwound in the symbolic representation (left). The DNA molecule takes on a twisted, double helix shape as shown in the space filling model on the right.

Related activities: DNA Molecules, Creating a DNA Molecule

A 1

Formation of a nucleotide

Condensation
(water removed)

H_2O

H_2O

A nucleotide is formed when phosphoric acid and a base are chemically bonded to a sugar molecule. In both cases, water is given off, and they are therefore condensation reactions. In the reverse reaction, a nucleotide is broken apart by the addition of water (**hydrolysis**).

Formation of a dinucleotide

H_2O

Two nucleotides are linked together by a condensation reaction between the phosphate of one nucleotide and the sugar of another.

Double-Stranded DNA

The **double-helix** structure of DNA is like a ladder twisted into a corkscrew shape around its longitudinal axis. It is 'unwound' here to show the relationships between the bases.

- The way the correct pairs of bases are attracted to each other to form hydrogen bonds is determined by the number of bonds they can form and the shape (length) of the base.

- The **template strand** the side of the DNA molecule that stores the information that is transcribed into mRNA. The template strand is also called the **antisense strand**.

- The other side (often called the **coding strand**) has the same nucleotide sequence as the mRNA except that T in DNA substitutes for U in mRNA. The coding strand is also called the **sense strand**.

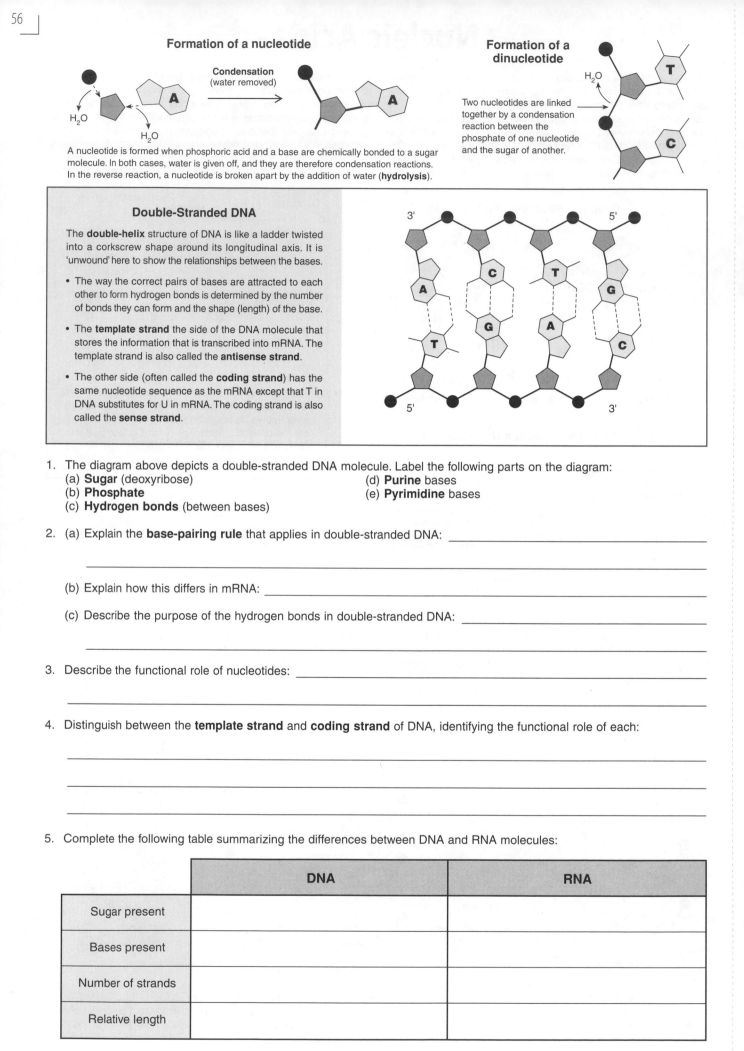

1. The diagram above depicts a double-stranded DNA molecule. Label the following parts on the diagram:
   (a) **Sugar** (deoxyribose)
   (b) **Phosphate**
   (c) **Hydrogen bonds** (between bases)
   (d) **Purine** bases
   (e) **Pyrimidine** bases

2. (a) Explain the **base-pairing rule** that applies in double-stranded DNA: _____

   _____

   (b) Explain how this differs in mRNA: _____

   (c) Describe the purpose of the hydrogen bonds in double-stranded DNA: _____

   _____

3. Describe the functional role of nucleotides: _____

   _____

4. Distinguish between the **template strand** and **coding strand** of DNA, identifying the functional role of each:

   _____

   _____

   _____

5. Complete the following table summarizing the differences between DNA and RNA molecules:

	DNA	RNA
Sugar present		
Bases present		
Number of strands		
Relative length		

Amino Acids

Amino acids are the basic units from which proteins are made. Plants can manufacture all the amino acids they require from simpler molecules, but animals must obtain a certain number of ready-made amino acids (called **essential amino acids**) from their diet. Which amino acids are essential varies from species to species, as different metabolisms are able to synthesize different substances. The distinction between essential and non-essential amino acids is somewhat unclear though, as some amino acids can be produced from others and some are interconvertible by the urea cycle. Amino acids can combine to form peptide chains in a **condensation reaction**. The reverse reaction, which breaks up peptide chains uses water and is called **hydrolysis**.

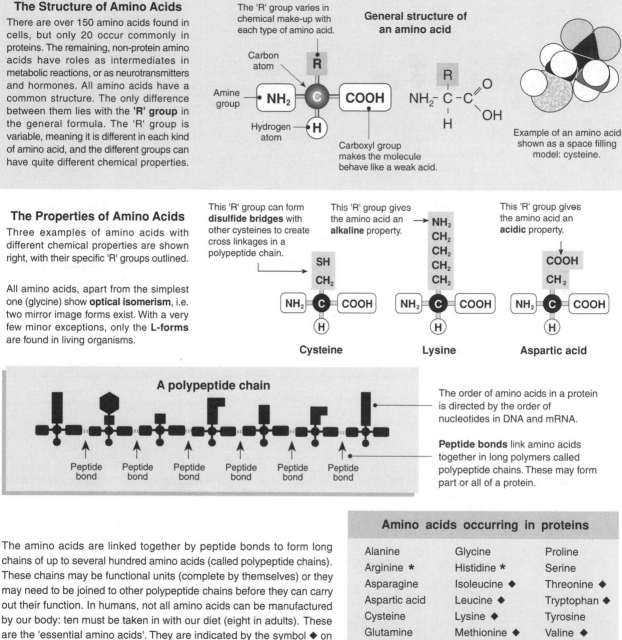

The Structure of Amino Acids

There are over 150 amino acids found in cells, but only 20 occur commonly in proteins. The remaining, non-protein amino acids have roles as intermediates in metabolic reactions, or as neurotransmitters and hormones. All amino acids have a common structure. The only difference between them lies with the **'R' group** in the general formula. The 'R' group is variable, meaning it is different in each kind of amino acid, and the different groups can have quite different chemical properties.

General structure of an amino acid

The 'R' group varies in chemical make-up with each type of amino acid.

Carbon atom
Amine group → NH₂
Hydrogen atom → H
Carboxyl group makes the molecule behave like a weak acid.

$$NH_2 - \overset{R}{\underset{H}{C}} - C \overset{O}{\underset{OH}{}}$$

Example of an amino acid shown as a space filling model: cysteine.

The Properties of Amino Acids

Three examples of amino acids with different chemical properties are shown right, with their specific 'R' groups outlined.

All amino acids, apart from the simplest one (glycine) show **optical isomerism**, i.e. two mirror image forms exist. With a very few minor exceptions, only the **L-forms** are found in living organisms.

This 'R' group can form **disulfide bridges** with other cysteines to create cross linkages in a polypeptide chain.

Cysteine

This 'R' group gives the amino acid an **alkaline** property.

Lysine

This 'R' group gives the amino acid an **acidic** property.

Aspartic acid

A polypeptide chain

The order of amino acids in a protein is directed by the order of nucleotides in DNA and mRNA.

Peptide bonds link amino acids together in long polymers called polypeptide chains. These may form part or all of a protein.

Peptide bond (×6)

The amino acids are linked together by peptide bonds to form long chains of up to several hundred amino acids (called polypeptide chains). These chains may be functional units (complete by themselves) or they may need to be joined to other polypeptide chains before they can carry out their function. In humans, not all amino acids can be manufactured by our body: ten must be taken in with our diet (eight in adults). These are the 'essential amino acids'. They are indicated by the symbol ◆ on the right. Those indicated with as asterisk are also required by infants.

Amino acids occurring in proteins

Alanine	Glycine	Proline
Arginine *	Histidine *	Serine
Asparagine	Isoleucine ◆	Threonine ◆
Aspartic acid	Leucine ◆	Tryptophan ◆
Cysteine	Lysine ◆	Tyrosine
Glutamine	Methionine ◆	Valine ◆
Glutamic acid	Phenylalanine ◆	

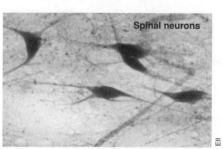

Spinal neurons

Several amino acids act as neurotransmitters in the central nervous system, Glutamic acid and ABA (gamma amino butyric acid) are the most common neurotransmitters in the brain. Others, such as glycine, are restricted to the spinal cord.

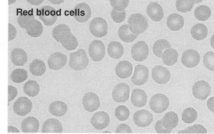

Red blood cells

Amino acids tend to stabilize the pH of solutions in which they are present (e.g. blood and tissue fluid) because they will remove excess H⁺ or OH⁻ ions. They retain this buffer capacity even when incorporated into peptides and proteins.

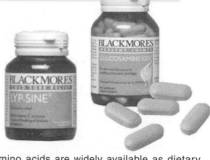

Amino acids are widely available as dietary supplements for specific purposes. Lysine is sold as relief for herpes infections and glucosamine supplements are used for alleviating the symptoms of arthritis and other joint disorders.

Related activities: Proteins, Translation
Web links: Amino Acids and Proteins

A 2

Condensation and Hydrolysis Reactions

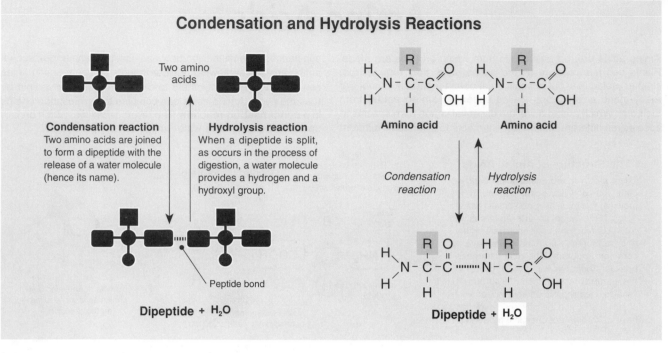

Dipeptide + H₂O

Condensation reaction
Two amino acids are joined to form a dipeptide with the release of a water molecule (hence its name).

Hydrolysis reaction
When a dipeptide is split, as occurs in the process of digestion, a water molecule provides a hydrogen and a hydroxyl group.

Peptide bond

Amino acid   Amino acid

Condensation reaction   *Hydrolysis reaction*

Dipeptide + H₂O

1. Discuss the various biological roles of amino acids: _____

_____

_____

_____

_____

_____

_____

2. Describe what makes each of the 20 amino acids found in proteins unique: _____

_____

_____

3. Describe the process that determines the sequence in which amino acids are linked together to form polypeptide chains:

_____

_____

4. Explain how the chemistry of amino acids enables them to act as buffers in biological tissues: _____

_____

_____

5. Giving examples, explain what is meant by an **essential amino acid**: _____

_____

_____

6. Describe the processes by which amino acids are joined together and broken down: _____

_____

_____

_____

_____

Proteins

The precise folding up of a protein into its **tertiary structure** creates a three dimensional arrangement of the active 'R' groups. The way each 'R' group faces with respect to the others gives the protein its unique chemical properties. If a protein loses this precise structure (denaturation), it is usually unable to carry out its biological function. Proteins are often classified on the basis of structure (globular vs fibrous). Some of the properties used for the basis of structural classification are outlined over the page.

Primary Structure - 1° *(amino acid sequence)*
Strings of hundreds of amino acids link together with peptide bonds to form molecules called polypeptide chains. There are 20 different kinds of amino acids that can be linked together in a vast number of different combinations. This sequence is called the **primary structure**. It is the arrangement of attraction and repulsion points in the amino acid chain that determines the higher levels of organization in the protein and its biological function.

Secondary Structure - 2° *(α-helix or β-pleated sheet)*
Polypeptides become folded in various ways, referred to as the secondary (2°) structure. The most common types of 2° structures are a coiled α-**helix** and a β-**pleated sheet**. Secondary structures are maintained with hydrogen bonds between neighboring CO and NH groups. H-bonds, although individually weak, provide considerable strength when there are a large number of them. The example, right, shows the two main types of secondary structure. In both, the **'R' side groups** (not shown) project out from the structure. Most globular proteins contain regions of α-helices together with β-sheets. Keratin (a fibrous protein) is composed almost entirely of α-helices. Fibroin (silk protein), is another fibrous protein, almost entirely in β-sheet form.

Tertiary Structure - 3° *(folding)*
Every protein has a precise structure formed by the folding of the secondary structure into a complex shape called the **tertiary structure**. The protein folds up because various points on the secondary structure are attracted to one another. The strongest links are caused by bonding between neighboring **cysteine** amino acids which form disulfide bridges. Other interactions that are involved in folding include weak ionic and hydrogen bonds as well as hydrophobic interactions.

Quaternary Structure - 4°
Some proteins (such as enzymes) are complete and functional with a tertiary structure only. However, many complex proteins exist as aggregations of polypeptide chains. The arrangement of the polypeptide chains into a functional protein is termed the **quaternary structure**. The example (right) shows a molecule of hemoglobin, a globular protein composed of 4 polypeptide subunits joined together; two identical **beta chains** and two identical **alpha chains**. Each has a heme (iron containing) group at the centre of the chain, which binds oxygen. Proteins containing non-protein material are **conjugated proteins**. The non-protein part is the **prosthetic group**.

Denaturation of Proteins
Denaturation refers to the loss of the three-dimensional structure (and usually also the biological function) of a protein. Denaturation is often, although not always, permanent. It results from an alteration of the bonds that maintain the secondary and tertiary structure of the protein, even though the sequence of amino acids remains unchanged. Agents that cause denaturation are:
- **Strong acids and alkalis**: Disrupt ionic bonds and result in coagulation of the protein. Long exposure also breaks down the primary structure of the protein.
- **Heavy metals**: May disrupt ionic bonds, form strong bonds with the carboxyl groups of the R groups, and reduce protein charge. The general effect is to cause the precipitation of the protein.
- **Heat and radiation** (e.g. UV): Cause disruption of the bonds in the protein through increased energy provided to the atoms.
- **Detergents and solvents**: Form bonds with the non-polar groups in the protein, thereby disrupting hydrogen bonding.

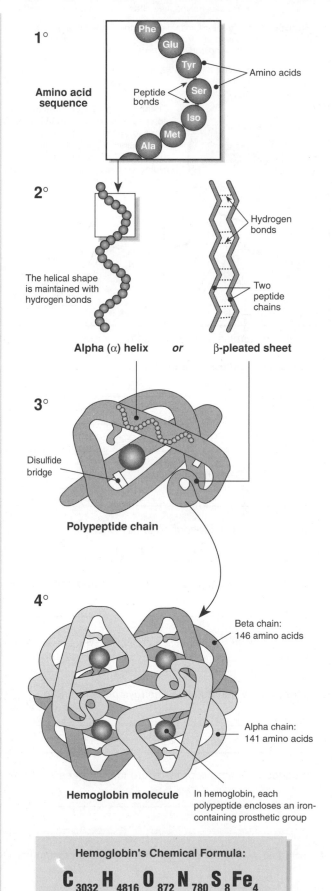

1°
Amino acid sequence
Amino acids
Peptide bonds

2°
The helical shape is maintained with hydrogen bonds
Hydrogen bonds
Two peptide chains
Alpha (α) helix *or* **β-pleated sheet**

3°
Disulfide bridge
Polypeptide chain

4°
Beta chain: 146 amino acids
Alpha chain: 141 amino acids
Hemoglobin molecule
In hemoglobin, each polypeptide encloses an iron-containing prosthetic group

Hemoglobin's Chemical Formula:
$$C_{3032}H_{4816}O_{872}N_{780}S_8Fe_4$$

Periodicals:
What is tertiary structure?
Modelling protein folding

Related activities: Enzymes

RA 2

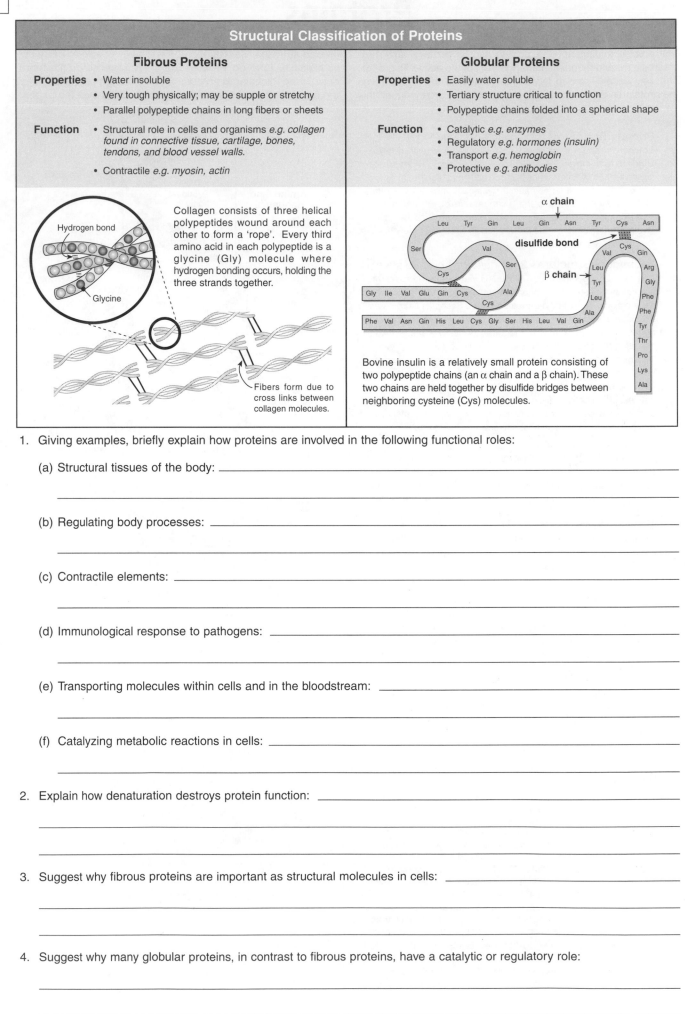

Structural Classification of Proteins

Fibrous Proteins

Properties • Water insoluble
• Very tough physically; may be supple or stretchy
• Parallel polypeptide chains in long fibers or sheets

Function • Structural role in cells and organisms *e.g. collagen found in connective tissue, cartilage, bones, tendons, and blood vessel walls.*
• Contractile *e.g. myosin, actin*

Collagen consists of three helical polypeptides wound around each other to form a 'rope'. Every third amino acid in each polypeptide is a glycine (Gly) molecule where hydrogen bonding occurs, holding the three strands together.

Fibers form due to cross links between collagen molecules.

Globular Proteins

Properties • Easily water soluble
• Tertiary structure critical to function
• Polypeptide chains folded into a spherical shape

Function • Catalytic *e.g. enzymes*
• Regulatory *e.g. hormones (insulin)*
• Transport *e.g. hemoglobin*
• Protective *e.g. antibodies*

Bovine insulin is a relatively small protein consisting of two polypeptide chains (an α chain and a β chain). These two chains are held together by disulfide bridges between neighboring cysteine (Cys) molecules.

1. Giving examples, briefly explain how proteins are involved in the following functional roles:

(a) Structural tissues of the body: _____

_____

(b) Regulating body processes: _____

_____

(c) Contractile elements: _____

_____

(d) Immunological response to pathogens: _____

_____

(e) Transporting molecules within cells and in the bloodstream: _____

_____

(f) Catalyzing metabolic reactions in cells: _____

_____

2. Explain how denaturation destroys protein function: _____

_____

_____

3. Suggest why fibrous proteins are important as structural molecules in cells: _____

_____

_____

4. Suggest why many globular proteins, in contrast to fibrous proteins, have a catalytic or regulatory role:

_____

_____

Enzymes

Most enzymes are proteins. They are capable of catalyzing (speeding up) biochemical reactions and are therefore called biological **catalysts**. Enzymes act on one or more compounds (called the **substrate**). They may break a single substrate molecule down into simpler substances, or join two or more substrate molecules chemically together. The enzyme itself is unchanged in the reaction; its presence merely allows the reaction to take place more rapidly. When the substrate attains the required **activation energy** to enable it to change into the product, there is a 50% chance that it will proceed forward to form the product, otherwise it reverts back to a stable form of

the reactant again. The part of the enzyme's surface into which the substrate is bound and undergoes reaction is known as the **active site**. This is made of different parts of polypeptide chain folded in a specific shape so they are closer together. For some enzymes, the complexity of the binding sites can be very precise, allowing only a single kind of substrate to bind to it. Some other enzymes have lower **specificity** and will accept a wide range of substrates of the same general type (e.g. lipases break up any fatty acid chain length of lipid). This is because the enzyme is specific for the type of chemical bond involved and not an exact substrate.

Enzyme Structure

The model on the right is of an enzyme called *Ribonuclease S*, which breaks up RNA molecules. It is a typical enzyme, being a globular protein and composed of up to several hundred atoms. The darkly shaded areas are called **active sites** and make up the **cleft**; the region into which the substrate molecule(s) are drawn. The correct positioning of these sites is critical for the catalytic reaction to occur. The substrate (RNA in this case) is drawn into the cleft by the active sites. By doing so, it puts the substrate molecule under stress, causing the reaction to proceed more readily.

Substrate molecule: Substrate molecules are the chemicals that an enzyme acts on. They are drawn into the cleft of the enzyme.

Active sites: These attraction points draw the substrate to the enzyme's surface. Substrate molecule(s) are positioned in a way to promote a reaction: either joining two molecules together or splitting up a larger one (as in this case).

Enzyme molecule: The complexity of the active site is what makes each enzyme so specific (i.e. precise in terms of the substrate it acts on).

Source: *Biochemistry*, (1981) by Lubert Stryer

How Enzymes Work

The **lock and key** model proposed earlier this century suggested that the substrate was simply drawn into a closely matching cleft on the enzyme molecule. More recent studies have revealed that the process more likely involves an **induced fit** (see diagram on the right), where the enzyme or the reactants change their shape slightly. The reactants become bound to enzymes by weak chemical bonds. This binding can weaken bonds within the reactants themselves, allowing the reaction to proceed more readily.

The presence of an enzyme simply makes it easier for a reaction to take place. All **catalysts** speed up reactions by influencing the stability of bonds in the reactants. They may also provide an alternative reaction pathway, thus lowering the activation energy needed for a reaction to take place (see the graph below).

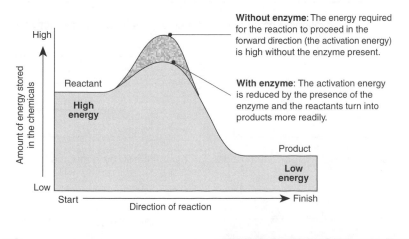

Without enzyme: The energy required for the reaction to proceed in the forward direction (the activation energy) is high without the enzyme present.

With enzyme: The activation energy is reduced by the presence of the enzyme and the reactants turn into products more readily.

Induced Fit Model

An enzyme fits to its substrate somewhat like a lock and key. The shape of the enzyme changes when the substrate fits into the cleft (called the **induced fit**):

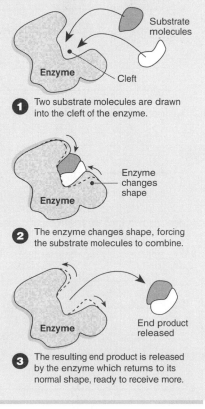

1. Two substrate molecules are drawn into the cleft of the enzyme.
2. The enzyme changes shape, forcing the substrate molecules to combine.
3. The resulting end product is released by the enzyme which returns to its normal shape, ready to receive more.

© Biozone International 2001-2010
Photocopying Prohibited

Periodicals: Enzymes: nature's catalytic machines

Related activities: Enzyme Reaction Rates, Gene Mutations
Web links: How Enzymes Work

RA 2

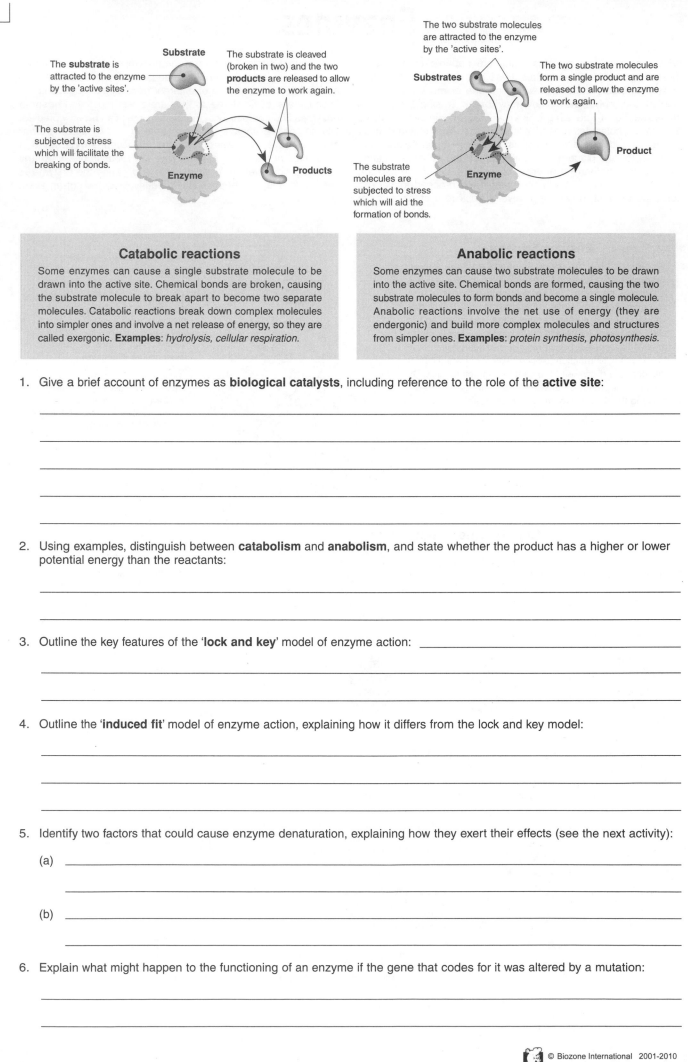

The **substrate** is attracted to the enzyme by the 'active sites'.

Substrate

The substrate is cleaved (broken in two) and the two **products** are released to allow the enzyme to work again.

The substrate is subjected to stress which will facilitate the breaking of bonds.

Enzyme

Products

The two substrate molecules are attracted to the enzyme by the 'active sites'.

Substrates

The two substrate molecules form a single product and are released to allow the enzyme to work again.

The substrate molecules are subjected to stress which will aid the formation of bonds.

Enzyme

Product

Catabolic reactions

Some enzymes can cause a single substrate molecule to be drawn into the active site. Chemical bonds are broken, causing the substrate molecule to break apart to become two separate molecules. Catabolic reactions break down complex molecules into simpler ones and involve a net release of energy, so they are called exergonic. **Examples**: *hydrolysis, cellular respiration*.

Anabolic reactions

Some enzymes can cause two substrate molecules to be drawn into the active site. Chemical bonds are formed, causing the two substrate molecules to form bonds and become a single molecule. Anabolic reactions involve the net use of energy (they are endergonic) and build more complex molecules and structures from simpler ones. **Examples**: *protein synthesis, photosynthesis*.

1. Give a brief account of enzymes as **biological catalysts**, including reference to the role of the **active site**:

   _____

   _____

   _____

   _____

   _____

2. Using examples, distinguish between **catabolism** and **anabolism**, and state whether the product has a higher or lower potential energy than the reactants:

   _____

   _____

3. Outline the key features of the '**lock and key**' model of enzyme action: _____

   _____

   _____

4. Outline the '**induced fit**' model of enzyme action, explaining how it differs from the lock and key model:

   _____

   _____

   _____

5. Identify two factors that could cause enzyme denaturation, explaining how they exert their effects (see the next activity):

   (a) _____

   _____

   (b) _____

   _____

6. Explain what might happen to the functioning of an enzyme if the gene that codes for it was altered by a mutation:

   _____

   _____

Enzyme Reaction Rates

Enzymes are sensitive molecules. They often have a narrow range of conditions under which they operate properly. For most of the enzymes associated with plant and animal metabolism, there is little activity at low temperatures. As the temperature increases, so too does the enzyme activity, until the point is reached where the temperature is high enough to damage the enzyme's structure. At this point, the enzyme ceases to function; a phenomenon called enzyme or protein **denaturation**.

Extremes in acidity (pH) can also cause the protein structure of enzymes to denature. Poisons often work by denaturing enzymes or occupying the enzyme's active site so that it does not function. In some cases, enzymes will not function without cofactors, such as vitamins or trace elements. In the four graphs below, the rate of reaction or degree of enzyme activity is plotted against each of four factors that affect enzyme performance. Answer the questions relating to each graph:

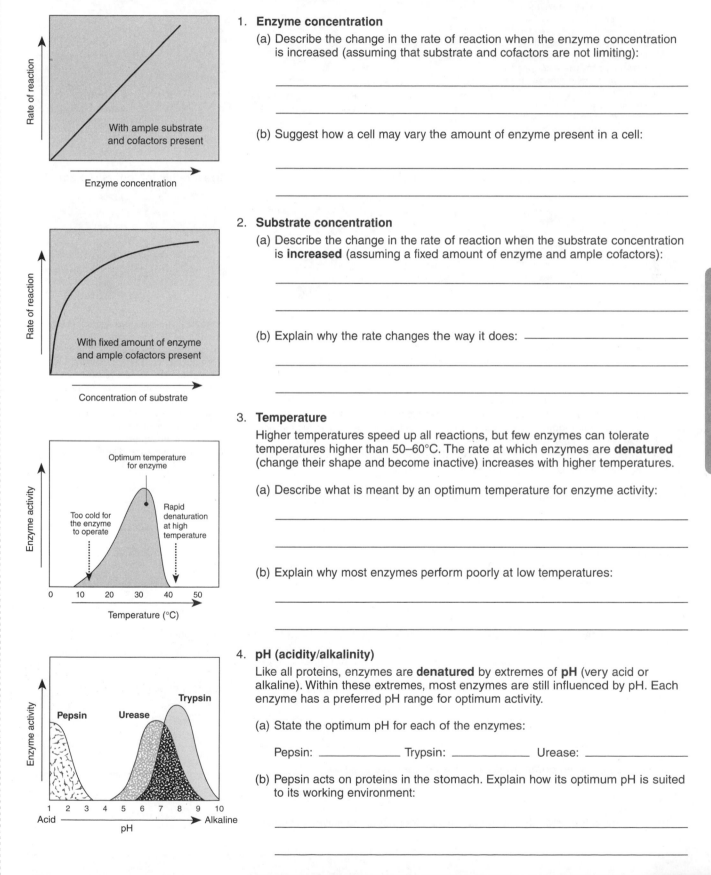

1. **Enzyme concentration**

   (a) Describe the change in the rate of reaction when the enzyme concentration is increased (assuming that substrate and cofactors are not limiting):

   _____

   _____

   (b) Suggest how a cell may vary the amount of enzyme present in a cell:

   _____

   _____

2. **Substrate concentration**

   (a) Describe the change in the rate of reaction when the substrate concentration is **increased** (assuming a fixed amount of enzyme and ample cofactors):

   _____

   _____

   (b) Explain why the rate changes the way it does: _____

   _____

   _____

3. **Temperature**

   Higher temperatures speed up all reactions, but few enzymes can tolerate temperatures higher than 50–60°C. The rate at which enzymes are **denatured** (change their shape and become inactive) increases with higher temperatures.

   (a) Describe what is meant by an optimum temperature for enzyme activity:

   _____

   _____

   (b) Explain why most enzymes perform poorly at low temperatures:

   _____

   _____

4. **pH (acidity/alkalinity)**

   Like all proteins, enzymes are **denatured** by extremes of **pH** (very acid or alkaline). Within these extremes, most enzymes are still influenced by pH. Each enzyme has a preferred pH range for optimum activity.

   (a) State the optimum pH for each of the enzymes:

   Pepsin: _____ Trypsin: _____ Urease: _____

   (b) Pepsin acts on proteins in the stomach. Explain how its optimum pH is suited to its working environment:

   _____

   _____

The Chemistry of Life

© Biozone International 2001-2010
Photocopying Prohibited

Periodicals:
Enzymes: fast and flexible

Related activities: Enzyme Cofactors and Inhibitors

RDA 2

Enzyme Cofactors and Inhibitors

Enzyme activity is often influenced by the presence of other chemicals. Some of these may enhance an enzyme's activity. Called **cofactors**, they are a nonprotein component of an enzyme and may be organic molecules (**coenzymes**) or inorganic ions (e.g. Ca^{2+}, Zn^{2+}). Enzymes may also be deactivated, temporarily or permanently, by chemicals called enzyme **inhibitors**.

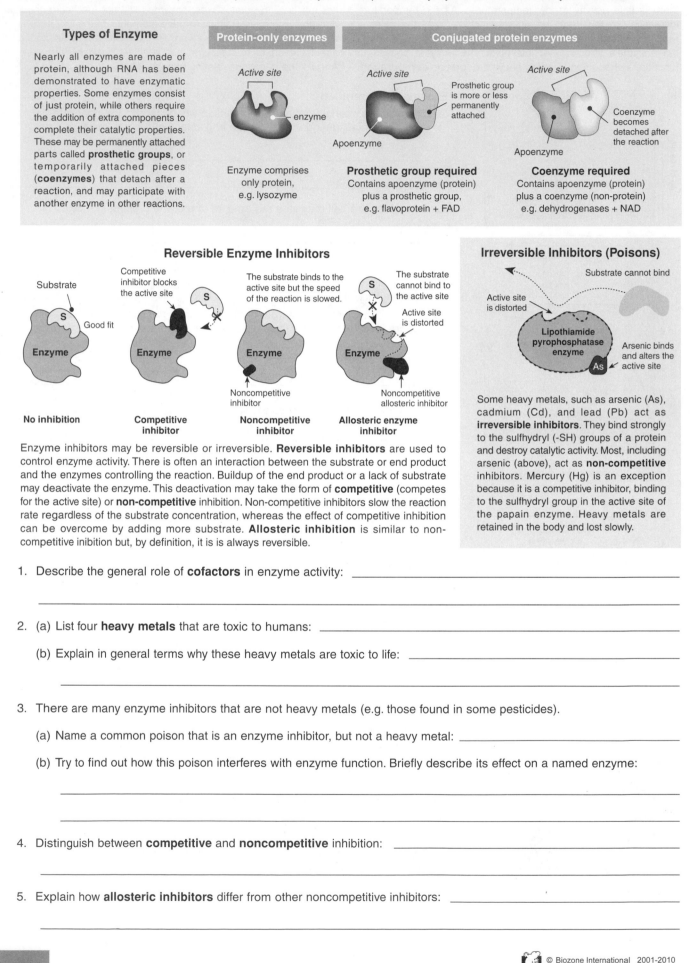

Types of Enzyme

Nearly all enzymes are made of protein, although RNA has been demonstrated to have enzymatic properties. Some enzymes consist of just protein, while others require the addition of extra components to complete their catalytic properties. These may be permanently attached parts called **prosthetic groups**, or temporarily attached pieces (**coenzymes**) that detach after a reaction, and may participate with another enzyme in other reactions.

Protein-only enzymes

Active site

enzyme

Enzyme comprises only protein, e.g. lysozyme

Conjugated protein enzymes

Active site

Prosthetic group is more or less permanently attached

Apoenzyme

Prosthetic group required
Contains apoenzyme (protein) plus a prosthetic group, e.g. flavoprotein + FAD

Active site

Coenzyme becomes detached after the reaction

Apoenzyme

Coenzyme required
Contains apoenzyme (protein) plus a coenzyme (non-protein) e.g. dehydrogenases + NAD

Reversible Enzyme Inhibitors

Substrate

S

Good fit

Enzyme

No inhibition

Competitive inhibitor blocks the active site

S

Enzyme

Competitive inhibitor

The substrate binds to the active site but the speed of the reaction is slowed.

Enzyme

Noncompetitive inhibitor

Noncompetitive inhibitor

The substrate cannot bind to the active site

S

Active site is distorted

Enzyme

Noncompetitive allosteric inhibitor

Allosteric enzyme inhibitor

Enzyme inhibitors may be reversible or irreversible. **Reversible inhibitors** are used to control enzyme activity. There is often an interaction between the substrate or end product and the enzymes controlling the reaction. Buildup of the end product or a lack of substrate may deactivate the enzyme. This deactivation may take the form of **competitive** (competes for the active site) or **non-competitive** inhibition. Non-competitive inhibitors slow the reaction rate regardless of the substrate concentration, whereas the effect of competitive inhibition can be overcome by adding more substrate. **Allosteric inhibition** is similar to non-competitive inibition but, by definition, it is is always reversible.

Irreversible Inhibitors (Poisons)

Substrate cannot bind

Active site is distorted

Lipothiamide pyrophosphatase enzyme

As

Arsenic binds and alters the active site

Some heavy metals, such as arsenic (As), cadmium (Cd), and lead (Pb) act as **irreversible inhibitors**. They bind strongly to the sulfhydryl (-SH) groups of a protein and destroy catalytic activity. Most, including arsenic (above), act as **non-competitive** inhibitors. Mercury (Hg) is an exception because it is a competitive inhibitor, binding to the sulfhydryl group in the active site of the papain enzyme. Heavy metals are retained in the body and lost slowly.

1. Describe the general role of **cofactors** in enzyme activity: _____

_____

2. (a) List four **heavy metals** that are toxic to humans: _____

(b) Explain in general terms why these heavy metals are toxic to life: _____

_____

3. There are many enzyme inhibitors that are not heavy metals (e.g. those found in some pesticides).

(a) Name a common poison that is an enzyme inhibitor, but not a heavy metal: _____

(b) Try to find out how this poison interferes with enzyme function. Briefly describe its effect on a named enzyme:

_____

_____

4. Distinguish between **competitive** and **noncompetitive** inhibition: _____

_____

5. Explain how **allosteric inhibitors** differ from other noncompetitive inhibitors: _____

_____

Applications of Enzymes

Microbes are ideal organisms for the industrial production of enzymes because of their high productivity, ease of culture in industrial fermenters, and the ease with which they can be genetically modified to produce particular products. In addition, because there is an enormous diversity in microbial metabolism, the variety of enzymes available for exploitation is very large. Some of the microorganisms involved in industrial fermentations, and their enzymes and their applications are described below.

Enzymes are used in various stages of **cheese production**, e.g. chymosin from GE microbes now replaces the rennin previously obtained from calves.

In **beer brewing**, **proteases** (from bacteria) are added to prevent cloudiness. Amyloglucosidases are used to produce low calorie beers.

Citric acid is used in **jam production** and is synthesized by a mutant strain of the fungus *Aspergillus niger*, which produces the enzyme citrate synthase.

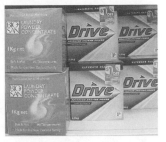

Biological detergents use **proteases**, **lipases**, and **amylases** extracted from fungi and thermophilic bacteria to break down organic material in stains.

Fungal ligninases are used in **pulp and paper industries** to remove lignin from wood pulp and treat wood waste.

In **soft centered chocolates**, **invertase** from yeast breaks down the solid filling to produce the soft center.

Medical treatment of blood clots employs protease enzymes such as streptokinase from *Streptomyces* spp.

Some of the many applications of microbial enzymes in medicine, industry, and food manufacture.

Bacterial proteases are used to break down the wheat protein (gluten) in flour, to produce low gluten breads.

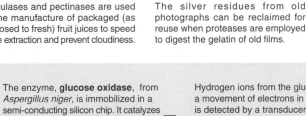

Cellulases and pectinases are used in the manufacture of packaged (as opposed to fresh) fruit juices to speed juice extraction and prevent cloudiness.

The silver residues from old photographs can be reclaimed for reuse when proteases are employed to digest the gelatin of old films.

The lactase from bacteria is used to convert lactose to glucose and galactose in the production of low-lactose and lactose free milk products.

Tanning industries now use proteases from *Bacillus subtilis* instead of toxic chemicals, such as sulfide pastes, to remove hairs and soften hides.

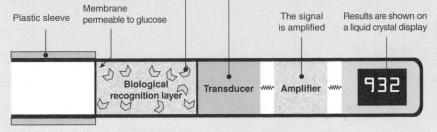

The enzyme, **glucose oxidase**, from *Aspergillus niger*, is immobilized in a semi-conducting silicon chip. It catalyzes the conversion of glucose (from the blood sample) to gluconic acid.

Hydrogen ions from the gluconic acid cause a movement of electrons in the silicon, which is detected by a transducer. The strength of the electric current is directly proportional to the blood glucose concentration.

Plastic sleeve

Membrane permeable to glucose

The signal is amplified

Results are shown on a liquid crystal display

Biological recognition layer

Transducer ⋀⋀⋀ Amplifier ⋀⋀⋀

932

Biosensors are electronic monitoring devices that use biological material to detect the presence or concentration of a particular substance. Enzymes are ideally suited for use in biosensors because of their specificity and sensitivity. This example illustrates how **glucose oxidase** from the fungus *Aspergillus niger* is used in a biosensor to measure blood glucose level in diabetics.

The Chemistry of Life

1. Identify two probable consequences of the absence of enzymes from a chemical reaction that normally uses them:

(a) _____

(b) _____

2. Identify three properties of microbial enzymes that make them highly suitable as industrial catalysts. For each, explain why the property is important:

(a) _____

_____

(b) _____

_____

(c) _____

_____

3. Choose one example from those described in the diagram on the previous page and, in more detail, identify:

(a) The enzyme and its specific microbial source: _____

_____

(b) The application of the enzyme in industry and the specific reaction it catalyzes: _____

_____

_____

_____

4. (a) Outline the basic principle of enzyme-based biosensors: _____

_____

(b) Suggest how a biosensor could be used to monitor blood alcohol level: _____

_____

_____

_____

5. For each of the examples described below, suggest how the use of microbial enzymes has improved the efficiency, cost effectiveness, and/or safety of processing compared with traditional methods:

(a) Use of microbial proteases to treat hides in the tanning industry: _____

_____

_____

_____

(b) Use of microbial chymosin in cheese production: _____

_____

_____

_____

(c) Use of fungal ligninases to treat wood waste: _____

_____

_____

_____

Biochemical Tests

Biochemical tests are used to detect the presence of nutrients such as lipids, proteins, and carbohydrates (sugar and starch) in various foods. These simple tests are useful for detecting nutrients when large quantities are present. A more accurate technique by which to separate a mixture of compounds involves **chromatography**. Chromatography is used when only a small sample is available or when you wish to distinguish between nutrients. Simple biochemical food tests will show whether sugar is present, whereas chromatography will distinguish between the different types of sugars (e.g. fructose or glucose).

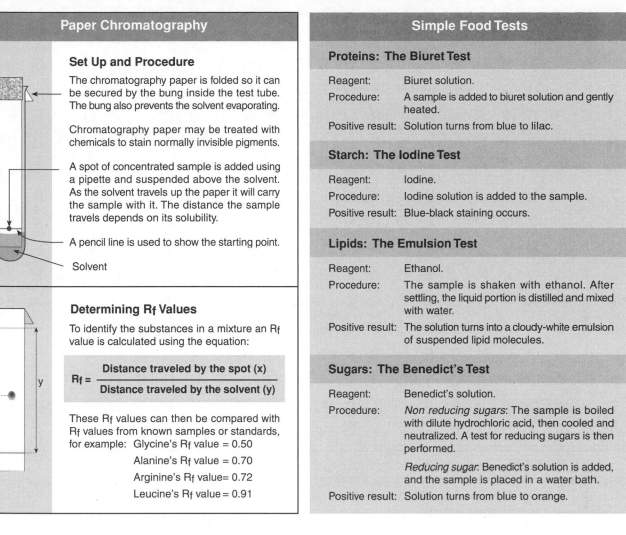

Paper Chromatography

Set Up and Procedure

The chromatography paper is folded so it can be secured by the bung inside the test tube. The bung also prevents the solvent evaporating.

Chromatography paper may be treated with chemicals to stain normally invisible pigments.

A spot of concentrated sample is added using a pipette and suspended above the solvent. As the solvent travels up the paper it will carry the sample with it. The distance the sample travels depends on its solubility.

A pencil line is used to show the starting point.

Solvent

Determining R$_f$ Values

To identify the substances in a mixture an R$_f$ value is calculated using the equation:

$$R_f = \frac{\text{Distance traveled by the spot (x)}}{\text{Distance traveled by the solvent (y)}}$$

These R$_f$ values can then be compared with R$_f$ values from known samples or standards, for example: Glycine's R$_f$ value = 0.50

Alanine's R$_f$ value = 0.70

Arginine's R$_f$ value = 0.72

Leucine's R$_f$ value = 0.91

Simple Food Tests

Proteins: The Biuret Test

Reagent:　　　Biuret solution.

Procedure:　　A sample is added to biuret solution and gently heated.

Positive result:　Solution turns from blue to lilac.

Starch: The Iodine Test

Reagent:　　　Iodine.

Procedure:　　Iodine solution is added to the sample.

Positive result:　Blue-black staining occurs.

Lipids: The Emulsion Test

Reagent:　　　Ethanol.

Procedure:　　The sample is shaken with ethanol. After settling, the liquid portion is distilled and mixed with water.

Positive result:　The solution turns into a cloudy-white emulsion of suspended lipid molecules.

Sugars: The Benedict's Test

Reagent:　　　Benedict's solution.

Procedure:　　*Non reducing sugars*: The sample is boiled with dilute hydrochloric acid, then cooled and neutralized. A test for reducing sugars is then performed.

Reducing sugar: Benedict's solution is added, and the sample is placed in a water bath.

Positive result:　Solution turns from blue to orange.

The Chemistry of Life

1. Calculate the R$_f$ value for the example given above (show your working): _____

   _____

2. Explain why the R$_f$ value of a substance is always less than 1: _____

   _____

3. Discuss when it is appropriate to use chromatography instead of a simple food test: _____

   _____

   _____

4. Predict what would happen if a sample was immersed in the chromatography solvent, instead of suspended above it:

   _____

   _____

5. With reference to their R$_f$ values, rank the four amino acids (listed above) in terms of their solubility: _____

   _____

6. Outline why lipids must be mixed in ethanol before they will form an emulsion in water: _____

   _____

Related activities: Carbohydrates, Lipids

RDA 1

KEY TERMS: Word Find

Use the clues below to find the relevant key terms in the WORD FIND grid

```
G L Y C O G E N C N S X A C T I V E S I T E L C I
S T G P A F L I P I D S X P E R S P C D Z P I M S
A M K L R R H E E X E R G O N I C H E A N Y Q O O
T P J H O I W R R A W E Z T A V M O L M K A N L M
U E R B V B M U I R O N Z K N O S S L I O I E E E
R V N D I N U A O P S B Y T A R J P U N D U U C R
A B L Z Z O S L R N A N O S B G V H L O D M T U S
T C V U Y J S U A Y W F S T O A X O O A B I R L K
E H S J U M S E O R S D X G L N R L S C W N A A O
D I T R D P E Y N Q T T I M I I T I E I A D L R K
Z T A N Z N J S G S X P R P C C Z P Z D T U F J F
M I R A A K L S Z E O I W U O C U I S S E C A E T
E N C H F I B R O U S R Q J C L E D C C R E T L H
Q H H C O N D E N S A T I O N T A S S O N D S F D
W D C O F A C T O R S D F D J T U R S F P F X O H
P V D D I P E P T I D E B U X U U R T W Z I C E V
D E N A T U R A T I O N E L D E L Z E Y F T Z R E
```

Known as the universal solvent.

Carbon-based compounds are known as this.

These proteins are very tough and often have a structural role in cells.

The loss of a protein's three dimensional functional structure is called this.

The formula that describes the number of atoms in a molecule.

This inorganic ion is a component of hemoglobin.

A molecule, like water, in which the opposite ends are oppositely charged.

The most abundant lipids in living things (2 words).

The emulsion tests detects these.

A carbohydrate storage molecule found in muscle and liver tissue.

A polysaccharide found in the exoskeleton of arthropods.

An important structural polysaccharide in plants.

A storage polymer in plants made up of long chains of alpha-glucose.

A general term for a reaction in which water is released.

These lipid molecules naturally form bilayers.

The building blocks of proteins (2 words).

The sequence of amino acids in a protein is called this (2 words).

The region of an enzyme responsible for substrate binding and reaction catalysis (2 words).

Substances required by an enzyme to enable its catalytic function.

Reactions that release energy are called this.

Reactions that build larger molecules from smaller ones are called this.

An electronic monitoring device that uses biological material to detect or measure a substance.

Currently accepted model for enzyme function (2 words).

A fatty acid containing the maximum number of hydrogen atoms is called this.

These proteins are water soluble and have catalytic and regulatory roles in cells.

The product of a condensation reaction between two amino acids
The forms a molecule can take are called this.

These are biological catalysts.

RA 2

Cell Structure

KEY CONCEPTS

▶ Cells are the fundamental units of life.

▶ Cells share many of the same components.

▶ There are distinguishing differences between the cells of different kingdoms.

▶ Microscopy can be used to understand cellular structure.

▶ Centrifugation is a simple technique by which to separate the components of cells.

KEY TERMS

cell wall
centrioles
chloroplast
cilia
contractile vacuole
cytoplasm
cytoskeleton
desmosome
electron microscope
endoplasmic reticulum
(ER)
eukaryotic cell
flagellum
gap junction
Golgi apparatus
lysosome
magnification
mitochondrion
nuclear envelope
nuclear pore
nucleolus
nucleus
optical microscope
organelles
plasma membrane
plasmodesmata
prokaryotic cell
Protista
resolution
ribosome
rough ER
SEM
smooth ER
stain
TEM
tight junction
vacuole

Periodicals:
*listings for this
chapter are on page 379*

Weblinks:
*www.thebiozone.com/
weblink/SB1-2597.html*

*Teacher Resource
CD-ROM:*

The Cell Theory

OBJECTIVES

☐ 1. Use the **KEY TERMS** to help you understand and complete these objectives.

Cells as Living Entities
pages 70-80, 93-94, 224

☐ 2. Describe the features of a generalized cell. Explain why cells are considered to be living entities.

☐ 3. Compare and contrast features of different cell types including **eukaryotic** vs **prokaryotic cells**, and protistan, fungal, plant, and animal cells.

☐ 4. Describe the structure and function of named cellular **organelles**. Describe the structure of cellular membranes, including the **plasma membrane**.

☐ 5. Describe the range of cell sizes. Express cell sizes in different units of measurement (mm, μm, nm).

☐ 6. Describe the non-cellular nature of viruses.

Microscopy
pages 82-88

☐ 7. Appreciate the significance of the **cell theory** to biology.

☐ 8. Recognize the contribution of microscopy to our modern understanding of cell structure and function.

☐ 9. Compare and contrast the structure and basic principles of optical and electron microscopes. Distinguish between **magnification** and **resolution**.

☐ 10. Distinguish between the structure and use of compound and stereo light microscopes.

☐ 11. PRACTICAL: Use both stereo and compound light microscopes to locate material and focus images.

☐ 12. Distinguish between the principles and uses of TEM (transmission electron microscopy) and SEM (scanning electron microscopy).

☐ 13. Interpret drawings and photomicrographs of typical **plant** and **animal cells** as seen using light microscopy and electron microscopy.

☐ 14. PRACTICAL: Demonstrate an ability to prepare a temporary mount for viewing with a light microscope.

☐ 15. PRACTICAL: Demonstrate an ability use simple **staining techniques** to show features of cells.

Separating Cell Fractions
page 81

☐ 16. Describe the principles of **cell fractionation**. Identify the components of the four fractions obtained from differential centrifugation: the **nuclear fraction**, **mitochondrial fraction**, **microsomal fraction**, and **soluble fraction**.

Types of Living Things

Living things are called organisms and **cells** are the functioning unit structure from which organisms are made. Under the five kingdom system, cells can be divided into two basic kinds: the **prokaryotes**, which are simple cells without a distinct, membrane-bound nucleus, and the more complex **eukaryotes**. The eukaryotes can be further organized into broad groups according to their basic cell type: the protists, fungi, plants, and animals. Viruses are non-cellular and have no cellular machinery of their own. All cells must secure a source of energy if they are to survive and carry out metabolic processes. **Autotrophs** can meet their energy requirements using light or chemical energy from the physical environment. Other types of cell, called **heterotrophs**, obtain their energy from other living organisms or their dead remains.

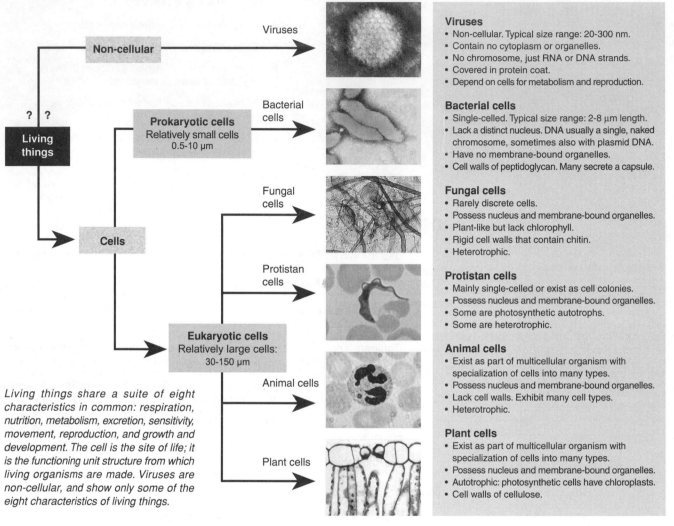

Living things share a suite of eight characteristics in common: respiration, nutrition, metabolism, excretion, sensitivity, movement, reproduction, and growth and development. The cell is the site of life; it is the functioning unit structure from which living organisms are made. Viruses are non-cellular, and show only some of the eight characteristics of living things.

Viruses
- Non-cellular. Typical size range: 20-300 nm.
- Contain no cytoplasm or organelles.
- No chromosome, just RNA or DNA strands.
- Covered in protein coat.
- Depend on cells for metabolism and reproduction.

Bacterial cells
- Single-celled. Typical size range: 2-8 µm length.
- Lack a distinct nucleus. DNA usually a single, naked chromosome, sometimes also with plasmid DNA.
- Have no membrane-bound organelles.
- Cell walls of peptidoglycan. Many secrete a capsule.

Fungal cells
- Rarely discrete cells.
- Possess nucleus and membrane-bound organelles.
- Plant-like but lack chlorophyll.
- Rigid cell walls that contain chitin.
- Heterotrophic.

Protistan cells
- Mainly single-celled or exist as cell colonies.
- Possess nucleus and membrane-bound organelles.
- Some are photosynthetic autotrophs.
- Some are heterotrophic.

Animal cells
- Exist as part of multicellular organism with specialization of cells into many types.
- Possess nucleus and membrane-bound organelles.
- Lack cell walls. Exhibit many cell types.
- Heterotrophic.

Plant cells
- Exist as part of multicellular organism with specialization of cells into many types.
- Possess nucleus and membrane-bound organelles.
- Autotrophic: photosynthetic cells have chloroplasts.
- Cell walls of cellulose.

1. List the cell types above according to the way in which they obtain their energy. Include viruses in your answer as well:

   (a) Autotrophic: _____

   (b) Heterotrophic: _____

2. Consult the diagram above and determine the two main features distinguishing **eukaryotic** cells from **prokaryotic** cells:

   (a) _____

   (b) _____

3. (a) Suggest why fungi were once classified as belonging to the plant kingdom: _____

   _____

   (b) Explain why, in terms of the distinguishing features of fungi, this classification was erroneous: _____

   _____

4. Suggest why the Protista have traditionally been a difficult group to classify: _____

   _____

   _____

Related activities: Bacterial Cells, Unicellular Eukaryotes, Fungal Cells, Plant cells, Animal Cells   **Web links**: Types of Microbes

Periodicals: *The living dead Are viruses alive?*

© Biozone International 2001-2010
Photocopying Prohibited

Bacterial Cells

Bacterial (prokaryotic) cells are much smaller and simpler than the cells of eukaryotes. They lack many eukaryotic features (e.g. a distinct nucleus and membrane-bound cellular organelles). The bacterial cell wall is an important feature. It is a complex, multi-layered structure and often has a role in virulence. These pages illustrate some features of bacterial structure and diversity.

Structure of a Generalized Bacterial Cell

Plasmids: Small, circular DNA molecules (accessory chromosomes) which can reproduce independently of the main chromosome. They can move between cells, and even between species, by **conjugation**. This property accounts for the transmission of antibiotic resistance between bacteria. Plasmids are also used as vectors in recombinant DNA technology.

Single, circular main chromosome: Makes them haploid for most genes. It is possible for some genes to be found on both the plasmid and chromosome and there may be several copies of a gene on a group of plasmids.

The cell lacks a nuclear membrane, so there is no distinct nucleus and the chromosome is in direct contact with the cytoplasm. It is possible for free ribosomes to attach to mRNA while the mRNA is still in the process of being transcribed from the DNA.

Fimbriae: Hairlike structures that are shorter, straighter, and thinner than flagella. They are used for attachment, not movement. Pili are similar to fimbriae, but are longer and less numerous. They are involved in bacterial conjugation (below) and as phage receptors (opposite).

1 µm

Cytoplasm

Cell surface membrane: Similar in composition to eukaryotic membranes, although less rigid.

Glycocalyx. A viscous, gelatinous layer outside the cell wall. It is composed of polysaccharide and/or polypeptide. If it is firmly attached to the wall, it is called a **capsule**. If loosely attached, it is called a **slime layer**. Capsules may contribute to virulence in pathogenic species, e.g. by protecting the bacteria from the host's immune attack. In some species, the glycocalyx allows attachment to substrates.

Cell wall. A complex, semi-rigid structure that gives the cell shape, prevents rupture, and serves as an anchorage point for flagella. The cell wall is composed of a macromolecule called **peptidoglycan**; repeating disaccharides attached by polypeptides to form a lattice. The wall also contains varying amounts of lipopolysaccharides and lipoproteins. The amount of peptidoglycan present in the wall forms the basis of the diagnostic **gram stain**. In many species, the cell wall contributes to their virulence (disease-causing ability).

Flagellum (pl. flagella). Some bacteria have long, filamentous appendages, called flagella, that are used for locomotion. There may be a single polar flagellum (monotrichous), one or more flagella at each end of the cell, or the flagella may be distributed over the entire cell (peritrichous).

Cell Structure

Bacterial cell shapes

Most bacterial cells range between 0.20-2.0 µm in diameter and 2-10 µm length. Although they are a very diverse group, much of this diversity is in their metabolism. In terms of gross morphology, there are only a few basic shapes found (illustrated below). The way in which members of each group aggregate after division is often characteristic and is helpful in identifying certain species.

Bacilli
Rod-shape,
e.g. *E. coli*

Bacilli: Rod-shaped bacteria that divide only across their short axis. Most occur as single rods, although pairs and chains are also found. The term bacillus can refer to shape. It may also denote a genus, e.g. *Bacillus anthracis*.

Cocci
Ball-shaped
e.g. *Staphylococcus*

Cocci: usually round, but sometimes oval or elongated. When they divide, the cells stay attached to each other and remain in aggregates e.g. pairs (diplococci) or clusters (staphylococci), that are usually a feature of the genus.

Spirilla
Spiral-shaped
e.g. *Leptospira*

Spirilla and vibrio: Bacteria with one or more twists. Spirilla bacteria have a helical (corkscrew) shape which may be rigid or flexible (as in spirochetes). Bacteria that look like curved rods (comma shaped) are called vibrios.

Bacterial conjugation

The two bacteria illustrated below are involved in 'pseudo sex'. This involves a one-way exchange of genetic information from a donor cell to a recipient cell. The plasmid, which must be of the 'conjugative' type, passes through a tube called a **sex pilus** to the other cell. Which is donor and which is recipient appears to be genetically determined.

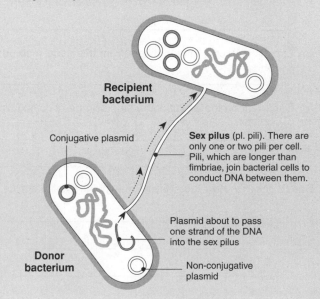

Recipient bacterium

Conjugative plasmid

Sex pilus (pl. pili). There are only one or two pili per cell. Pili, which are longer than fimbriae, join bacterial cells to conduct DNA between them.

Plasmid about to pass one strand of the DNA into the sex pilus

Donor bacterium

Non-conjugative plasmid

Periodicals:
Bacteria

Related activities: Antibiotic Resistance, Features of Taxonomic Groups
Web links: Gram Stain Animation, Bacterial Conjugation Animation

RA 2

72

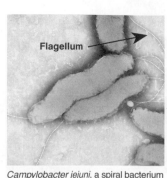

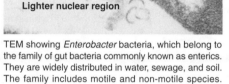

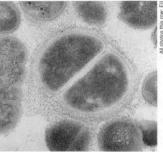

Campylobacter jejuni, a spiral bacterium responsible for foodborne intestinal disease. Note the single flagellum at each end (amphitrichous arrangement).

Helicobacter pylori, a comma-shaped vibrio bacterium that causes stomach ulcers in humans. This bacterium moves by means of multiple polar flagella.

A species of *Spirillum*, a spiral shaped bacterium with a tuft of polar flagella. Most of the species in this genus are harmless aquatic organisms.

Bacteria usually divide by binary fission. During this process, DNA is copied and the cell splits into two cells, as in these gram positive cocci.

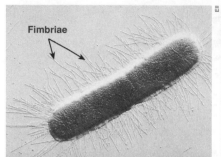

Escherichia coli, a common gut bacterium with **peritrichous** (around the entire cell) **fimbriae**. *E. coli* is a gram negative rod; it does not take up the gram stain but can be counter stained with safranin.

TEM showing *Enterobacter* bacteria, which belong to the family of gut bacteria commonly known as enterics. They are widely distributed in water, sewage, and soil. The family includes motile and non-motile species.

SEM showing a large rod-shaped bacterium with an approaching bacteriophage (viral particle). The bacterium has hair-like **pili** (not visible) protruding from the surface which act as phage receptors.

1. (a) Describe the function of flagella in bacteria: _____

   (b) Explain how fimbriae differ structurally and functionally from flagella: _____

2. (a) Describe the location and general composition of the bacterial cell wall: _____

   (b) Describe how the glycocalyx differs from the cell wall: _____

3. (a) Describe the main method by which bacteria reproduce: _____

   (b) Explain how conjugation differs from this usual method: _____

   (c) Comment on the evolutionary significance of conjugation: _____

4. Briefly describe how the artificial manipulation of plasmids has been used for technological applications:

Unicellular Eukaryotes

Unicellular (single-celled) **eukaryotes** comprise the majority of the diverse kingdom, Protista. They are found almost anywhere there is water, including within larger organisms (as parasites or symbionts). The protists are a very diverse group, exhibiting some features typical of generalized eukaryotic cells, as well as specialized features, which may be specific to one genus. Note that even within the genera below there is considerable variation in size and appearance. *Amoeba* and *Paramecium* are both **heterotrophic**, ingesting food, which accumulates inside a **vacuole**. *Euglena* and *Chlamydomonas* are autotrophic algae, although *Euglena* is heterotrophic when deprived of light. Other protists include the marine foraminiferans and radiolarians, specialized intracellular parasites such as *Plasmodium*, and zooflagellates such as the parasites *Trypanosoma* and *Giardia*.

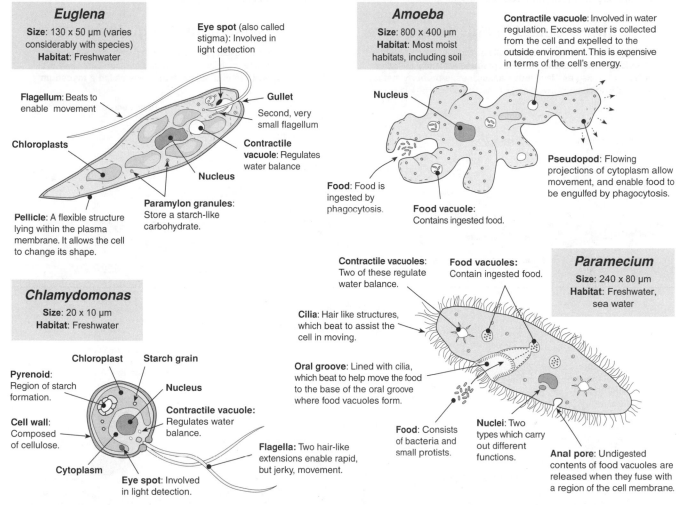

Euglena

Size: 130 x 50 μm (varies considerably with species)
Habitat: Freshwater

Eye spot (also called stigma): Involved in light detection

Flagellum: Beats to enable movement

Gullet
Second, very small flagellum

Chloroplasts

Contractile vacuole: Regulates water balance

Nucleus

Pellicle: A flexible structure lying within the plasma membrane. It allows the cell to change its shape.

Paramylon granules: Store a starch-like carbohydrate.

Amoeba

Size: 800 x 400 μm
Habitat: Most moist habitats, including soil

Contractile vacuole: Involved in water regulation. Excess water is collected from the cell and expelled to the outside environment. This is expensive in terms of the cell's energy.

Nucleus

Pseudopod: Flowing projections of cytoplasm allow movement, and enable food to be engulfed by phagocytosis.

Food: Food is ingested by phagocytosis.

Food vacuole: Contains ingested food.

Chlamydomonas

Size: 20 x 10 μm
Habitat: Freshwater

Chloroplast
Starch grain

Pyrenoid: Region of starch formation.

Nucleus

Cell wall: Composed of cellulose.

Contractile vacuole: Regulates water balance.

Cytoplasm

Eye spot: Involved in light detection.

Flagella: Two hair-like extensions enable rapid, but jerky, movement.

Paramecium

Size: 240 x 80 μm
Habitat: Freshwater, sea water

Contractile vacuoles: Two of these regulate water balance.

Food vacuoles: Contain ingested food.

Cilia: Hair like structures, which beat to assist the cell in moving.

Oral groove: Lined with cilia, which beat to help move the food to the base of the oral groove where food vacuoles form.

Food: Consists of bacteria and small protists.

Nuclei: Two types which carry out different functions.

Anal pore: Undigested contents of food vacuoles are released when they fuse with a region of the cell membrane.

1. Fill in the table below to summarize differences in some of the features and life functions of the protists shown above:

Organism	Nutrition	Movement	Osmoregulation	Eye spot present / absent	Cell wall present / absent
Amoeba					
Paramecium					
Euglena					
Chlamydomonas					

2. List the four organisms shown above in order of size (largest first): _____

_____

3. Suggest why an autotroph would have an eye spot: _____

_____

Related activities: *Features of Taxonomic Groups*
Web links: *Paramecium Animation*

EA 1

Cell Structure

Fungal Cells

The fungi are a large, successful group of eukaryotes that includes the yeasts, moulds, and fleshy fungi. The study of fungi is called **mycology**. All fungi are chemoheterotrophs: they lack chlorophyll and require organic compounds for a source of energy and carbon. Most fungi are also **saprophytic**, feeding on dead material, although some are parasitic or mutualistic. Fungal nutrition is absorptive and digestion is extracellular and takes place outside the fungal body. Of more than 100 000 fungal species, only about 100 are pathogenic to humans or other animals. However, many are plant pathogens and virtually every economically important plant species is attacked by one or more fungi. Note that the **lichens** have been reclassified into the fungal kingdom. They are dual organisms, formed by a mutualistic association between a green alga or a cyanobacterium, and a fungus (usually an ascomycete). Features of two fungal groups: yeasts and moulds are described below.

Single Celled Fungi: Yeasts

Yeasts are nonfilamentous, unicellular fungi that are typically spherical or oval shaped. Yeasts reproduce asexually by fission or budding. They are facultative anaerobes, a property that is exploited in the brewing, wine making, and bread making industries.

Storage granules · Rough endoplasmic reticulum · **Budding cell** · Nucleolus · **Parent cell** · Nucleus · Golgi apparatus · Nuclear pore · Vacuole · Mitochondrion

Filamentous Fungi: Molds

Moulds are multicellular, filamentous fungi often divided by septa into uni-nucleate, cell-like units. When conditions are favorable, hyphae grow to form a filamentous mass called a **mycelium.**

Rigid, chitinous cell wall · Rough endoplasmic reticulum · Nucleus · Vesicles · Vacuole · Some species have hyphae divided by crosswalls (**septa**) · Golgi apparatus · Mitochondrion

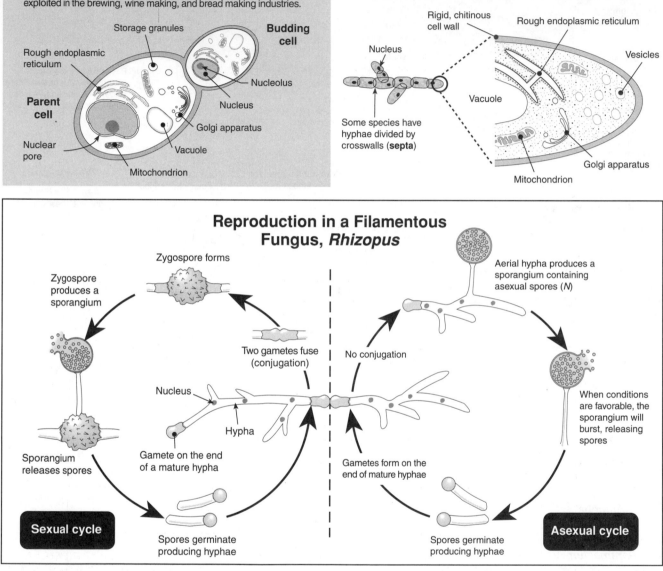

Reproduction in a Filamentous Fungus, *Rhizopus*

Zygospore forms · Zygospore produces a sporangium · Two gametes fuse (conjugation) · No conjugation · Aerial hypha produces a sporangium containing asexual spores (*N*) · Nucleus · When conditions are favorable, the sporangium will burst, releasing spores · Sporangium releases spores · Gamete on the end of a mature hypha · Hypha · Gametes form on the end of mature hyphae · **Sexual cycle** · Spores germinate producing hyphae · Spores germinate producing hyphae · **Asexual cycle**

1. List three distinguishing features of fungi: _____

   _____

2. Outline the key differences in the reproductive strategies of yeasts and molds: _____

   _____

   _____

3. Identify two commonly exploited fungal species and state how they are used:

   (a) _____

   (b) _____

Plant Cells

Plant cells are enclosed in a cellulose cell wall. The cell wall protects the cell, maintains its shape, and prevents excessive water uptake. It does not interfere with the passage of materials into and out of the cell. The diagram below shows the structure and function of a typical plant cell and its organelles. Also see the following pages where further information is provided on the organelles listed here but not described.

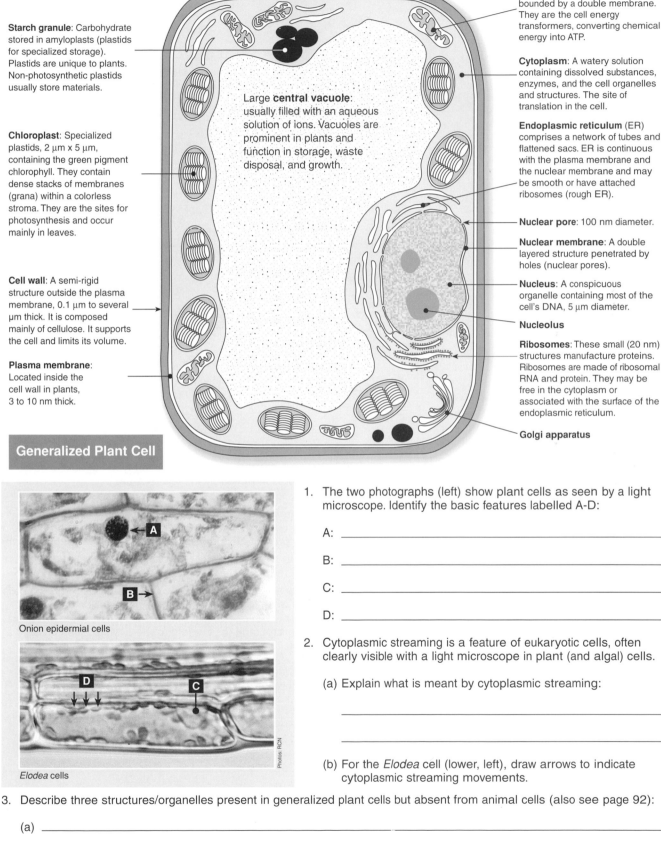

Starch granule: Carbohydrate stored in amyloplasts (plastids for specialized storage). Plastids are unique to plants. Non-photosynthetic plastids usually store materials.

Chloroplast: Specialized plastids, 2 μm x 5 μm, containing the green pigment chlorophyll. They contain dense stacks of membranes (grana) within a colorless stroma. They are the sites for photosynthesis and occur mainly in leaves.

Cell wall: A semi-rigid structure outside the plasma membrane, 0.1 μm to several μm thick. It is composed mainly of cellulose. It supports the cell and limits its volume.

Plasma membrane: Located inside the cell wall in plants, 3 to 10 nm thick.

Large **central vacuole**: usually filled with an aqueous solution of ions. Vacuoles are prominent in plants and function in storage, waste disposal, and growth.

Mitochondrion: 1.5 μm X 2–8 μm. Mitochondria are ovoid structures bounded by a double membrane. They are the cell energy transformers, converting chemical energy into ATP.

Cytoplasm: A watery solution containing dissolved substances, enzymes, and the cell organelles and structures. The site of translation in the cell.

Endoplasmic reticulum (ER) comprises a network of tubes and flattened sacs. ER is continuous with the plasma membrane and the nuclear membrane and may be smooth or have attached ribosomes (rough ER).

Nuclear pore: 100 nm diameter.

Nuclear membrane: A double layered structure penetrated by holes (nuclear pores).

Nucleus: A conspicuous organelle containing most of the cell's DNA, 5 μm diameter.

Nucleolus

Ribosomes: These small (20 nm) structures manufacture proteins. Ribosomes are made of ribosomal RNA and protein. They may be free in the cytoplasm or associated with the surface of the endoplasmic reticulum.

Golgi apparatus

Generalized Plant Cell

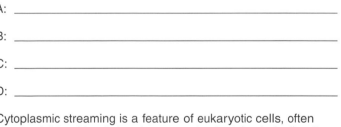

Onion epidermial cells

Elodea cells

Photos: RCN

1. The two photographs (left) show plant cells as seen by a light microscope. Identify the basic features labelled A-D:

A: _____

B: _____

C: _____

D: _____

2. Cytoplasmic streaming is a feature of eukaryotic cells, often clearly visible with a light microscope in plant (and algal) cells.

(a) Explain what is meant by cytoplasmic streaming:

_____

_____

(b) For the *Elodea* cell (lower, left), draw arrows to indicate cytoplasmic streaming movements.

3. Describe three structures/organelles present in generalized plant cells but absent from animal cells (also see page 92):

(a) _____

(b) _____

(c) _____

Related activities: Animal Cells, Cell Structures and Organelles
Web links: Review of Eukaryotic Cells

RA 2

Cell Structure

Animal Cells

Animal cells, unlike plant cells, do not have a regular shape. In fact, some animal cells (such as phagocytes) are able to alter their shape for various purposes (e.g. engulfment of foreign material). The diagram below shows the structure and function of a typical animal cell and its organelles. Note the differences between this cell and the generalized plant cell. Also see the previous page and following two pages, where further information is provided on the organelles listed here but not described.

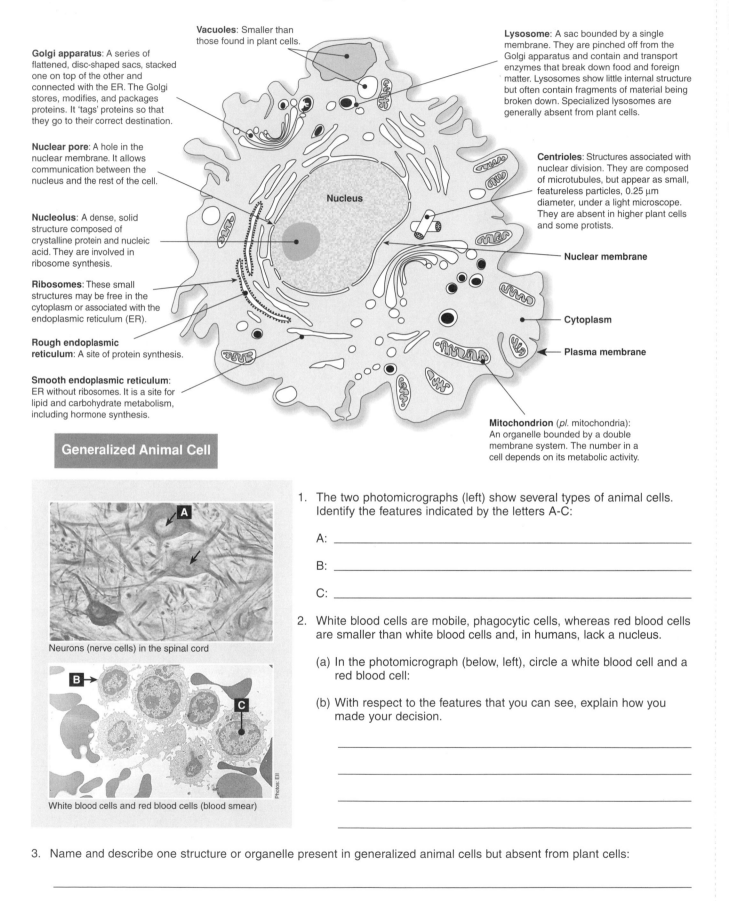

Vacuoles: Smaller than those found in plant cells.

Golgi apparatus: A series of flattened, disc-shaped sacs, stacked one on top of the other and connected with the ER. The Golgi stores, modifies, and packages proteins. It 'tags' proteins so that they go to their correct destination.

Nuclear pore: A hole in the nuclear membrane. It allows communication between the nucleus and the rest of the cell.

Nucleolus: A dense, solid structure composed of crystalline protein and nucleic acid. They are involved in ribosome synthesis.

Ribosomes: These small structures may be free in the cytoplasm or associated with the endoplasmic reticulum (ER).

Rough endoplasmic reticulum: A site of protein synthesis.

Smooth endoplasmic reticulum: ER without ribosomes. It is a site for lipid and carbohydrate metabolism, including hormone synthesis.

Lysosome: A sac bounded by a single membrane. They are pinched off from the Golgi apparatus and contain and transport enzymes that break down food and foreign matter. Lysosomes show little internal structure but often contain fragments of material being broken down. Specialized lysosomes are generally absent from plant cells.

Centrioles: Structures associated with nuclear division. They are composed of microtubules, but appear as small, featureless particles, 0.25 μm diameter, under a light microscope. They are absent in higher plant cells and some protists.

Nuclear membrane

Cytoplasm

Plasma membrane

Mitochondrion (*pl.* mitochondria): An organelle bounded by a double membrane system. The number in a cell depends on its metabolic activity.

Nucleus

Generalized Animal Cell

Neurons (nerve cells) in the spinal cord

White blood cells and red blood cells (blood smear)

Photos: Ell

1. The two photomicrographs (left) show several types of animal cells. Identify the features indicated by the letters A-C:

   A: _____

   B: _____

   C: _____

2. White blood cells are mobile, phagocytic cells, whereas red blood cells are smaller than white blood cells and, in humans, lack a nucleus.

   (a) In the photomicrograph (below, left), circle a white blood cell and a red blood cell:

   (b) With respect to the features that you can see, explain how you made your decision.

   _____

   _____

   _____

   _____

3. Name and describe one structure or organelle present in generalized animal cells but absent from plant cells:

   _____

   _____

Related activities: Plant Cells, Cell Structures and Organelles
Web links: Review of Eukaryotic Cells

Cell Sizes

Cells are extremely small and they can only be seen properly when viewed through the magnifying lenses of a microscope. The diagrams and photographs below show a variety of cell types, together with a virus and a microscopic animal for comparison. For each of these images, note the scale and relate this to the type of microscopy used.

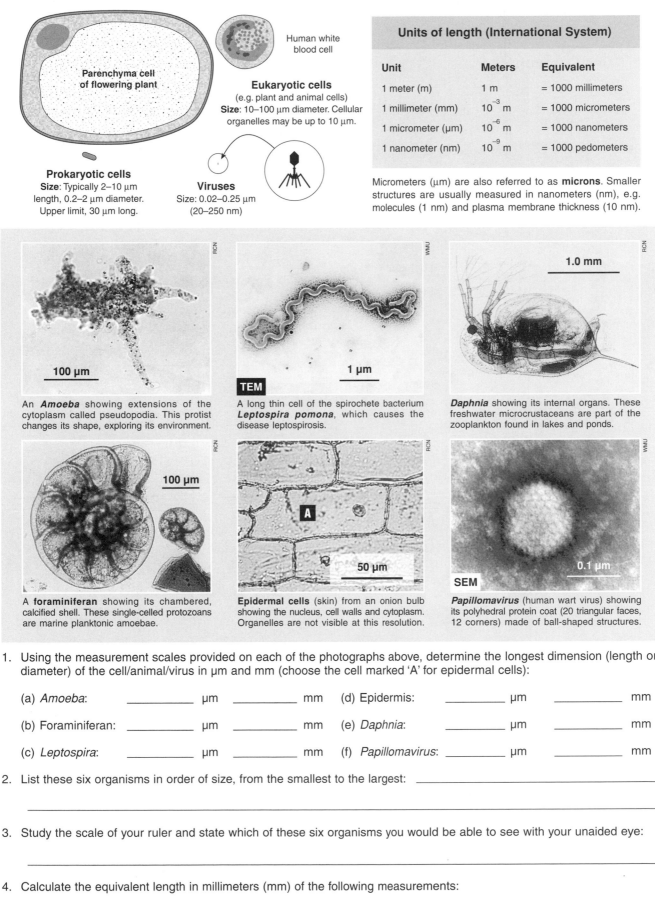

Parenchyma cell of flowering plant

Human white blood cell

Eukaryotic cells
(e.g. plant and animal cells)
Size: 10–100 µm diameter. Cellular organelles may be up to 10 µm.

Prokaryotic cells
Size: Typically 2–10 µm length, 0.2–2 µm diameter. Upper limit, 30 µm long.

Viruses
Size: 0.02–0.25 µm
(20–250 nm)

Units of length (International System)

Unit	Meters	Equivalent
1 meter (m)	1 m	= 1000 millimeters
1 millimeter (mm)	10^{-3} m	= 1000 micrometers
1 micrometer (µm)	10^{-6} m	= 1000 nanometers
1 nanometer (nm)	10^{-9} m	= 1000 pedometers

Micrometers (µm) are also referred to as **microns**. Smaller structures are usually measured in nanometers (nm), e.g. molecules (1 nm) and plasma membrane thickness (10 nm).

100 µm

An **Amoeba** showing extensions of the cytoplasm called pseudopodia. This protist changes its shape, exploring its environment.

TEM **1 µm**

A long thin cell of the spirochete bacterium **Leptospira pomona**, which causes the disease leptospirosis.

1.0 mm

Daphnia showing its internal organs. These freshwater microcrustaceans are part of the zooplankton found in lakes and ponds.

100 µm

A **foraminiferan** showing its chambered, calcified shell. These single-celled protozoans are marine planktonic amoebae.

A **50 µm**

Epidermal cells (skin) from an onion bulb showing the nucleus, cell walls and cytoplasm. Organelles are not visible at this resolution.

0.1 µm **SEM**

Papillomavirus (human wart virus) showing its polyhedral protein coat (20 triangular faces, 12 corners) made of ball-shaped structures.

1. Using the measurement scales provided on each of the photographs above, determine the longest dimension (length or diameter) of the cell/animal/virus in µm and mm (choose the cell marked 'A' for epidermal cells):

(a) *Amoeba*: _____ µm _____ mm (d) Epidermis: _____ µm _____ mm

(b) Foraminiferan: _____ µm _____ mm (e) *Daphnia*: _____ µm _____ mm

(c) *Leptospira*: _____ µm _____ mm (f) *Papillomavirus*: _____ µm _____ mm

2. List these six organisms in order of size, from the smallest to the largest: _____

_____

3. Study the scale of your ruler and state which of these six organisms you would be able to see with your unaided eye:

_____

4. Calculate the equivalent length in millimeters (mm) of the following measurements:

(a) 0.25 µm: _____ (b) 450 µm: _____ (c) 200 nm: _____

Cell Structure

Periodicals:
Size does matter

Related activities: Optical Microscopes, Electron Microscopes

DA 2

Cell Structures and Organelles

The table below, and the following page, provides a format to summarize information about the structures and organelles of typical eukaryotic cells. Complete the table using the list provided and by referring to other pages in this topic. The first cell component has been completed for you as a guide and

the log scale of measurements (top of next page) illustrates the relative sizes of some cellular structures. **List of structures and organelles**: *cell wall, mitochondrion, chloroplast, cell junctions, centrioles, ribosome, flagella, endoplasmic reticulum, Golgi apparatus, nucleus, flagella, cytoskeleton and vacuoles.*

Cell Component	Details	Present in		Visible under light microscope
		Plant cells	Animal cells	
(a) Double layer of phospholipids (called the lipid bilayer); Proteins	Name: Plasma (cell surface) membrane Location: Surrounding the cell Function: Gives the cell shape and protection. It also regulates the movement of substances into and out of the cell.	YES	YES	YES *(but not at the level of detail shown in the diagram)*
(b)	Name: Location: Function:			
(c) Outer membrane, Inner membrane, Matrix, Cristae	Name: Location: Function:			
(d) Secretory vesicles budding off, Cisternae, Transfer vesicles from the smooth endoplasmic reticulum	Name: Location: Function:			
(e) Ribosomes, Transport pathway, Rough, Smooth, Vesicles budding off, Flattened membrane sacs	Name: Location: Function:			
(f) Grana comprise stacks of thylakoids, Stroma, Lamellae	Name: Location: Function:			

Related activities: *Plant Cells, Animal Cells*
Web links: *Eukaryotic Cells Interactive Animation*

Periodicals:
Cellular factories, Chloroplasts: biosynthetic powerhouses

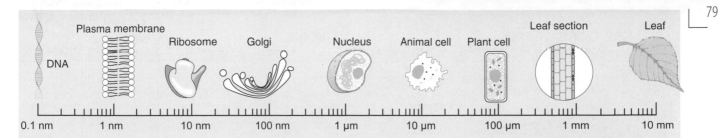

Cell Component	Details	Present in		Visible under light microscope
		Plant cells	Animal cells	
(g)	Name: Lysosome and food vacuole Location: Function:			
(h)	Name: Location: Function:			
(i)	Name: Location: Function:			
(j)	Name: Cilia and flagella (some eukaryotic cells) Location: Function:			

Cell Structure

Cell Component	Details	Present in		Visible under light microscope
		Plant cells	Animal cells	
(k) 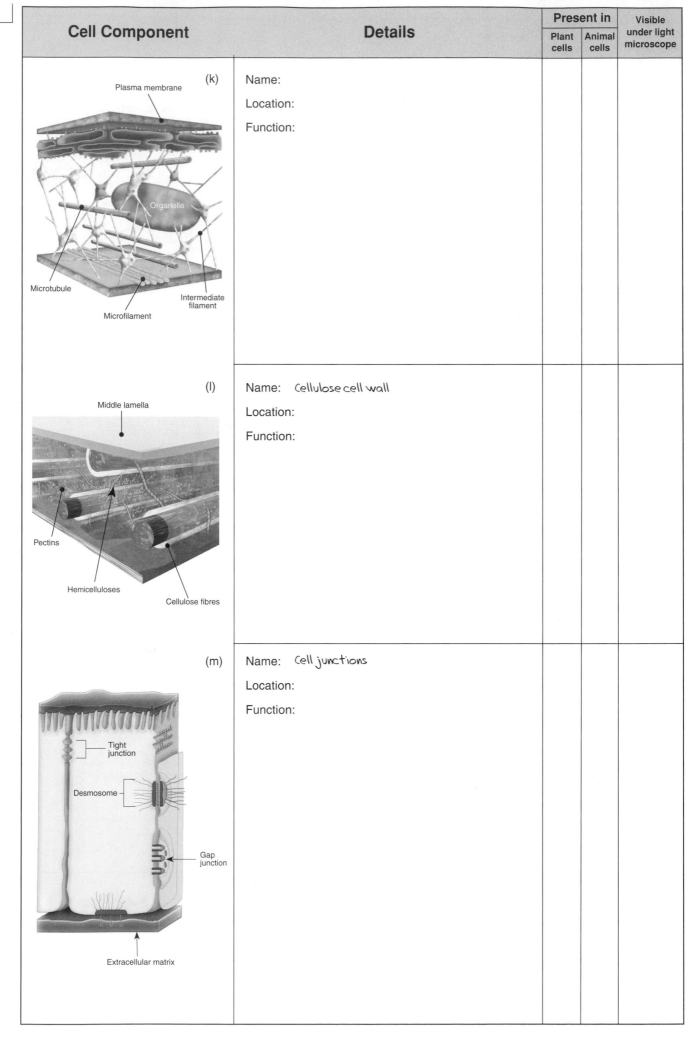 Plasma membrane / Organelle / Microtubule / Intermediate filament / Microfilament	Name: Location: Function:			
(l) Middle lamella / Pectins / Hemicelluloses / Cellulose fibres	Name: Cellulose cell wall Location: Function:			
(m) Tight junction / Desmosome / Gap junction / Extracellular matrix	Name: Cell junctions Location: Function:			

Differential Centrifugation

Differential centrifugation (also called **cell fractionation**) is a technique used to extract organelles from cells so that they can be studied. The aim is to extract undamaged intact organelles. Samples must be kept very cool so that metabolism is slowed and self digestion of the organelles is prevented. The samples must also be kept in a buffered, isotonic solution so that the organelles do not change volume and the enzymes are not denatured by changes in pH.

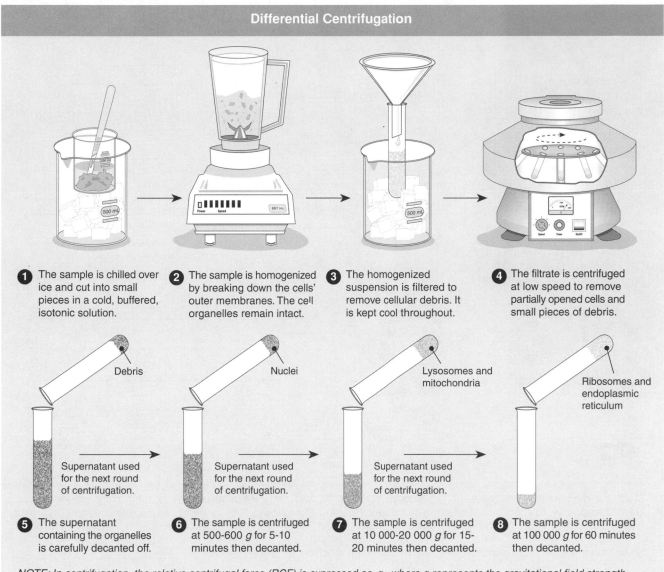

Differential Centrifugation

1. The sample is chilled over ice and cut into small pieces in a cold, buffered, isotonic solution.

2. The sample is homogenized by breaking down the cells' outer membranes. The cell organelles remain intact.

3. The homogenized suspension is filtered to remove cellular debris. It is kept cool throughout.

4. The filtrate is centrifuged at low speed to remove partially opened cells and small pieces of debris.

Debris

Nuclei

Lysosomes and mitochondria

Ribosomes and endoplasmic reticulum

Supernatant used for the next round of centrifugation.

Supernatant used for the next round of centrifugation.

Supernatant used for the next round of centrifugation.

5. The supernatant containing the organelles is carefully decanted off.

6. The sample is centrifuged at 500-600 *g* for 5-10 minutes then decanted.

7. The sample is centrifuged at 10 000-20 000 *g* for 15-20 minutes then decanted.

8. The sample is centrifuged at 100 000 *g* for 60 minutes then decanted.

NOTE: In centrifugation, the relative centrifugal force (RCF) is expressed as g , where g represents the gravitational field strength.

Cell Structure

1. Explain why it is possible to separate cell organelles using centrifugation: _____

_____

_____

2. Suggest why the sample is homogenized before centrifugation: _____

_____

3. Explain why the sample must be kept in a solution that is:

(a) Isotonic: _____

(b) Cool: _____

(c) Buffered: _____

(a)

(b)

(c)

(d) *Cellular debris*

4. **Density gradient centrifugation** (right) is another method of cell fractionation. Sucrose is added to a sample, which is then centrifuged at high speed. The organelles will form layers according to their specific densities. Using the information above, label the centrifuge tube on the right with the organelles you would find in each layer.

Related activities: Enzyme Reaction Rates

A 1

Identifying Cell Structures

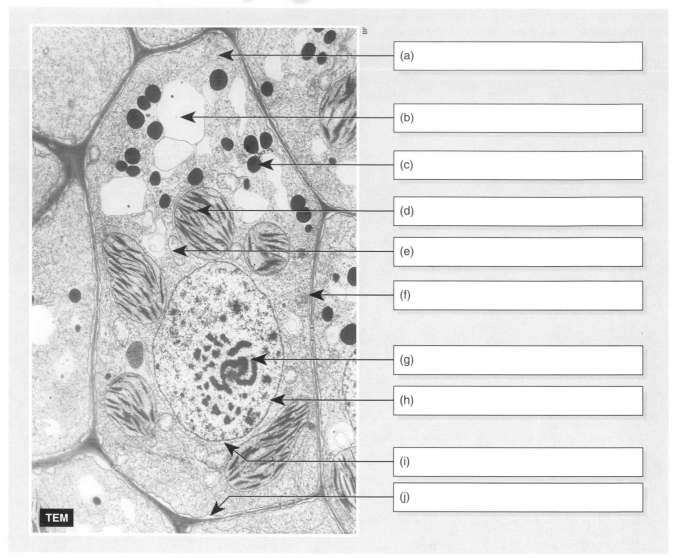

(a)

(b)

(c)

(d)

(e)

(f)

(g)

(h)

(i)

(j)

TEM

1. Study the diagrams on the previous pages to become familiar with the various structures found in plant and animal cells. Identify and label the ten structures in the cell above using the following list of terms: *nuclear membrane, cytoplasm, endoplasmic reticulum, mitochondrion, starch granules, chromosome, vacuole, plasma membrane, cell wall, chloroplast*

2. State how many cells, or parts of cells, are visible in the electron micrograph above: _____

3. Identify the **type** of cell illustrated above (bacterial cell, plant cell, or animal cell). Explain your answer:

   _____

4. (a) Explain where cytoplasm is found in the cell: _____

   _____

   (b) Describe what cytoplasm is made up of: _____

   _____

   (c) Explain why nucleoplasm is only found in eukaryotic cells: _____

   _____

5. Describe two structures, pictured in the cell above, that are associated with storage:

   (a) _____

   _____

   (b) _____

   _____

Related activities: *Electron Microscopes*

Optical Microscopes

The light microscope is one of the most important instruments used in biology practicals, and its correct use is a basic and essential skill of biology. High power light microscopes use a combination of lenses to magnify objects up to several hundred times. They are called **compound microscopes** because there are two or more separate lenses involved. A typical compound light microscope (bright field) is shown below (top photograph). The specimens viewed with these microscopes must be thin and mostly transparent. Light is focused up through the condenser and specimen; if the specimen is thick or opaque, little or no detail will be visible. The microscope below has two eyepieces (**binocular**), although monocular microscopes, with a mirror rather than an internal light source, may still be encountered. Dissecting microscopes (lower photograph) are a type of binocular microscope used for observations at low total magnification (x4 to x50), where a large working distance between objectives and stage is required. A dissecting microscope has two separate lens systems, one for each eye. Such microscopes produce a 3-D view of the specimen and are sometimes called stereo microscopes for this reason.

(a)

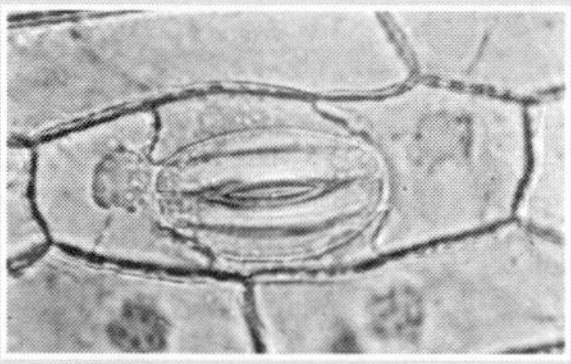

Stoma in leaf epidermis

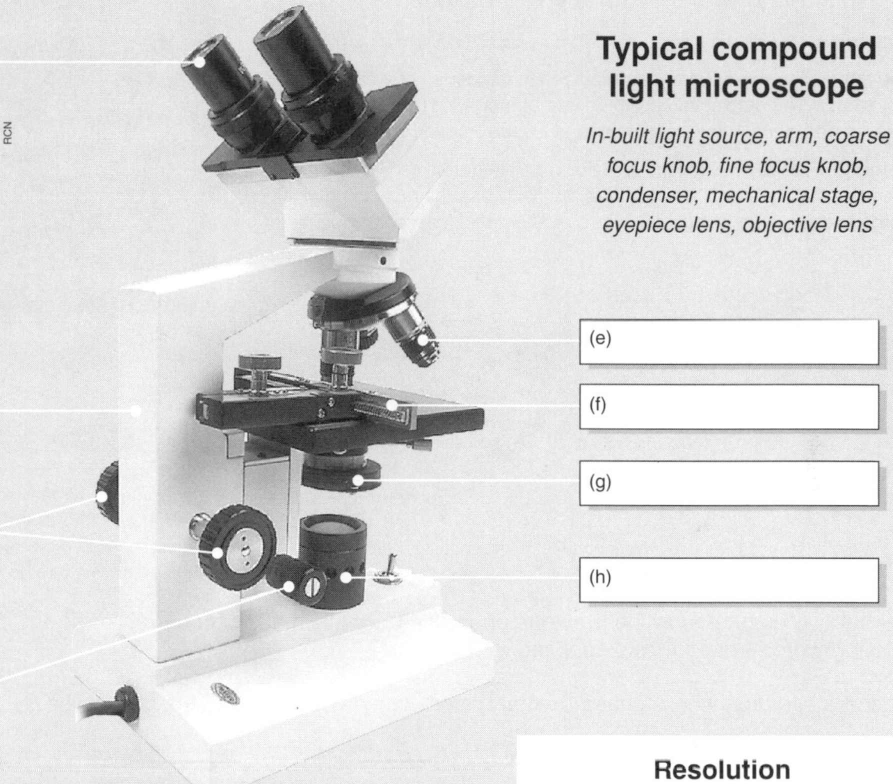

(b)

(c)

(d)

Typical compound light microscope

In-built light source, arm, coarse focus knob, fine focus knob, condenser, mechanical stage, eyepiece lens, objective lens

(e)

(f)

(g)

(h)

(i)

(j)

(k)

(l)

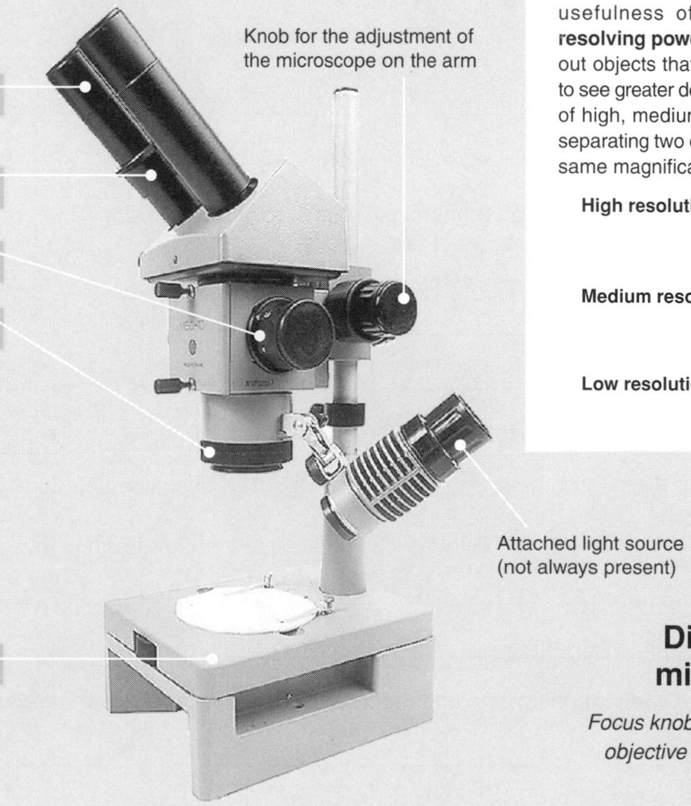

Knob for the adjustment of the microscope on the arm

Drosophila

(m)

Attached light source
(not always present)

Resolution

One important factor that determines the usefulness of a microscope is its **resolving power**; the ability to separate out objects that are close together and to see greater detail. Below is an example of high, medium and low resolution for separating two objects viewed under the same magnification.

High resolution

Medium resolution

Low resolution

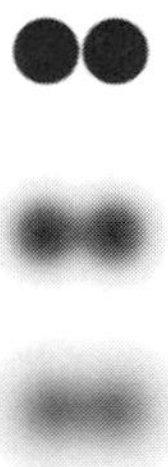

Dissecting microscope

Focus knob, stage, eyepiece lens, objective lens, eyepiece focus

Cell Structure

Periodicals:
Light microscopy

Related activities: Plant Cells, Animal Cells, Unicellular Eukaryotes
Web links: Introduction to the Microscope

RDA 2

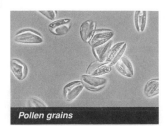

Pollen grains

Phase contrast illumination increases contrast of transparent specimens by producing interference effects.

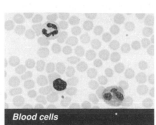

Blood cells

Leishman s stain is used to show red blood cells as red/pink, while staining the nucleus of white blood cells blue.

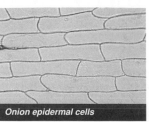

Onion epidermal cells

Standard **bright field** lighting shows cells with little detail; only cell walls, with the cell nuclei barely visible.

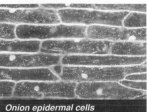

Onion epidermal cells

Dark field illumination is excellent for viewing near transparent specimens. The nucleus of each cell is visible.

Photos: Eli

Making a temporary wet mount

1. **Sectioning**: Very thin sections of fresh material are cut with a razorblade.

2. **Mounting**: The thin section(s) are placed in the center of a clean glass microscope slide and covered with a drop of mounting liquid (e.g. water, glycerol or stain). A coverslip is placed on top to exclude air (below).

3. **Staining**: Dyes can be applied to stain some structures and leave others unaffected. The stains used in dyeing living tissues are called **vital stains** and they can be applied before or after the specimen is mounted.

Commonly used temporary stains

Stain	Final color	Used for
Iodine solution	blue-black	Starch
Aniline sulfate	yellow	Lignin
Schultz's solution	blue	Starch
	blue or violet	Cellulose
	yellow	Protein, cutin, lignin, suberin
Methylene blue	blue	Nuclei

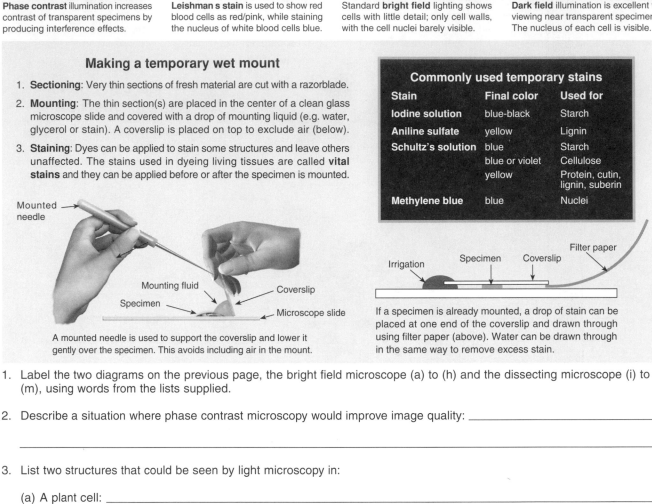

Mounted needle

Mounting fluid

Specimen

Coverslip

Microscope slide

A mounted needle is used to support the coverslip and lower it gently over the specimen. This avoids including air in the mount.

Irrigation   Specimen   Coverslip   Filter paper

If a specimen is already mounted, a drop of stain can be placed at one end of the coverslip and drawn through using filter paper (above). Water can be drawn through in the same way to remove excess stain.

1. Label the two diagrams on the previous page, the bright field microscope (a) to (h) and the dissecting microscope (i) to (m), using words from the lists supplied.

2. Describe a situation where phase contrast microscopy would improve image quality: _____

_____

3. List two structures that could be seen by light microscopy in:

   (a) A plant cell: _____

   (b) An animal cell: _____

4. Name one cell structure that can not be seen by light microscopy: _____

5. Identify a stain that would be appropriate for improving definition of the following:

   (a) Blood cells: _____   (d) Lignin: _____

   (b) Starch: _____   (e) Nuclei and DNA: _____

   (c) Protein: _____   (f) Cellulose: _____

6. Determine the magnification of a microscope using:

   (a) 15 X eyepiece and 40 X objective lens: _____   (b) 10 X eyepiece and 60 X objective lens: _____

7. Describe the main difference between a bright field light microscope and a dissecting microscope. _____

_____

_____

8. Explain the difference between magnification and resolution (resolving power) with respect to microscope use:

_____

_____

Electron Microscopes

Electron microscopes (EMs) use a beam of electrons, instead of light, to produce an image. The higher resolution of EMs is due to the shorter wavelengths of electrons. There are two basic types of electron microscope: **scanning electron microscopes** (SEM) and **transmission electron microscopes** (TEM). In SEMs, the electrons are bounced off the surface of an object to produce detailed images of the external appearance. TEMs produce very clear images of specially prepared thin sections.

Transmission Electron Microscope (TEM)

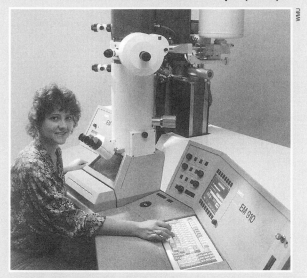

The transmission electron microscope is used to view extremely thin sections of material. Electrons pass through the specimen and are scattered. Magnetic lenses focus the image onto a fluorescent screen or photographic plate. The sections are so thin that they have to be prepared with a special machine, called an **ultramicrotome**, that can cut wafers to just 30 thousandths of a millimeter thick. It can magnify several hundred thousand times.

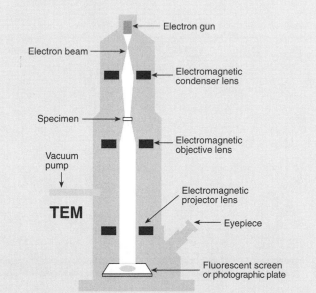

TEM diagram labels: Electron gun, Electron beam, Electromagnetic condenser lens, Specimen, Vacuum pump, Electromagnetic objective lens, Electromagnetic projector lens, Eyepiece, Fluorescent screen or photographic plate, **TEM**

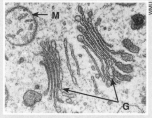

TEM photo showing the Golgi (**G**) and a mitochondrion (**M**).

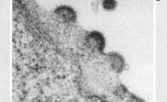

Three HIV viruses budding out of a human lymphocyte (TEM).

Scanning Electron Microscope (SEM)

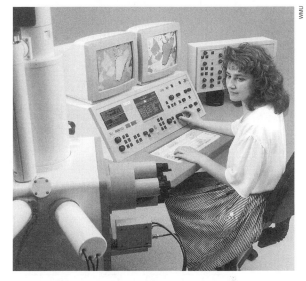

The scanning electron microscope scans a sample with a beam of primary electrons that knock electrons from its surface. These secondary electrons are picked up by a collector, amplified, and transmitted onto a viewing screen or photographic plate, producing a superb 3-D image. A microscope of this power can easily obtain clear pictures of organisms as small as bacteria and viruses. The image produced is of the outside surface only.

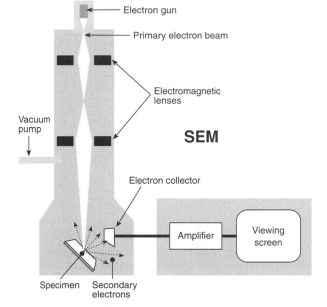

SEM diagram labels: Electron gun, Primary electron beam, Electromagnetic lenses, Vacuum pump, **SEM**, Electron collector, Amplifier, Viewing screen, Specimen, Secondary electrons

SEM photo of stoma and epidermal cells on the upper surface of a leaf.

Image of hair louse clinging to two hairs on a Hooker's sealion (SEM).

Cell Structure

A 2

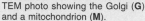
Periodicals: Transmission electron microscopy

Related activities: Optical Microscopes, Interpreting Electron Micrographs

	Light Microscope	Transmission Electron Microscope (TEM)	Scanning Electron Microscope (SEM)
Radiation source:	light	electrons	electrons
Wavelength:	400-700 nm	0.005 nm	0.005 nm
Lenses:	glass	electromagnetic	electromagnetic
Specimen:	living or non-living supported on glass slide	non-living supported on a small copper grid in a vacuum	non-living supported on a metal disc in a vacuum
Maximum resolution:	200 nm	1 nm	10 nm
Maximum magnification:	1500 x	250 000 x	100 000 x
Stains:	colored dyes	impregnated with heavy metals	coated with carbon or gold
Type of image:	colored	monochrome (black & white)	monochrome (black & white)

1. Explain why electron microscopes are able to resolve much greater detail than a light microscope:

   _____

   _____

   _____

2. Describe two typical applications for each of the following types of microscope:

   (a) Transmission electron microscope (TEM): _____

   _____

   (b) Scanning electron microscope (SEM): _____

   _____

   (c) Bright field microscope (thin section): _____

   _____

   (d) Dissecting microscope: _____

   _____

3. Identify which type of electron microscope (SEM or TEM) or optical microscope (compound light (bright field) microscope or dissecting microscope) was used to produce each of the images in the photos below (A-H):

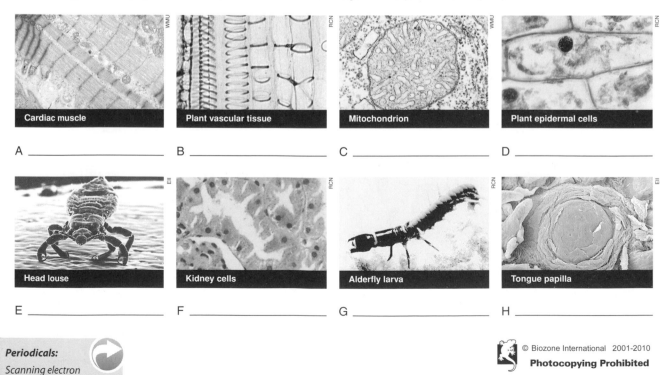

Cardiac muscle

Plant vascular tissue

Mitochondrion

Plant epidermal cells

A _____   B _____   C _____   D _____

Head louse

Kidney cells

Alderfly larva

Tongue papilla

E _____   F _____   G _____   H _____

Interpreting Electron Micrographs

The photographs below were taken using a transmission electron microscope (TEM). They show some of the cell organelles in great detail. Remember that these photos are showing only **parts of cells, not whole cells**. Some of the photographs show more than one type of organelle. The questions refer to the main organelle in the center of the photo.

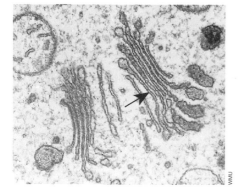

1. (a) Name this organelle (arrowed): _____

   (b) State which kind of cell(s) this organelle would be found in:

   _____

   (c) Describe the function of this organelle: _____

   _____

   _____

   (d) Label two structures that can be seen inside this organelle.

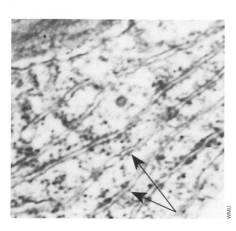

2. (a) Name this organelle (arrowed): _____

   (b) State which kind of cell(s) this organelle would be found in:

   _____

   (c) Describe the function of this organelle: _____

   _____

   _____

   _____

   _____

3. (a) Name the large, circular organelle: _____

   (b) State which kind of cell(s) this organelle would be found in:

   _____

   (c) Describe the function of this organelle: _____

   _____

   _____

   (d) Label two regions that can be seen inside this organelle.

4. (a) Name and label the ribbon-like organelle in this photograph (arrowed):

   _____

   (b) State which kind of cell(s) this organelle is found in:

   _____

   (c) Describe the function of these organelles: _____

   _____

   _____

   (d) Name the dark 'blobs' attached to the organelle you have labeled:

   _____

Cell Structure

© Biozone International 2001-2010
Photocopying Prohibited

Periodicals:
The power behind
an electron microscopist

Related activities: Electron Microscopes, Plant Cells, Animal Cells,
Cell Structures and Organelles

RA 2

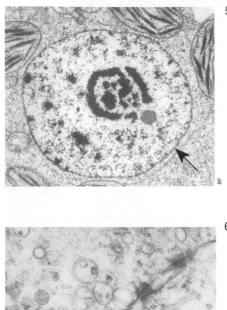

5. (a) Name this large circular structure (arrowed): _____

(b) State which kind of cell(s) this structure would be found in:

_____

(c) Describe the function of this structure: _____

_____

_____

(d) Label three features relating to this structure in the photograph.

6. The four dark structures shown in this photograph are called **desmosomes**. They cause the plasma membranes of neighboring cells to stick together. Without desmosomes, animal cells would not combine together to form tissues.

(a) Describe the functions of the plasma membrane:

_____

_____

_____

(b) Label the plasma membrane and the four desmosomes in the photograph.

7. In the space below, draw a simple, labeled diagram of a **generalized cell** to show the relative size and location of these six structures and organelles (simple outlines of the organelles will do):

KEY TERMS: Memory Card Game

The cards below have a keyword or term printed on one side and its definition printed on the opposite side. The aim is to win as many cards as possible from the table. To play the game.....

1) Cut out the cards and lay them definition side down on the desk. You will need one set of cards between two students.

2) Taking turns, choose a card and, BEFORE you pick it up, state your own best definition of the keyword to your opponent.

3) Check the definition on the opposite side of the card. If both you and your opponent agree that your stated definition matches, then keep the card. If your definition does not match then return the card to the desk.

4) Once your turn is over, your opponent may choose a card.

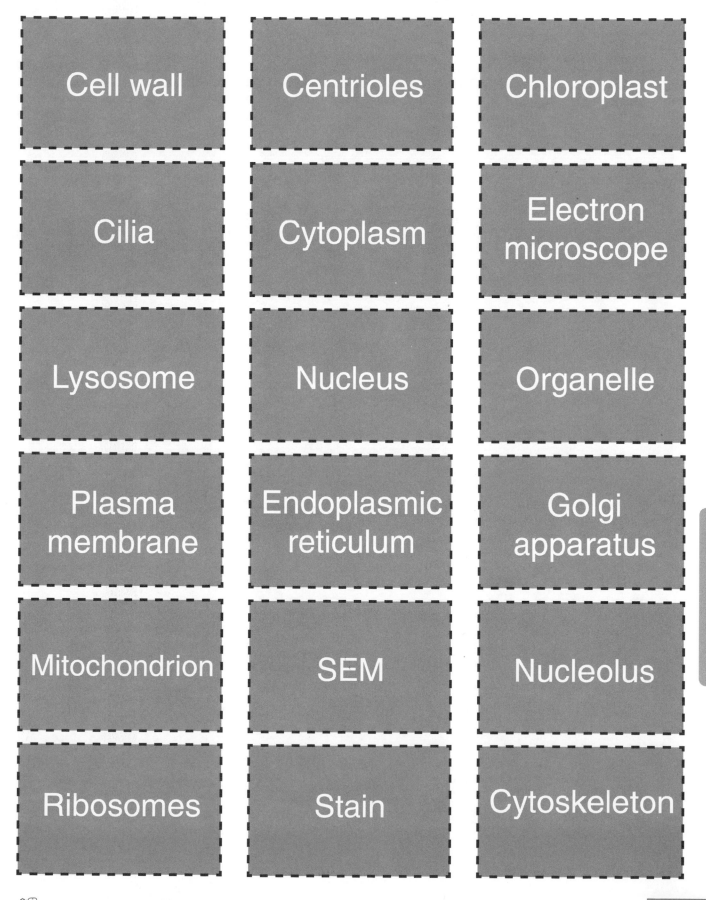

Cell wall	Centrioles	Chloroplast
Cilia	Cytoplasm	Electron microscope
Lysosome	Nucleus	Organelle
Plasma membrane	Endoplasmic reticulum	Golgi apparatus
Mitochondrion	SEM	Nucleolus
Ribosomes	Stain	Cytoskeleton

Cell Structure

R 2

When you've finished the game keep these cutouts and use them as flash cards!

An organelle found in plants that contains chlorophyll and in which the reactions of photosynthesis take place.

Cell structures comprising a hollow cylinder of microtubules. During cell division they produce spindle fibres that pull apart the chromosomes.

A structure, present in plants and bacteria, which is found outside the plasma membrane and gives rigidity to the cell.

An advanced microscope that uses electron beams to produce high resolution images.

The watery contents of the cell within the plasma membrane, but excluding the contents of the nucleus.

Microtubular extensions of the plasma membrane that allow for cell locomotion in some organisms or for moving particles about in others.

A structural and functional part of the cell usually bound within its own membrane. Examples include the mitochondria and chloroplasts.

Membrane bound area within a eukaryotic cell where the chromosomes are found.

Membrane bound vacuolar organelles that contain enzymes that form part of the intracellular digestive system.

Organelle that resembles a series of flattened stacks. It modifies and packages proteins and also performs a secretory function by budding off vesicles.

An organelle comprising a convoluted membranous stack and divided into rough and smooth regions. It plays a part in protein synthesis and membrane synthesis.

Lipid bilayered membrane surrounding the cell. Proteins are embedded in it and are responsible for the passage of material into and out of the cell.

An organelle found within the nucleus that contains ribosomal RNA and is associated with the part of the chromosome that codes for rRNA.

Scanning electron microscope (or microscopy). An EM that produces images of the surface features of objects. An image produced by a scanning electron microscope.

Organelle responsible for producing the cell's energy. It appears oval in shape with an outer double membrane and a convoluted interior membrane. Contains its own circular DNA.

A network of actin filaments and microtubles within the cytosol that provide structure and assist the movement of materials within the cell.

A chemical that binds to parts of the cell and allows those parts to be seen more easily under a microscope.

Small structures comprising RNA and protein that are found in all cells. They function in translation of the genetic code (genes into proteins).

Processes in
Cells

KEY CONCEPTS

▶ Cellular metabolism depends on the transport of substances across cellular membranes.

▶ Cell size is limited by surface area to volume ratio.

▶ New cells arise through cell division.

▶ Cellular diversity arises through specialization from stem cell progenitors.

▶ Regulation of cell division is important in development.

▶ Emergent properties are a feature of increasing complexity in biological systems.

KEY TERMS

active transport
amphipathic
anaphase
cancer
cellular differentiation
concentration gradient
cytokinesis
diffusion
emergent property
endocytosis
exocytosis
facilitated diffusion
fluid mosaic model
gametic meiosis
glycolipid
glycoprotein
hypertonic
hypotonic
interphase
ion pump
isotonic
meiosis
metaphase
mitosis
oncogene
osmosis
partially permeable
passive transport
phagocytosis
phospholipid
pinocytosis
plasma membrane
plasmolysis
potency
prophase
proto-oncogene
sporic meiosis
stem cell
surface area: volume ratio
tumor suppressor gene
turgor
water potential

Periodicals: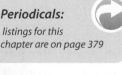

listings for this chapter are on page 379

Weblinks:
www.thebiozone.com/
weblink/SB1-2597.html

OBJECTIVES

☐ 1. Use the **KEY TERMS** to help you understand and complete these objectives.

Cellular Transport Processes   pages 92-107

☐ 2. Describe the structure and function of the **plasma membrane**, including the significance of the **amphipathic** character the phospholipids and the role of transmembrane proteins, glycoproteins, and glycolipids.

☐ 3. Explain **passive transport** across membranes by **diffusion** and **osmosis**. If required, use **water potential** to explain net movement of water by osmosis.

☐ 4. Explain the terms **hypotonic**, **isotonic**, and **hypertonic** with reference to the movement of water in and out of cells.

☐ 5. Using examples, explain active transport processes in cells, including **ion pumps**, **endocytosis**, and **exocytosis**.

☐ 6. Explain the role of **surface area: volume ratio** in limiting the size of cells. Describe factors affecting the rates of transport processes in cells.

The Nucleus and Cell Division   pages 108-118

☐ 7. Recall the general structure of nucleic acids. Describe the structure and role of the nucleus and its contents, including the **DNA** and **chromosomes**.

☐ 8. Describe the **cell cycle** in eukaryotes, including reference to: **mitosis**, **growth** (G1 and G2), and DNA replication (S).

☐ 9. Recognize and describe stages in mitosis: **prophase**, **metaphase**, **anaphase**, and **telophase**. Describe **cytokinesis** in plant and animal cells.

☐ 10. Describe the role of mitosis in growth and repair, and asexual reproduction. Recognize mitotic cell division as a prelude to **cellular differentiation**.

☐ 11. Distinguish between **meiosis** and **mitosis** in terms of their cellular outcomes.

☐ 12. Explain the role of **tumor suppressor genes** and **proto-oncogenes** in regulating cell division. Explain how **carcinogens** can upset these regulatory mechanisms and trigger tumor formation (**cancer**).

☐ 13. Describe **apoptosis** (programmed cell death) and explain its role.

☐ 14. Describe the properties of **stem cells** and explain the role of stem cells in multicellular organisms. Describe the potential of stem cells in medicine.

Organization of Life   pages 119-124

☐ 15. Describe the hierarchy of organization in multicellular organisms. Appreciate that each step in the hierarchy of biological order is associated with the emergence of properties not present at simpler levels of organization.

☐ 16. Explain **cellular differentiation** and the role of specialized cells in tissues.

Cell Processes

All of the organelles and other structures in the cell have functions. The cell can be compared to a factory with an assembly line. Organelles in the cell provide the equivalent of the power supply, assembly line, packaging department, repair and maintenance, transport system, and the control center. The sum total of all the processes occurring in a cell is known as **metabolism**. Some of these processes store energy in molecules (anabolism) while others release that stored energy (catabolism). Below is a summary of the major processes that take place in a cell.

Autolysis
Lysosomes contain powerful digestive enzymes that destroy unwanted cell organelles and foreign objects brought into the cell.

Transport in and out of the cell
Simple diffusion and active transport move substances into and out of the cell across the plasma (cell surface) membrane.

Protein synthesis
Chromosomes in the nucleus store genetic instructions for the production of specific proteins. These proteins are put together by ribosomes on the endoplasmic reticulum.

Cell division
Centrioles control the movement of chromosomes during cell division.

Secretion
The Golgi apparatus is the packaging department of the cell. It produces secretory vesicles (small membrane-bound sacs) that are used to store useful chemicals, prepare substances for movement out of the cell (e.g. hormones), or to package digestive enzymes.

Phagocytosis
The plasma membrane can engulf objects (such as food particles or bacteria). The membrane pinches off to become a vesicle, and the particles are then digested by powerful enzymes.

Pinocytosis
The plasma membrane can pinch off to trap some of the surrounding watery fluid in a vesicle.

Cellular respiration
Respiration is a complex chemical process which starts in the cell cytoplasm but is completed in the mitochondria. This process supplies the cell with energy to carry out the many other chemical reactions of metabolism.

Chloroplasts are found only in plant cells. The diagram above is of an animal cell.

Photosynthesis
Chloroplasts (found only in plant cells) carry out the chemical process of photosynthesis. This captures light energy and transfers it into useful chemical energy.

1. State which organelles or structures are associated with each of the processes listed below:

   (a) Secretion: _____

   (b) Respiration: _____

   (c) Pinocytosis: _____

   (d) Phagocytosis: _____

   (e) Protein synthesis: _____

   (f) Photosynthesis: _____

   (g) Cell division: _____

   (h) Autolysis: _____

   (i) Transport in/out of cell: _____

2. Explain what is meant by **metabolism** and describe an example of a metabolic process: _____
   _____
   _____

Related activities: Cell Structures and Organelles

Periodicals:
Cellular factories

The Structure of Membranes

All cells have a plasma membrane that forms the outer limit of the cell. Bacteria, fungi, and plant cells have a cell wall outside this, but it is quite distinct and outside the plasma membrane. Membranes are also found inside eukaryotic cells as part of membranous **organelles**. Present day knowledge of membrane structure has been built up as a result of many observations and experiments.

The original model of membrane structure, proposed by Davson and Danielli, was the unit membrane (a lipid bilayer coated with protein). This model was later modified after the discovery that the protein molecules were embedded *within* the bilayer rather than coating the outside. The now-accepted model of membrane structure is the **fluid mosaic model** described below.

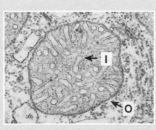
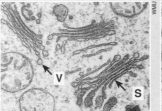
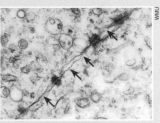

The **nuclear membrane** that surrounds the nucleus helps to control the passage of genetic information to the cytoplasm. It may also serve to protect the DNA.

Mitochondria have an outer membrane (**O**) which controls the entry and exit of materials involved in aerobic respiration. Inner membranes (**I**) provide attachment sites for enzyme activity.

The **Golgi apparatus** comprises stacks of membrane-bound sacs (**S**). It is involved in packaging materials for transport or export from the cell as secretory vesicles (**V**).

The cell is surrounded by a **plasma membrane** which controls the movement of most substances into and out of the cell. This photo shows two neighboring cells (arrows).

The Fluid Mosaic Model of Membrane Structure

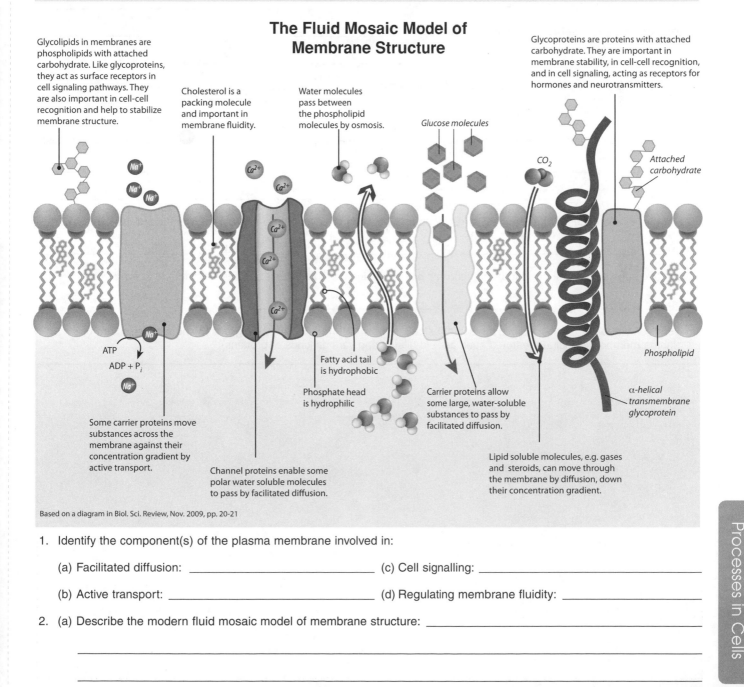

Glycolipids in membranes are phospholipids with attached carbohydrate. Like glycoproteins, they act as surface receptors in cell signaling pathways. They are also important in cell-cell recognition and help to stabilize membrane structure.

Cholesterol is a packing molecule and important in membrane fluidity.

Water molecules pass between the phospholipid molecules by osmosis.

Glucose molecules

Glycoproteins are proteins with attached carbohydrate. They are important in membrane stability, in cell-cell recognition, and in cell signaling, acting as receptors for hormones and neurotransmitters.

Attached carbohydrate

ATP
ADP + P$_i$

Some carrier proteins move substances across the membrane against their concentration gradient by active transport.

Channel proteins enable some polar water soluble molecules to pass by facilitated diffusion.

Fatty acid tail is hydrophobic

Phosphate head is hydrophilic

Carrier proteins allow some large, water-soluble substances to pass by facilitated diffusion.

Lipid soluble molecules, e.g. gases and steroids, can move through the membrane by diffusion, down their concentration gradient.

Phospholipid

α-helical transmembrane glycoprotein

Based on a diagram in Biol. Sci. Review, Nov. 2009, pp. 20-21

1. Identify the component(s) of the plasma membrane involved in:

   (a) Facilitated diffusion: _____ (c) Cell signalling: _____

   (b) Active transport: _____ (d) Regulating membrane fluidity: _____

2. (a) Describe the modern fluid mosaic model of membrane structure: _____

   _____

   _____

Processes in Cells

Periodicals:
Border control
The fluid mosaic model

Related activities: The Role of Membranes in Cells
Web links: Membrane Structure Tutorial

RA 2

(b) Explain how the fluid mosaic model accounts for the observed properties of cellular membranes:

_____

_____

_____

_____

_____

_____

_____

3. Discuss the various functional roles of membranes in cells: _____

_____

_____

_____

4. (a) Name a cellular organelle that possesses a membrane: _____

(b) Describe the membrane's purpose in this organelle: _____

_____

5. (a) Describe the purpose of cholesterol in plasma membranes: _____

_____

(b) Suggest why marine organisms living in polar regions have a very high proportion of cholesterol in their membranes:

_____

_____

6. List three substances that need to be transported **into** all kinds of animal cells, in order for them to survive:

(a) _____ (b) _____ (c) _____

7. List two substances that need to be transported **out** of all kinds of animal cells, in order for them to survive:

(a) _____ (b) _____

8. Use the symbol for a phospholipid molecule (below) to draw a **simple labelled diagram** to show the structure of a plasma membrane (include features such as lipid bilayer and various kinds of proteins):

The Role of Membranes in Cells

Many of the important structures and organelles in cells are composed of, or are enclosed by, membranes. These include: the endoplasmic reticulum, mitochondria, nucleus, Golgi body, chloroplasts, lysosomes, vesicles and the cell plasma membrane itself. All membranes within eukaryotic cells share the same basic structure as the plasma membrane that encloses the entire cell. They perform a number of critical functions in the cell: serving to compartmentalise regions of different function within the cell, controlling the entry and exit of substances, and fulfilling a role in recognition and communication between cells. Some of these roles are described below and electron micrographs of the organelles involved are on the following page.

Isolation of enzymes Membrane-bound lysosomes contain enzymes for the destruction of wastes and foreign material. Peroxisomes are the site for destruction of the toxic and reactive molecule, hydrogen peroxide (formed as a result of some cellular reactions).

Role in lipid synthesis
The smooth ER is the site of lipid and steroid synthesis.

Containment of DNA
The nucleus is surrounded by a nuclear envelope of two membranes, forming a separate compartment for the cell's genetic material.

Role in protein and membrane synthesis
Some protein synthesis occurs on free ribosomes, but much occurs on membrane-bound ribosomes on the rough endoplasmic reticulum. Here, the protein is synthesized directly into the space within the ER membranes. The rough ER is also involved in membrane synthesis, growing in place by adding proteins and phospholipids.

Entry and export of substances The plasma membrane may take up fluid or solid material and form membrane-bound vesicles (or larger vacuoles) within the cell. Membrane-bound transport vesicles move substances to the inner surface of the cell where they can be exported from the cell by exocytosis.

Cell communication and recognition
The proteins embedded in the membrane act as receptor molecules for hormones and neurotransmitters. Glycoproteins and glycolipids stabilize the plasma membrane and act as cell identity markers, helping cells to organize themselves into tissues, and enabling foreign cells to be recognized.

Packaging and secretion
The Golgi apparatus is a specialized membrane-bound organelle which produces lysosomes and compartmentalizes the modification, packaging and secretion of substances such as proteins and hormones.

Transport processes
Channel and carrier proteins are involved in selective transport across the plasma membrane. Cholesterol in the membrane can help to prevent ions or polar molecules from passing through the membrane (acting as a plug).

Energy transfer The reactions of cellular respiration (and photosynthesis in plants) take place in the membrane-bound energy transfer systems occurring in mitochondria and chloroplasts respectively. See the example explained below.

Compartmentation within Membranes

Membranes play an important role in separating regions within the cell (and within organelles) where particular reactions occur. Specific enzymes are therefore often located in particular organelles. Reaction rates are controlled by controlling the rate at which substrates enter the organelle. This regulates the availability of the raw materials required for the metabolic reactions.

Example: *The enzymes involved in cellular respiration are arranged in different parts of the mitochondria. The various reactions are localized and separated by membrane systems.*

Amine oxidases and other enzymes on the outer membrane surface

Adenylate kinase and other phosphorylases between the membranes

Respiratory assembly enzymes embedded in the membrane (ATPase)

Many soluble enzymes of the Krebs cycle floating in the matrix, as well as enzymes for fatty acid degradation.

Matrix

Cross-section of a mitochondrion

1. Discuss the importance of membrane systems and organelles in providing compartments within the cell:

_____

_____

_____

_____

Related activities: Cell Structures and Organelles, Packaging Macromolecules
Web links: Cell Membranes

A 2

Functional Roles of Membranes in Cells

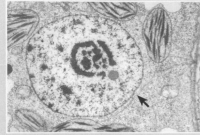

The **nuclear membrane**, which surrounds the nucleus, regulates the passage of genetic information to the cytoplasm and may also protect the DNA from damage.

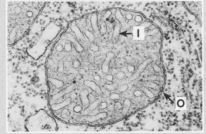

Mitochondria have an outer membrane (**O**) which controls the entry and exit of materials involved in aerobic respiration. Inner membranes (**I**) provide attachment sites for enzyme activity.

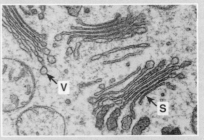

The **Golgi apparatus** comprises stacks of membrane-bound sacs (**S**). It is involved in packaging materials for transport or export from the cell as secretory vesicles (**V**).

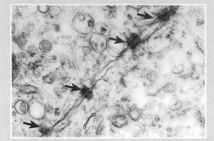

The **plasma membrane** surrounds the cell. In this photo, intercellular junctions called **desmosomes**, which connect neighbouring cells, are indicated with arrows.

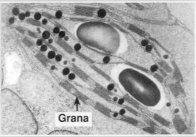

Chloroplasts are large organelles found in plant cells. The stacked membrane systems of chloroplasts (grana) trap light energy which is then used to fix carbon into 6-C sugars.

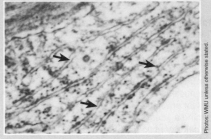

This EM shows stacks of rough endoplasmic reticulum (arrows). The membranes are studded with ribosomes, which synthesize proteins into the intermembrane space.

2. Match each of the following organelles with the correct description of its functional role in the cell:

chloroplast, rough endoplasmic reticulum, lysosome, smooth endoplasmic reticulum, mitochondrion, Golgi apparatus

(a) Active in synthesis, sorting, and secretion of cell products: _____

(b) Digestive organelle where macromolecules are hydrolyzed: _____

(c) Organelle where most cellular respiration occurs and most ATP is generated: _____

(d) Active in membrane synthesis and synthesis of secretory proteins: _____

(e) Active in lipid and hormone synthesis and secretion: _____

(f) Photosynthetic organelle converts light energy to chemical energy stored in sugar molecules: _____

3. Explain how the membrane surface area is increased within cells and organelles: _____

_____

4. Discuss the importance of each of the following to cellular function:

(a) High membrane surface area: _____

_____

(b) Channel proteins and carrier proteins in the plasma membrane: _____

_____

5. Non-polar (lipid-soluble) molecules diffuse more rapidly through membranes than polar (lipid-insoluble) molecules:

(a) Explain the reason for this: _____

_____

(b) Discuss the implications of this to the transport of substances into the cell through the plasma membrane:

_____

_____

How Do We Know? Membrane Structure

Cellular membranes play many extremely important roles in cells, and understanding their structure is central to understanding cellular function. Moreover, understanding the structure and function of membrane proteins is essential to understanding cellular transport processes, and cell recognition and signaling. Cellular membranes are far too small to be seen clearly using light microscopy, and certainly any detail is impossible to resolve. Since early last century, scientists have known that membranes comprised a lipid bilayer with associated proteins. But how did they elucidate just how these molecules were organized?

The answers were provided with electron microscopy, and one technique in particular – **freeze fracture**. As the name implies, freeze fracture, at its very simplest level, is the freezing of a cell and then cleaving it so that it fractures in a certain way. Scientists can then use electron microscopy to see the indentations and outlines of the structures remaining after cleavage. Membranes are composed of two layers held together by weak intermolecular bonds, so they cleave into two halves when fractured. This provides views of the inner surfaces of the membrane.

The procedure involves several steps:

▶ The tissue is prefixed using a cross linking agent. This alters the strength of the internal and external parts of the membrane.

▶ The cell is fixed to immobilise any mobile macromolecules.

▶ The specimen is passed through a sequential series of glycerol solutions of increasing strength. This protects the cells from bursting when placed into the cryomaterial.

▶ The specimen is frozen using liquid propane cooled by liquid nitrogen. The specimens are mounted on gold supports and cooled briefly before transfer to the freeze-etch machine.

▶ Specimen is cleaved in a helium-vented vacuum at -150°C. A razor blade cooled to -170°C acts as both a cold trap for water and the cleaving instrument.

▶ At this stage the specimen may be evaporated a little to produce some relief in the surface of the fracture (known as etching) so that a 3-dimensional effect occurs.

▶ For viewing under EM, a replica of the specimen is made and coated in gold or platinum to ~3 nm thick. This produces a shadow effect that allows structures to be seen clearly. A layer of carbon around 30 nm thick is used to stabilize the specimen.

▶ The samples are then raised to room temperature and placed into distilled water or digestive enzymes, which allows the replica to separate from the sample. The replica is then rinsed several times in distilled water before it is ready for viewing.

The freeze fracture technique provided the necessary supporting evidence for the current fluid mosaic model of membrane structure. When cleaved, proteins in the membrane left impressions that showed they were embedded into the membrane and not a continuous layer on the outside as earlier models proposed.

Cleaving the membrane

Razor blade

Proteins leave bumps and holes in the membrane when it is cleaved

— 50 nm   Photo: Louisa Howard and Chuck Daghlian, Dartmouth College

1. Describe the principle of freeze fracture and explain why it is such a useful technique for studying membrane structure:

_____

_____

_____

2. Explain how this freeze-fracture studies provided evidence for our current model of membrane structure: _____

_____

_____

3. An earlier model of membrane structure was the unit membrane; a phospholipid bilayer with a protein coat. Explain how the freeze-fracture studies showed this model to be flawed:

_____

_____

_____

Processes in Cells

RA 2

Modification of Proteins

Proteins may be modified after they have been produced by ribosomes. After they pass into the interior of rough endoplasmic reticulum, some proteins may have carbohydrates added to them to form **glycoproteins**. Proteins may be further altered in the Golgi apparatus. The **Golgi apparatus** functions principally as a system for processing, sorting, and modifying proteins. Proteins that are to be secreted from the cell are synthesized by ribosomes on the rough endoplasmic reticulum and transported to the Golgi apparatus. At this stage, carbohydrates may be removed or added in a step-wise process. Some of the possible functions of glycoproteins are illustrated below. Other proteins may have fatty acids added to them to form **lipoproteins**. These modified proteins transport lipids in the plasma between various organs in the body (e.g. gut, liver, and adipose tissue).

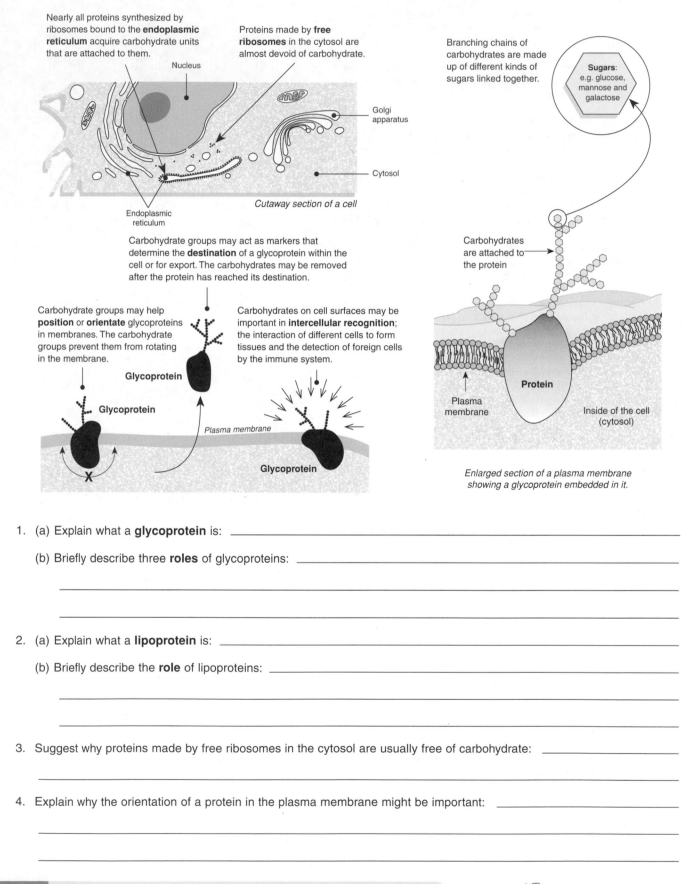

Nearly all proteins synthesized by ribosomes bound to the **endoplasmic reticulum** acquire carbohydrate units that are attached to them.

Proteins made by **free ribosomes** in the cytosol are almost devoid of carbohydrate.

Nucleus

Golgi apparatus

Cytosol

Cutaway section of a cell

Endoplasmic reticulum

Branching chains of carbohydrates are made up of different kinds of sugars linked together.

Sugars: e.g. glucose, mannose and galactose

Carbohydrate groups may act as markers that determine the **destination** of a glycoprotein within the cell or for export. The carbohydrates may be removed after the protein has reached its destination.

Carbohydrate groups may help **position** or **orientate** glycoproteins in membranes. The carbohydrate groups prevent them from rotating in the membrane.

Carbohydrates on cell surfaces may be important in **intercellular recognition**; the interaction of different cells to form tissues and the detection of foreign cells by the immune system.

Glycoprotein

Glycoprotein

Plasma membrane

Glycoprotein

Carbohydrates are attached to the protein

Protein

Plasma membrane

Inside of the cell (cytosol)

Enlarged section of a plasma membrane showing a glycoprotein embedded in it.

1. (a) Explain what a **glycoprotein** is: _____

   (b) Briefly describe three **roles** of glycoproteins: _____

   _____

   _____

2. (a) Explain what a **lipoprotein** is: _____

   (b) Briefly describe the **role** of lipoproteins: _____

   _____

   _____

3. Suggest why proteins made by free ribosomes in the cytosol are usually free of carbohydrate: _____

   _____

4. Explain why the orientation of a protein in the plasma membrane might be important: _____

   _____

   _____

Packaging Macromolecules

Cells produce a range of organic polymers made up of repeating units of smaller molecules. The synthesis, packaging and movement of these **macromolecules** inside the cell involves a number of membrane bound organelles, as indicated below. These organelles provide compartments where the enzyme systems involved can be isolated.

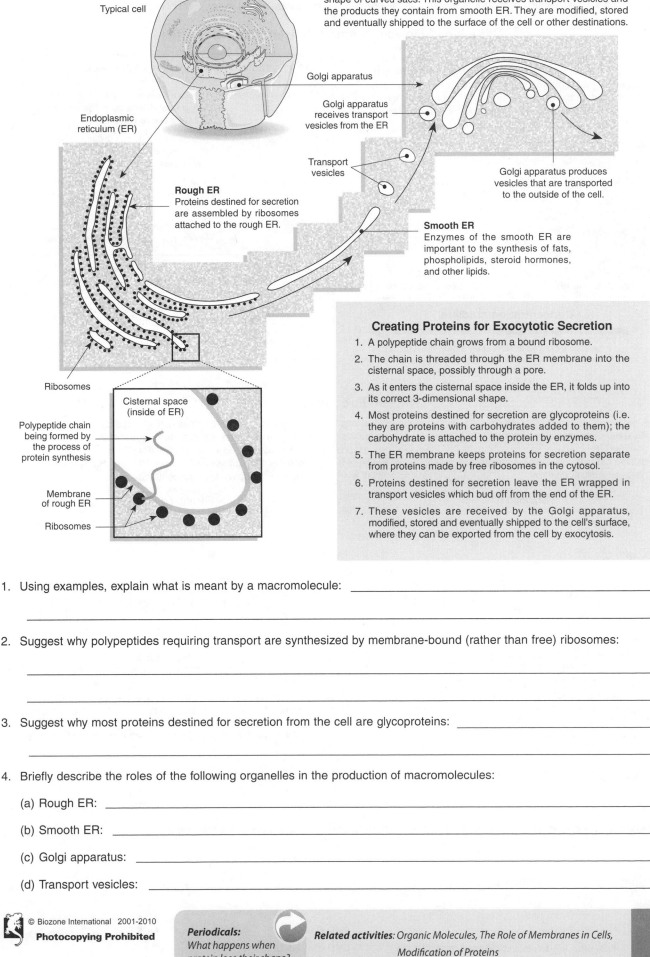

Golgi apparatus
The Golgi apparatus comprises stacks of flattened membranes in the shape of curved sacs. This organelle receives transport vesicles and the products they contain from smooth ER. They are modified, stored and eventually shipped to the surface of the cell or other destinations.

Typical cell

Endoplasmic reticulum (ER)

Golgi apparatus

Golgi apparatus receives transport vesicles from the ER

Transport vesicles

Golgi apparatus produces vesicles that are transported to the outside of the cell.

Rough ER
Proteins destined for secretion are assembled by ribosomes attached to the rough ER.

Smooth ER
Enzymes of the smooth ER are important to the synthesis of fats, phospholipids, steroid hormones, and other lipids.

Ribosomes

Cisternal space (inside of ER)

Polypeptide chain being formed by the process of protein synthesis

Membrane of rough ER

Ribosomes

Creating Proteins for Exocytotic Secretion

1. A polypeptide chain grows from a bound ribosome.
2. The chain is threaded through the ER membrane into the cisternal space, possibly through a pore.
3. As it enters the cisternal space inside the ER, it folds up into its correct 3-dimensional shape.
4. Most proteins destined for secretion are glycoproteins (i.e. they are proteins with carbohydrates added to them); the carbohydrate is attached to the protein by enzymes.
5. The ER membrane keeps proteins for secretion separate from proteins made by free ribosomes in the cytosol.
6. Proteins destined for secretion leave the ER wrapped in transport vesicles which bud off from the end of the ER.
7. These vesicles are received by the Golgi apparatus, modified, stored and eventually shipped to the cell's surface, where they can be exported from the cell by exocytosis.

1. Using examples, explain what is meant by a macromolecule: _____

_____

2. Suggest why polypeptides requiring transport are synthesized by membrane-bound (rather than free) ribosomes:

_____

_____

3. Suggest why most proteins destined for secretion from the cell are glycoproteins: _____

_____

4. Briefly describe the roles of the following organelles in the production of macromolecules:

   (a) Rough ER: _____

   (b) Smooth ER: _____

   (c) Golgi apparatus: _____

   (d) Transport vesicles: _____

Processes in Cells

Periodicals:
What happens when protein lose their shape?

Related activities: Organic Molecules, The Role of Membranes in Cells, Modification of Proteins

RDA 2

Diffusion

The molecules that make up substances are constantly moving about in a random way. This random motion causes molecules to disperse from areas of high to low concentration; a process called **diffusion**. The molecules move along a **concentration gradient**. Diffusion and osmosis (diffusion of water molecules across a partially permeable membrane) are **passive** processes, and use no energy. Diffusion occurs freely across membranes, as long as the membrane is permeable to that molecule (partially permeable membranes allow the passage of some molecules but not others). Each type of molecule diffuses along its own concentration gradient. Diffusion of molecules in one direction does not hinder the movement of other molecules. Diffusion is important in allowing exchanges with the environment and in the regulation of cell water content.

Diffusion of Molecules Along Concentration Gradients

Diffusion is the movement of particles from regions of high to low concentration (the **concentration gradient**), with the end result being that the molecules become evenly distributed. In biological systems, diffusion often occurs across partially permeable membranes. Various factors determine the rate at which this occurs (see right).

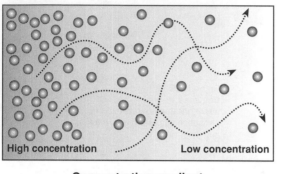

High concentration   **Low concentration**

Concentration gradient

If molecules are free to move, they move from high to low concentration until they are evenly dispersed.

Factors affecting rates of diffusion

Concentration gradient:	Diffusion rates will be higher when there is a greater difference in concentration between two regions.
The distance involved:	Diffusion over shorter distances occurs at a greater rate than diffusion over larger distances.
The area involved:	The larger the area across which diffusion occurs, the greater the rate of diffusion.
Barriers to diffusion:	Thicker barriers slow diffusion rate. Pores in a barrier enhance diffusion.

These factors are expressed in **Fick's law**, which governs the rate of diffusion of substances within a system. It is described by:

$$\frac{\text{Surface area of membrane} \quad X \quad \text{Difference in concentration across the membrane}}{\text{Length of the diffusion path (thickness of the membrane)}}$$

Diffusion through Membranes

Each type of diffusing molecule (gas, solvent, solute) moves **along its own concentration gradient**. Two-way diffusion (below) is common in biological systems, e.g. at the lung surface, carbon dioxide diffuses out and oxygen diffuses into the blood. Facilitated diffusion (below, right) increases the diffusion rate selectively and is important for larger molecules (e.g. glucose, amino acids) where a higher diffusion rate is desirable (e.g. transport of glucose into skeletal muscle fibers, transport of ADP into mitochondria). Neither type of diffusion requires energy expenditure because the molecules are not moving against their concentration gradient.

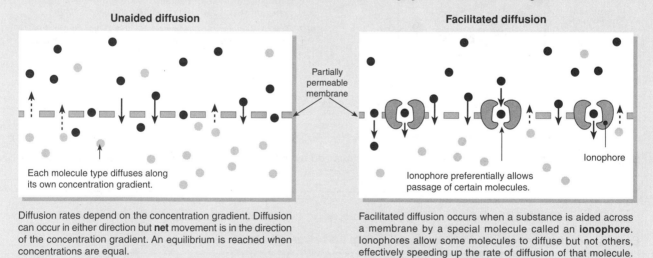

Unaided diffusion

Each molecule type diffuses along its own concentration gradient.

Partially permeable membrane

Facilitated diffusion

Ionophore preferentially allows passage of certain molecules.

Ionophore

Diffusion rates depend on the concentration gradient. Diffusion can occur in either direction but **net** movement is in the direction of the concentration gradient. An equilibrium is reached when concentrations are equal.

Facilitated diffusion occurs when a substance is aided across a membrane by a special molecule called an **ionophore**. Ionophores allow some molecules to diffuse but not others, effectively speeding up the rate of diffusion of that molecule.

1. Describe two properties of an exchange surface that would facilitate rapid diffusion rates:

   (a) _____ (b) _____

2. Identify one way in which organisms maintain concentration gradients across membranes: _____

   _____

3. State how facilitated diffusion is achieved: _____

Osmosis and Water Potential

Osmosis is the term describing the diffusion of water along its concentration gradient across a partially permeable membrane. It is the principal mechanism by which water enters and leaves cells in living organisms. As it is a type of diffusion, the rate at which osmosis occurs is affected by the same factors that affect all diffusion rates (see earlier). The tendency for water to move in any particular direction can be calculated on the basis of the

water potential (ψ) of the cell sap relative to its surrounding environment. The use of water potential to express the water relations of cells has replaced the terms osmotic potential and osmotic pressure although these are still frequently used in areas of animal physiology and medicine. An alternative version which does not use this terminology is available on the *TRC: Osmosis and Diffusion* (or see web links below).

Osmosis and the Water Potential of Cells

Osmosis is simply the diffusion of water molecules from high concentration to lower concentration, across a partially permeable membrane. The direction of this movement can be predicted on the basis of the water potential of the solutions involved. The **water potential** of a solution (denoted with the symbol ψ) is the term given to the tendency for water molecules to enter or leave a solution by osmosis. Pure water has the highest water potential, set at zero. Dissolving any solute into

pure water lowers the water potential (makes it more negative). *Water always diffuses from regions of less negative to more negative water potential.* Water potential is determined by two components: the **solute potential**, ψs (of the cell sap) and the **pressure potential**, ψp. This is expressed as a simple equation:

$$\psi cell \ = \ \psi s \ + \ \psi p$$

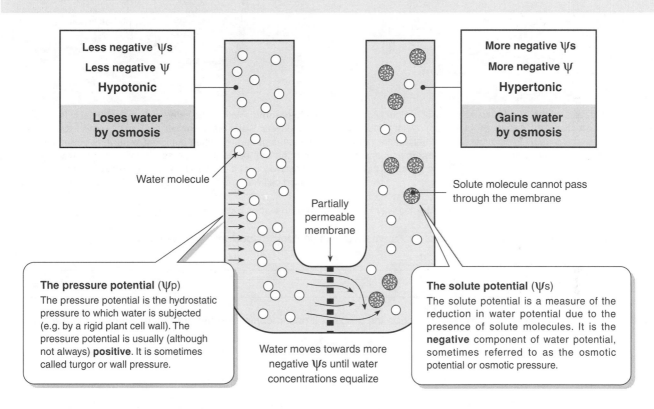

Less negative ψs
Less negative ψ
Hypotonic

Loses water by osmosis

Water molecule

The pressure potential (ψp)
The pressure potential is the hydrostatic pressure to which water is subjected (e.g. by a rigid plant cell wall). The pressure potential is usually (although not always) **positive**. It is sometimes called turgor or wall pressure.

Partially permeable membrane

Water moves towards more negative ψs until water concentrations equalize

More negative ψs
More negative ψ
Hypertonic

Gains water by osmosis

Solute molecule cannot pass through the membrane

The solute potential (ψs)
The solute potential is a measure of the reduction in water potential due to the presence of solute molecules. It is the **negative** component of water potential, sometimes referred to as the osmotic potential or osmotic pressure.

1. State the water potential of pure water at standard temperature and pressure: _____

2. The three diagrams below show the solute and pressure potential values for three hypothetical situations where two solutions are separated by a selectively permeable membrane. For each example (a) - (c) calculate ψ for the solutions on each side of the membrane, as indicated:

3. Draw arrows on each diagram to indicate the direction of net flow of water:

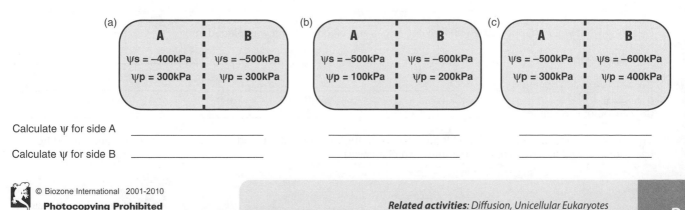

(a)

A	B
ψs = –400kPa	ψs = –500kPa
ψp = 300kPa	ψp = 300kPa

(b)

A	B
ψs = –500kPa	ψs = –600kPa
ψp = 100kPa	ψp = 200Pta

(c)

A	B
ψs = –500kPa	ψs = –600kPa
ψp = 300kPa	ψp = 400kPa

Calculate ψ for side A _____ _____ _____

Calculate ψ for side B _____ _____ _____

Processes in Cells

Related activities: Diffusion, Unicellular Eukaryotes
Web links: Cellular Transport, Osmosis and Diffusion

DA 2

Water Relations in Plant Cells

The plasma membrane of cells is a partially permeable membrane and osmosis is the principal mechanism by which water enters and leaves the cell. When the external water potential is the same as that of the cell there is no net movement of water. Two systems (cell and environment) with the same water potential are termed **isotonic**. The diagram below illustrates two different situations: when the external water potential is less negative than the cell (**hypotonic**) and when it is more negative than the cell (**hypertonic**).

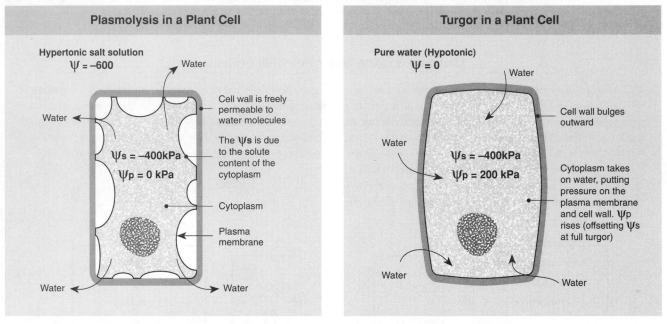

Plasmolysis in a Plant Cell

Hypertonic salt solution
$\Psi = -600$

Water

Water

$\Psi s = -400kPa$
$\Psi p = 0$ kPa

Cell wall is freely permeable to water molecules

The Ψs is due to the solute content of the cytoplasm

Cytoplasm

Plasma membrane

Water

Water

Turgor in a Plant Cell

Pure water (Hypotonic)
$\Psi = 0$

Water

$\Psi s = -400kPa$
$\Psi p = 200$ kPa

Cell wall bulges outward

Cytoplasm takes on water, putting pressure on the plasma membrane and cell wall. Ψp rises (offsetting Ψs at full turgor)

Water

Water

Water

In a **hypertonic** solution, the external water potential is more negative than the water potential of the cell ($\Psi cell = \Psi s + \Psi p$). Water leaves the cell and, because the cell wall is rigid, the plasma membrane shrinks away from the cell wall. This process is termed **plasmolysis** and the cell becomes **flaccid** ($\Psi p = 0$). Full plasmolysis is irreversible; the cell cannot recover by taking up water.

In a **hypotonic** solution, the external water potential is less negative than the $\Psi cell$. Water enters the cell causing it to swell tight. A pressure potential is generated when sufficient water has been taken up to cause the cell contents to press against the cell wall. Ψp rises progressively until it offsets Ψs. Water uptake stops when $\Psi cell = 0$. The rigid cell wall prevents cell rupture. Cells in this state are **turgid**.

4. Fluid replacements are usually provided for heavily perspiring athletes after endurance events.

    (a) Identify the preferable tonicity of these replacement drinks (isotonic, hypertonic, or hypotonic): _____

    (b) Give a reason for your answer: _____

    _____

5. *Paramecium* is a freshwater protozoan. Describe the problem it has in controlling the amount of water inside the cell:

    _____

    _____

6. (a) Explain the role of pressure potential in generating cell turgor in plants: _____

    _____

    (b) Explain the purpose of cell turgor to plants: _____

    _____

7. Explain how animal cells differ from plant cells with respect to the effects of net water movements: _____

    _____

8. Describe what would happen to an animal cell (e.g. a red blood cell) if it was placed into:

    (a) Pure water: _____

    (b) A hypertonic solution: _____

    (c) A hypotonic solution: _____

9. The malarial parasite lives in human blood. Relative to the tonicity of the blood, the parasite's cell contents would be hypertonic / isotonic / hypotonic (circle the correct answer).

Surface Area and Volume

When an object (e.g. a cell) is small it has a large surface area in comparison to its volume. In this case diffusion will be an effective way to transport materials (e.g. gases) into the cell. As an object becomes larger, its surface area compared to its volume is smaller. Diffusion is no longer an effective way to transport materials to the inside. For this reason, there is a physical limit for the size of a cell, with the effectiveness of diffusion being the controlling factor.

Diffusion in Organisms of Different Sizes

Respiratory gases and some other substances are exchanged with the surroundings by diffusion or active transport across the plasma membrane.

The **plasma membrane**, which surrounds every cell, functions as a selective barrier that regulates the cell's chemical composition. For each square micrometer of membrane, only so much of a particular substance can cross per second.

The surface area of an elephant is increased, for radiating body heat, by large flat ears.

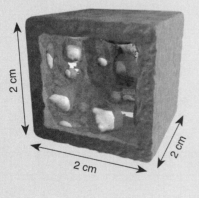

The nucleus can control a smaller cell more efficiently.

Oxygen

Food

Carbon dioxide

Wastes

A specialized gas exchange surface (lungs) and circulatory (blood) system are required to speed up the movement of substances through the body.

Respiratory gases cannot reach body tissues by diffusion alone.

Amoeba: The small size of single-celled protists, such as *Amoeba,* provides a large surface area relative to the cell s volume. This is adequate for many materials to be moved into and out of the cell by diffusion or active transport.

Multicellular organisms: To overcome the problems of small cell size, plants and animals became multicellular. They provide a small surface area compared to their volume but have evolved various adaptive features to improve their effective surface area.

Smaller is Better for Diffusion

One large cube

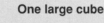

2 cm

2 cm

2 cm

Volume: = 8 cm³

Surface area: = 24 cm²

Eight small cubes

1 cm

1 cm

1 cm

Volume: = 8 cm³ for 8 cubes

Surface area: = 6 cm² for 1 cube

= 48 cm² for 8 cubes

The eight small cells and the single large cell have the same total volume, but their surface areas are different. The small cells together have twice the total surface area of the large cell, because there are more exposed (inner) surfaces. Real organisms have complex shapes, but the same principles apply.

The surface-area volume relationship has important implications for processes involving transport into and out of cells across membranes. For activities such as gas exchange, the surface area available for diffusion is a major factor limiting the rate at which oxygen can be supplied to tissues.

Processes in Cells

Periodicals:
Getting in and out

Related activities: Diffusion

DA 1

The diagram below shows four hypothetical cells of different sizes (cells do not actually grow to this size, their large size is for the sake of the exercise). They range from a small 2 cm cube to a larger 5 cm cube. This exercise investigates the effect of cell size on the efficiency of diffusion.

2 cm cube   **3 cm cube**   **4 cm cube**   **5 cm cube**

1. Calculate the volume, surface area and the ratio of surface area to volume for each of the four cubes above (the first has been done for you). When completing the table below, show your calculations.

Cube size	Surface area	Volume	Surface area to volume ratio
2 cm cube	$2 \times 2 \times 6 = 24\ cm^2$ (2 cm x 2 cm x 6 sides)	$2 \times 2 \times 2 = 8\ cm^3$ (height x width x depth)	$24\ to\ 8 = 3{:}1$
3 cm cube			
4 cm cube			
5 cm cube			

2. Create a graph, plotting the surface area against the volume of each cube, on the grid on the right. Draw a line connecting the points and label axes and units.

3. State which increases the fastest with increasing size: the **volume** or **surface area**.

_____

4. Explain what happens to the ratio of surface area to volume with increasing size:

_____

_____

_____

_____

5. (a) Diffusion of substances into and out of a cell occurs across the cell surface. Describe how increasing the size of a cell affects the ability of diffusion to transport materials into and out of a cell:

_____

_____

(b) Describe how this places constraints on cell size and explain how multicellular organisms have overcome this:

_____

_____

_____

Ion Pumps

Diffusion alone cannot supply the cell's entire requirements for molecules (and ions). Some molecules (e.g. glucose) are required by the cell in higher concentrations than occur outside the cell. Others (e.g. sodium) must be removed from the cell in order to maintain fluid balance. These molecules must be moved across the plasma membrane by active transport mechanisms. **Active transport** requires the expenditure of energy because the molecules (or ions) must be moved **against** their concentration gradient. The work of active transport is performed by specific carrier proteins in the membrane. These transport proteins harness the energy of ATP to pump molecules from a low to a high concentration. When ATP transfers a phosphate group to the carrier protein, the protein changes its shape in such a way as to move the bound molecule across the membrane. Three types of membrane pump are illustrated below. The sodium-potassium pump (below, centre) is almost universal in animal cells and is common in plant cells also. The concentration gradient created by ion pumps such as this and the proton pump (left) is frequently coupled to the transport of molecules such as sucrose (in phloem) and glucose (e.g. in the intestine) as shown below right.

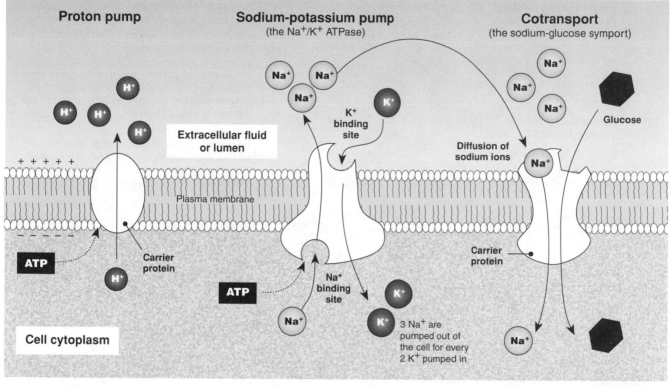

Proton pumps
ATP driven proton pumps use energy to remove hydrogen ions (H^+) from inside the cell to the outside. This creates a large difference in the proton concentration either side of the membrane, with the inside of the plasma membrane being negatively charged. This potential difference can be coupled to the transport of other molecules.

Sodium-potassium pump
The sodium-potassium pump is a specific protein in the membrane that uses energy in the form of ATP to exchange sodium ions (Na^+) for potassium ions (K^+) across the membrane. The unequal balance of Na^+ and K^+ across the membrane creates large concentration gradients that can be used to drive transport of other substances (e.g. cotransport of glucose).

Cotransport (coupled transport)
A gradient in sodium ions drives the active transport of **glucose** in intestinal epithelial cells. The specific transport protein couples the return of Na^+ down its concentration gradient to the transport of glucose into the intestinal epithelial cell. A low intracellular concentration of Na^+ (and therefore the concentration gradient) is maintained by a sodium-potassium pump.

1. Explain why the ATP is required for membrane pump systems to operate: _____

_____

2. (a) Explain what is meant by cotransport: _____

_____

(b) Explain how cotransport is used to move glucose into the intestinal epithelial cells: _____

_____

_____

(c) Explain what happens to the glucose that is transported into the intestinal epithelial cells: _____

_____

3. Describe two consequences of the extracellular accumulation of sodium ions: _____

_____

_____

Periodicals:
How biological membranes
achieve selective transport

Related activities: Active and Passive Transport
Web links: Cellular Transport, Symport

A 2

Processes in Cells

Exocytosis and Endocytosis

Most cells carry out **cytosis**: a form of **active transport** involving the in- or outfolding of the plasma membrane. The ability of cells to do this is a function of the flexibility of the plasma membrane. Cytosis results in the bulk transport into or out of the cell and is achieved through the localized activity of microfilaments and microtubules in the cell cytoskeleton. Engulfment of material is termed **endocytosis.** Endocytosis typically occurs in protozoans and certain white blood cells of the mammalian defense system (e.g. neutrophils, macrophages). **Exocytosis** is the reverse of endocytosis and involves the release of material from vesicles or vacuoles that have fused with the plasma membrane. Exocytosis is typical of cells that export material (secretory cells).

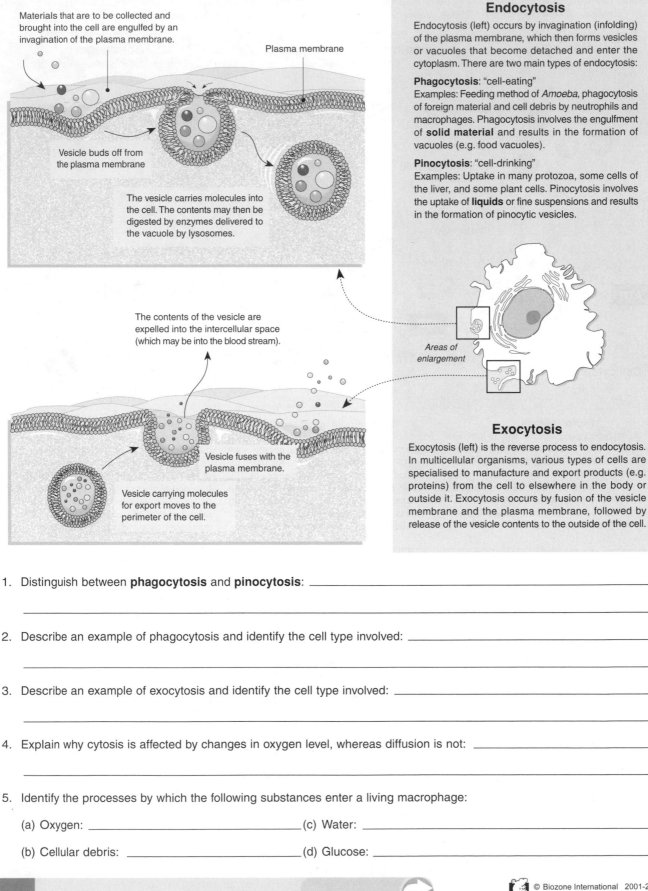

Materials that are to be collected and brought into the cell are engulfed by an invagination of the plasma membrane.

Plasma membrane

Vesicle buds off from the plasma membrane

The vesicle carries molecules into the cell. The contents may then be digested by enzymes delivered to the vacuole by lysosomes.

The contents of the vesicle are expelled into the intercellular space (which may be into the blood stream).

Vesicle fuses with the plasma membrane.

Vesicle carrying molecules for export moves to the perimeter of the cell.

Areas of enlargement

Endocytosis

Endocytosis (left) occurs by invagination (infolding) of the plasma membrane, which then forms vesicles or vacuoles that become detached and enter the cytoplasm. There are two main types of endocytosis:

Phagocytosis: "cell-eating"
Examples: Feeding method of *Amoeba*, phagocytosis of foreign material and cell debris by neutrophils and macrophages. Phagocytosis involves the engulfment of **solid material** and results in the formation of vacuoles (e.g. food vacuoles).

Pinocytosis: "cell-drinking"
Examples: Uptake in many protozoa, some cells of the liver, and some plant cells. Pinocytosis involves the uptake of **liquids** or fine suspensions and results in the formation of pinocytic vesicles.

Exocytosis

Exocytosis (left) is the reverse process to endocytosis. In multicellular organisms, various types of cells are specialised to manufacture and export products (e.g. proteins) from the cell to elsewhere in the body or outside it. Exocytosis occurs by fusion of the vesicle membrane and the plasma membrane, followed by release of the vesicle contents to the outside of the cell.

1. Distinguish between **phagocytosis** and **pinocytosis**: _____

_____

2. Describe an example of phagocytosis and identify the cell type involved: _____

_____

3. Describe an example of exocytosis and identify the cell type involved: _____

_____

4. Explain why cytosis is affected by changes in oxygen level, whereas diffusion is not: _____

_____

5. Identify the processes by which the following substances enter a living macrophage:

   (a) Oxygen: _____   (c) Water: _____

   (b) Cellular debris: _____   (d) Glucose: _____

Related activities: Active and Passive Transport

Periodicals:
What is endocytosis?

Active and Passive Transport

Cells have a need to move materials both into and out of the cell. Raw materials and other molecules necessary for metabolism must be accumulated from outside the cell. Some of these substances are scarce outside of the cell and some effort is required to accumulate them. Waste products and molecules for use in other parts of the body must be 'exported' out of the cell.

Some materials (e.g. gases and water) move into and out of the cell by **passive transport** processes, without the expenditure of energy on the part of the cell. Other molecules (e.g. sucrose) are moved into and out of the cell using **active transport**. Active transport processes involve the expenditure of energy in the form of ATP, and therefore use oxygen.

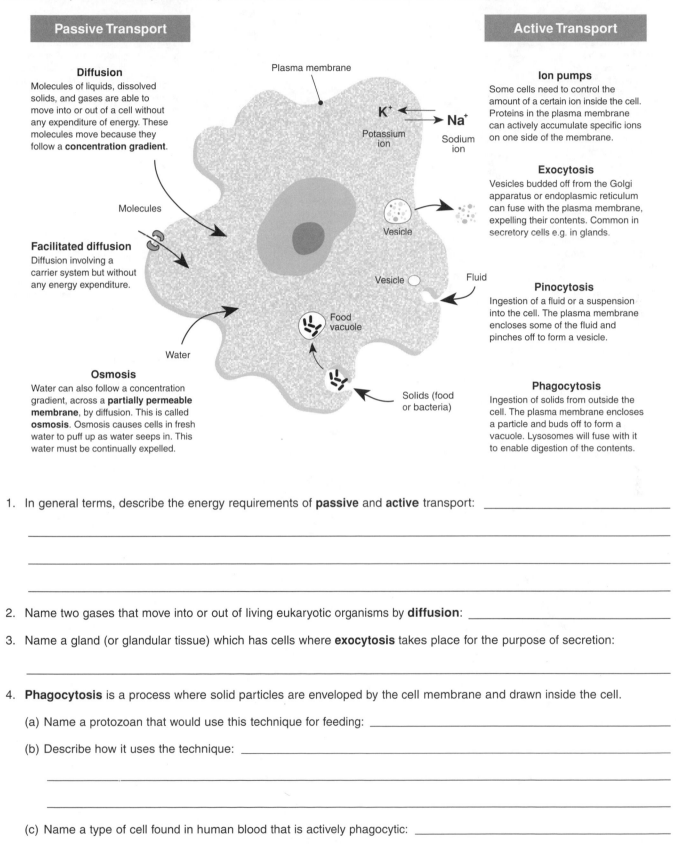

Passive Transport

Diffusion
Molecules of liquids, dissolved solids, and gases are able to move into or out of a cell without any expenditure of energy. These molecules move because they follow a **concentration gradient**.

Facilitated diffusion
Diffusion involving a carrier system but without any energy expenditure.

Osmosis
Water can also follow a concentration gradient, across a **partially permeable membrane**, by diffusion. This is called **osmosis**. Osmosis causes cells in fresh water to puff up as water seeps in. This water must be continually expelled.

Plasma membrane

K⁺ → Na⁺

Potassium ion   Sodium ion

Molecules

Vesicle

Vesicle   Fluid

Food vacuole

Water

Solids (food or bacteria)

Active Transport

Ion pumps
Some cells need to control the amount of a certain ion inside the cell. Proteins in the plasma membrane can actively accumulate specific ions on one side of the membrane.

Exocytosis
Vesicles budded off from the Golgi apparatus or endoplasmic reticulum can fuse with the plasma membrane, expelling their contents. Common in secretory cells e.g. in glands.

Pinocytosis
Ingestion of a fluid or a suspension into the cell. The plasma membrane encloses some of the fluid and pinches off to form a vesicle.

Phagocytosis
Ingestion of solids from outside the cell. The plasma membrane encloses a particle and buds off to form a vacuole. Lysosomes will fuse with it to enable digestion of the contents.

1. In general terms, describe the energy requirements of **passive** and **active** transport: _____

   _____

   _____

   _____

2. Name two gases that move into or out of living eukaryotic organisms by **diffusion**: _____

3. Name a gland (or glandular tissue) which has cells where **exocytosis** takes place for the purpose of secretion:

   _____

4. **Phagocytosis** is a process where solid particles are enveloped by the cell membrane and drawn inside the cell.

   (a) Name a protozoan that would use this technique for feeding: _____

   (b) Describe how it uses the technique: _____

       _____

       _____

   (c) Name a type of cell found in human blood that is actively phagocytic: _____

   (d) Describe the functional role of the cell that is related to this feature: _____

Related activities: Unicellular Eukaryotes, Human Cell Specialization
Web links: Cellular Transport

RA 1

Processes in Cells

Cell Division

The life cycle of a diploid sexually reproducing organism, such as a human, with **gametic meiosis** is illustrated below. In this life cycle, **gametogenesis** involves meiotic division to produce male and female gametes for the purpose of sexual reproduction. The life cycle in flowering plants is different in that the gametes are produced through mitosis in haploid gametophytes. The male gametes are produced inside the pollen grain and the female gametes are produced inside the embryo sac of the ovule. The gametophytes develop and grow from haploid spores, which are produced from meiosis.

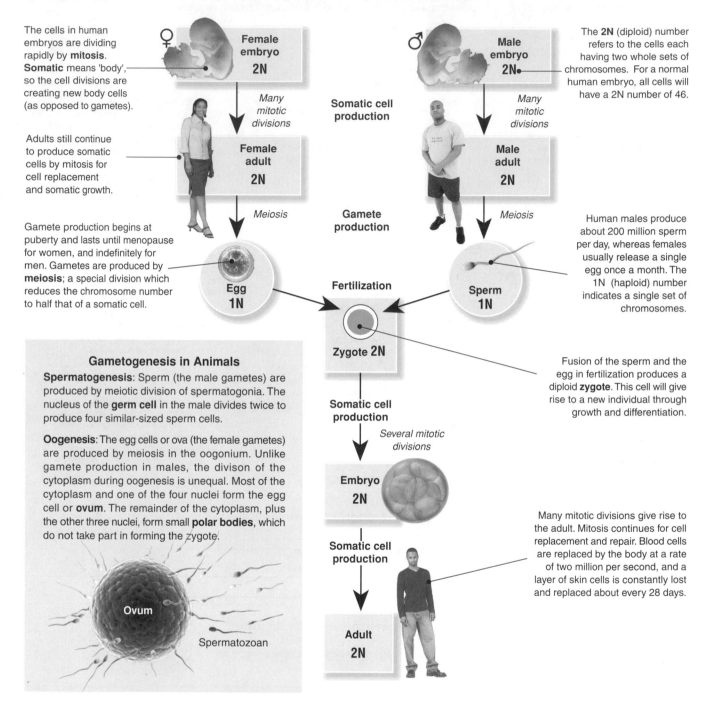

The cells in human embryos are dividing rapidly by **mitosis**. **Somatic** means 'body', so the cell divisions are creating new body cells (as opposed to gametes).

Adults still continue to produce somatic cells by mitosis for cell replacement and somatic growth.

Gamete production begins at puberty and lasts until menopause for women, and indefinitely for men. Gametes are produced by **meiosis**; a special division which reduces the chromosome number to half that of a somatic cell.

The **2N** (diploid) number refers to the cells each having two whole sets of chromosomes. For a normal human embryo, all cells will have a 2N number of 46.

Human males produce about 200 million sperm per day, whereas females usually release a single egg once a month. The 1N (haploid) number indicates a single set of chromosomes.

Fusion of the sperm and the egg in fertilization produces a diploid **zygote**. This cell will give rise to a new individual through growth and differentiation.

Many mitotic divisions give rise to the adult. Mitosis continues for cell replacement and repair. Blood cells are replaced by the body at a rate of two million per second, and a layer of skin cells is constantly lost and replaced about every 28 days.

Female embryo 2N — Many mitotic divisions — **Female adult 2N** — Meiosis — **Egg 1N**

Somatic cell production

Gamete production

Male embryo 2N — Many mitotic divisions — **Male adult 2N** — Meiosis — **Sperm 1N**

Fertilization — **Zygote 2N** — **Somatic cell production** — Several mitotic divisions — **Embryo 2N** — **Somatic cell production** — **Adult 2N**

Gametogenesis in Animals

Spermatogenesis: Sperm (the male gametes) are produced by meiotic division of spermatogonia. The nucleus of the **germ cell** in the male divides twice to produce four similar-sized sperm cells.

Oogenesis: The egg cells or ova (the female gametes) are produced by meiosis in the oogonium. Unlike gamete production in males, the divison of the cytoplasm during oogenesis is unequal. Most of the cytoplasm and one of the four nuclei form the egg cell or **ovum**. The remainder of the cytoplasm, plus the other three nuclei, form small **polar bodies**, which do not take part in forming the zygote.

Ovum — Spermatozoan

1. Describe the purpose of the following types of cell division in an organism with a life cycle involving gametic meiosis:

   (a) Mitosis: _____

   (b) Meiosis: _____

2. Describe the basic difference between the cell divisions involved in spermatogenesis and oogenesis:

   _____

   _____

3. Explain how gametogenesis differs between humans and flowering plants: _____

   _____

Related activities: Mitosis and the Cell Cycle

Mitosis and the Cell Cycle

Mitosis is part of the 'cell cycle' in which an existing cell (the parent cell) divides into two (the daughter cells). Mitosis does not result in a change of chromosome numbers (unlike meiosis) and the daughter cells are identical to the parent cell. Although mitosis is part of a continuous cell cycle, it is divided into stages (below). The example below illustrates the cell cycle in a plant cell. Note that in animal cells, **cytokinesis** involves the formation of a constriction that divides the cell in two. It is usually well underway by the end of telophase and does not involve the formation of a cell plate.

The Cell Cycle and Stages of Mitosis

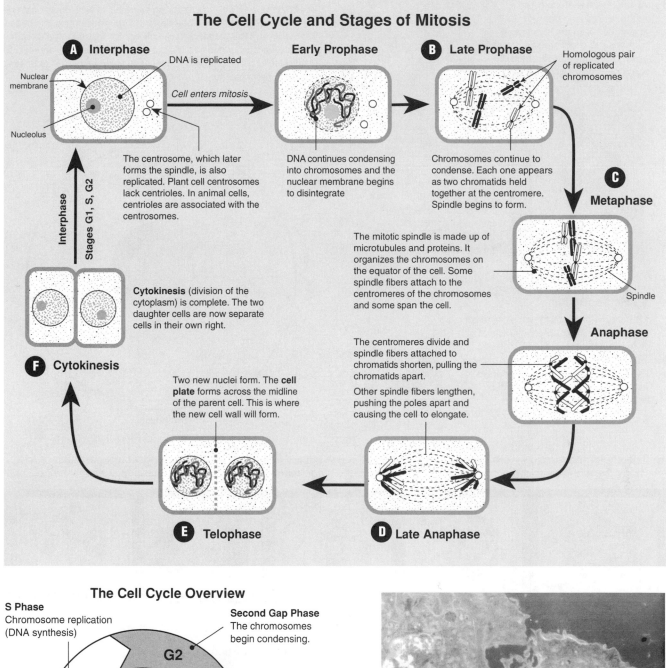

A Interphase

DNA is replicated

Nuclear membrane

Nucleolus

Cell enters mitosis

The centrosome, which later forms the spindle, is also replicated. Plant cell centrosomes lack centrioles. In animal cells, centrioles are associated with the centrosomes.

Interphase

Stages G1, S, G2

Early Prophase

DNA continues condensing into chromosomes and the nuclear membrane begins to disintegrate

B Late Prophase

Homologous pair of replicated chromosomes

Chromosomes continue to condense. Each one appears as two chromatids held together at the centromere. Spindle begins to form.

C Metaphase

The mitotic spindle is made up of microtubules and proteins. It organizes the chromosomes on the equator of the cell. Some spindle fibers attach to the centromeres of the chromosomes and some span the cell.

Spindle

F Cytokinesis

Cytokinesis (division of the cytoplasm) is complete. The two daughter cells are now separate cells in their own right.

Anaphase

The centromeres divide and spindle fibers attached to chromatids shorten, pulling the chromatids apart.

Other spindle fibers lengthen, pushing the poles apart and causing the cell to elongate.

Two new nuclei form. The **cell plate** forms across the midline of the parent cell. This is where the new cell wall will form.

E Telophase

D Late Anaphase

The Cell Cycle Overview

S Phase
Chromosome replication (DNA synthesis)

Second Gap Phase
The chromosomes begin condensing.

G2

S

The Cell Cycle

M

Mitosis
Nuclear division

C

G1

Cytokinesis
Division of the cytoplasm and separation of the two cells. Cytokinesis is distinct from nuclear division.

First Gap Phase
Cell growth and development

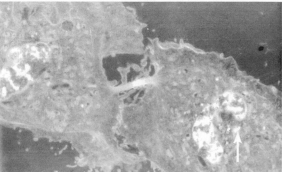

Animal cell cytokinesis (above) begins shortly after the sister chromatids have separated in anaphase of mitosis. A contractile ring of microtubular elements assembles in the middle of the cell, next to the plasma membrane, constricting it to form a **cleavage furrow**. In an energy-using process, the cleavage furrow moves inwards, forming a region of abscission where the two cells will separate. In the photograph above, an arrow points to a centrosome, which is still visible near the nucleus.

Processes in Cells

Periodicals:
The cell cycle and mitosis

Related activities: *Cancer: Cells Out of Control, Root Cell Development*
Web links: *Mitosis in an Animal Cell*

A 2

Mitotic cell division has several purposes (below left). In multicellular organisms, mitosis repairs damaged cells and tissues, and produces the growth in an organism that allows it to reach its adult size. In unicellular organisms, and some small multicellular organisms, cell division allows organisms to reproduce asexually (as in the budding yeast cell cycle below).

The Functions of Mitosis

❶ Growth

In plants, cell division occurs in regions of **meristematic tissue**. In the plant root tip (right), the cells in the root apical meristem are dividing by mitosis to produce new cells. This elongates the root, resulting in **plant growth**.

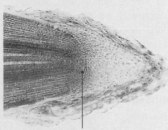

Root apical meristem

❷ Repair

Some animals, such as this skink (left), detach their limbs as a defence mechanism in a process called **autotomy**. The limbs can be **regenerated** via the mitotic process, although the tissue composition of the new limb differs slightly from that of the original.

Photo: AB Sheldon

❸ Reproduction

Mitotic division enables some animals to reproduce **asexually**. The cells of this Hydra (left) undergo mitosis, forming a 'bud' on the side of the parent organism. Eventually the bud, which is genetically identical to its parent, detaches to continue the life cycle.

Parent

The Budding Yeast Cell Cycle

Yeasts can reproduce asexually through **budding**. In *Saccharomyces cerevisiae* (baker's yeast), budding involves mitotic division in the parent cell, with the formation of a daughter cell (or bud). As budding begins, a ring of chitin stabilizes the area where the bud will appear and enzymatic activity and turgor pressure act to weaken and extrude the cell wall. New cell wall material is incorporated during this phase. The nucleus of the parent cell also divides in two, to form a daughter nucleus, which migrates into the bud. The daughter cell is genetically identical to its parent cell and continues to grow, eventually separating from the parent cell.

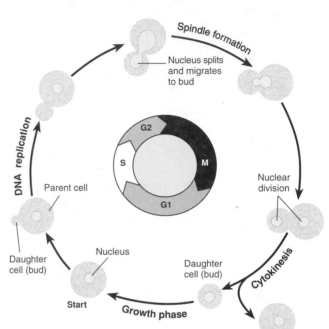

Spindle formation

Nucleus splits and migrates to bud

DNA replication

Parent cell

Daughter cell (bud)

Nucleus

Start

Growth phase

Nuclear division

Cytokinesis

Daughter cell (bud)

Parent cell

G2

S

M

G1

1. The photographs below were taken at various stages through mitosis in a plant cell. They are not in any particular order. Study the diagram on the previous page and determine the stage represented in each photograph (e.g. anaphase).

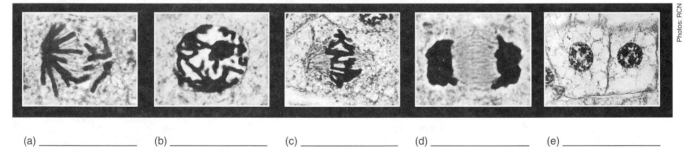

Photos: RCN

(a) _____   (b) _____   (c) _____   (d) _____   (e) _____

2. State two important changes that chromosomes must undergo before cell division can take place: _____

_____

3. Briefly summarise the stages of the cell cycle by describing what is happening at the points (**A-F**) in the diagram on the previous page:

A. _____

B. _____

C. _____

D. _____

E. _____

F. _____

Apoptosis: Programmed Cell Death

Apoptosis or programmed cell death (PCD) is a normal and necessary mechanism in multicellular organisms to trigger the death of a cell. Apoptosis has a number of crucial roles in the body, including the maintenance of adult cell numbers, and defence against damaged or dangerous cells, such as virus-infected cells and cells with DNA damage. Apoptosis also has a role in "sculpting" embryonic tissue during its development, e.g. in the formation of fingers and toes in a developing human embryo. Programmed cell death involves an orderly series of biochemical events that result in set changes in cell morphology and end in cell death. The process is carried out in such a way as to safely dispose of cell remains and fragments. This is in contrast to another type of cell death, called **necrosis**, in which traumatic damage to the cell results in spillage of cell contents. Apoptosis is tightly regulated by a balance between the factors that promote cell survival and those that trigger cell death. An imbalance between these regulating factors leads to defective apoptotic processes and is implicated in an extensive variety of diseases. For example, low rates of apoptosis result in uncontrolled proliferation of cells and cancers.

Stages in Apoptosis

Apoptosis is a normal cell suicide process in response to particular cell signals. It characterized by an overall compaction (shrinking) of the cell and its nucleus, and the orderly dissection of chromatin by endonucleases. Death is finalized by a rapid engulfment of the dying cell by phagocytosis. The cell contents remain membrane-bound and there is no inflammation.

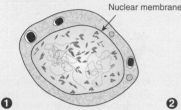

Nuclear membrane

Chromatin

❶ The cell shrinks and loses contact with neighboring cells. The chromatin condenses and begins to degrade.

❷ The nuclear membrane degrades. The cell loses volume. The chromatin clumps into **chromatin bodies**.

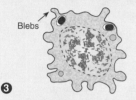

Blebs

Organelle

❸ **Zeiosis**: The plasma membrane forms bubble like **blebs** on its surface.

❹ The nucleus collapses, but many membrane-bound organelles are unaffected.

Nucleus

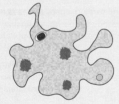

Apoptotic body

❺ The nucleus breaks up into spheres and the DNA breaks up into small fragments.

❻ The cell breaks into numerous **apoptotic bodies**, which are quickly resorbed by phagocytosis.

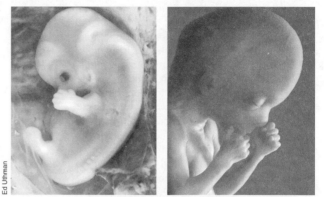

Ed Uhman

In humans, the mesoderm initially formed between the fingers and toes is removed by apoptosis. 41 days after fertilization (top left), the digits of the hands and feet are webbed, making them look like small paddles. Apoptosis selectively destroys this superfluous webbing and, later in development, each of the digits can be individually seen (right).

Regulating Apoptosis

Apoptosis is a complicated and tightly controlled process, distinct from cell necrosis (uncontrolled cell death), when the cell contents are spilled. Apoptosis is regulated through both:

Positive signals, which prevent apoptosis and allow a cell to function normally. They include:
▶ interleukin-2
▶ bcl-2 protein and growth factors

Interleukin-2 is a positive signal for cell survival. Like other signalling molecules, it binds to surface receptors on the cell to regulate metabolism.

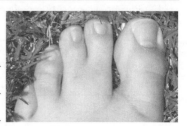

Negative signals (death activators), which trigger the changes leading to cell death. They include:
▶ inducer signals generated from within the cell itself in response to stress, e.g. DNA damage or cell starvation.
▶ signalling proteins and peptides such as lymphotoxin.

1. The photograph (right) depicts a condition called syndactyly. Explain what might have happened during development to result in this condition:

_____

_____

2. Describe one difference between apoptosis and necrosis: _____

_____

3. Describe two situations, other than digit formation in development, in which apoptosis plays a crucial role:

(a) _____

_____

(b) _____

_____

Periodicals:
What is cell suicide?

Related activities: Cancer: Cells Out of Control
Web links: Apoptosis: Dance of Death

Processes in Cells

A 2

Cancer: Cells out of Control

Normal cells do not live forever. Under certain circumstances, cells are programed to die, particularly during development. Cells that become damaged beyond repair will normally undergo this programed cell death (called **apoptosis** or **cell suicide**). Cancer cells evade this control and become immortal, continuing to divide regardless of any damage incurred. **Carcinogens** are agents capable of causing cancer. Roughly 90% of carcinogens are also mutagens, i.e. they damage DNA. Chronic exposure to carcinogens accelerates the rate at which dividing cells make errors. Susceptibility to cancer is also influenced by genetic make-up. Any one or a number of cancer-causing factors (including defective genes) may interact to induce cancer.

Cancer: Cells out of Control

Cancerous transformation results from changes in the genes controlling normal cell growth and division. The resulting cells become immortal and no longer carry out their functional role. Two types of gene are normally involved in controlling the cell cycle: **proto-oncogenes**, which start the cell division process and are essential for normal cell development, and **tumor-suppressor** genes, which switch off cell division. In their normal form, both kinds of genes work as a team, enabling the body to perform vital tasks such as repairing defective cells and replacing dead ones. However mutations in these genes can disrupt these checks and balances. Proto-oncogenes, through mutation, can give rise to **oncogenes**; genes that lead to uncontrollable cell division. Mutations to tumor-suppressor genes initiate most human cancers. The best studied tumor-suppressor gene is **p53**, which encodes a protein that halts the cell cycle so that DNA can be repaired before division.

The panel, right, shows the mutagenic action of some selected carcinogens on four of five codons of the **p53 gene**.

Features of Cancer Cells

The diagram right shows a single **lung cell** that has become cancerous. It no longer carries out the role of a lung cell, and instead takes on a parasitic lifestyle, taking from the body what it needs in the way of nutrients and contributing nothing in return. The rate of cell division is greater than in normal cells in the same tissue because there is no *resting phase* between divisions.

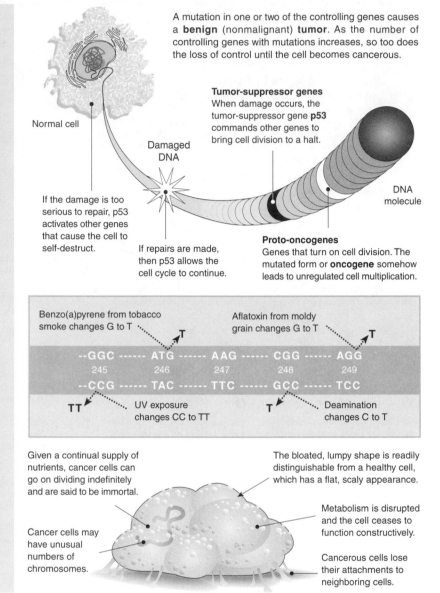

A mutation in one or two of the controlling genes causes a **benign** (nonmalignant) tumor. As the number of controlling genes with mutations increases, so too does the loss of control until the cell becomes cancerous.

Normal cell

Damaged DNA

If the damage is too serious to repair, p53 activates other genes that cause the cell to self-destruct.

If repairs are made, then p53 allows the cell cycle to continue.

Tumor-suppressor genes
When damage occurs, the tumor-suppressor gene **p53** commands other genes to bring cell division to a halt.

DNA molecule

Proto-oncogenes
Genes that turn on cell division. The mutated form or **oncogene** somehow leads to unregulated cell multiplication.

Benzo(a)pyrene from tobacco smoke changes G to T

Aflatoxin from moldy grain changes G to T

--GGC ------ ATG ------ AAG ------ CGG ------ AGG
245   246   247   248   249
--CCG ------ TAC ------ TTC ------ GCC ------ TCC

UV exposure changes CC to TT

Deamination changes C to T

Given a continual supply of nutrients, cancer cells can go on dividing indefinitely and are said to be immortal.

Cancer cells may have unusual numbers of chromosomes.

The bloated, lumpy shape is readily distinguishable from a healthy cell, which has a flat, scaly appearance.

Metabolism is disrupted and the cell ceases to function constructively.

Cancerous cells lose their attachments to neighboring cells.

1. Explain how cancerous cells differ from normal cells: _____

_____

2. Explain how the cell cycle is normally controlled, including reference to the role of **tumor-suppressor genes**:

_____

_____

_____

_____

3. With reference to the role of **oncogenes**, explain how the normal controls over the cell cycle can be lost:

_____

_____

_____

Related activities: The Causes of Mutation

Periodicals:
Rebels without a cause

Estrogen, Transcription, and Cancer

Estrogen, a hormone found in high levels in premenopausal women, has long been implicated in immune system function. **Autoimmune diseases**, in which the body attacks its own tissues, tend to be more common in women than in men and normal immune responses to infection are slightly faster in women than in men. Both of these responses have been linked to levels of **estrogen** in the blood. We now know that estrogen plays an important role in the immune system by acting as a switch to turn on the gene involved in antibody production. As it happens, activation of that gene is also linked to immune system cancer.

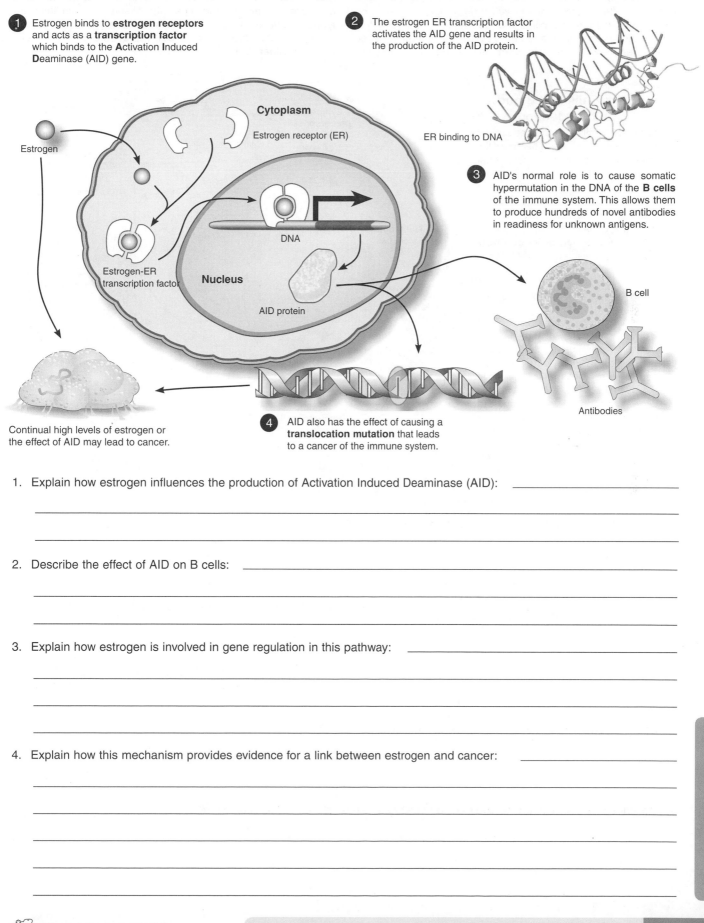

1 Estrogen binds to **estrogen receptors** and acts as a **transcription factor** which binds to the **A**ctivation **I**nduced **D**eaminase (AID) gene.

2 The estrogen ER transcription factor activates the AID gene and results in the production of the AID protein.

ER binding to DNA

3 AID's normal role is to cause somatic hypermutation in the DNA of the **B cells** of the immune system. This allows them to produce hundreds of novel antibodies in readiness for unknown antigens.

Cytoplasm

Estrogen receptor (ER)

Estrogen

DNA

Estrogen-ER transcription factor

Nucleus

AID protein

B cell

Antibodies

Continual high levels of estrogen or the effect of AID may lead to cancer.

4 AID also has the effect of causing a **translocation mutation** that leads to a cancer of the immune system.

1. Explain how estrogen influences the production of Activation Induced Deaminase (AID): _____
   _____
   _____

2. Describe the effect of AID on B cells: _____
   _____
   _____

3. Explain how estrogen is involved in gene regulation in this pathway: _____
   _____
   _____
   _____

4. Explain how this mechanism provides evidence for a link between estrogen and cancer: _____
   _____
   _____
   _____
   _____

Related activities: Cancer: Cells Out of Control

EA 3

Processes in Cells

Stem Cells

Stem cells are undifferentiated cells found in multicellular organisms. They are characterized by two features. The first, **self renewal**, is the ability to undergo numerous cycles of cell division while maintaining an unspecialized state. The second, **potency**, is the ability to differentiate into specialized cells. **Totipotent** cells, produced in the first few divisions of a fertilized egg, can differentiate into any cell type, embryonic or extra-embryonic. **Pluripotent cells** are descended from totipotent cells and can give rise to any of the cells derived from the three germ layers (endoderm, mesoderm, and ectoderm). Embryonic stem cells at the blastocyst stage and fetal stem cells are pluripotent. Adult (somatic) stem cells are termed **multipotent**. They are undifferentiated cells found among differentiated cells in a tissue or organ. These cells can give rise to several other cell types, but those types are limited mainly to the cells of the blood, heart, muscle and nerves. The primary roles of adult stem cells are to maintain and repair the tissue in which they are found. A potential use of stem cells is making cells and tissues for medical therapies, such as **cell replacement therapy** and **tissue engineering**.

Stem Cells and Blood Cell Production

New blood cells are produced in the red bone marrrow, which becomes the main site of blood production after birth, taking over from the fetal liver. All types of blood cells develop from a single cell type: called a **multipotent stem cell** or hemocytoblast. These cells are capable of mitosis and of differentiation into 'committed' precursors of each of the main types of blood cell.

Each of the different cell lines is controlled by a specific **growth factor**. When a stem cell divides, one of its daughters remains a stem cell, while the other becomes a precursor cell, either a **lymphoid cell** or **myeloid cell**. These cells continue to mature into the various type of blood cells, developing their specialized features and characteristic roles as they do so.

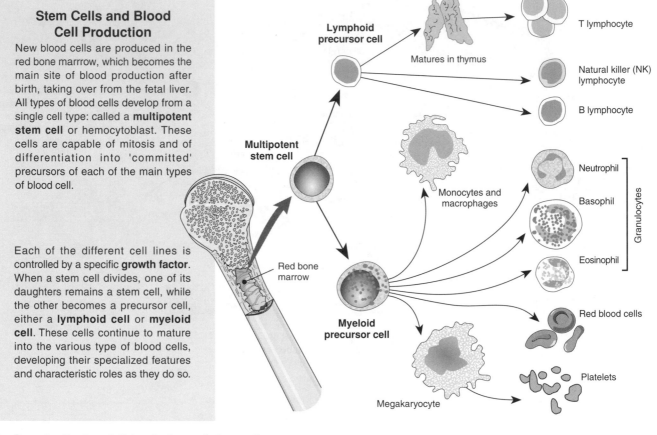

1. Describe the two defining features of stem cells:

    (a) _____

    (b) _____

2. Distinguish between embryonic stem cells and adult stem cells with respect to their **potency** and their potential applications in medical technologies:

    _____

    _____

    _____

3. Using an example, explain the purpose of stem cells in an adult: _____

    _____

    _____

4. Describe one potential advantage of using embryonic stem cells for tissue engineering technology: _____

    _____

    _____

    _____

Differentiation of Human Cells

A zygote commences development by dividing into a small ball of a few dozen identical cells called **embryonic stem cells**. These cells start to take different developmental paths to become specialized cells such as nerve stem cells which means they can no longer produce any other type of cell. **Differentiation** is cell specialization that occurs at the end of a developmental pathway.

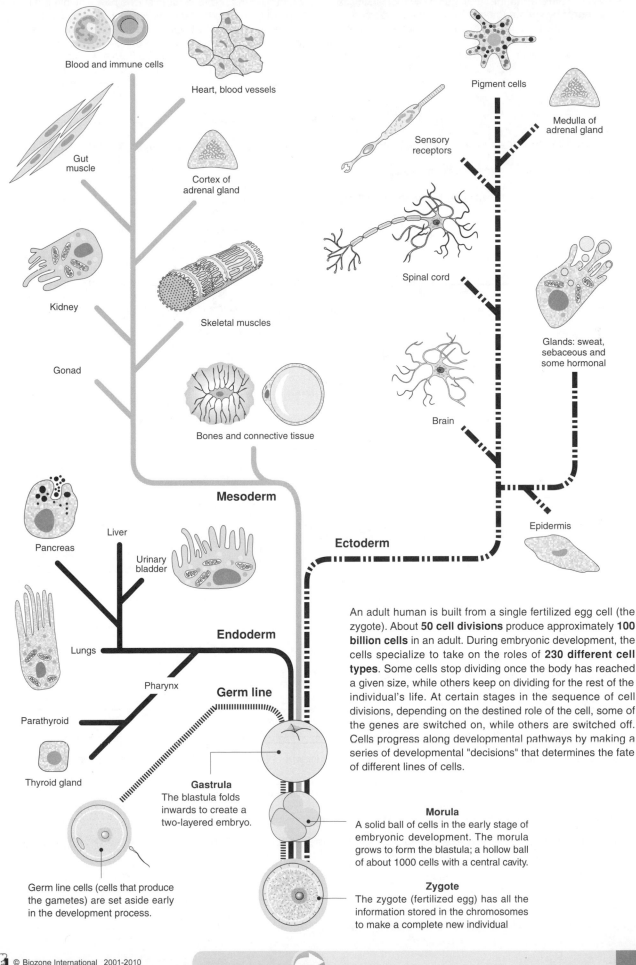

Blood and immune cells

Heart, blood vessels

Pigment cells

Medulla of adrenal gland

Gut muscle

Cortex of adrenal gland

Sensory receptors

Kidney

Skeletal muscles

Spinal cord

Glands: sweat, sebaceous and some hormonal

Gonad

Bones and connective tissue

Brain

Mesoderm

Pancreas

Liver

Urinary bladder

Ectoderm

Epidermis

Endoderm

Lungs

Pharynx

Germ line

Parathyroid

Thyroid gland

Gastrula
The blastula folds inwards to create a two-layered embryo.

An adult human is built from a single fertilized egg cell (the zygote). About **50 cell divisions** produce approximately **100 billion cells** in an adult. During embryonic development, the cells specialize to take on the roles of **230 different cell types**. Some cells stop dividing once the body has reached a given size, while others keep on dividing for the rest of the individual's life. At certain stages in the sequence of cell divisions, depending on the destined role of the cell, some of the genes are switched on, while others are switched off. Cells progress along developmental pathways by making a series of developmental "decisions" that determines the fate of different lines of cells.

Morula
A solid ball of cells in the early stage of embryonic development. The morula grows to form the blastula; a hollow ball of about 1000 cells with a central cavity.

Germ line cells (cells that produce the gametes) are set aside early in the development process.

Zygote
The zygote (fertilized egg) has all the information stored in the chromosomes to make a complete new individual

Periodicals:
Cell differentiation

Related activities: Stem Cells, Human Cell Specialization,
Plant Cell Specialization

Processes in Cells

RA 2

Development is the process of progressive change through the lifetime of an organism. Part of this process involves growth (increase in size) and cell division (to generate the multicellular body). Cellular **differentiation** (the generation of specialized cells) and **morphogenesis** (the creation of the shape and form of the body) are also part of development. Differentiation defines the specific structure and function of a cell. As development proceeds, the possibilities available to individual cells become fewer, until each cell's **fate** is determined. The tissues and organs making up the body form from the aggregation and

organization of these differentiated cells. In animals, the final body form is the result of cell migration and the programed death of certain cells (**apoptosis**) during embryonic development. The diagram on the previous page shows how a single fertilized egg (zygote) gives rise to the large number of specialized cell types that make up the adult human body. The morula, blastula, and gastrula stages mentioned at the bottom of the diagram show the early development of the embryo from the zygote. The gastrula gives rise to the three layers of cells (ectoderm, mesoderm, and endoderm), from which specific cell types develop.

1. State how many different types of cell are found in the human body: _____

2. State approximately how many cell divisions take place from fertilized egg (zygote) to produce an adult: _____

3. State approximately how many cells make up an adult human body: _____

4. Name one cell type that continues to divide throughout a person's lifetime: _____

_____

5. Name one cell type that does not continue to divide throughout a person's lifetime: _____

_____

6. Germ line cells diverge (become isolated) from other cells at a very early stage in embryonic development.

   (a) Explain what the **germ line** is: _____

   _____

   _____

   (b) Explain why it is necessary for the germ line to become separated at such an early stage of development:

   _____

   _____

   _____

7. Cloning whole new organisms is possible by taking a nucleus from a cell during the blastula stage of embryonic development and placing it into an egg cell that has had its own nucleus removed.

   (a) Explain what a **clone** is: _____

   _____

   _____

   (b) Explain why the cell required for cloning needs to be taken at such an early stage of embryonic development:

   _____

   _____

   _____

8. Cancer cells are particularly damaging to organisms. Explain what has happened to a cell that has become cancerous:

   _____

   _____

9. Explain the genetic events that enable so many different cell types to arise from one unspecialized cell (the zygote):

   _____

   _____

   _____

Stem Cells and Tissue Engineering

The properties of self renewal and potency that characterize stem cells make them highly suitable for use in many areas of cell and tissue research. Potentially, they could be used to study human development and gene regulation, to test the safety and effectiveness of new drugs and vaccines, to produce monoclonal antibodies, and to supply cells to treat diseased and damaged tissue. Engineered skin tissue is already widely used in treating the victims of burns and other tissue injuries. These technologies require a disease-free and plentiful supply of cells of specific types. The greatest potential for stem cell therapies comes from embryonic stem cells, because they are pluripotent, but there are many ethical issues to address when blastocysts are used as a source of cells. However, even adult (somatic) stem cells, obtained from umbilical cord blood and bone marrow, have great potential, and could be used to treat a variety of diseases including leukemia, lymphomas, anemia, and a range of congenital disorders. Although therapeutic **stem cell cloning** is still in its very early stages, it has the potential to provide specific cell types to treat any number of tissue disorders and injuries. The recent decision of President Barack Obama's government to lift the ban on stem cell research has been welcomed by many as a first step in developing new therapies.

Embryonic Stem Cells

Embryonic stem cells (ESC) are stem cells taken from **blastocysts** (below). Blastocysts are embryos which are about five days old and consist of a hollow ball containing 50-150 cells. In general, ESCs come from embryos which have been fertilized *in vitro* at fertilization clinics and have then been donated for research.

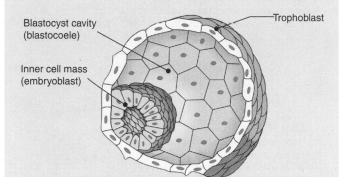

Blastocyst cavity (blastocoele)

Trophoblast

Inner cell mass (embryoblast)

Cells derived from the inner cell mass are **pluripotent**; they can become any cells of the body, with the exception of placental cells. When grown *in vitro*, without any stimulation for differentiation, these cells retain their pluripotency through multiple cell divisions. As a consequence of this, **ESC therapies** have potential use in regenerative medicine and tissue replacement. By manipulating the culture conditions, scientists are able to select and control the type of cells grown (e.g. heart cells). This ability could allow for specific cell types to be grown to treat specific diseases or replace damaged tissue. However, no ESC treatments have been approved to date due to ethical issues.

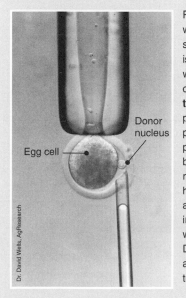

Egg cell

Donor nucleus

Dr. David Wells, AgResearch

Poor **histocompatibility**, in which the recipient's immune system rejects foreign cells, is one of the major difficulties with transplantation. Stem cell cloning (also called **therapeutic cloning**) provides a way around this problem. Stem cell cloning produces cells that have been derived from the recipient and are therefore histocompatible. Such an approach would mean immunosuppressant drugs would no longer be required. Diseases such as leukemia and Parkinson's could be treated using this technique.

Embryonic Stem Cell Cloning

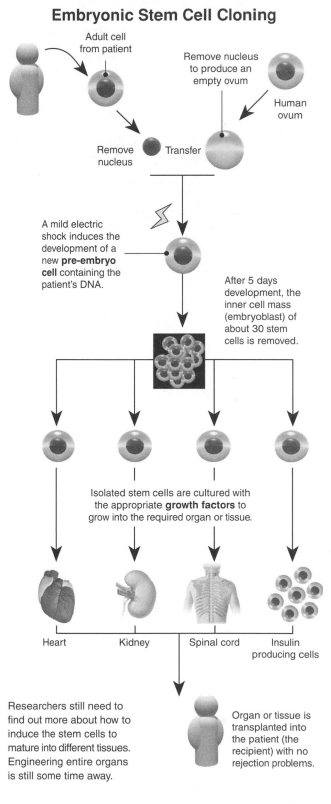

Adult cell from patient

Remove nucleus to produce an empty ovum

Human ovum

Remove nucleus

Transfer

A mild electric shock induces the development of a new **pre-embryo cell** containing the patient's DNA.

After 5 days development, the inner cell mass (embryoblast) of about 30 stem cells is removed.

Isolated stem cells are cultured with the appropriate **growth factors** to grow into the required organ or tissue.

Heart

Kidney

Spinal cord

Insulin producing cells

Researchers still need to find out more about how to induce the stem cells to mature into different tissues. Engineering entire organs is still some time away.

Organ or tissue is transplanted into the patient (the recipient) with no rejection problems.

© Biozone International 2001-2010

Periodicals:
Grown to order,
Embryonic stem cells

Related activities: *Stem Cells*
Web links: *Stem Cell Resources*

RA 3

Processes in Cells

Engineering a Living Skin

New technologies such as cell replacement therapy and tissue engineering require a disease-free and plentiful supply of cells of specific types. Tissue engineering, for example, involves inducing living cells to grow on a scaffold of natural or synthetic material to produce a three-dimensional tissue such as bone or skin.

In 1998, an artificial skin called **Apligraf** became the first product of this type to be approved for use as a biomedical device. It is now widely used in place of skin grafts to treat diabetic ulcers and burns, with the patient's own cells and tissues helping to complete the biological repair. Producing Apligraf is a three stage process (right), which results in a bilayered, living structure capable of stimulating wound repair through its own growth factors and proteins. The cells used to start the culture are usually obtained from discarded neonatal foreskins collected after circumcision. The key to future tissue engineering will be the developments in stem cell research. The best source of stem cells is from very early embryos, but some adult tissues (e.g. bone marrow) also contain stem cells.

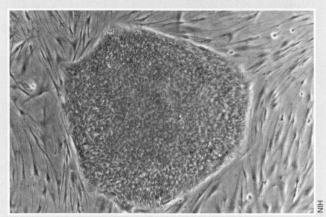

Human embryonic stem cells (ESCs) growing on mouse embryonic fibroblasts. The mouse fibroblasts act as feeder cells for the culture, releasing nutrients and providing a surface for the ESCs to grow on.

Human dermal cells

Collagen

Day 0
Undifferentiated human dermal cells (fibroblasts) are combined with a gel containing **collagen**, the primary protein of skin. The dermal cells move through the gel, rearranging the collagen and producing a fibrous, living matrix similar to the natural dermis.

Step 1
Form the lower dermal layer

Human epidermal cells

Day 6
Human epidermal cells (called **keratinocytes**) are placed on top of the dermal layer. These cells multiply to cover the dermal layer.

Step 2
Form the upper epidermal layer

Air exposure

Day 10
Exposing the culture to air prompts the epidermal cells to form the outer protective (keratinized) layer of skin. The final size of the Apligraf product is about 75 mm and, from this, tens of thousands of pieces can be made.

Step 3
Form the outer layer

1. Describe the benefits of using a tissue engineered skin product, such as Apligraf, to treat wounds that require grafts:

_____

_____

2. (a) Describe one of the major difficulties with transplantation of cells: _____

_____

(b) Explain one way in which this problem could be overcome: _____

_____

_____

(c) Describe one of the ethical concerns associated with this solution: _____

_____

3. Discuss some of the difficulties which must be overcome when growing *in vitro* tissue cultures:

_____

_____

_____

_____

_____

Human Cell Specialization

Animal cells are often specialized to perform particular functions. The eight specialized cell types shown below are representative of some 230 different cell types in humans. Each has specialized features that suit it to performing a specific role.

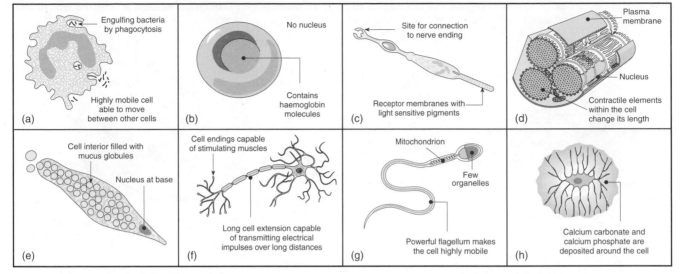

(a) Engulfing bacteria by phagocytosis / Highly mobile cell able to move between other cells

(b) No nucleus / Contains haemoglobin molecules

(c) Site for connection to nerve ending / Receptor membranes with light sensitive pigments

(d) Plasma membrane / Nucleus / Contractile elements within the cell change its length

(e) Cell interior filled with mucus globules / Nucleus at base

(f) Cell endings capable of stimulating muscles / Long cell extension capable of transmitting electrical impulses over long distances

(g) Mitochondrion / Few organelles / Powerful flagellum makes the cell highly mobile

(h) Calcium carbonate and calcium phosphate are deposited around the cell

1. Identify each of the cells (b) to (h) pictured above, and describe their **specialized features** and **role** in the body:

   (a) Type of cell: _Phagocytic white blood cell (neutrophil)_

   Specialized features: _Engulfs bacteria and other foreign material by phagocytosis_

   Role of cell within body: _Destroys pathogens and other foreign material as well as cellular debris_

   (b) Type of cell: _____

   Specialized features: _____

   Role of cell within body: _____

   (c) Type of cell: _____

   Specialized features: _____

   Role of cell within body: _____

   (d) Type of cell: _____

   Specialized features: _____

   Role of cell within body: _____

   (e) Type of cell: _____

   Specialized features: _____

   Role of cell within body: _____

   (f) Type of cell: _____

   Specialized features: _____

   Role of cell within body: _____

   (g) Type of cell: _____

   Specialized features: _____

   Role of cell within body: _____

   (h) Type of cell: _____

   Specialized features: _____

   Role of cell within body: _____

Related activities: Differentiation of Human Cells, Stem Cells

RA 2

Processes in Cells

Plant Cell Specialization

Plants show a wide variety of cell types. The vegetative plant body consists of three organs: stems, leaves, and roots. Flowers, fruits, and seeds comprise additional organs that are concerned with reproduction. The eight cell types illustrated below are representatives of these plant organ systems. Each has structural or physiological features that set it apart from the other cell types. The differentiation of cells enables each specialized type to fulfill a specific role in the plant.

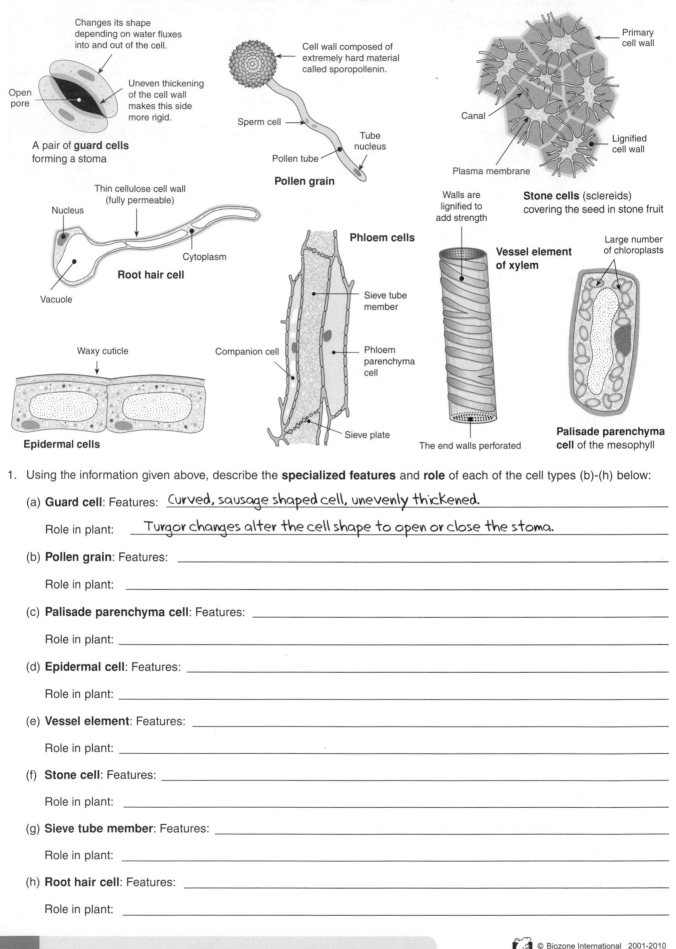

Changes its shape depending on water fluxes into and out of the cell.

Open pore

Uneven thickening of the cell wall makes this side more rigid.

A pair of **guard cells** forming a stoma

Cell wall composed of extremely hard material called sporopollenin.

Sperm cell

Tube nucleus

Pollen tube

Pollen grain

Primary cell wall

Canal

Lignified cell wall

Plasma membrane

Stone cells (sclereids) covering the seed in stone fruit

Thin cellulose cell wall (fully permeable)

Nucleus

Cytoplasm

Root hair cell

Vacuole

Walls are lignified to add strength

Phloem cells

Sieve tube member

Companion cell

Phloem parenchyma cell

Sieve plate

Vessel element of xylem

The end walls perforated

Large number of chloroplasts

Waxy cuticle

Epidermal cells

Palisade parenchyma cell of the mesophyll

1. Using the information given above, describe the **specialized features** and **role** of each of the cell types (b)-(h) below:

   (a) **Guard cell**: Features: *Curved, sausage shaped cell, unevenly thickened.*

   Role in plant: *Turgor changes alter the cell shape to open or close the stoma.*

   (b) **Pollen grain**: Features: _____

   Role in plant: _____

   (c) **Palisade parenchyma cell**: Features: _____

   Role in plant: _____

   (d) **Epidermal cell**: Features: _____

   Role in plant: _____

   (e) **Vessel element**: Features: _____

   Role in plant: _____

   (f) **Stone cell**: Features: _____

   Role in plant: _____

   (g) **Sieve tube member**: Features: _____

   Role in plant: _____

   (h) **Root hair cell**: Features: _____

   Role in plant: _____

Related activities: Root Cell Development

Levels of Organization

Organization and the emergence of novel properties in complex systems are two of the defining features of living organisms. Organisms are organized according to a hierarchy of structural levels (below), each level building on the one below it. At each level, novel properties emerge that were not present at the simpler level. Hierarchical organization allows specialized cells to group together into tissues and organs to perform a particular function. This improves efficiency of function in the organism.

In the spaces provided for each question below, assign each of the examples listed to one of the levels of organisation as indicated.

1. **Animals**: *adrenaline, blood, bone, brain, cardiac muscle, cartilage, collagen, DNA, heart, leukocyte, lysosome, mast cell, nervous system, neuron, phospholipid, reproductive system, ribosomes, Schwann cell, spleen, squamous epithelium.*

(a) Molecular level: _____

_____

(b) Organelles: _____

_____

(c) Cells: _____

_____

(d) Tissues: _____

_____

(e) Organs: _____

_____

(f) Organ system: _____

_____

_____

2. **Plants**: *cellulose, chloroplasts, collenchyma, companion cells, DNA, epidermal cell, fibers, flowers, leaf, mesophyll, parenchyma, pectin, phloem, phospholipid, ribosomes, roots, sclerenchyma, tracheid.*

(a) Molecular level: _____

_____

(b) Organelles: _____

_____

(c) Cells: _____

_____

(d) Tissues: _____

_____

(e) Organs: _____

_____

_____

MOLECULAR LEVEL

Atoms and molecules form the most basic level of organization. This level includes all the chemicals essential for maintaining life e.g. water, ions, fats, carbohydrates, amino acids, proteins, and nucleic acids.

ORGANELLE LEVEL

Many diverse molecules may associate together to form complex, highly specialized structures within cells called cellular organelles e.g. mitochondria, Golgi apparatus, endoplasmic reticulum, chloroplasts.

Golgi apparatus

Mitochondria

CELLULAR LEVEL

Cells are the basic structural and functional units of an organism. Each cell type has a different structure and function; the result of cellular differentiation during development.

Animal examples *include: epithelial cells, osteoblasts, muscle fibers.*

Plant examples *include: sclereids, xylem vessels, sieve tubes.*

Epithelial cells

TISSUE LEVEL

Tissues are composed of groups of cells of similar structure that perform a particular, related function.

Animal examples *include: epithelial tissue, bone, muscle.*

Plant examples *include: phloem, chlorenchyma, endodermis, xylem.*

Epithelial tissue of the glomerulus

ORGAN LEVEL

Organs are structures of definite form and structure, made up of two or more tissues.

Animal examples *include: heart, lungs, brain, stomach, kidney.*

Plant examples *include: leaves, roots, storage organs, ovary.*

Kidney

ORGAN SYSTEM LEVEL

In animals, organs form parts of even larger units known as **organ systems**. An organ system is an association of organs with a common function, e.g. digestive system, cardiovascular system, and the urinary system. In all, eleven organ systems make up the **organism**.

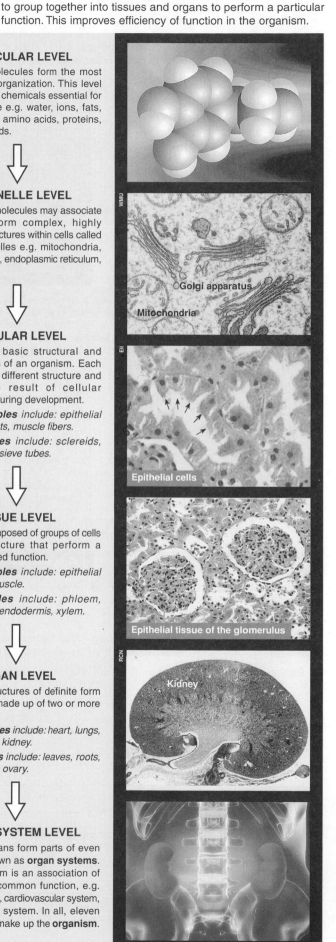

Processes in Cells

A 2

Animal Tissues

The study of tissues (plant or animal) is called **histology**. The cells of a tissue, and their associated extracellular substances, e.g. collagen, are grouped together to perform particular functions. Tissues improve the efficiency of operation because they enable tasks to be shared amongst various specialized cells. **Animal tissues** can be divided into four broad groups: **epithelial tissues**, **connective tissues**, **muscle**, and **nervous**

tissues. Organs usually consist of several types of tissue. The heart mostly consists of cardiac muscle tissue, but also has epithelial tissue, which lines the heart chambers to prevent leaking, connective tissue for strength and elasticity, and nervous tissue, in the form of neurons, which direct the contractions of the cardiac muscle. The features of some of the more familiar animal tissues are described below.

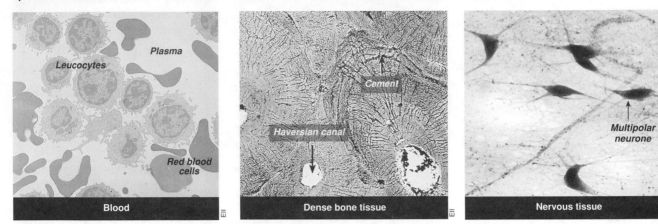

Blood   Dense bone tissue   Nervous tissue

Connective tissue is the major supporting tissue of the animal body. It comprises cells, widely dispersed in a semi-fluid matrix. Connective tissues bind other structures together and provide support, and protection against damage, infection, or heat loss. Connective tissues include dentine (teeth), adipose (fat) tissue, bone (above) and cartilage, and the tissues around the body's organs and blood vessels. Blood (above, left) is a special type of liquid tissue, comprising cells floating in a liquid matrix.

Nervous tissue contains densely packed nerve cells (neurones) which are specialized for the transmission of nerve impulses. Associated with the neurones there may also be supporting cells and connective tissue containing blood vessels.

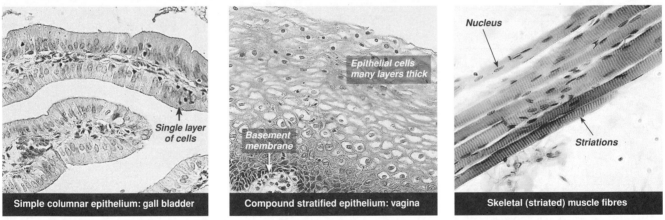

Simple columnar epithelium: gall bladder   Compound stratified epithelium: vagina   Skeletal (striated) muscle fibres

Epithelial tissue is organized into single (above, left) or layered (above) sheets. It lines internal and external surfaces (e.g. blood vessels, ducts, gut lining) and protects the underlying structures from wear, infection, and/or pressure. Epithelial cells rest on a basement membrane of fibres and collagen and are held together by a carbohydrate-based "glue". The cells may also be specialized for absorption, secretion, or excretion. Examples: stratified (compound) epithelium of vagina, ciliated epithelium of respiratory tract, cuboidal epithelium of kidney ducts, and the columnar epithelium of the intestine.

Muscle tissue consists of very highly specialized cells called fibres, held together by connective tissue. The three types of muscle in the body are cardiac muscle, skeletal muscle (above), and smooth muscle. Muscles bring about both voluntary and involuntary (unconscious) body movements.

1. Explain how the development of tissues improves functional efficiency: _____

_____

_____

2. Describe the general functional role of each of the following broad tissue types:

    (a) Epithelial tissue: _____ (c) Muscle tissue: _____

    (b) Nervous tissue: _____ (d) Connective tissue: _____

3. Identify the particular features that contribute to the particular functional role of each of the following tissue types:

    (a) Muscle tissue: _____

       _____

    (b) Nervous tissue: _____

       _____

Plant Tissues

Plant tissues are divided into two groups: simple and complex. **Simple tissues** contain only one cell type and form packing and support tissues. **Complex tissues** contain more than one cell type and form the conducting and support tissues of plants. Tissues are in turn grouped into tissue systems which make up the plant body. Vascular plants have three systems; the dermal, vascular, and ground tissue systems. The **dermal system** is the outer covering of the plant providing protection and reducing water loss. **Vascular tissue** provides the transport system by which water and nutrients are moved through the plant. The **ground tissue** system, which makes up the bulk of a plant, is made up mainly of simple tissues such as parenchyma, and carries out a wide variety of roles within the plant including photosynthesis, storage, and support.

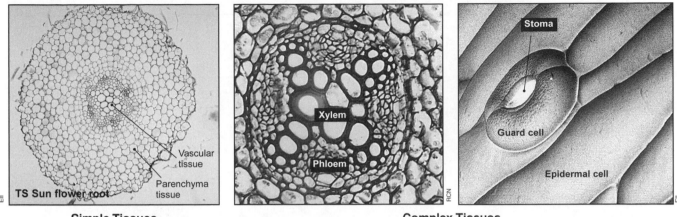

TS Sun flower root — Vascular tissue, Parenchyma tissue

Xylem, Phloem

Stoma, Guard cell, Epidermal cell

Simple Tissues

Simple tissues consists of only one or two cell types. **Parenchyma tissue** is the most common and involved in storage, photosynthesis, and secretion. **Collenchyma tissue** comprises thick-walled collenchyma cells alternating with layers of intracellular substances (pectin and cellulose) to provide flexible support. The cells of **sclerenchyma** tissue (fibers and sclereids) have rigid cell walls which provide support.

Complex Tissues

Xylem and phloem tissue (above left), which together make up the plant **vascular tissue** system, are complex tissues. Each comprises several tissue types including tracheids, vessel members, parenchyma and fibers in xylem, and sieve tube members, companion cells, parenchyma and sclerenchyma in phloem. **Dermal tissue** is also complex tissue and covers the outside of the plant. The composition of dermal tissue varies depending upon its location on the plant. Root epidermal tissue consist of epidermal cells which extend to root hairs (**trichomes**) for increasing surface area. In contrast, the epidermal tissue of leaves (above right) are covered by a waxy cuticle to reduce water loss, and specialized guard cells regulate water intake via the stomata (pores in the leaf through which gases enter and leave the leaf tissue).

1. The table below lists the major types of simple and complex plant tissue. Complete the table by filling in the role each of the tissue types plays within the plant. The first example has been completed for you.

Simple Tissue	Cell Type(s)	Role within the Plant
Parenchyma	Parenchyma cells	Involved in respiration, photosynthesis, storage and secretion.
Collenchyma		
Sclerenchyma		
Root endodermis	Endodermal cells	
Pericycle		
Complex Tissue		
Leaf mesophyll	Spongy mesophyll cells, palisade mesophyll cells	
Xylem		
Phloem		
Epidermis		

Related activities: Levels of Organization, Xylem, Phloem
Web links: Photographic Atlas of Plant Anatomy

RA 2

Processes in Cells

Root Cell Development

In plants, cell division for growth (mitosis) is restricted to growing tips called **meristematic** tissue. These are located at the tips of every stem and root. This is unlike mitosis in a growing animal where cell divisions can occur all over the body. The diagram below illustrates the position and appearance of developing and growing cells in a plant root. Similar zones of development occur in the growing stem tips, which may give rise to specialized structures such as leaves and flowers.

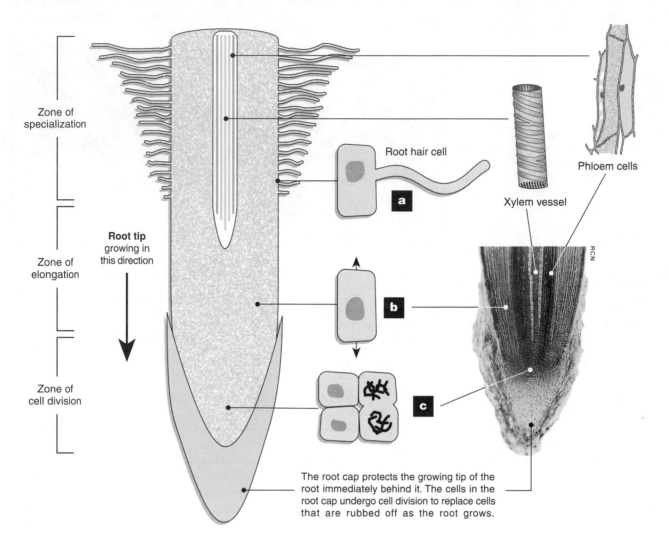

Zone of specialization

Root hair cell

a

Xylem vessel

Phloem cells

Zone of elongation

Root tip
growing in this direction

b

Zone of cell division

c

The root cap protects the growing tip of the root immediately behind it. The cells in the root cap undergo cell division to replace cells that are rubbed off as the root grows.

1. Briefly describe what is happening to the plant cells at each of the points labelled **a** to **c** in the diagram above:

    (a) _____

    (b) _____

    (c) _____

2. The light micrograph (below) shows a section of the cells of an onion root tip, stained to show up the chromosomes.

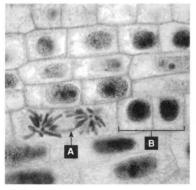

   (a) State the mitotic stage of the cell labeled **A** and explain your answer:

       _____

   (b) State the mitotic stage just completed in the cells labeled **B** and explain:

       _____

   (c) If, in this example, 250 cells were examined and 25 were found to be in the process of mitosis, state the proportion of the cell cycle occupied by mitosis:

       _____

       _____

3. Identify the cells that divide and specialize when a tree increases its girth (diameter): _____

 © Biozone International 2001-2010
Photocopying Prohibited

KEY TERMS: Mix and Match

INSTRUCTIONS: Test your vocab by matching each term to its correct definition, as identified by its preceding letter code.

ACTIVE TRANSPORT

AMPHIPATHIC

CELLULAR DIFFERENTIATION

CONCENTRATION GRADIENT

CYTOKINESIS

DIFFUSION

EMERGENT PROPERTY

ENDOCYTOSIS

EXOCYTOSIS

FACILITATED DIFFUSION

FLUID MOSAIC MODEL

GAMETE

GLYCOLIPIDS

GLYCOPROTEINS

HYPERTONIC

HYPOTONIC

ION PUMP

ISOTONIC

MITOSIS

OSMOSIS

ONCOGENE

PARTIALLY PERMEABLE MEMBRANE

PASSIVE TRANSPORT

PHAGOCYTOSIS

PLASMA MEMBRANE

PLASMOLYSIS

SURFACE AREA: VOLUME RATIO

TURGOR

WATER POTENTIAL

A Passive movement of water molecules across a partially permeable membrane down a concentration gradient.

B The model for membrane structure which proposes a double phospholipid bilayer in which proteins and cholesterol are embedded.

C A type of passive transport, facilitated by transport proteins.

D The potential energy of water per unit volume relative to pure water in reference conditions.

E The process in plant cells where the plasma membrane pulls away from the cell wall due to the loss of water through osmosis.

F The energy-requiring movement of substances across a biological membrane against a concentration gradient.

G A solution with lower solute concentration relative to another solution (across a membrane)

H Active transport in which molecules are engulfed by the plasma membrane, forming a phagosome or food vacuole within the cell.

I A novel property that arises as a result of a complex of relatively simple interactions.

J The force exerted outward on a plant cell wall by the water contained in the cell.

K Lipids with attached carbohydrates which serve as markers for cellular recognition.

L Solutions of equal solute concentration are termed this.

M This relationship determines capacity for effective diffusion in a cell.

N The phase of a cell cycle resulting in nuclear division.

O The passive movement of molecules from high to low concentration.

P The movement of substances across a biological membrane without energy expenditure.

Q A solution with higher solute concentration relative to another solution (across a membrane).

R A regulatory gene that has become mutated so that the controls it normally have over cell division are lost.

S A partially-permeable phospholipid bilayer forming the boundary of all cells.

T A transmembrane protein that moves ions across a plasma membrane against their concentration gradient.

U Active transport process by which cells take in molecules (such as proteins) from outside the cell by engulfing it with their plasma membrane.

V Gradual change in the concentration of solutes as a function of distance through the solution. In biology, this is usually results from unequal distribution of ions across a membrane.

W The process by which a less specialized cell becomes a more specialized cell type.

X Division of the cytoplasm of a eukaryotic cell to form two daughter cells.

Y Active transport process by which membrane-bound secretory vesicles fuse with the plasma membrane and release the vesicle contents into the external environment.

Z A membrane that acts selectively to allow some substances, but not others, to pass.

AA Membrane-bound proteins with attached carbohydrates, involved in cell to cell interactions.

BB Possessing both hydrophilic and hydrophobic (lipophilic) properties.

CC A haploid cell produced for the purposes of sexual reproduction.

Processes in Cells

RA 1

Cellular Energetics

KEY CONCEPTS

► ATP is the universal energy currency in cells.

► Cellular respiration and photosynthesis are important energy transformation processes.

► Both cellular respiration and photosynthesis involve the use of ATP and electron carriers.

► Cellular respiration involves the stepwise oxidation of glucose in the mitochondria.

► Photosynthesis uses light energy to fix carbon in organic compounds. It occurs in the chloroplasts.

KEY TERMS

absorption spectrum
accessory pigment
acetyl coA
action spectrum
anaerobic metabolism
ATP
bundle sheath cells
Calvin cycle
cellular respiration
chemiosmosis
chloroplast
chlorophyll
cristae
cyclic photophosphorylation
electron transport chain
ethanol
fermentation
glycolysis
grana
H$^+$ acceptor
Hatch and Slack pathway
Krebs cycle
lactic acid
light dependent phase
matrix
mitochondrion
NAD/NADP
non-cyclic photophosphorylation
oxidative phosphorylation
photolysis
photophosphorylation
photosynthesis
respiratory quotient (RQ)
respiratory substrate
ribulose bisphosphate
stroma
substrate level phosphorylation
thylakoid discs

Periodicals:
listings for this chapter are on page 380

Weblinks:
www.thebiozone.com/
weblink/SB1-2597.html

Teacher Resource CD-ROM:
Events in Biochemistry

OBJECTIVES

□ 1. Use the **KEY TERMS** to help you understand and complete these objectives.

Role of ATP pages 126-127

□ 2. Explain the universal role of **ATP** in metabolism, as illustrated by examples, e.g. active transport, anabolic reactions, movement, and thermoregulation.

□ 3. Describe the structure of ATP. Describe its synthesis from ADP and inorganic phosphate (Pi) and explain how it stores and releases its energy.

□ 4. Compare and contrast cellular respiration and photosynthesis as energy transformation processes.

Cellular Respiration pages 129-133

□ 5. Describe the structure of a mitochondrion, identifying the **matrix** and **cristae**. Identify the location of the main steps in cellular respiration: **glycolysis**, **Krebs cycle**, and **electron transport chain**.

□ 6 Identify glucose as the primary **respiratory substrate**. Explain the use of **respiratory quotient** in identifying respiratory substrates.

□ 7. Describe glycolysis and recognize it as the major anaerobic pathway in cells. State the net yield of ATP and NADH$_2$ from glycolysis.

□ 8. Describe the complete oxidation of glucose to CO$_2$, including reference to:
  • The conversion of pyruvate to acetyl-coenzyme A.
  • The stepwise oxidation of intermediates in the Krebs cycle.
  • Generation of ATP by **chemiosmosis** in the electron transport chain.
  • The role of oxygen as the terminal electron acceptor.

□ 9. Describe **fermentation** in mammalian muscle and in yeast, identifying the H$^+$ **acceptor** to each case. Compare and explain the differences in the yields of ATP from aerobic respiration (#8) and from fermentation.

Photosynthesis pages 134-139

□ 10. Describe the structure of a **chloroplast** and explain the role of **chlorophyll a** and **b**, and **accessory pigments** in light capture by green plants.

□ 11. Describe and explain **photosynthesis** in a C$_3$ plant, including reference to:
  • The generation of ATP and NADPH$_2$ in the **light dependent phase**.
  • The **Calvin cycle** and the fixation of carbon dioxide using ATP and NADPH$_2$ in the **light independent phase**.

□ 12. Describe and explain factors affecting **photosynthetic rate**.

□ 13. Describe photosynthesis in C$_4$ plants via the **Hatch and Slack** pathway. Explain the advantage of C$_4$ metabolism under conditions of drought, high temperatures and CO$_2$ limitation.

The Role of ATP in Cells

127

All organisms require energy for their metabolism. The universal energy carrier for the cell is the molecule ATP (**adenosine triphosphate**). ATP transports chemical energy within the cell for use in metabolic processes such as biosynthesis, cell division, cell signaling, thermoregulation, cell motility, and active transport. ATP can release its energy quickly; only one chemical reaction (hydrolysis of the terminal phosphate) is required. This reaction is catalyzed by the enzyme ATPase. Once ATP has released its energy, it becomes ADP (adenosine diphosphate), a low energy molecule that can be recharged by adding a phosphate. This requires energy, which is supplied by the controlled breakdown of respiratory substrates (commonly glucose) in **cellular respiration**.

Cellular Energetics

Adenosine Triphosphate (ATP)

The ATP molecule consists of three components; a purine base (**adenine**), a pentose sugar (**ribose**), and **three phosphate groups** which attach to the 5' carbon of the pentose sugar. The three dimensional structure of ATP is described below.

ATP acts as a store of energy within the cell. The bonds between the phosphate groups are **high-energy bonds**, meaning that a large amount of free energy is released when they are hydrolyzed. Typically, this hydrolysis involves the removal of one phosphate group from the ATP molecule resulting in the formation of adenosine diphosphate (ADP).

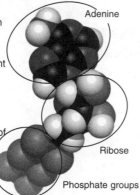

Adenine

Ribose

Phosphate groups

The Mitochondrion

Cellular respiration and ATP production occur in the mitochondria. A mitochondrion is bound by a double membrane. The inner and outer membranes are separated by an intermembrane space, compartmentalising the regions of the mitochondrion in which the different reactions of cellular respiration take place.

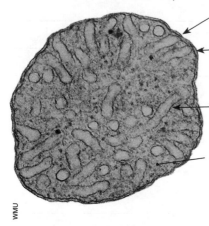

WMU

Amine oxidases on the outer membrane surface

Phosphorylases between the inner and outer membranes

ATPases on the inner membranes (the cristae)

Soluble enzymes for the Krebs cycle and fatty acid degradation floating in the matrix

ATP Powers Metabolism

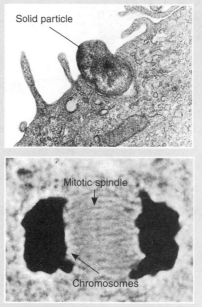

Solid particle

The energy released from the removal of a phosphate group of ATP is used to actively transport molecules and substances across the cellular membrane. **Phagocytosis** (left), which involves the engulfment of solid particles, is one such example.

Mitotic spindle

Chromosomes

Cell division (mitosis), as observed in this onion cell, requires ATP to proceed. Formation of the mitotic spindle and chromosome separation are two aspects of cell division which require energy from ATP hydrolysis to occur.

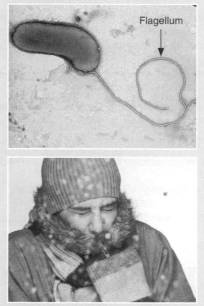

Flagellum

The hydrolysis of ATP provides the energy for motile cells to achieve movement via a tail-like structure called a flagellum. For example, the bacterium, *Helicobacter pylori* (left), is motile. Likewise, mammalian sperm must be able to move to the ovum to fertilise it.

The maintenance of body temperature requires energy. To maintain body heat, muscular activity increases (e.g. shivering, erection of body hairs). Cooling requires expenditure of energy too. For example, sweating is an energy requiring process involving secretion from glands in the skin.

In the presence of the enzyme **ATPase**, the ATP loses a phosphate.

P*i*

30.7 kJ

The energy released from the loss of a phosphate is available for work inside the cell.

Adenosine P P P

Adenosine triphosphate

Adenosine P P

Adenosine diphosphate

ATP
A high energy compound able to supply energy for metabolic activity.

Inorganic phosphate P*i*

ADP
A low energy compound that stores less energy than ATP.

Mitochondrion

© Biozone International 2001-2010
Photocopying Prohibited

Periodicals:
The role of ATP in cells
The double life of ATP

Related activities: Photosynthesis, Cellular Respiration

RA 2

Energy Transformations in a Photosynthetic Plant Cell

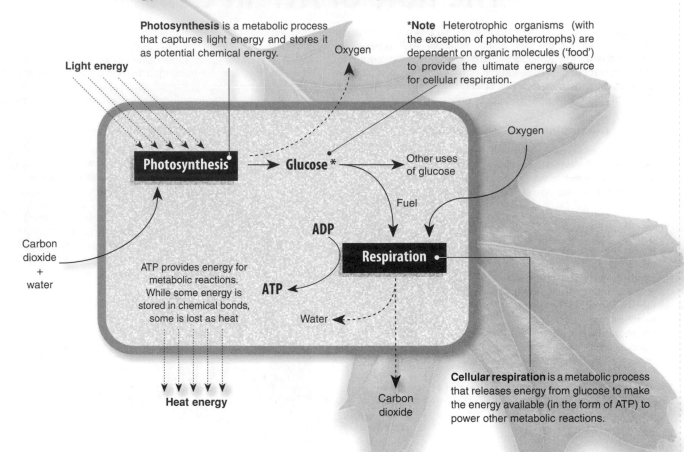

Photosynthesis is a metabolic process that captures light energy and stores it as potential chemical energy.

*Note Heterotrophic organisms (with the exception of photoheterotrophs) are dependent on organic molecules ('food') to provide the ultimate energy source for cellular respiration.

Light energy

Oxygen

Oxygen

Photosynthesis → Glucose * → Other uses of glucose

Carbon dioxide + water

ATP provides energy for metabolic reactions. While some energy is stored in chemical bonds, some is lost as heat

Fuel

ADP

Respiration

ATP

Water

Heat energy

Carbon dioxide

Cellular respiration is a metabolic process that releases energy from glucose to make the energy available (in the form of ATP) to power other metabolic reactions.

A summary of the flow of energy within a plant cell is illustrated above. Animal cells have a similar flow except the glucose is supplied by feeding rather than by photosynthesis. The energy not immediately stored in chemical bonds is lost as heat. Note the role of ATP; it is made in cellular respiration and provides the energy for metabolic reactions, including photosynthesis.

1. Describe how ATP acts as a supplier of energy to power metabolic reactions: _____

    _____

2. Name the immediate source of energy used to reform ATP from ADP molecules: _____

3. Name the process of re-energizing ADP into ATP molecules: _____

4. Name the ultimate source of energy for plants: _____

5. Name the ultimate source of energy for animals: _____

6. Explain in what way the ADP/ATP system can be likened to a rechargeable battery: _____

    _____

7. In the following table, use brief statements to contrast photosynthesis and respiration in terms of the following:

Feature	Photosynthesis	Cellular respiration
Starting materials		
Waste products		
Role of hydrogen carriers: NAD, NADP		
Role of ATP		
Overall biological role		

Measuring Respiration

Cellular Energetics

In small animals or germinating seeds, the rate of cellular respiration can be measured using a simple respirometer: a sealed unit where the carbon dioxide produced by the respiring tissues is absorbed by soda lime and the volume of oxygen consumed is detected by fluid displacement in a manometer. Germinating seeds are also often used to calculate the **respiratory quotient** (RQ): the ratio of the amount of carbon dioxide produced during cellular respiration to the amount of oxygen consumed. RQ provides a useful indication of the respiratory substrate being used.

Respiratory Substrates and RQ

The respiratory quotient (RQ) can be expressed simply as:

$$RQ = \frac{CO_2 \text{ produced}}{O_2 \text{ consumed}}$$

When pure carbohydrate is oxidized in cellular respiration, the RQ is 1.0; more oxygen is required to oxidize fatty acids (RQ = 0.7). The RQ for protein is about 0.9. Organisms usually respire a mix of substrates, giving RQ values of between 0.8 and 0.9 (see table 1, below).

Table 1: RQ values for the respiration of various substrates

RQ	Substrate
> 1.0	Carbohydrate with some anaerobic respiration
1.0	Carbohydrates e.g. glucose
0.9	Protein
0.7	Fat
0.5	Fat with associated carbohydrate synthesis
0.3	Carbohydrate with associated organic acid synthesis

Using RQ to determine respiratory substrate

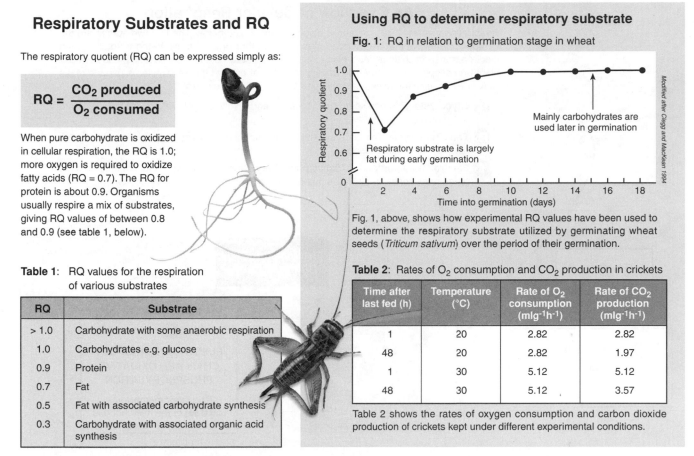

Fig. 1: RQ in relation to germination stage in wheat

Modified after Clegg and MacKean 1994

Fig. 1, above, shows how experimental RQ values have been used to determine the respiratory substrate utilized by germinating wheat seeds (*Triticum sativum*) over the period of their germination.

Table 2: Rates of O_2 consumption and CO_2 production in crickets

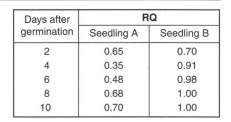

Time after last fed (h)	Temperature (°C)	Rate of O_2 consumption ($mlg^{-1}h^{-1}$)	Rate of CO_2 production ($mlg^{-1}h^{-1}$)
1	20	2.82	2.82
48	20	2.82	1.97
1	30	5.12	5.12
48	30	5.12	3.57

Table 2 shows the rates of oxygen consumption and carbon dioxide production of crickets kept under different experimental conditions.

1. Table 2 above shows the results of an experiment to measure the rates of oxygen consumption and carbon dioxide production of crickets 1 hour and 48 hours after feeding at different temperatures:

   (a) Calculate the RQ of a cricket kept at 20°C, 48 hours after feeding (show working): _____

   (b) Compare this RQ to the RQ value obtained for the cricket 1 hour after being fed (20°C). Explain the difference:

   _____

   _____

2. The RQs of two species of seeds were calculated at two day intervals after germination. Results are tabulated to the right:

   (a) Plot the change in RQ of the two species during early germination:

   (b) Explain the values in terms of the possible substrates being respired:

   _____

   _____

   _____

   _____

   _____

   _____

   _____

   _____

Days after germination	RQ	
	Seedling A	Seedling B
2	0.65	0.70
4	0.35	0.91
6	0.48	0.98
8	0.68	1.00
10	0.70	1.00

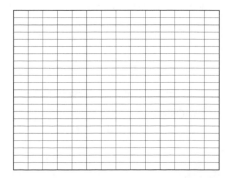

Cellular Respiration

Cellular respiration is the process by which organisms break down energy rich molecules (e.g. glucose) to release the energy in a usable form (ATP). All living cells respire in order to exist, although the substrates they use may vary. **Aerobic respiration** requires oxygen. Forms of cellular respiration that do not require oxygen are said to be **anaerobic**. Some plants and animals can generate ATP anaerobically for short periods of time. Other organisms use only anaerobic respiration and live in oxygen-free environments. For these organisms, there is some other final electron acceptor other than oxygen (e.g. nitrate or Fe^{2+}).

An Overview of Cellular Respiration

Respiration involves three metabolic stages, summarized below. The first two stages are the catabolic pathways that decompose glucose and other organic fuels. In the third stage, the electron transport chain accepts electrons from the first two stages and passes these from one electron acceptor to another. The energy released at each stepwise transfer is used to make ATP. The final electron acceptor in this process is molecular oxygen.

1 **Glycolysis**. This occurs in the cytoplasm and involves the breakdown of glucose into two molecules of pyruvate.

2 **The Krebs cycle**. This occurs in the mitochondrial matrix, and decomposes a derivative of pyruvate to carbon dioxide.

3 **Electron transport and oxidative phosphorylation**. This occurs in the inner membranes of the mitochondrion and accounts for almost 90% of the ATP generated by respiration.

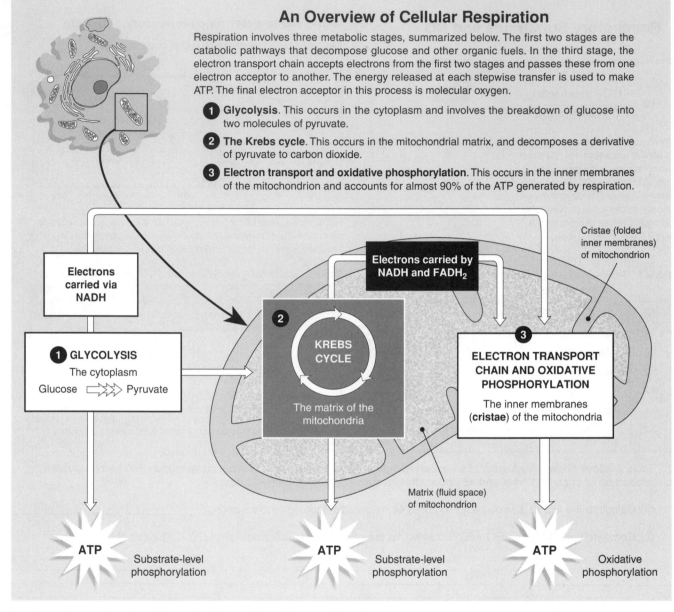

1. Describe precisely in which part of the cell the following take place:

    (a) Glycolysis: _____

    (b) Krebs cycle reactions: _____

    (c) Electron transport chain: _____

2. Provide a clear explanation of what is involved in each of the following processes:

    (a) Substrate-level phosphorylation: _____

    _____

    _____

    (b) Oxidative phosphorylation: _____

    _____

    _____

Periodicals:
Fuelled for Life

The Biochemistry of Respiration

Cellular respiration is a catabolic, energy yielding pathway. The breakdown of glucose and other organic fuels (such as fats and proteins) to simpler molecules is **exergonic** and releases energy for the synthesis of ATP. As summarized in the previous activity, respiration involves glycolysis, the Krebs cycle, and electron transport. The diagram below provides a more detailed overview of the events in each of these stages. Glycolysis and the Krebs cycle supply electrons (via NADH) to the electron transport chain, which drives **oxidative phosphorylation**. Glycolysis nets two ATP, produced by **substrate-level phosphorylation**.

The conversion of pyruvate (the end product of glycolysis) to **acetyl CoA** links glycolysis to the Krebs cycle. One "turn" of the cycle releases carbon dioxide, forms one ATP by substrate level phosphorylation, and passes electrons to three NAD^+ and one FAD. Most of the ATP generated in cellular respiration is produced by oxidative phosphorylation when NADH and $FADH_2$ donate electrons to the series of electron carriers in the electron transport chain. At the end of the chain, electrons are passed to molecular oxygen, reducing it to water. Electron transport is coupled to ATP synthesis by **chemiosmosis** (following page).

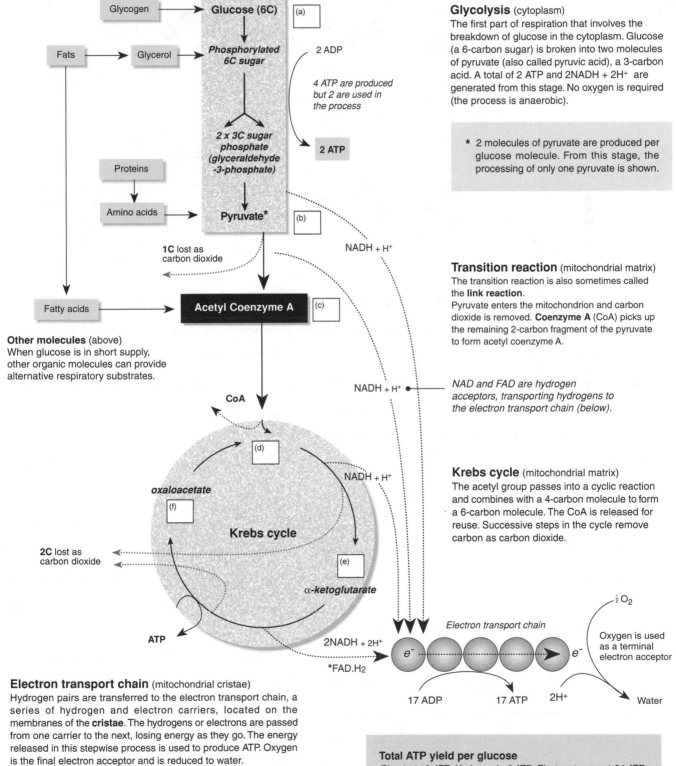

Glycolysis (cytoplasm)
The first part of respiration that involves the breakdown of glucose in the cytoplasm. Glucose (a 6-carbon sugar) is broken into two molecules of pyruvate (also called pyruvic acid), a 3-carbon acid. A total of 2 ATP and 2NADH + $2H^+$ are generated from this stage. No oxygen is required (the process is anaerobic).

* 2 molecules of pyruvate are produced per glucose molecule. From this stage, the processing of only one pyruvate is shown.

Transition reaction (mitochondrial matrix)
The transition reaction is also sometimes called the **link reaction**.
Pyruvate enters the mitochondrion and carbon dioxide is removed. **Coenzyme A** (CoA) picks up the remaining 2-carbon fragment of the pyruvate to form acetyl coenzyme A.

NAD and FAD are hydrogen acceptors, transporting hydrogens to the electron transport chain (below).

Krebs cycle (mitochondrial matrix)
The acetyl group passes into a cyclic reaction and combines with a 4-carbon molecule to form a 6-carbon molecule. The CoA is released for reuse. Successive steps in the cycle remove carbon as carbon dioxide.

Other molecules (above)
When glucose is in short supply, other organic molecules can provide alternative respiratory substrates.

Electron transport chain (mitochondrial cristae)
Hydrogen pairs are transferred to the electron transport chain, a series of hydrogen and electron carriers, located on the membranes of the **cristae**. The hydrogens or electrons are passed from one carrier to the next, losing energy as they go. The energy released in this stepwise process is used to produce ATP. Oxygen is the final electron acceptor and is reduced to water.
*Note FAD enters the electron transport chain at a lower energy level than NAD, and only 2ATP are generated per FAD.H2.

Total ATP yield per glucose
Glycolysis: 2 ATP, *Krebs cycle*: 2 ATP, *Electron transport*: 34 ATP

Periodicals:
AcetylCoA: a central metabolite

Related activities: Cellular Respiration
Web links: Glycolysis and the Krebs Cycle

A 3

Chemiosmosis

Chemiosmosis is the process whereby the synthesis of ATP is coupled to electron transport and the movement of protons (H⁺ ions). **Electron transport carriers** are arranged over the inner membrane of the mitochondrion and oxidize NADH + H⁺ and FADH₂. Energy from this process forces protons to move, against their concentration gradient, from the mitochondrial matrix into the space between the two membranes. Eventually the protons flow back into the matrix via ATP synthetase molecules in the membrane. As the protons flow down their concentration gradient, energy is released and ATP is synthesized. Chemiosmotic theory also explains the generation of ATP in the light dependent phase of photosynthesis.

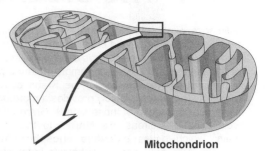

Mitochondrion

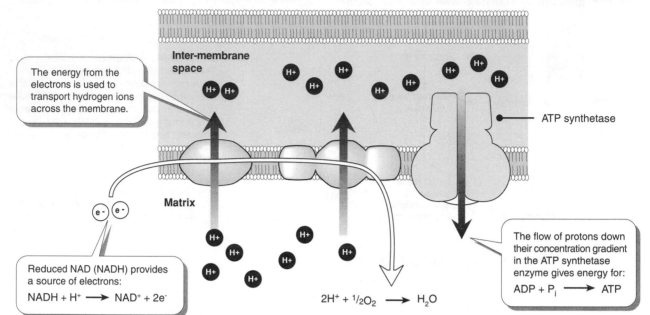

The energy from the electrons is used to transport hydrogen ions across the membrane.

Inter-membrane space

ATP synthetase

Matrix

Reduced NAD (NADH) provides a source of electrons:

$$NADH + H^+ \longrightarrow NAD^+ + 2e^-$$

$$2H^+ + \tfrac{1}{2}O_2 \longrightarrow H_2O$$

The flow of protons down their concentration gradient in the ATP synthetase enzyme gives energy for:

$$ADP + P_i \longrightarrow ATP$$

1. On the diagram of cellular respiration (previous page), state the number of carbon atoms in each of the molecules (a) – (f):

2. Determine how many ATP molecules **per molecule of glucose** are generated during the following stages of respiration:

   (a) Glycolysis: _____ (b) Krebs cycle: _____ (c) Electron transport chain: _____(d) Total: _____

3. Explain what happens to the carbon atoms lost during respiration: _____

   _____

4. Describe the role of the following in aerobic cellular respiration:

   (a) Hydrogen atoms: _____

   _____

   _____

   (b) Oxygen: _____

   _____

   _____

5. (a) Identify the process by which ATP is synthesized in respiration: _____

   (b) Briefly summarize this process: _____

   _____

   _____

   _____

   _____

   _____

Anaerobic Pathways

All organisms can metabolize glucose anaerobically (without oxygen) using glycolysis in the cytoplasm, but the energy yield from this process is low and few organisms can obtain sufficient energy for their needs this way. In the absence of oxygen, glycolysis soon stops unless there is an alternative acceptor for the electrons produced from the glycolytic pathway. In yeasts and the root cells of higher plants this acceptor is ethanal, and the pathway is called alcoholic fermentation. In the skeletal muscle of mammals, the acceptor is pyruvate itself and the end product is lactic acid. In both cases, the duration of the fermentation is limited by the toxic effects of the organic compound produced. Although fermentation is often used synonymously with anaerobic respiration, they are not the same. Respiration always involves hydrogen ions passing down a chain of carriers to a terminal acceptor, and this does not occur in fermentation. In anaerobic respiration, the terminal H^+ acceptor is a molecule other than oxygen, e.g. Fe^{2+} or nitrate.

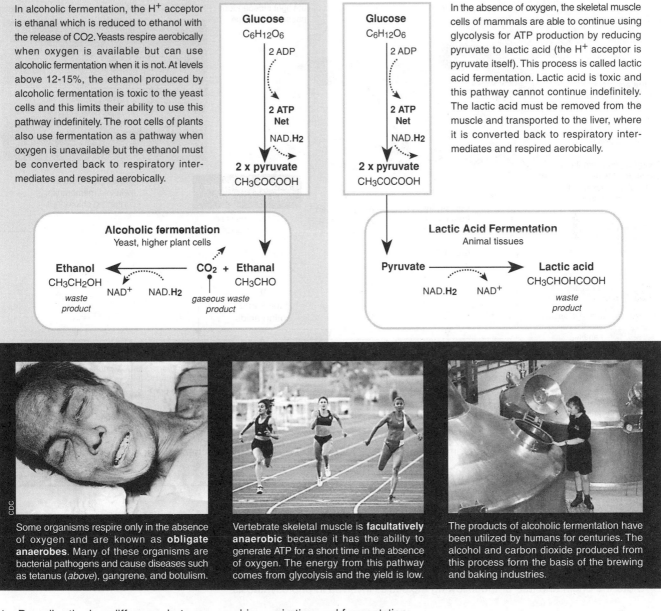

Alcoholic Fermentation

In alcoholic fermentation, the H^+ acceptor is ethanal which is reduced to ethanol with the release of CO_2. Yeasts respire aerobically when oxygen is available but can use alcoholic fermentation when it is not. At levels above 12-15%, the ethanol produced by alcoholic fermentation is toxic to the yeast cells and this limits their ability to use this pathway indefinitely. The root cells of plants also use fermentation as a pathway when oxygen is unavailable but the ethanol must be converted back to respiratory intermediates and respired aerobically.

Glucose
$C_6H_{12}O_6$
2 ADP
2 ATP Net
NAD.H_2
2 x pyruvate
$CH_3COCOOH$

Alcoholic fermentation
Yeast, higher plant cells

Ethanol
CH_3CH_2OH
waste product
NAD$^+$ NAD.H_2
CO_2 + Ethanal
CH_3CHO
gaseous waste product

Lactic Acid Fermentation

In the absence of oxygen, the skeletal muscle cells of mammals are able to continue using glycolysis for ATP production by reducing pyruvate to lactic acid (the H^+ acceptor is pyruvate itself). This process is called lactic acid fermentation. Lactic acid is toxic and this pathway cannot continue indefinitely. The lactic acid must be removed from the muscle and transported to the liver, where it is converted back to respiratory intermediates and respired aerobically.

Glucose
$C_6H_{12}O_6$
2 ADP
2 ATP Net
NAD.H_2
2 x pyruvate
$CH_3COCOOH$

Lactic Acid Fermentation
Animal tissues

Pyruvate → **Lactic acid**
$CH_3CHOHCOOH$
waste product
NAD.H_2 NAD$^+$

Some organisms respire only in the absence of oxygen and are known as **obligate anaerobes**. Many of these organisms are bacterial pathogens and cause diseases such as tetanus (*above*), gangrene, and botulism.

Vertebrate skeletal muscle is **facultatively anaerobic** because it has the ability to generate ATP for a short time in the absence of oxygen. The energy from this pathway comes from glycolysis and the yield is low.

The products of alcoholic fermentation have been utilized by humans for centuries. The alcohol and carbon dioxide produced from this process form the basis of the brewing and baking industries.

1. Describe the key difference between aerobic respiration and fermentation: _____

_____

2. (a) Refer to the previous activity and determine the efficiency of fermentation compared to aerobic respiration: _____%

   (b) In simple terms, explain why the efficiency of anaerobic pathways is so low: _____

   _____

3. Explain why fermentation cannot go on indefinitely: _____

_____

_____

Photosynthesis

Photosynthesis is of fundamental importance to living things because it transforms sunlight energy into chemical energy stored in molecules, releases free oxygen gas, and absorbs carbon dioxide (a waste product of cellular metabolism). Photosynthetic organisms use special pigments, called **chlorophylls**, to absorb light of specific wavelengths and thereby capture the light energy.

Visible light is a small fraction of the total **electromagnetic radiation** reaching Earth from the sun. Of the visible spectrum, only certain wavelengths (red and blue) are absorbed for photosynthesis. Other wavelengths, particularly green, are reflected or transmitted. The diagram below summarizes the process of photosynthesis..

Summary of Photosynthesis in a C₃ Plant

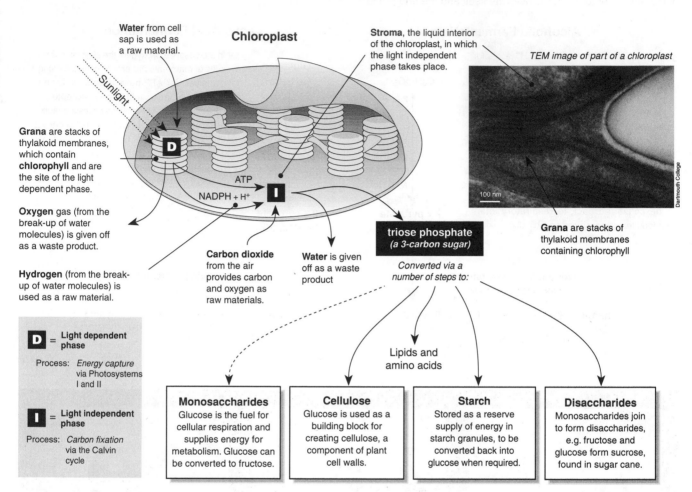

Water from cell sap is used as a raw material.

Chloroplast

Stroma, the liquid interior of the chloroplast, in which the light independent phase takes place.

TEM image of part of a chloroplast

Sunlight

Grana are stacks of thylakoid membranes, which contain **chlorophyll** and are the site of the light dependent phase.

D

ATP

NADPH + H⁺

I

100 nm

Dartmouth College

Oxygen gas (from the break-up of water molecules) is given off as a waste product.

Carbon dioxide from the air provides carbon and oxygen as raw materials.

Water is given off as a waste product

triose phosphate *(a 3-carbon sugar)*

Grana are stacks of thylakoid membranes containing chlorophyll

Hydrogen (from the break-up of water molecules) is used as a raw material.

Converted via a number of steps to:

D = **Light dependent phase**

Process: *Energy capture via Photosystems I and II*

I = **Light independent phase**

Process: *Carbon fixation via the Calvin cycle*

Lipids and amino acids

Monosaccharides
Glucose is the fuel for cellular respiration and supplies energy for metabolism. Glucose can be converted to fructose.

Cellulose
Glucose is used as a building block for creating cellulose, a component of plant cell walls.

Starch
Stored as a reserve supply of energy in starch granules, to be converted back into glucose when required.

Disaccharides
Monosaccharides join to form disaccharides, e.g. fructose and glucose form sucrose, found in sugar cane.

1. Distinguish between the two different regions of a chloroplast and describe the biochemical processes that occur in each:

(a) _____

_____

(b) _____

_____

2. State the origin and fate of the following molecules involved in photosynthesis:

(a) Carbon dioxide: _____

(b) Oxygen: _____

(c) Hydrogen: _____

3. Suggest how scientists might have determined the fate of these molecules: _____

_____

4. Discuss the potential uses for the end products of photosynthesis: _____

_____

_____

Related activities: The Biochemistry of Photosynthesis

Periodicals: Photosynthesis: most hated topic?

Pigments and Light Absorption

As light meets matter, it may be reflected, transmitted, or absorbed. Substances that absorb visible light are called **pigments**, and different pigments absorb light of different wavelengths. The ability of a pigment to absorb particular wavelengths of light can be measured with a spectrophotometer. The light absorption vs the wavelength is called the **absorption spectrum** of that pigment. The absorption spectrum of different photosynthetic pigments provides clues to their role in photosynthesis, since light can only perform work if it is absorbed. An **action spectrum** profiles the effectiveness of different wavelength light in fuelling photosynthesis. It is obtained by plotting wavelength against some measure of photosynthetic rate (e.g. CO_2 production). Some features of photosynthetic pigments and their light absorbing properties are outlined below.

The Electromagnetic Spectrum

Light is a form of energy known as electromagnetic radiation. The segment of the electromagnetic spectrum most important to life is the narrow band between about 380 and 750 nanometres (nm). This radiation is known as visible light because it is detected as colors by the human eye (although some other animals, such as insects, can see in the ultraviolet range). It is the visible light that drives photosynthesis.

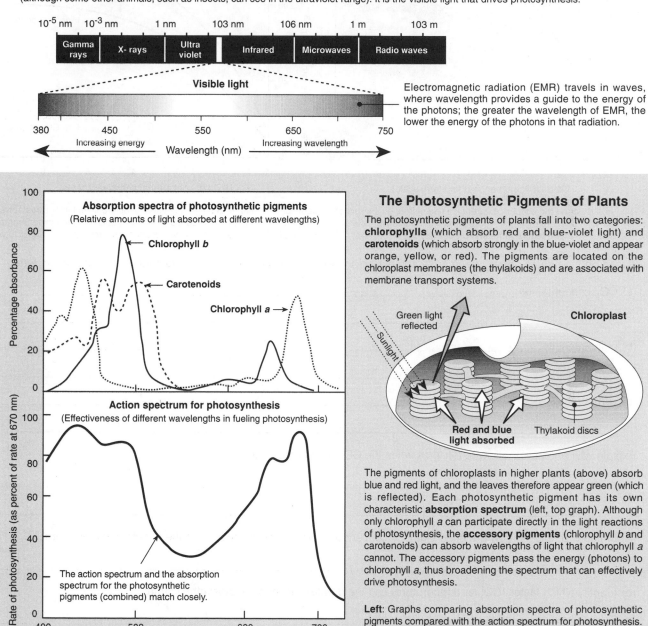

Electromagnetic radiation (EMR) travels in waves, where wavelength provides a guide to the energy of the photons; the greater the wavelength of EMR, the lower the energy of the photons in that radiation.

The Photosynthetic Pigments of Plants

The photosynthetic pigments of plants fall into two categories: **chlorophylls** (which absorb red and blue-violet light) and **carotenoids** (which absorb strongly in the blue-violet and appear orange, yellow, or red). The pigments are located on the chloroplast membranes (the thylakoids) and are associated with membrane transport systems.

The pigments of chloroplasts in higher plants (above) absorb blue and red light, and the leaves therefore appear green (which is reflected). Each photosynthetic pigment has its own characteristic **absorption spectrum** (left, top graph). Although only chlorophyll *a* can participate directly in the light reactions of photosynthesis, the **accessory pigments** (chlorophyll *b* and carotenoids) can absorb wavelengths of light that chlorophyll *a* cannot. The accessory pigments pass the energy (photons) to chlorophyll *a*, thus broadening the spectrum that can effectively drive photosynthesis.

Left: Graphs comparing absorption spectra of photosynthetic pigments compared with the action spectrum for photosynthesis.

1. Explain what is meant by the absorption spectrum of a pigment: _____

_____

2. Explain why the action spectrum for photosynthesis does not exactly match the absorption spectrum of chlorophyll *a*:

_____

_____

A 2

Photosynthetic Rate

The rate at which plants make carbohydrate (the photosynthetic rate) is dependent on environmental factors, the most important of which are the availability of **light** and **carbon dioxide** (CO_2). Temperature is important, but its influence is less clear because it depends on the availability of the other two **limiting factors** (CO_2 and light) and the temperature tolerance of the plant. The effect of these factors can be tested experimentally by altering one of the factors while holding the others constant (a controlled experiment). In reality, a plant in its natural environment is subjected to variations in many different environmental factors, all of which will influence, directly or indirectly, the rate at which photosynthesis can occur.

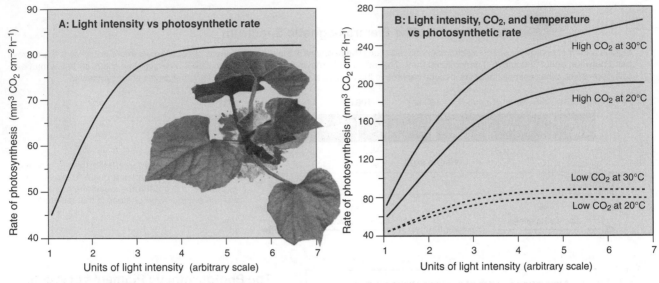

These figures illustrate the effect of different limiting factors on the rate of photosynthesis in cucumber plants. Figure A shows the effect of different light intensities when the temperature and carbon dioxide ($CO2$) level are kept constant. Figure B shows the effect of different light intensities at two temperatures and two $CO2$ concentrations. In each of these experiments, either $CO2$ level or temperature was changed at each light intensity in turn.

1. Based on the figures above, summarise and explain the effect of each of the following factors on photosynthetic rate:

    (a) CO_2 concentration: _____

    _____

    (b) Light intensity: _____

    _____

    (c) Temperature: _____

    _____

2. Explain why photosynthetic rate declines when the CO_2 level is reduced: _____

    _____

    _____

3. (a) In figure B, explain how the effects of CO_2 concentration were distinguished from the effects of temperature:

    _____

    (b) Identify which factor (CO_2 or temperature) had the greatest effect on photosynthetic rate: _____

    (c) Explain how you can tell this from the graph:_____

    _____

4. Explain how glasshouses can be used to create an environment in which photosynthetic rates are maximised:

    _____

    _____

    _____

5. Design an experiment to demonstrate the effect of temperature on photosynthetic rate. You should include a hypothesis, list of equipment, and methods. Staple your experiment to this page.

Related activities: The Biochemistry of Photosynthesis

The Biochemistry of Photosynthesis

Like cellular respiration, photosynthesis is a redox process, but in photosynthesis, water is split and electrons are transferred together with hydrogen ions from water to CO_2, reducing it to sugar. The electrons increase in potential energy as they move from water to sugar. The energy to do this is provided by light. Photosynthesis comprises two phases. In the **light dependent phase**, light energy is converted to chemical energy (ATP and

reducing power). In the **light independent phase** (**Calvin cycle**), the chemical energy is used to synthesise carbohydrate. The light dependent phase illustrated below shows **non-cyclic phosphorylation**, which produces ATP and NADPH in roughly equal quantities. In **cyclic phosphorylation**, the electrons lost from photosystem II are replaced by those from photosystem I. ATP is generated, but not NADPH.

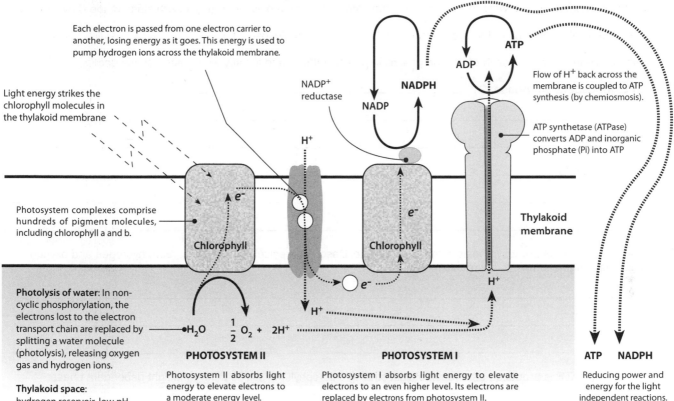

Each electron is passed from one electron carrier to another, losing energy as it goes. This energy is used to pump hydrogen ions across the thylakoid membrane.

Light energy strikes the chlorophyll molecules in the thylakoid membrane

Photosystem complexes comprise hundreds of pigment molecules, including chlorophyll a and b.

Flow of H$^+$ back across the membrane is coupled to ATP synthesis (by chemiosmosis).

ATP synthetase (ATPase) converts ADP and inorganic phosphate (Pi) into ATP

Thylakoid membrane

Photolysis of water: In non-cyclic phosphorylation, the electrons lost to the electron transport chain are replaced by splitting a water molecule (photolysis), releasing oxygen gas and hydrogen ions.

$H_2O \quad \frac{1}{2} O_2 + 2H^+$

Thylakoid space: hydrogen reservoir, low pH

PHOTOSYSTEM II

Photosystem II absorbs light energy to elevate electrons to a moderate energy level.

PHOTOSYSTEM I

Photosystem I absorbs light energy to elevate electrons to an even higher level. Its electrons are replaced by electrons from photosystem II.

ATP   NADPH

Reducing power and energy for the light independent reactions.

Light Dependent Reactions:
Energy Capture (above)

The light dependent reactions in the thylakoid of the chlorplasts capture light energy to produce ATP and reducing power (as NADPH) which is used in the light independent reactions. This diagram depicts non-cyclic phosphorylation in which both ATP and NADPH are generated. In cyclic phosphorylation the electrons lost from photosystem II are replaced by electrons from photosystem I and no NADPH is generated.

Light Independent Reactions:
Carbon Fixation (right)

The light independent reactions of photosynthesis (the **Calvin cycle**) occur in the stroma of the chloroplast, and do not require light to proceed. Here, hydrogen (H$^+$) is added to CO_2 and a 5C intermediate to make carbohydrate. The H$^+$ and ATP are supplied by the light dependent phase. The Calvin cycle uses more ATP than NADPH, but the cell uses cyclic phosphorylation (which does not produce NADPH) when it runs low on ATP to make up the difference.

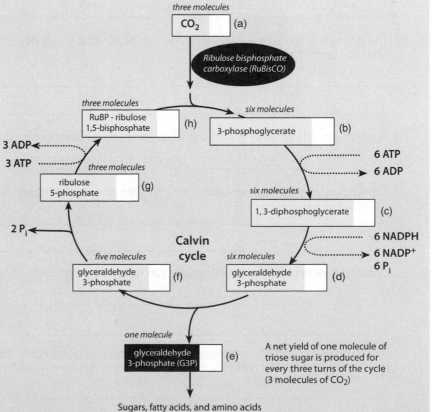

A net yield of one molecule of triose sugar is produced for every three turns of the cycle (3 molecules of CO_2)

1. Describe the role of the carrier molecule **NADP** in photosynthesis: _____

_____

2. Explain the role of chlorophyll molecules in the process of photosynthesis: _____

_____

_____

3. On the previous diagram, write the number of carbon atoms of each molecule (a)-(h)at each stage of the Calvin cycle:

4. State the number of turns of the cycle to produce one molecule of glucose (6C): _____

5. Summarize the events in each of the two phases in photosynthesis and identify where each phase occurs:

   (a) **Light dependent phase (D)**: _____

   _____

   _____

   (b) **Calvin cycle**: _____

   _____

   _____

6. The final product of photosynthesis is triose phosphate. Describe precisely where the carbon, hydrogen and oxygen molecules originate from to make this molecule:

   _____

   _____

   _____

7. Explain how ATP is produced as a result of light striking chlorophyll molecules during the light dependent phase:

   _____

   _____

   _____

8. (a) The diagram of the light dependent phase (top of previous page) describes **non-cyclic phosphorylation**. Explain what you understand by this term:

   _____

   _____

   (b) Suggest why this process is also known as non-cyclic **photo**phosphorylation: _____

   _____

   (c) Explain how photophosphorylation differs from the oxidative phosphorylation occurring in cellular respiration:

   _____

   _____

9. (a) Describe how **cyclic photophosphorylation** differs from non-cyclic photophosphorylation: _____

   _____

   _____

   (b) Both cyclic and noncyclic pathways operate to varying degrees during photosynthesis. Since the non-cyclic pathway produces both ATP and NADPH, explain the purpose of the cyclic pathway of electron flow:

   _____

   _____

Photosynthesis in C₄ Plants

When photosynthesis takes place, the first detectable compound which is made by a plant is usually a 3-carbon compound called GP (glycerate 3-phosphate). Plants which do this are called C₃ plants. In some plants, however, a 4-carbon molecule called oxaloacetate, is the first to be made. Such plants, which include cereals and tropical grasses, are called C₄ plants. These plants have a high rate of photosynthesis, thriving in environments with high light levels and warm temperatures. Their yield of photosynthetic products is higher than that of C₃ plants, giving them a competitive advantage in tropical climates. The high productivity of the C₄ system is also an important property of crop plants such as sugar cane and maize.

Structure of a Leaf from a C₄ Plant

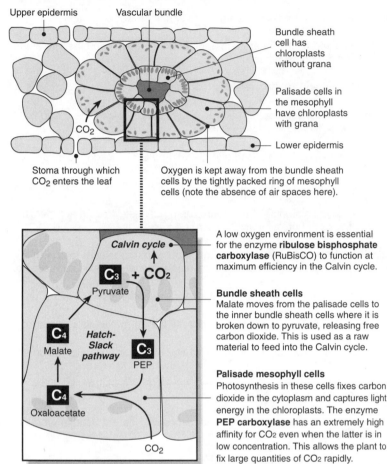

Upper epidermis
Vascular bundle
Bundle sheath cell has chloroplasts without grana
Palisade cells in the mesophyll have chloroplasts with grana
CO₂
Lower epidermis
Stoma through which CO₂ enters the leaf
Oxygen is kept away from the bundle sheath cells by the tightly packed ring of mesophyll cells (note the absence of air spaces here).

Calvin cycle
C₃ + **CO₂**
Pyruvate
C₄
Malate
Hatch-Slack pathway
C₃
PEP
C₄
Oxaloacetate
CO₂

A low oxygen environment is essential for the enzyme **ribulose bisphosphate carboxylase** (RuBisCO) to function at maximum efficiency in the Calvin cycle.

Bundle sheath cells
Malate moves from the palisade cells to the inner bundle sheath cells where it is broken down to pyruvate, releasing free carbon dioxide. This is used as a raw material to feed into the Calvin cycle.

Palisade mesophyll cells
Photosynthesis in these cells fixes carbon dioxide in the cytoplasm and captures light energy in the chloroplasts. The enzyme **PEP carboxylase** has an extremely high affinity for CO₂ even when the latter is in low concentration. This allows the plant to fix large quantities of CO₂ rapidly.

Examples of C₄ plants
- Sugar cane *(Saccharum officinale)*
- Maize *(Zea mays)*
- Sorghum *(Sorghum bicolor)*
- Sun plant *(Portulaca grandifolia)*

Distribution of grasses using C₄ mechanism in North America

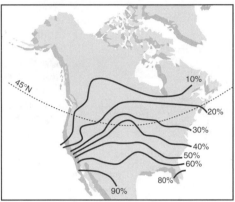

45°N
10%
20%
30%
40%
50%
60%
80%
90%

The photosynthetic strategy that a plant possesses is an important factor in determining where it lives. Because many of the enzymes of C₄ plants have optimum temperatures well above 25°C, they thrive in hot tropical and sub-tropical climates. Under these conditions, they can out-compete most C₃ plants because they achieve faster rates of photosynthesis. The proportion of grasses using the C₄ mechanism in North America is greatest near the tropics and diminishes northwards.

1. Explain why C₄ plants have a competitive advantage over C₃ plants in the tropics: _____

_____

_____

2. Explain why the bundle sheath cells are arranged in a way that keeps them isolated from air spaces in the leaf:

_____

_____

3. Study the map of North America above showing the distribution of C₄ plants. Explain the distribution pattern in terms of their competitive advantage and the environmental conditions required for this advantage:

_____

_____

_____

4. In C₃ plants, the rate of photosynthesis is enhanced by higher atmospheric CO₂ concentrations. Explain why this is not the case for C₄ plants:

_____

_____

Related activities: The Biochemistry of Photosynthesis

A 3

KEY TERMS: Mix and Match

INSTRUCTIONS: Test your vocab by matching each term to its correct definition, as identified by its preceding letter code.

ABSORPTION SPECTRUM

ACCESSORY PIGMENT

ACETYL COENZYME A

ACTION SPECTRUM

ATP

CALVIN CYCLE

CELLULAR RESPIRATION

CHEMIOSMOSIS

CHLOROPLAST

CHLOROPHYLL

CRISTAE

ELECTRON TRANSPORT CHAIN

FERMENTATION

GLYCOLYSIS

HATCH AND SLACK PATHWAY

KREBS CYCLE

LACTIC ACID

MATRIX

MITOCHONDRION

NON-CYCLIC PHOTO-PHOSPHORYLATION

OXIDATIVE PHOSPHORYLATION

PHOTOLYSIS

PHOTOSYNTHESIS

RESPIRATORY QUOTIENT

RIBULOSE BISPHOSPHATE

STROMA

SUBSTRATE LEVEL PHOSPHORYLATION

THYLAKOID DISCS

A   A type of photophosphorylation in which light energy is used to generate both ATP and NADPH2.

B   The internal compartments formed by the inner membrane of a mitochondrion.

C   The principal photosynthetic pigment found in green plants and algae.

D   A series of electron carriers which transfer electrons in a series of redox reactions.

E   The colorless material enclosed by the inner membrane of a chloroplast and the site of the light independent reactions of photosynthesis.

F   The light absorption (of a pigment) vs the wavelength of light.

G   A key series of metabolic reactions in aerobic cellular respiration in which CO_2 is oxidized to various carboxylic acid and NAD and FAD are reduced.

H   The ratio between the volume of CO_2 produced and the volume of O_2 used in respiration.

I   Catabolic process by which a cell produces ATP through the stepwise oxidation of glucose.

J   Membranous compartments in chloroplasts, which are arranged in stacks and are the site of the light dependent reactions of photosynthesis.

K   The anaerobic breakdown of organic compounds (e.g. carbohydrates) in cells to one of a variety of end-products (e.g. lactic acid or ethanol), using an endogenous electron acceptor.

L   The inner region in a mitochondrion enclosed by the inner mitochondrial membrane.

M   Formation of ATP by transfer of a phosphate to ADP directly from a metabolic substrate.

N   An organelle in eukaryotic cells in which the Krebs cycle and oxidative phosphorylation occur.

O   A series of biochemical reactions occurring in the stroma of chloroplasts in which carbon dioxide is fixed in organic compounds.

P   Anabolic process by which sunlight energy, water, and carbon dioxide are used to produce oxygen and chemical energy (sugar).

Q   An alternative photosynthetic pathway in some plants in which CO_2 is temporarily fixed as a four-carbon acid, which is then broken down to supply CO_2 to the normal Calvin cycle.

R   A process in aerobic organisms in which ATP is generated by electron transfer along an electron transport chain to oxygen.

S   The primary CO_2 acceptor in photosynthesis.

T   A waste product of fermentation in mammalian muscle.

U   The generation of ATP using a proton gradient across a selectively permeable membrane.

V   The relative effectiveness of different wavelengths of light at generating electrons from a pigment.

W   An important intermediary molecule, which transfers carbon atoms within the acetyl group to the Krebs cycle to be oxidized for energy production.

X   The main energy storage and transfer molecule in the cell.

Y   An organelle in photosynthetic eukaryotes which light capture and carbon fixation take place.

Z   Splitting a molecule using light energy, as in the splitting of water into hydrogen and oxygen.

AA The anaerobic breakdown of glucose to pyruvate in living cells.

BB Plant pigments that absorb wavelengths of light that chlorophyll a does not absorb.

Important in this section ...

- *Understand how meiosis and sexual reproduction produce variation and how this is inherited.*
- *Understand the structure and role of DNA.*
- *Develop an understanding of the techniques and applications of nucleic acid technology.*

Chromosomes and Meiosis

Meiosis	• Meiosis and variation • Linkage, crossing over, & recombination • Meiosis vs mitosis
Chromosomes & karyotypes	• The genome • Eukaryotic chromosomes • Karyotypes and karyotyping

Heredity

Variation	• Genetic sources of variation • Environmental effects on phenotype • Continuous and discontinuous variation
Mendelian genetics	• Mendel's laws • Dominance: Mono-and dihybrid crosses • Departures from expected Mendelian ratios - lethal alleles
Various aspects of inheritance	• Sex determination • Sex linked inheritance • Epigenetics and gene silencing • Pedigrees and genetic counseling
Gene interactions	• Simple gene interactions: collaboration • Pleiotropy and epistasis • Polygeny and continuous variation

Meiosis creates new allele combinations in the offspring. A karyotype displays its genetic makeup.

Both genes and environment contribute to final expression of the phenotype.

Part 3

Heredity and Molecular Genetics

Mutation and sexual reproduction contribute to the heritable variation in organisms.

DNA, through the genetic code, provides the blueprint for life.

Humans have developed the ability to alter the genome of organisms to diagnose and correct disorders and produce valuable commodities.

The genetic code is universal among living things. Changes to DNA can create new alleles.

The genome of organisms can be both analyzed and manipulated to meet human needs.

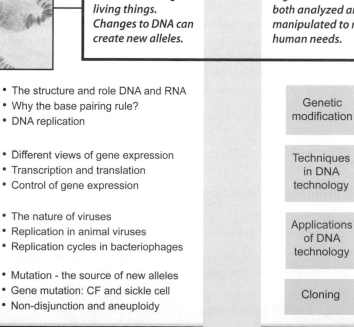

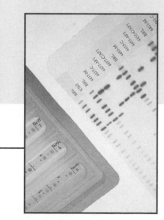

Nucleic acids	• The structure and role DNA and RNA • Why the base pairing rule? • DNA replication
Gene expression	• Different views of gene expression • Transcription and translation • Control of gene expression
Replication in viruses	• The nature of viruses • Replication in animal viruses • Replication cycles in bacteriophages
Mutation	• Mutation - the source of new alleles • Gene mutation: CF and sickle cell • Non-disjunction and aneuploidy

Molecular Genetics & Mutation

Genetic modification	• Adding novel genes • Switching genes off • Altering existing genes
Techniques in DNA technology	• Restriction digestion and ligation • Gel electrophoresis • DNA amplification and gene cloning • DNA sequencing
Applications of DNA technology	• Profiling for diagnostics and forensics • Use of transgenics as biofactories • Genome sequencing projects • Gene therapy and genetic screening
Cloning	• Animal and plant cloning • Rapid dissemination of transgenics

Nucleic Acid Technology

Chromosomes and Meiosis

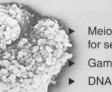

KEY CONCEPTS

▶ Meiosis is a reduction division and is essential for sexual reproduction.

▶ Gametic meiosis produces haploid gametes.

▶ DNA is packaged into chromosomes with organizing histone proteins.

▶ The genome describes the entire genetic content of a haploid organism or cell.

▶ A karyotype is the number and appearance of chromosomes in the nucleus of a eukaryote cell.

KEY TERMS

allele
amniocentesis
anaphase
autosome
bivalent
chorionic villus sampling
chromatid
chromosome
crossing over
crossover frequency
diploid (2N)
fertilization
gamete
haploid (1N)
histone
independent assortment
interphase
karyotype
karyotyping
linkage
maternal chromosome
meiosis
metaphase
nuclear division
nucleosome
paternal chromosome
prophase
recombination
sex chromosome
somatic cell
synapsis
telophase
variation

Periodicals:
listings for this chapter are on page 380

Weblinks:
www.thebiozone.com/
weblink/SB1-2597.html

Teacher Resource CD-ROM:
Chromosome Mapping

OBJECTIVES

☐ 1. Use the **KEY TERMS** to help you understand and complete these objectives.

Sex and Meiosis   pages 93-96

☐ 2. Discuss the possible reasons for the evolution of **sex** in organisms.

☐ 3. Know that **meiosis**, like mitosis, involves DNA replication during interphase in the parent cell, but that this is followed by two cycles of nuclear division.

☐ 4. Recall how chromosome numbers vary between **somatic cells** (**diploid 2N**) and **gametes** (**haploid 1N**). Compare the outcomes of meiosis and mitosis.

☐ 5. In more detail than #3, summarize the principal events in meiosis, including:
   (a) **Synapsis** and formation of **bivalents**.
   (b) **Chiasma** formation and exchange of genetic material between **chromatids** in the first, (reduction) division.
   (c) Separation of chromatids (second division); production of haploid cells.
   (d) Identification of the names of the main stages.

☐ 6. Describe the behavior of **homologous chromosomes** (and their associated **alleles**) during meiosis and fertilization, with reference to:
   • The **independent assortment** of **maternal** and **paternal chromosomes**.
   • The recombination of segments of maternal and paternal homologous chromosomes in **crossing over**.
   • The random fusion of gametes during **fertilization**.
   Explain how these events create new allele combinations in the gametes.

☐ 7. Explain the significance of **linkage** and **recombination** to the inheritance of alleles.

☐ 8. EXTENSION: Explain what is meant by the **crossover frequency**. Demonstrate the use of crossover values in chromosome (gene) mapping.

Chromosome Structure   pages 97-98

☐ 9. Describe the structure and morphology of eukaryotic chromosomes, identifying the role of **histone proteins** in packaging the DNA in the nucleus.

☐ 10. EXTENSION: With respect to structure and organization, distinguish between prokaryotic and eukaryotic chromosomes.

☐ 11. Define the terms: **karyotype**, **autosome**, and **sex chromosome**. Explain the basis of **karyotyping** and describe one application of this process.

☐ 12. Know that karyotyping uses cells collected by **chorionic villus sampling** or **amniocentesis**, for **prenatal diagnosis** of chromosome abnormalities.

Meiosis

Meiosis is a special type of division, called a reduction division. In animals, it always produces gametes, but in some other organisms it gives rise to haploid spores. Meiosis involves a chromosomal duplication followed by two nuclear divisions, and results in a halving of the diploid chromosome number. Meiosis occurs in the sex organs of plants and animals and is essential for sexual reproduction. If genetic mistakes (**mutations**) occur here, they will be passed on to the offspring (they will be inherited).

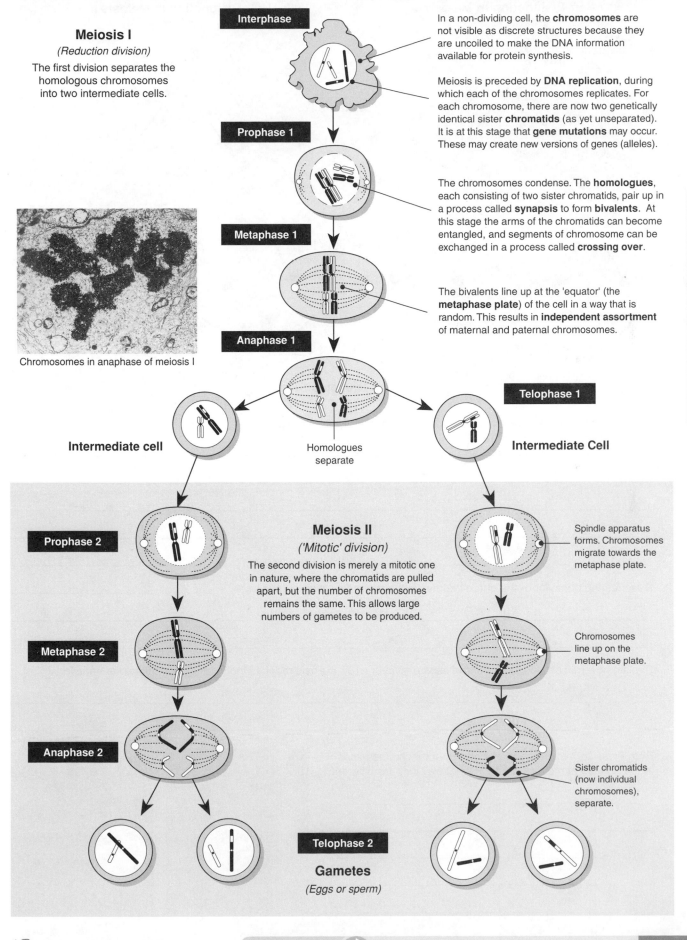

Meiosis I
(Reduction division)

The first division separates the homologous chromosomes into two intermediate cells.

Interphase

In a non-dividing cell, the **chromosomes** are not visible as discrete structures because they are uncoiled to make the DNA information available for protein synthesis.

Prophase 1

Meiosis is preceded by **DNA replication**, during which each of the chromosomes replicates. For each chromosome, there are now two genetically identical sister **chromatids** (as yet unseparated). It is at this stage that **gene mutations** may occur. These may create new versions of genes (alleles).

Metaphase 1

The chromosomes condense. The **homologues**, each consisting of two sister chromatids, pair up in a process called **synapsis** to form **bivalents**. At this stage the arms of the chromatids can become entangled, and segments of chromosome can be exchanged in a process called **crossing over**.

The bivalents line up at the 'equator' (the **metaphase plate**) of the cell in a way that is random. This results in **independent assortment** of maternal and paternal chromosomes.

Anaphase 1

Chromosomes in anaphase of meiosis I

Telophase 1

Intermediate cell

Intermediate Cell

Homologues separate

Meiosis II
('Mitotic' division)

The second division is merely a mitotic one in nature, where the chromatids are pulled apart, but the number of chromosomes remains the same. This allows large numbers of gametes to be produced.

Prophase 2

Spindle apparatus forms. Chromosomes migrate towards the metaphase plate.

Metaphase 2

Chromosomes line up on the metaphase plate.

Anaphase 2

Sister chromatids (now individual chromosomes), separate.

Telophase 2

Gametes
(Eggs or sperm)

Periodicals:
Mechanisms of meiosis

Related activities: Crossing Over, Mitosis vs Meiosis
Web links: Meiosis Tutorial

RA 2

The meiotic spindle normally distributes chromosomes to daughter cells without error. However, mistakes can occur in which the homologous chromosomes fail to separate properly at anaphase during meiosis I, or sister chromatids fail to separate during meiosis II. In these cases, one gamete receives two of the same type of chromosome and the other gamete receives no copy. This mishap, called **non-disjunction**, results in abnormal numbers of chromosomes passing to the gametes. If either of the aberrant gametes unites with a normal one at fertilization, the offspring will have an abnormal chromosome number, known as an **aneuploidy**.

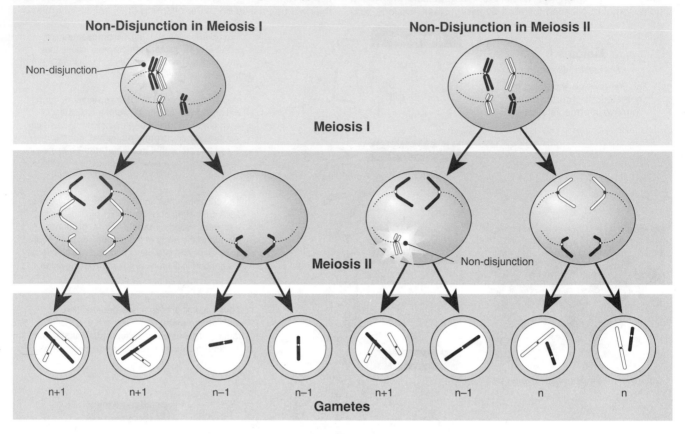

1. Describe the behavior of the chromosomes in the first division of meiosis: _____

_____

_____

_____

_____

_____

2. Describe the behavior of the chromosomes in the second division of meiosis: _____

_____

_____

3. Explain how mitosis conserves chromosome number while meiosis reduces the number from diploid to haploid:

_____

_____

4. Both these light micrographs (A and B) show chromosomes in metaphase of meiosis. State in what way they are different:

_____

_____

_____

_____

_____

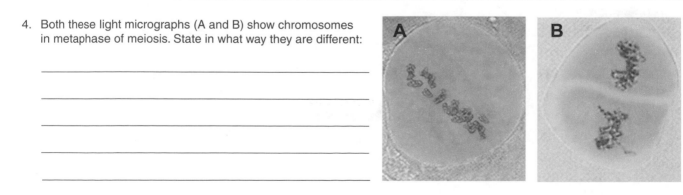

Crossing Over

Crossing over refers to the mutual exchange of pieces of chromosome and involves the swapping of whole groups of genes between the **homologous** chromosomes. This process can occur only during the first division of **meiosis**. Errors in crossing over can result in **block mutations** (see activity *Chromosome Mutations*), which can be very damaging to development. Crossing over can upset expected frequencies of offspring in dihybrid crosses. The frequency of crossing over (COV) for different genes (as followed by inherited, observable traits) can be used to determine the relative positions of genes on a chromosome and provide a genetic map. There has been a recent suggestion that crossing over may be necessary to ensure accurate cell division.

Pairing of Homologous Chromosomes
Every somatic cell contains a pair of each type of chromosome, one from each parent. These are called **homologous pairs** or **homologues**. In prophase of meiosis I, the homologues pair up to form **bivalents**. This process is called **synapsis** and it brings the chromatids of the homologues into close contact.

Chiasma Formation and Crossing Over
Synapsis allows the homologous, non-sister chromatids to become entangled and the chromosomes exchange segments. This exchange occurs at regions called **chiasmata** (sing. chiasma). In the diagram (center), a chiasma is forming and the exchange of pieces of chromosome has not yet taken place. Numerous chiasmata may develop between homologues.

Separation
Crossing over produces new allele combinations, a phenomenon known as **recombination**. When the homologues separate in anaphase of meiosis I, each of the chromosomes pictured will have new mix of alleles that will be passed into the gametes soon to be formed. Recombination is an important source of variation in population gene pools.

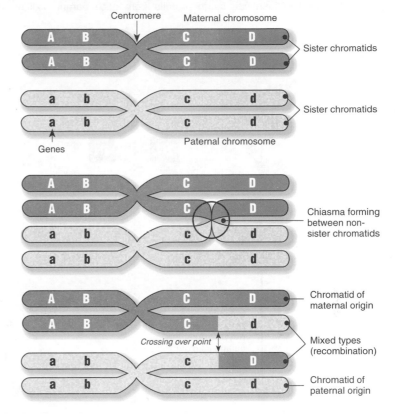

Chromosomes and Meiosis

Gamete Formation
Once the final division of meiosis is complete, the two chromatids that made up each replicated chromosome become separated and are now referred to as chromosomes. As a result of the crossing over, **four** genetically different chromosomes are produced. If no crossing over had occurred, there would have been only two parental types (two copies of each).

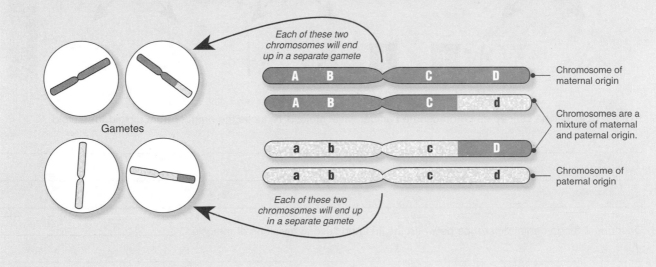

1. Briefly explain how the process of crossing over is going to alter the genotype of gametes: _____

_____

2. Describe the importance of crossing over in the process of evolution: _____

_____

Related activities: Meiosis
Web links: Crossing Over Problems

A 2

Mitosis vs Meiosis

Cell division is fundamental to all life, as cells arise only by the division of existing cells. All types of cell division begin with replication of the cell's DNA. In eukaryotes, this is followed by division of the nucleus. There are two forms of nuclear division: **mitosis** and **meiosis**, and they have quite different purposes and outcomes. Mitosis is the simpler of the two and produces two identical daughter cells from each parent cell. Mitosis is responsible for growth and repair processes in multicellular organisms and reproduction in single-celled and asexual eukaryotes. Meiosis involves a **reduction division** in which haploid gametes are produced for the purposes of sexual reproduction. Fusion of haploid gametes in fertilization restores the diploid cell number in the **zygote**. These two fundamentally different types of cell division are compared below.

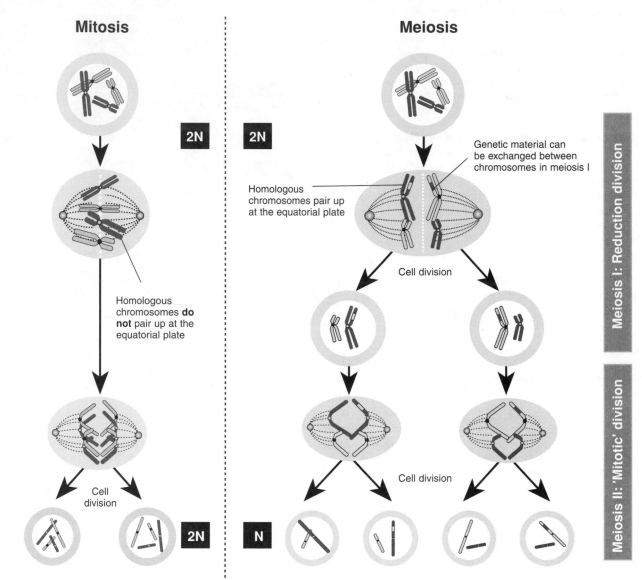

1. Explain how mitosis conserves chromosome number while meiosis reduces the number from diploid to haploid:

_____

_____

_____

2. Describe a fundamental difference between the first and second divisions of meiosis: _____

_____

_____

3. Explain how meiosis introduces genetic variability into gametes and offspring (following gamete fusion in fertilization):

_____

_____

_____

Linkage

Linkage refers to genes that are located on the same chromosome. Linked genes tend to be inherited together and fewer genetic combinations of their alleles are possible. Linkage reduces the variety of offspring that can be produced (contrast this with recombination). In genetic crosses, linkage is indicated when a greater proportion of the progeny resulting from a cross are of the parental type (than would be expected if the alleles were assorting independently). If the genes in question had been on separate chromosomes, there would have been more genetic variation in the gametes and therefore in the offspring. Note that in the example below, wild type alleles are dominant and are denoted by an upper case symbol of the mutant phenotype (Cu or Eb). This symbology used for *Drosophila* departs from the convention of using the dominant gene to provide the symbol. This is necessary because there are many mutant alternative phenotypes to the wild type (e.g. curled and vestigial wings). A lower case symbol of the wild type (e.g. ss for straight wing), would not indicate the mutant phenotype involved. Alternatively, the wild type is sometimes denoted with a raised plus sign e.g. cu^+cu^+ and all symbols are in lower case.

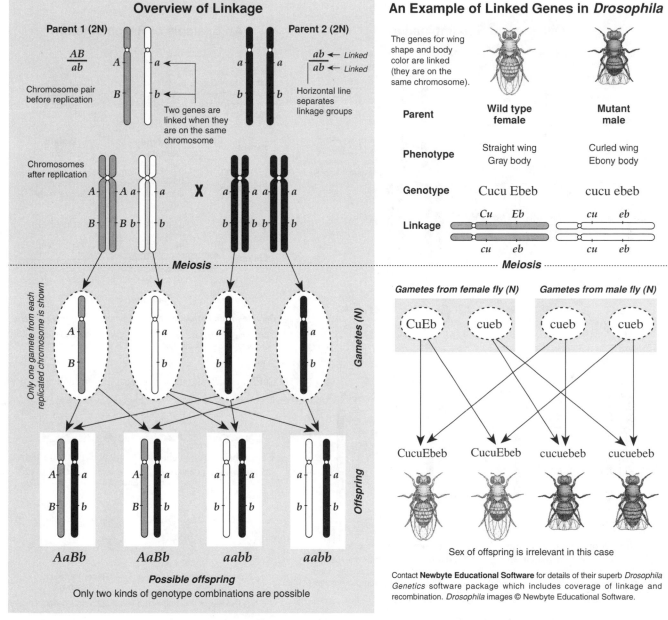

Overview of Linkage

Possible offspring
Only two kinds of genotype combinations are possible

An Example of Linked Genes in *Drosophila*

Sex of offspring is irrelevant in this case

Contact **Newbyte Educational Software** for details of their superb *Drosophila Genetics* software package which includes coverage of linkage and recombination. *Drosophila* images © Newbyte Educational Software.

1. Describe the effect of **linkage** on the inheritance of genes: _____

_____

2. (a) List the possible genotypes in the offspring (above, left) if genes A and B had been on **separate chromosomes**:

_____

(b) If the female *Drosophila* had been homozygous for the dominant wild type alleles (CuCu EbEb), state:

The genotype(s) of the F$_1$: _____ The phenotype(s) of the F$_1$: _____

3. Explain how linkage decreases the amount of genetic variation in the offspring: _____

_____

Related activities: Recombination

A 3

Recombination

Genetic recombination refers to the exchange of alleles between homologous chromosomes as a result of **crossing over**. The alleles of parental linkage groups separate and new associations of alleles are formed in the gametes. Offspring formed from these gametes show new combinations of characteristics and are known as **recombinants** (they are offspring with genotypes unlike either parent). The proportion of recombinants in the offspring can be used to calculate the frequency of recombination (crossover value). These values are fairly constant for any given pair of alleles and can be used to

produce gene maps indicating the relative positions of genes on a chromosome. In contrast to linkage, recombination increases genetic variation. Recombination between the alleles of parental linkage groups is indicated by the appearance of recombinants in the offspring, although not in the numbers that would be expected had the alleles been on separate chromosomes (independent assortment). The example below uses the same genotypes as the previous activity, *Linkage*, but in this case crossing over occurs between the alleles in a linkage group in one parent. The symbology is the same for both activities.

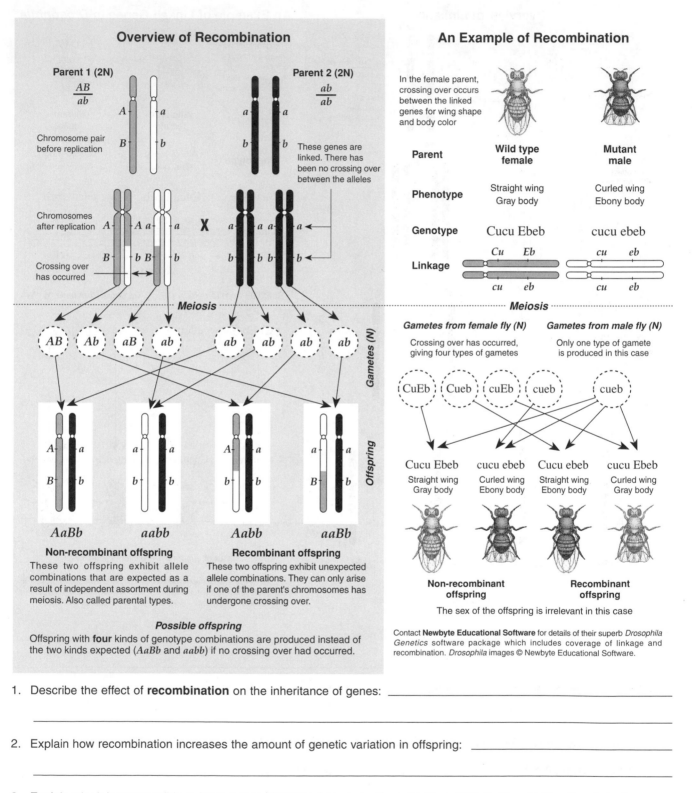

Overview of Recombination

Parent 1 (2N)

Chromosome pair before replication

Chromosomes after replication

Crossing over has occurred

Meiosis

Gametes (N)

AaBb   aabb   Aabb   aaBb

Non-recombinant offspring
These two offspring exhibit allele combinations that are expected as a result of independent assortment during meiosis. Also called parental types.

Recombinant offspring
These two offspring exhibit unexpected allele combinations. They can only arise if one of the parent's chromosomes has undergone crossing over.

Possible offspring
Offspring with **four** kinds of genotype combinations are produced instead of the two kinds expected (*AaBb* and *aabb*) if no crossing over had occurred.

Parent 2 (2N)

These genes are linked. There has been no crossing over between the alleles

An Example of Recombination

In the female parent, crossing over occurs between the linked genes for wing shape and body color

	Wild type female	Mutant male
Parent		
Phenotype	Straight wing Gray body	Curled wing Ebony body
Genotype	Cucu Ebeb	cucu ebeb
Linkage	*Cu* *Eb* / *cu* *eb*	*cu* *eb* / *cu* *eb*

Meiosis

Gametes from female fly (N)
Crossing over has occurred, giving four types of gametes

CuEb   Cueb   cuEb   cueb

Gametes from male fly (N)
Only one type of gamete is produced in this case

cueb

Cucu Ebeb
Straight wing Gray body

cucu ebeb
Curled wing Ebony body

Cucu ebeb
Straight wing Ebony body

cucu Ebeb
Curled wing Gray body

Non-recombinant offspring

Recombinant offspring

The sex of the offspring is irrelevant in this case

Contact **Newbyte Educational Software** for details of their superb *Drosophila Genetics* software package which includes coverage of linkage and recombination. *Drosophila* images © Newbyte Educational Software.

1. Describe the effect of **recombination** on the inheritance of genes: _____

_____

2. Explain how recombination increases the amount of genetic variation in offspring: _____

_____

3. Explain why it is not possible to have a recombination frequency of greater than 50% (half recombinant progeny):

_____

Related activities: Linkage

The Advantages of Sex

Asexual and **sexual reproduction** have different outcomes in terms of the inheritance of characteristics. In asexually reproducing organisms, the offspring are **clones** of the parent and clones with favorable mutations compete with other clones. Sexually reproducing organisms produce many variants and at any one time, the best suited "allele combinations" will be more successful and produce more offspring. Many scientists have questioned why sexual reproduction persists despite its higher energetic costs. One hypothesis (called the red Queen hypothesis) is that sexual reproduction enables constant "trialling" of variants against the ever-changing adaptations of pathogens and parasites (below left).

Chromosomes and Meiosis

Larger Organisms Need Sex

From the moment you're born, you are exposed to parasites and pathogens in the environment. The Red Queen hypothesis proposes that sexual reproduction persists because it enables host species to evolve new genetic defenses against rapidly evolving pathogens and parasites in the prevailing environment. Sex increases the rate at which adaptation can occur; it recombines alleles and increases the likelihood that advantageous pairings of alleles will occur. Sex also provides a way to "store" unfavorable genes and continually try them in combination, waiting for the time when the focus of disadvantage has moved elsewhere.

There is certainly anecdotal evidence for this hypothesis. The **topminnow**, (below right), is under constant attack by a parasite that causes black-spot disease. It sometimes crossbreeds with another similar fish to produce an asexual hybrid, and these asexually producing topminnows harbor many more black-spot worms than those producing sexually. This finding fits the Red Queen hypothesis: the sexual topminnows could devise new defenses faster by recombination than the asexually producing ones.

Above:
Deer tick on toddler's scalp

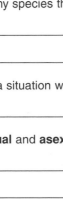

Left:
Flea bites

Photo: Seotaro

Changing Reproductive Strategies

Aphids can switch their reproductive strategy to suit the prevailing conditions. During the summer, aphid populations comprise only females, which produce live nymphs (all females), without the involvement of meiosis or male fertilization. This asexual reproduction maximizes population numbers to take advantage of the abundant food supply.

In autumn, they switch to sexual reproduction, producing winged males and females. These produce gametes by meiosis and mate. The eggs produced survive winter in diapause. These "sexual" eggs increase variation through recombination of alleles. Note that if an advantageous mutation occurs in an asexual line, it is impossible for that mutation to spread without wiping out all other lines, which may have different advantageous mutations of their own.

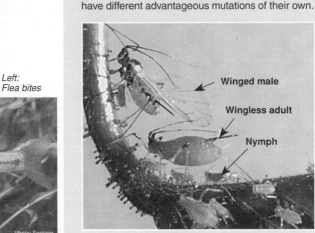

Winged male

Wingless adult

Nymph

1. Explain how sexual reproduction confers an advantage over asexual reproduction in the long term:

_____

_____

2. (a) Explain why species that reproduce asexually can out-perform sexually reproducing organisms in some situations:

_____

_____

(b) Describe a situation where species that reproduces asexually might not be favoured: _____

_____

3. Contrast **sexual** and **asexual** reproduction with respect to the inheritance of favorable mutations: _____

_____

_____

4. Explain how a reproductive strategy involving **alternation** of sexual and asexual phases might benefit a species:

_____

_____

Related activities: For Harm or Benefit?
Web links: The Advantage of Sex, The Red Queen

A 2

Genomes

Genome research has become an important field of genetics. A **genome** is the entire haploid complement of genetic material of a cell or organism. Each species has a unique genome, although there is a small amount of genetic variation between individuals within a species. For example, in humans the average genetic difference is one in every 500-1000 bases. Every cell in an individual has a complete copy of the genome. The base sequence shown below is the total DNA sequence for the genome of a virus. There are nine genes in the sequence, coding for nine different proteins. At least 2000 times this amount of DNA would be found in a single bacterial cell. Half a million times this quantity of DNA would be found in the genome of a single human cell. The first gene has been highlighted gray, while the start and stop codes are in black rectangles.

Genome for the φX174 bacterial virus

Start

The gray area represents the nucleotide sequence for a single gene

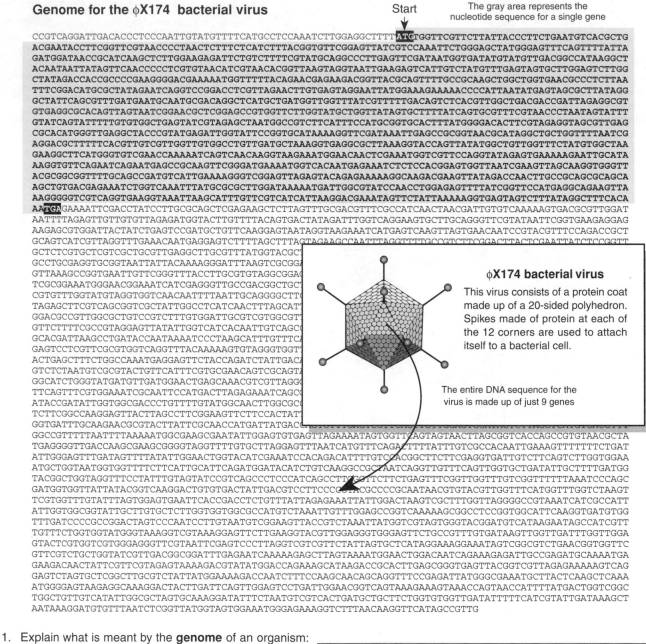

φX174 bacterial virus

This virus consists of a protein coat made up of a 20-sided polyhedron. Spikes made of protein at each of the 12 corners are used to attach itself to a bacterial cell.

The entire DNA sequence for the virus is made up of just 9 genes

1. Explain what is meant by the **genome** of an organism: _____

    _____

2. Determine the number of bases, kilobases, and megabases in this genome (100 bases in each row, except the last):

    1 kb = 1 kilobase = 1000 bases   **1 Mb** = 1 megabase = 1 000 000 bases

    (a) Bases: _____   (b) Kilobases: _____   (c) Megabases: _____

3. Determine how many bases are present in the gene shown above (in the gray area): _____

4. State whether the genome of the virus above is **small, average** or **large** in size compared to those viruses listed in the table on the earlier page *DNA Molecules* (in the topic Molecular Genetics):

    _____

Related activities: DNA Molecules, Genome Projects, The Human Genome Project

© Biozone International 2001-2010
Photocopying Prohibited

Eukaryotic Chromosome Structure

The chromosomes of eukaryotes are more complex than those of prokaryotes. Chromosomes are made up of a complex of DNA and protein called **chromatin**. The DNA is coiled at several levels so that the long DNA molecules can be packed into the nucleus. This **condensation** is achieved by wrapping the DNA around protein cores and then further folding and wrapping the chromatin fiber (as described below) around a protein scaffold. During the early stage of meiosis, a chromosome consists of two chromatids. A non-dividing cell would have chromosomes with the 'equivalent' of a single chromatid only.

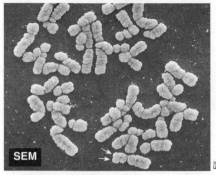

SEM

A cluster of human chromosomes seen during metaphase of cell division. Individual chromatids (arrowed) are difficult to discern on these double chromatid chromosomes.

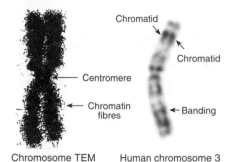

Chromatid

Chromatid

Centromere

Chromatin fibres

Banding

Chromosome TEM   Human chromosome 3

A human chromosome from a dividing white blood cell (above left). Note the compact organization of the chromatin in the two chromatids. The LM photograph (above right) shows the banding visible on human chromosome 3.

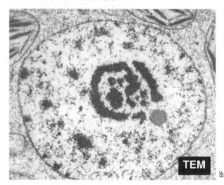

TEM

In non-dividing cells, chromosomes exist as single-armed structures. They are not visible as coiled structures, but are 'unwound' to make the genes accessible for transcription (above).

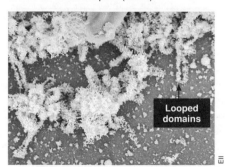

Looped domains

The evidence for the existence of looped domains comes from the study of giant lampbrush chromosomes in amphibian oocytes (above). Under electron microscopy, the lateral loops of the DNA-protein complex have a brushlike appearance.

The Packaging of Chromatin

Chromatin structure is based on successive levels of DNA packing. **Histone proteins** are responsible for packing the DNA into a compact form. Without them, the DNA could not fit into the nucleus. Five types of histone proteins form a complex with DNA, in a way that resembles "beads on a string". These beads, or **nucleosomes**, form the basic unit of DNA packing.

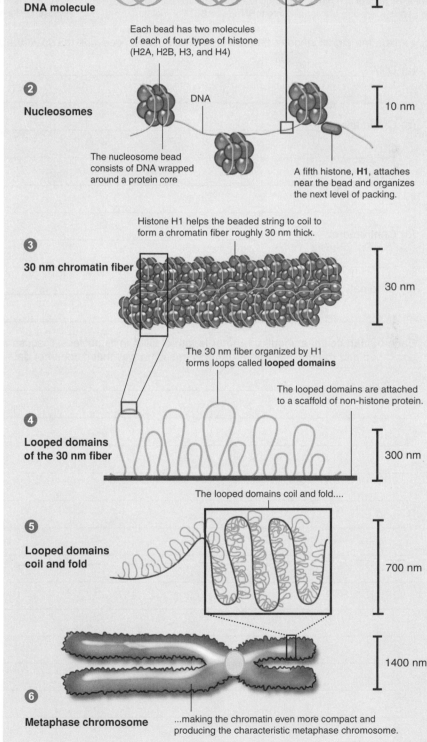

1

DNA molecule

2 nm

Each bead has two molecules of each of four types of histone (H2A, H2B, H3, and H4)

2

Nucleosomes

DNA

10 nm

The nucleosome bead consists of DNA wrapped around a protein core

A fifth histone, **H1**, attaches near the bead and organizes the next level of packing.

Histone H1 helps the beaded string to coil to form a chromatin fiber roughly 30 nm thick.

3

30 nm chromatin fiber

30 nm

The 30 nm fiber organized by H1 forms loops called **looped domains**

The looped domains are attached to a scaffold of non-histone protein.

4

Looped domains of the 30 nm fiber

300 nm

The looped domains coil and fold....

5

Looped domains coil and fold

700 nm

6

Metaphase chromosome

1400 nm

...making the chromatin even more compact and producing the characteristic metaphase chromosome.

Chromosomes and Meiosis

Related activities: DNA Molecules
Web links: Chromosome Structure, Prokaryotic Chromosome Structure

A 2

152

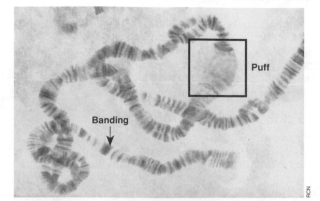

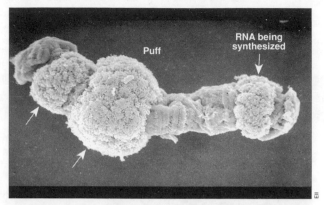

Banded chromosome: This light microscope photo is a view of the polytene chromosomes in a salivary gland cell of a sandfly. It shows a banding pattern that is thought to correspond to groups of genes. Regions of chromosome **puffing** are thought to occur where the genes are being transcribed into mRNA (see SEM on right).

A **polytene chromosome** viewed with a scanning electron microscope (SEM). The arrows indicate localized regions of the chromosome that are uncoiling to expose their genes (puffing) to allow transcription of those regions. Polytene chromosomes are a special type of chromosome consisting of a large bundle of chromatids bound tightly together.

1. Explain the significance of the following terms used to describe the structure of chromosomes:

(a) DNA: _____

_____

(b) Chromatin: _____

_____

(c) Histone: _____

_____

(d) Centromere: _____

_____

(e) Chromatid: _____

_____

2. Each human cell has about a 1 meter length of DNA in its nucleus. Discuss the mechanisms by which this DNA is packaged into the nucleus and organized in such a way that it does not get ripped apart during cell division:

_____

_____

_____

_____

_____

_____

_____

_____

_____

_____

_____

_____

_____

_____

_____

Karyotypes

The diagram below shows the **karyotype** of a normal human. Karyotypes are prepared from the nuclei of cultured white blood cells that are 'frozen' at the metaphase stage of mitosis (see the photo circled opposite). Photographs of the chromosomes are arranged on a grid so that the homologous pairs are placed together. Homologous pairs are identified by their general shape, length, and the pattern of banding produced by a special staining technique. Karyotypes for a human male and female are shown below. The **male karyotype** has 44 autosomes, a single X chromosome, and a Y chromosome (written as 44 + XY), whereas the **female karyotype** shows two X chromosomes (written as 44 + XX).

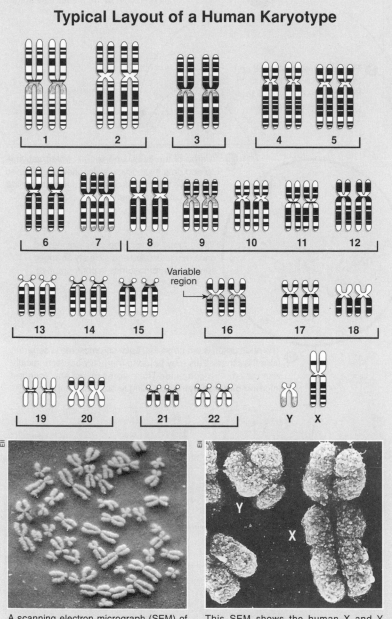

Typical Layout of a Human Karyotype

1 2 3 4 5

6 7 8 9 10 11 12

13 14 15 Variable region 16 17 18

19 20 21 22 Y X

A scanning electron micrograph (SEM) of human chromosomes clearly showing their double chromatids.

This SEM shows the human X and Y chromosomes. Although these two are the sex chromosomes, they are not homologous.

Karyotypes for different species

The term **karyotype** refers to the chromosome complement of a cell or a whole organism. In particular, it shows the number, size, and shape of the chromosomes as seen during metaphase of mitosis. The diagram on the left depicts the human karyotype. Chromosome numbers vary considerably among organisms and may differ markedly between closely related species:

Organism	Chromosome number (2N)
Vertebrates	
human	46
chimpanzee	48
gorilla	48
horse	64
cattle	60
dog	78
cat	38
rabbit	44
rat	42
turkey	82
goldfish	94
Invertebrates	
fruit fly, *Drosophila*	8
housefly	12
honey bee	32 or 16
Hydra	32
Plants	
cabbage	18
broad bean	12
potato	48
orange	18, 27 or 36
barley	14
garden pea	14
Ponderosa pine	24

NOTE: The number of chromosomes is not a measure of the quantity of genetic information.

1. Explain what a **karyotype** is and comment on the information it provides: _____

_____

_____

_____

2. Distinguish between **autosomes** and **sex chromosomes**: _____

_____

_____

_____

Related activities: Human Karyotype Exercise
Web links: Making a Karyotype

RA 1

Chromosomes and Meiosis

Preparing a Karyotype

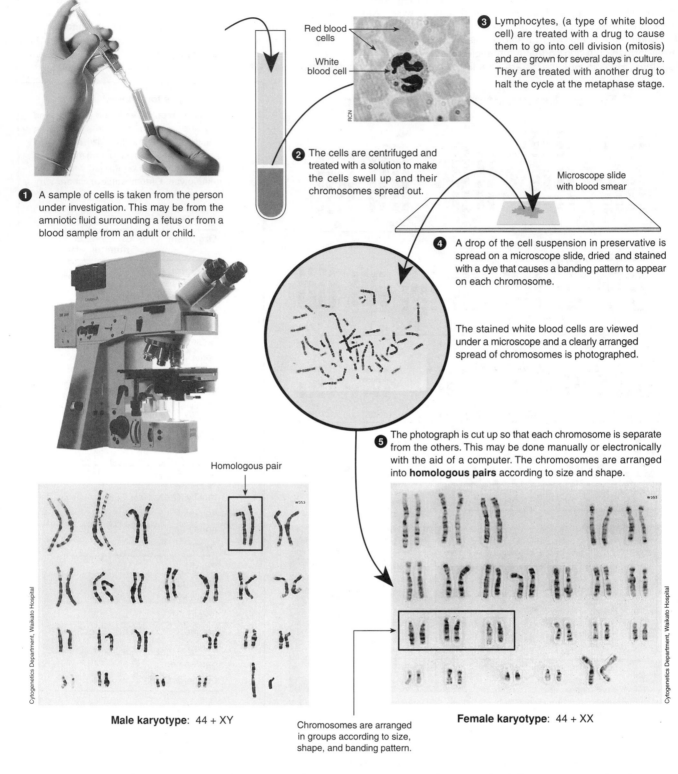

① A sample of cells is taken from the person under investigation. This may be from the amniotic fluid surrounding a fetus or from a blood sample from an adult or child.

② The cells are centrifuged and treated with a solution to make the cells swell up and their chromosomes spread out.

Red blood cells

White blood cell

③ Lymphocytes, (a type of white blood cell) are treated with a drug to cause them to go into cell division (mitosis) and are grown for several days in culture. They are treated with another drug to halt the cycle at the metaphase stage.

Microscope slide with blood smear

④ A drop of the cell suspension in preservative is spread on a microscope slide, dried and stained with a dye that causes a banding pattern to appear on each chromosome.

The stained white blood cells are viewed under a microscope and a clearly arranged spread of chromosomes is photographed.

⑤ The photograph is cut up so that each chromosome is separate from the others. This may be done manually or electronically with the aid of a computer. The chromosomes are arranged into **homologous pairs** according to size and shape.

Homologous pair

Male karyotype: 44 + XY

Chromosomes are arranged in groups according to size, shape, and banding pattern.

Female karyotype: 44 + XX

Cytogenetics Department, Waikato Hospital

3. On the male and female karyotype photographs *above* **number** each homologous pair of chromosomes using the diagram on the previous page as a guide.

4. **Circle** the sex chromosomes (**X** and **Y**) in the female karyotype and male karyotype.

5. Write down the number of *autosomes* and the arrangement of *sex chromosomes* for each sex:

(a) **Female**:   No. of autosomes: _____   Sex chromosomes: _____

(b) **Male**:   No. of autosomes: _____   Sex chromosomes: _____

6. State how many chromosomes are found in a:

(a) Normal human body (**somatic**) cell: _____   (b) Normal human sperm or egg (**gametic**) cell: _____

Human Karyotype Exercise

Each chromosome has distinctive features that enable it to be identified and distinguished from others. Chromosomes are stained in a special technique that gives them a banded appearance. The banding pattern represents regions of the chromosome that contain up to many hundreds of genes.

Determine the sex and chromosome condition of the individual whose chromosomes are displayed below. The karyotypes presented on the previous pages, and the hints on how to recognize chromosome pairs, can be used to help you complete this activity.

Distinguishing Characteristics of Chromosomes

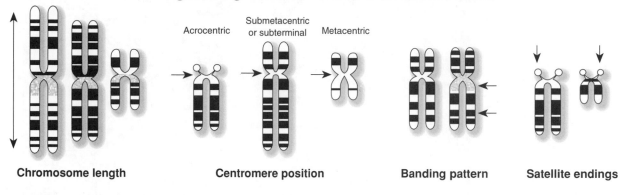

Chromosome length　　　**Centromere position**　　　**Banding pattern**　　　**Satellite endings**

Acrocentric　　Submetacentric or subterminal　　Metacentric

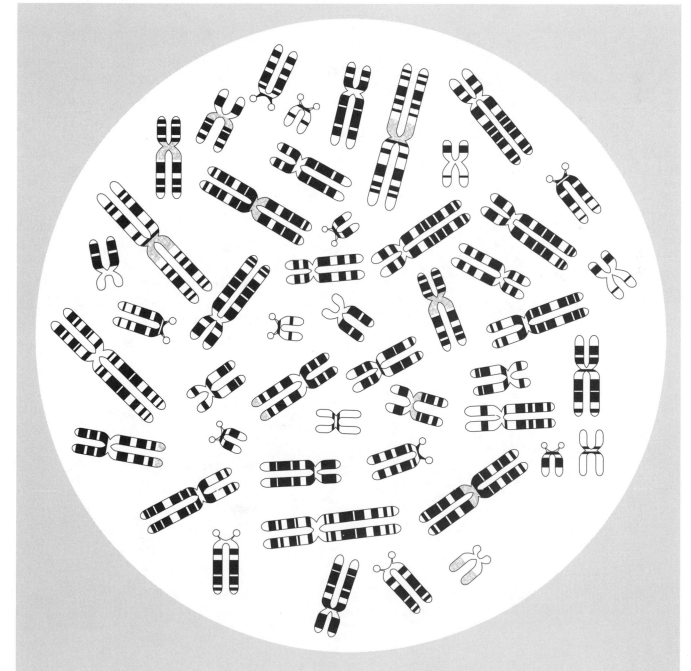

Chromosomes and Meiosis

This page is left blank deliberately

1. Cut out the chromosomes on page 155 and arrange them on the record sheet below in their homologous pairs.

2. (a) Determine the sex of this individual: **male** or **female** (circle one)

   (b) State whether the individual's *chromosome arrangement* is: **normal** or **abnormal** (circle one)

   (c) If the arrangement is *abnormal*, state in what way and name the syndrome displayed: _____

   _____

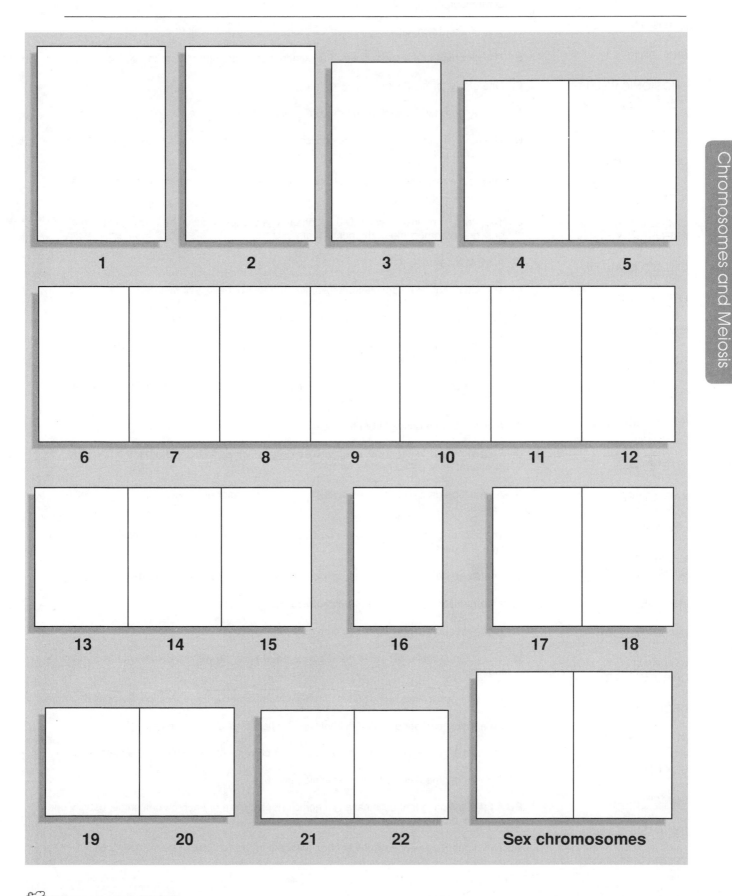

| 1 | 2 | 3 | 4 | 5 |

| 6 | 7 | 8 | 9 | 10 | 11 | 12 |

| 13 | 14 | 15 | 16 | 17 | 18 |

| 19 | 20 | 21 | 22 | **Sex chromosomes** |

Chromosomes and Meiosis

KEY TERMS: Mix and Match

INSTRUCTIONS: Test your vocab by matching each term to its correct definition, as identified by its preceding letter code.

ALLELE

AMNIOCENTESIS

ANAPHASE 1

AUTOSOME

BIVALENT

CHORIONIC VILLUS SAMPLING

CHROMATID

CHROMOSOME

CROSSING OVER

CROSSOVER FREQUENCY

DIPLOID (2N)

FERTILIZATION

GAMETE

HAPLOID (1N)

HISTONE PROTEIN

INDEPENDENT ASSORTMENT

INTERPHASE

KARYOTYPE

LINKAGE

MATERNAL CHROMOSOME

MEIOSIS

METAPHASE 1

PATERNAL CHROMOSOME

PROPHASE 1

RECOMBINATION

SEX CHROMOSOMES

SOMATIC CELL

SYNAPSIS

TELOPHASE 1

A A term denoting that cells have one copy of each chromosome, as in a gamete.

B An amniotic fluid test used in prenatal diagnosis of chromosomal abnormalities.

C A percentage value that provides a measure of the relative distance between genes on a chromosome.

D The union of two haploid gametes to reconstitute a diploid cell (the zygote).

E The chromosome responsible for the deterination of gender; in humans, X and Y.

F One of the forms a gene may take.

G The random assortment of chromosomes during meiosis.

H The stage in meiosis 1 when the duplicated chromatin condenses. Crossing-over can occur during the latter part of this stage.

I One of the two identical copies of DNA, joined at their centromeres, which make up a replicated chromosome.

J The stage in meiosis 1 when two daughter cells are formed with each daughter containing only one chromosome of the homologous pair.

K A haploid sex cell.

L The pairing of homologous chromosomes during prophase 1 of meiosis.

M Any cells forming the body of an organism, as opposed to the germline cells.

N The tendency of certain alleles to be inherited together because they are located close together on a chromosome.

O A prenatal diagnostic test for chromosome abnormalities which involves the extraction of a small amount of placental tissue.

P A chromosome derived from the male parent.

Q Proteins found in the nuclei of eukaryotic cells, which package and order the DNA into structural units called nucleosomes.

R A reduction division in eukaryotic cells in which the number of chromosomes per cell is halved.

S The exchange of alleles between homologous chromosomes during meiosis as a result of crossing over.

T The stage before cell division begins when the genetic material is duplicated.

U An organized structure of DNA and protein found in cells.

V A chromosome that is not a sex chromosome.

W The stage in meiosis 1 when homologous pairs separate with sister chromatids remaining together.

X A pair of associated homologous chromosomes formed after replication; also called a tetrad.

Y An exchange of genetic material between homologous chromosomes.

Z The stage in meiosis 1 when the homologous chromosomes align at the equatorial plate.

AA A chromosome derived from the female parent.

BB A term denoting that cells have two homologous copies of each chromosome, usually one from the mother and one from the father.

CC The number and appearance of chromosomes in the nucleus of a eukaryotic cell.

Heredity

KEY CONCEPTS

▶ Sexual reproduction introduces variation in the offspring: the raw material for natural selection.

▶ The dominance of alleles can be inferred from the genetic outcomes of crosses.

▶ Lethal alleles and linkage cause departures from expected offspring phenotype ratios.

▶ Gene interactions contribute to genetic variation.

▶ Environmental factors can modify the phenotype encoded by genes.

KEY TERMS

allele
autosome
back cross
chi-squared test
cline
codominance (of alleles)
collaboration
complementary genes
continuous variation
cross
dihybrid cross
discontinuous variation
dominant (of alleles)
epigenetics
epistasis
F1 / F2 generation
genetic counseling
genomic imprinting
genotype
heterozygous
homozygous
incomplete dominance
lethal allele
linkage
monohybrid cross
multiple alleles
pedigree analysis
phenotype
polygenes (=multiple genes)
Punnett square
pure (true)-breeding
recessive (of alleles)
recombination
reduction division
selfing
sex chromosome
sex linked gene
test cross

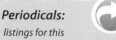

Periodicals:
listings for this chapter are on page 380

Weblinks:
www.thebiozone.com/
weblink/SB1-2597.html

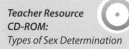

Teacher Resource CD-ROM:
Types of Sex Determination

OBJECTIVES

☐ 1. Use the **KEY TERMS** to help you understand and complete these objectives.

Variation and the Study of Inheritance   pages 160-180, 197-199

☐ 2. Recall the role of sexual reproduction in generating genetic variation. Explain how genotype and environment contribute to phenotypic **variation**.

☐ 3. Describe examples of **discontinuous** and **continuous variation** in the characteristics of organisms. Explain the genetic basis for each pattern (see #17)

☐ 4. Summarize Mendel's **principles of inheritance** and their importance to our understanding of heredity and evolution.

☐ 5. Demonstrate appropriate use of the terms commonly used in inheritance studies: **allele**, **locus**, **trait**, **heterozygous, homozygous, genotype, phenotype, cross, test cross, back cross, carrier, offspring, trait, F1 generation, F2 generation**.

☐ 6. Solve problems involving **monohybrid inheritance** with a simple **dominant-recessive** pattern.

☐ 7. Describe and explain inheritance involving: **codominance, incomplete dominance, multiple alleles, lethal alleles**, and **sex linked genes**.

☐ 8. Solve problems involving **dihybrid inheritance** of **unlinked, autosomal genes** for two independent characteristics.

☐ 9. EXTENSION: Explain how you could determine, from a cross, if genes are linked. Solve problems involving **dihybrid inheritance** of **linked genes**.

Sex Determination, Sex Linkage, and Epigenetics   pages 160,181-189

☐ 10. Describe and explain an example of **genomic imprinting**.

☐ 11. Explain the basis of **sex determination** in humans. Recognize humans as being of the XX / XY type. Distinguish **sex chromosomes** from **autosomes**.

☐ 12. Describe examples and solve problems involving different patterns of inheritance involving **sex linked genes** (e.g. red-green color-blindness or hemophilia).

☐ 13. Describe and explain the use of **pedigree analysis** to illustrate the inheritance of traits in a 'family tree'. Describe the principles and role of **genetic counseling**.

Gene Interactions   pages 190-196

☐ 14. Recognize and describe a simple interaction between two genes: e.g. **collaboration** in the determination of comb shape of domestic hens.

☐ 15. Describe and explain **pleiotropy**, e.g. sickle cell gene mutation or PKU.

☐ 16. Describe and explain **epistasis**. Examples include the inheritance of coat color in mammals and the control of flower color by **complementary genes** in sweet peas.

☐ 17. Describe and explain inheritance involving **polygenes (multiple genes)**.

A Gene That Can Tell Your Future

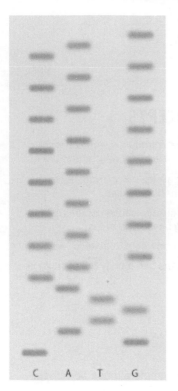

Huntington's disease (HD) is a genetic neuro-degenerative disease that normally does not affect people until about the age of 40. Its symptoms usually appear first as a shaking of the hands and an awkward gait. Later manifestations of the disease include serious loss of muscle control and mental function, often ending in dementia and ultimately death.

All humans have the huntingtin (**HTT**) gene, which in its normal state produces a protein with roles in gene transcription, synaptic transmission, and brain cell survival. The mutant gene (**mHTT**) causes changes to and death of the cells of the cerebrum, the hippocampus, and cerebellum, resulting in the atrophy (reduction) of brain matter. The gene was discovered by Nancy Wexler in 1983 after ten years of research working with cell samples and family histories of more than 10,000 people from the town of San Luis in Venezuela, where around 1% of the population have the disease (compared to about 0.01% in the rest of the world). Ten years later the exact location of the gene on the chromosome 4 was discovered.

The identification of the HD gene began by looking for a gene probe that would bind to the DNA of people who had HD, and not to those who didn't. Eventually a marker for HD, called **G8**, was found. The next step was to find which chromosome carried the marker and where on the chromosome it was. The researchers hybridised human cells with those of mice so that each cell contained only one human chromosome, a different chromosome in each cell. The hybrid cell with chromosome 4 was the one with the G8 marker. They then found a marker that overlapped G8 and then another marker that overlapped that marker. By repeating this many times, they produced a map of the genes on chromosome 4. The researchers then sequenced the genes and found people who had HD had one gene that was considerably longer than people who did not have HD. Moreover the increase in length was caused by the repetition of the base sequence CAG.

The HD mutation (mHTT) is called a trinucleotide repeat expansion. In the case of mHTT, the base sequence CAG is repeated multiple times on the short arm of chromosome 4. The normal number of CAG repeats is between 6 and 30. The mHTT gene causes the repeat number to be 35 or more and the size of the repeat often increases from generation to generation, with the severity of the disease increasing with the number of repeats. Individuals who have 27 to 35 CAG repeats in the HTT gene do not develop Huntington disease, but they are at risk of having children who will develop the disorder. The mutant allele, mHTT, is also dominant, so those who are homozygous or heterozygous for the allele are both at risk of developing HD.

New research has shown that the mHTT gene activates an enzyme called JNK3, which is expressed only in the neurones and causes a drop in nerve cell activity. While a person is young and still growing, the neurones can compensate for the accumulation of JNK3. However, when people get older and neurone growth stops, the effects of JNK3 become greater and the physical signs of HD become apparent. Because of mHTT's dominance, an affected person has a 50% chance of having offspring who are also affected. Genetic testing for the disease is relatively easy now that the genetic cause of the disease is known. While locating and counting the CAG repeats does not give a date for the occurrence of HD, it does provide some understanding of the chances of passing on the disease.

1. Describe the physical effects of Huntington's disease: _____
_____

2. Describe how the mHTT gene was discovered: _____
_____
_____
_____

3. Discuss the cause of Huntington's disease and its pattern of increasing severity with each generation: _____
_____
_____
_____
_____
_____

Variation

Variation is a characteristic of all living organisms; we see it not only between species but between individuals of the same species. The genetic variability within species is due mostly to a **shuffling** of the existing genetic material into new combinations as genetic information is passed from generation to generation. In addition to this, **mutation** creates new alleles in individuals. While most mutations are harmful, some are 'silent' (without visible effect on the phenotype), and some may even be beneficial. Depending on the nature of the inheritance pattern, variation in a population can be continuous or discontinuous. Traits determined by a single gene (e.g. ABO blood groups) show **discontinuous variation**, with a very limited number of variants present in the population. In contrast, traits determined by a large number of genes (e.g. skin color) show **continuous variation**, and the number of phenotypic variations is exceedingly large. Environmental influences (differences in diet for example) also contribute to the observable variation in a population, helping or hindering the expression of an individual's full genetic potential.

Albinism (above) is the result of the inheritance of recessive alleles for melanin production. Those with the albino phenotype lack melanin pigment in the eyes, skin, and hair.

Comb shape in poultry is a **qualitative trait** and birds have one of four phenotypes depending on which combination of four alleles they inherit. The dash (missing allele) indicates that the allele may be recessive or dominant.

Quantitative traits are characterised by **continuous variation**, with individuals falling somewhere on a normal distribution curve of the phenotypic range. Typical examples include skin colour and height in humans (left), grain yield in corn (above), growth in pigs (above, left), and milk production in cattle (far left). Quantitiative traits are determined by genes at many loci (polygenic) but most are also influenced by environmental factors.

Single comb	Walnut comb	Pea comb	Rose comb
rrpp	**R_P_**	**rrP_**	**R_pp**

Flower colour in snapdragons (right) is also a **qualitative trait** determined by two alleles. (red and white) The alleles show incomplete dominance and the heterozygote ($C^R C^W$) exhibits an intermediate phenotype between the two homozygotes.

$C^R C^R$

$C^W C^W$

Heredity

Sources of Variation in Organisms

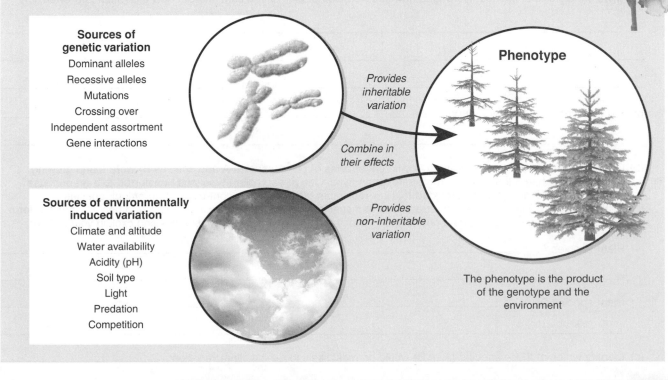

Sources of genetic variation
Dominant alleles
Recessive alleles
Mutations
Crossing over
Independent assortment
Gene interactions

Provides inheritable variation

Phenotype

Combine in their effects

Sources of environmentally induced variation
Climate and altitude
Water availability
Acidity (pH)
Soil type
Light
Predation
Competition

Provides non-inheritable variation

The phenotype is the product of the genotype and the environment

Periodicals:
What is variation?

Related activities: Descriptive Statistics, Interpreting Sample Variability

The Effects of Environment on Phenotype

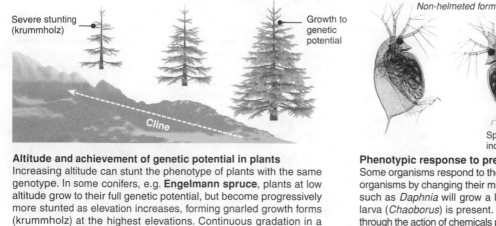

Altitude and achievement of genetic potential in plants

Increasing altitude can stunt the phenotype of plants with the same genotype. In some conifers, e.g. **Engelmann spruce**, plants at low altitude grow to their full genetic potential, but become progressively more stunted as elevation increases, forming gnarled growth forms (krummholz) at the highest elevations. Continuous gradation in a phenotypic character within a species, associated with a change in an environmental variable, is called a **cline**.

Phenotypic response to predation in zooplankton

Some organisms respond to the presence of other, potentially harmful, organisms by changing their morphology or body shape. Invertebrates such as *Daphnia* will grow a large helmet when a predatory midge larva (*Chaoborus*) is present. Such responses are usually mediated through the action of chemicals produced by the predator (or competitor), and are common in plants as well as animals.

1. Giving appropriate examples, distinguish clearly between **genotype** and **phenotype**: _____

_____

_____

2. Identify each of the following phenotypic traits as continuous (quantitative) or discontinuous (qualitative):

   (a) Wool production in sheep: _____ (d) Albinism in mammals: _____

   (b) Hand span in humans: _____ (e) Body weight in mice: _____

   (c) Blood groups in humans: _____ (f) Flower color in snapdragons: _____

3. In the examples above, identify those in which an environmental influence on phenotype could be expected:

_____

4. Identify some of the physical factors associated with altitude that could affect plant phenotype: _____

_____

5. The hydrangea is a plant that exhibits a change in the color of its flowers according to the condition of the soil. Identify the physical factor that causes hydrangea flowers to be blue or pink. If you can, find out how this effect is exerted:

_____

_____

_____

6. (a) Explain what is meant by a **cline**: _____

_____

   (b) On a windswept portion of a coast, two different species of plant (species A and species B) were found growing together. Both had a low growing (prostrate) phenotype. One of each plant type was transferred to a greenhouse where "ideal" conditions were provided to allow maximum growth. In this controlled environment, species B continued to grow in its original prostrate form, but species A changed its growing pattern and became erect in form. Identify the **cause** of the prostrate phenotype in each of the coastal grown plant species and explain your answer:

   Plant species A: _____

_____

   Plant species B: _____

_____

   (c) Identify which of these species (A or B) would be most likely to exhibit clinal variation: _____

Alleles

Sexually reproducing organisms in nearly all cases have paired sets of chromosomes, one set coming from each parent. The equivalent chromosomes that form a pair are termed **homologues**. They contain equivalent sets of genes on them. But there is the potential for different versions of a gene to exist in a population and these are termed **alleles**.

Homologous Chromosomes

In sexually reproducing organisms, most cells have a homologous pair of chromosomes (one coming from each parent). This diagram shows the position of three different genes on the same chromosome that control three different traits (A, B and C).

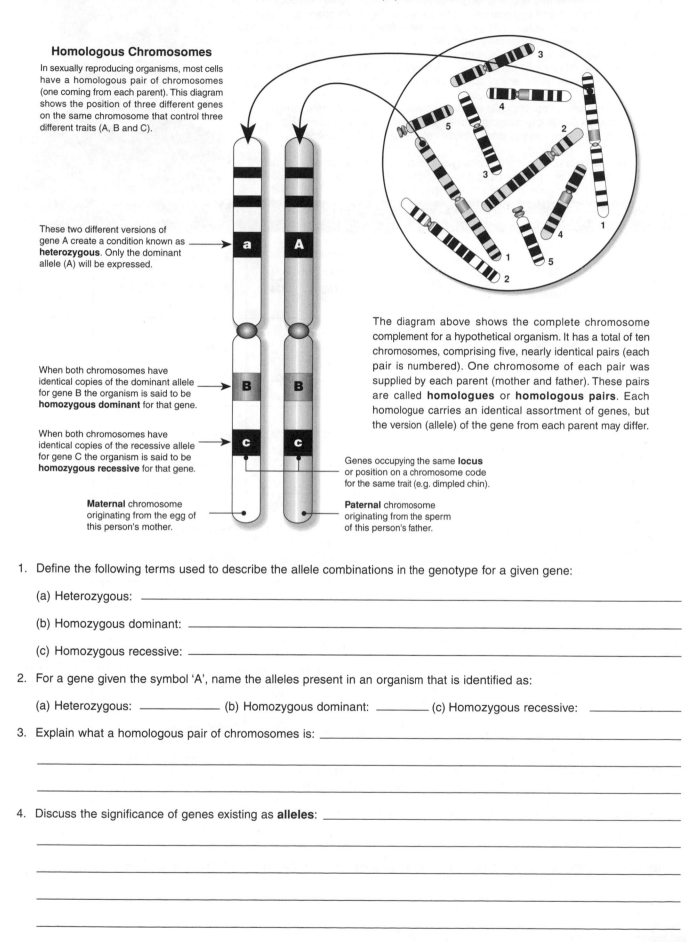

These two different versions of gene A create a condition known as **heterozygous**. Only the dominant allele (A) will be expressed.

When both chromosomes have identical copies of the dominant allele for gene B the organism is said to be **homozygous dominant** for that gene.

When both chromosomes have identical copies of the recessive allele for gene C the organism is said to be **homozygous recessive** for that gene.

Maternal chromosome originating from the egg of this person's mother.

The diagram above shows the complete chromosome complement for a hypothetical organism. It has a total of ten chromosomes, comprising five, nearly identical pairs (each pair is numbered). One chromosome of each pair was supplied by each parent (mother and father). These pairs are called **homologues** or **homologous pairs**. Each homologue carries an identical assortment of genes, but the version (allele) of the gene from each parent may differ.

Genes occupying the same **locus** or position on a chromosome code for the same trait (e.g. dimpled chin).

Paternal chromosome originating from the sperm of this person's father.

Heredity

1. Define the following terms used to describe the allele combinations in the genotype for a given gene:

   (a) Heterozygous: _____

   (b) Homozygous dominant: _____

   (c) Homozygous recessive: _____

2. For a gene given the symbol 'A', name the alleles present in an organism that is identified as:

   (a) Heterozygous: _____ (b) Homozygous dominant: _____ (c) Homozygous recessive: _____

3. Explain what a homologous pair of chromosomes is: _____

   _____

   _____

4. Discuss the significance of genes existing as **alleles**: _____

   _____

   _____

   _____

   _____

Mendel's Pea Plant Experiments

Gregor Mendel (1822-1884), pictured on the right, was an Austrian monk who is regarded as the 'father of genetics'. He carried out some pioneering work using pea plants to study the inheritance patterns of a number of **traits** (characteristics). Mendel observed that characters could be masked in one generation of peas but could reappear in later generations. He showed that inheritance involved the passing on to offspring of discrete units of inheritance; what we now call genes. Mendel examined a number of phenotypic traits and found that they were inherited in predictable ratios, depending on the phenotype of the parents. Below are some of his results from crossing heterozygous plants (e.g. tall plants that were the offspring of tall and dwarf parent plants: Tt x Tt). The numbers in the results column represent how many offspring had those phenotypic features.

1. Study the **results** for each of the six experiments below. Determine which of the two phenotypes is the dominant one, and which is the recessive. Place your answers in the spaces in the **dominance** column in the table below.

2. Calculate the ratio of dominant phenotypes to recessive phenotypes (to two decimal places). The first one (for seed shape) has been done for you (5474 ÷ 1850 = 2.96). Place your answers in the spaces provided in the table below:

Trait	Possible Phenotypes	Results	Dominance	Ratio
Seed shape	*Wrinkled* *Round*	Wrinkled 1850 Round 5474 **TOTAL 7324**	Dominant: Round Recessive: Wrinkled	2.96 : 1
Seed color	*Green* *Yellow*	Green 2001 Yellow 6022 **TOTAL 8023**	Dominant: Recessive:	
Pod color	*Green* *Yellow*	Green 428 Yellow 152 **TOTAL 580**	Dominant: Recessive:	
Flower position	*Axial* *Terminal*	Axial 651 Terminal 207 **TOTAL 858**	Dominant: Recessive:	
Pod shape	*Constricted* *Inflated*	Constricted 299 Inflated 882 **TOTAL 1181**	Dominant: Recessive:	
Stem length	*Tall* *Dwarf*	Tall 787 Dwarf 277 **TOTAL 1064**	Dominant: Recessive:	

3. Mendel's experiments identified that two heterozygous parents should produce offspring in the ratio of three times as many dominant offspring to those showing the recessive phenotype.

   (a) State which three of Mendel's experiments provided ratios closest to the theoretical 3:1 ratio:

   _____

   (b) Suggest a possible reason why these results deviated less from the theoretical ratio than the others:

   _____

   _____

Mendel's Laws of Inheritance

From his work on the inheritance of phenotypic traits in peas, Mendel formulated a number of ideas about the inheritance of characters. These were later given formal recognition as Mendel's laws of inheritance. These are outlined below.

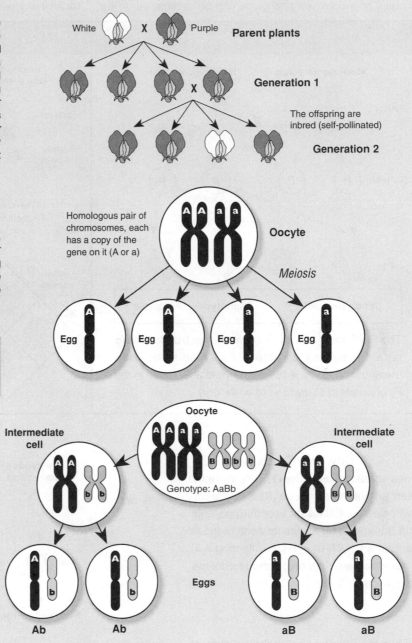

The Theory of Particulate Inheritance

Mendel recognized that characters are determined by discrete units that are inherited intact down through the generations. This model explained many observations that could not be explained by the idea of blending inheritance that was universally accepted prior to this. The diagram on the right illustrates this principle, showing that the trait for flower color appears to take on the appearance of only one parent plant in the first generation, but reappears in later generations.

Law of Segregation

The diagram on the right illustrates how, during meiosis, the two members of any pair of alleles segregate unchanged by passing into different gametes. These gametes are eggs (ova) and sperm cells. The allele in the gamete will be passed on to the offspring.

> NOTE: This diagram has been simplified, omitting the stage where the second chromatid is produced for each chromosome.

Law of Independent Assortment

The diagram on the right illustrates how genes are carried on chromosomes. There are two genes shown (A and B) that code for different traits. Each of these genes is represented twice, one copy (allele) on each of two homologous chromosomes. The genes A and B are located on different chromosomes and, because of this, they will be inherited independently of each other, i.e. the gametes may contain any combination of the parental alleles.

Heredity

1. Briefly state what **property of genetic inheritance** allows parent pea plants that differ in flower color to give rise to flowers of a single color in the first generation, with both parental flower colors reappearing in the following generation:

   _____

2. The oocyte is the egg producing cell in the ovary of an animal. In the diagram illustrating the **law of segregation** above:

   (a) State the genotype for the oocyte (adult organism): _____

   (b) State the genotype of each of the **four** gametes: _____

   (c) State how many different kinds of gamete can be produced by this oocyte: _____

3. The diagram illustrating the **law of independent assortment** (above) shows only one possible result of the random sorting of the chromosomes to produce: Ab and aB in the gametes.

   (a) List another possible combination of genes (on the chromosomes) ending up in gametes from the same oocyte:

   _____

   (b) State how many different gene combinations are possible for the oocyte: _____

Periodicals:
Mendel's legacy

Related activities: Alleles, Mendel's Pea Plant Experiments

A 2

Basic Genetic Crosses

For revision purposes, examine the diagrams below on monohybrid crosses and complete the exercise for dihybrid (two gene) inheritance. A **test cross** is also provided to show how the genotype of a dominant phenotype can be determined. A test cross will yield one of two different results, depending on the genotype of the dominant individual. A **back cross** (not shown) refers to any cross between an offspring and one of its parents (or an individual genetically identical to one of its parents).

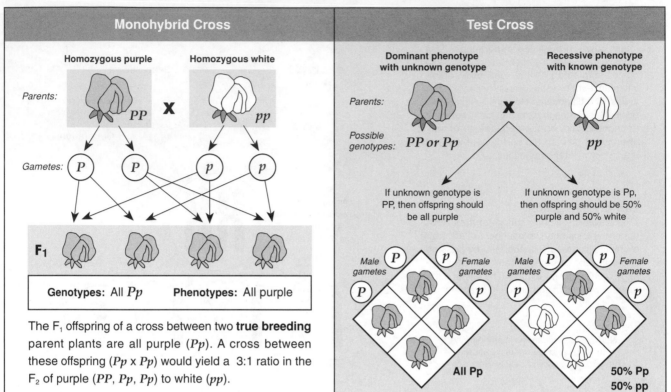

Monohybrid Cross	Test Cross

Monohybrid Cross

Homozygous purple PP **X** Homozygous white pp

Parents:

Gametes: P P p p

F_1

Genotypes: All Pp **Phenotypes:** All purple

The F_1 offspring of a cross between two **true breeding** parent plants are all purple (Pp). A cross between these offspring (Pp x Pp) would yield a 3:1 ratio in the F_2 of purple (PP, Pp, Pp) to white (pp).

Test Cross

Dominant phenotype with unknown genotype **X** Recessive phenotype with known genotype

Parents:

Possible genotypes: PP or Pp pp

If unknown genotype is PP, then offspring should be all purple

If unknown genotype is Pp, then offspring should be 50% purple and 50% white

Male gametes P p Female gametes
P p

All Pp

Male gametes P p Female gametes
p p

50% Pp
50% pp

Dihybrid Cross

In pea seeds, yellow color (Y) is dominant to green (y) and round shape (R) is dominant to wrinkled (r). Each **true breeding** parental plant has matching alleles for each of these characters ($YYRR$ or $yyrr$). F_1 offspring will all have the same genotype and phenotype (yellow-round: $YyRr$).

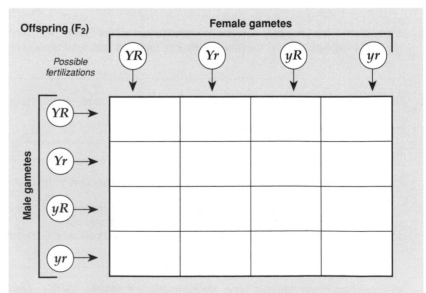

Homozygous yellow-round **X** Homozygous green-wrinkled

Parents:

Gametes: YR yr

F_1 all yellow-round $YyRr$ **X** $YyRr$ for the F_2

1. Fill in the Punnett square (below right) to show the genotypes of the F_2 generation.

2. In the boxes below, use fractions to indicate the numbers of each phenotype produced from this cross.

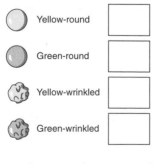

Yellow-round [ ]

Green-round [ ]

Yellow-wrinkled [ ]

Green-wrinkled [ ]

3. Express these numbers as a ratio:

_____

Offspring (F_2)

Possible fertilizations

Female gametes: YR Yr yR yr

Male gametes: YR Yr yR yr

A 2   *Related activities: Monohybrid Cross, Dihybrid Cross, The Test Cross*

The Test Cross

It is not always possible to determine an organism's genotype by its appearance because the expression of genes is complicated by patterns of dominance and by gene interactions. The **test cross** was developed by Gregor Mendel as a way to establish the genotype of an organism with the dominant phenotype for a particular trait. The principle of the test cross is simple. The individual with the unknown genotype is bred with a homozygous recessive individual for the trait(s) of interest. The homozygous recessive can produce only one type of allele (recessive), so the phenotypes of the resulting offspring will reveal the genotype of the unknown parent. For example, if the unknown individual is homozygous for the trait, all of the offspring will display the dominant phenotype. However, if the offspring display both dominant and recessive phenotypes, then the unknown must be heterozygous for that trait. The test cross can be used to determine the genotype of single genes or multiple genes.

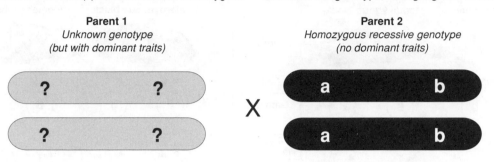

Parent 1
Unknown genotype
(but with dominant traits)

Parent 2
Homozygous recessive genotype
(no dominant traits)

The common fruit fly (*Drosophila melanogaster*) is often used to illustrate basic principles of inheritance because it has several genetic markers whose phenotypes are easily identified . Once such phenotype is body color. Wild type (normal) *Drosophila* have yellow-brown bodies. The allele for yellow-brown body color (E) is dominant. The allele for an ebony colored body (e) is recessive. The test crosses below show the possible outcomes for an individual with homozygous and heterozygous alleles for ebony body color.

A. A homozygous recessive female (ee) with an ebony body is crossed with a homozyogous dominant male (EE).

B. A homozygous recessive female (ee) with an ebony body is crossed with a heterozygous male (Ee).

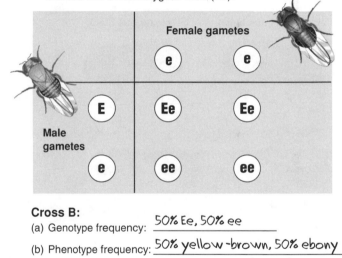

Cross A:
(a) Genotype frequency: _100% Ee_
(b) Phenotype frequency: _100% yellow-brown_

Cross B:
(a) Genotype frequency: _50% Ee, 50% ee_
(b) Phenotype frequency: _50% yellow-brown, 50% ebony_

1. In *Drosophila*, the allele for brown eyes (b) is recessive, while the red eye allele (B) is dominant. Explain how you would set up a **two gene test cross** to determine the genotype of a male who has a normal body color and red eyes:

   _____

   _____

   _____

2. List all of the **possible genotypes** for the male *Drosophila*: _____

   _____

3. 50% of the resulting progeny are yellow-brown bodies with red eyes, and 50% have ebony bodies with red eyes.

   (a) State the genotype of the male *Drosophila*: _____

   (b) Explain how you came to this conclusion: _____

   _____

   _____

Related activities: Monohybrid Cross

A 2

Monohybrid Cross

The study of **single-gene inheritance** is achieved by performing **monohybrid crosses**. The six basic types of matings possible among the three genotypes can be observed by studying a pair of alleles that govern coat color in the guinea pig. A dominant allele: given the symbol **B** produces **black** hair, and its recessive allele: **b**, produces white. Each of the parents can produce two types of gamete by the process of **meiosis** (in reality there are four, but you get identical pairs). Determine the **genotype** and **phenotype frequencies** for the crosses below (enter the frequencies in the spaces provided). For crosses 3 to 6, you must also determine gametes produced by each parent (write these in the circles), and offspring (F₁) genotypes and phenotypes (write in the genotype inside the offspring and state if black or white).

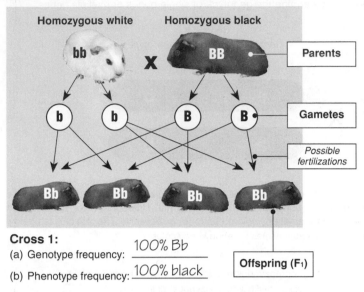

Cross 1:
(a) Genotype frequency: _100% Bb_

(b) Phenotype frequency: _100% black_

Parents

Gametes

Possible fertilizations

Offspring (F₁)

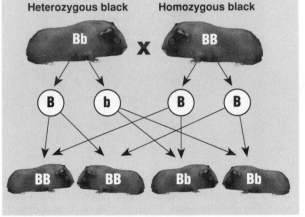

Cross 2:
(a) Genotype frequency: _____

(b) Phenotype frequency: _____

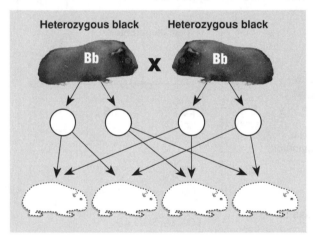

Cross 3:
(a) Genotype frequency: _____

(b) Phenotype frequency: _____

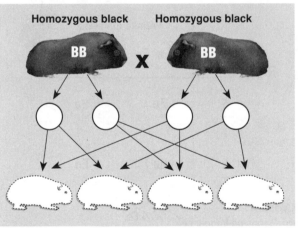

Cross 4:
(a) Genotype frequency: _____

(b) Phenotype frequency: _____

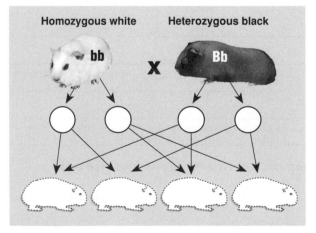

Cross 5:
(a) Genotype frequency: _____

(b) Phenotype frequency: _____

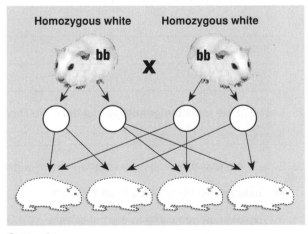

Cross 6:
(a) Genotype frequency: _____

(b) Phenotype frequency: _____

Related activities: Basic Genetic Crosses

Dominance of Alleles

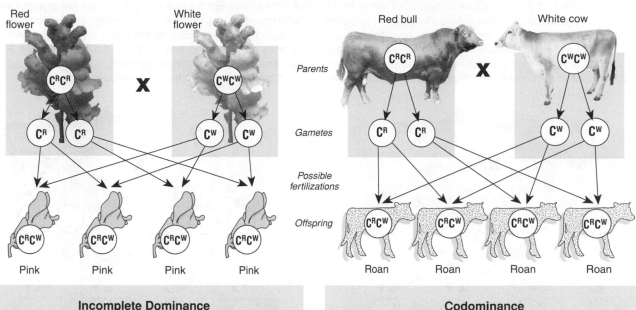

Red flower
CRCR X White flower CWCW

CR CR CW CW

CRCW Pink   CRCW Pink   CRCW Pink   CRCW Pink

Parents — Red bull CRCR X White cow CWCW

Gametes — CR CR CW CW

Possible fertilizations

Offspring — CRCW Roan   CRCW Roan   CRCW Roan   CRCW Roan

Incomplete Dominance

Incomplete dominance refers to the situation where the action of one allele does not completely mask the action of the other and neither allele has dominant control over the trait. The heterozygous offspring is **intermediate** in phenotype between the contrasting homozygous parental phenotypes. In crosses involving incomplete dominance the phenotype and genotype ratios are identical. Examples include snapdragons (*Antirrhinum*), where red and white-flowered parent plants are crossed to produce pink-flowered offspring. In this type of inheritance the phenotype of the offspring results from the partial influence of both alleles.

Codominance

Codominance refers to inheritance patterns when both alleles in a heterozygous organism contribute to the phenotype. Both alleles are **independently** and **equally expressed**. One example includes the human blood group AB which is the result of two alleles: A and B, both being equally expressed. Other examples include certain coat colors in horses and cattle. Reddish coat color is not completely dominant to white. Animals that have both alleles have coats that are **roan**-colored (coats with a mix of red and white hairs). The red hairs and white hairs are expressed equally and independently (not blended to produce pink).

1. In incomplete and codominance, two parents of differing phenotype produce offspring different from either parent. Explain the mechanism by which this occurs in:

   (a) Incomplete dominance: _____

   _____

   (b) Codominance: _____

   _____

2. For each situation below, explain how the heterozygous individuals differ in their phenotype from homozygous ones:

   (a) Incomplete dominance: _____

   (b) Codominance: _____

3. Describe the classical phenotypic ratio for a codominant gene resulting from the cross of two heterozygous parents (in the case of the cattle described above, this would be a cross between two roan cattle). Use the Punnett square (provided right) to help you:

   _____

   _____

   _____

Gametes from male

Gametes from female

4. A plant breeder wanted to produce flowers for sale that were only pink or white (i.e. no red). Determine the phenotypes of the two parents necessary to produce these desired offspring. Use the Punnett square (provided right) to help you:

   _____

   _____

Gametes from male

Gametes from female

Heredity

Related activities: Multiple Alleles in Blood Groups
Web links: Drag and Drop Genetics

A 2

In the shorthorn cattle breed coat color is inherited. White shorthorn parents always produce calves with white coats. Red parents always produce red calves. But when a red parent mates with a white one the calves have a coat color that is different from either parent, called roan (a mixture of red hairs and white hairs). Look at the example on the previous page for guidance and determine the offspring for the following two crosses. In the cross on the left, you are given the phenotype of the parents. From this information, their genotypes can be determined, and therefore the gametes and genotypes and phenotypes of the calves. In the cross on the right, only one parent's phenotype is known. Work out the genotype of the cow and calves first, then trace back to the unknown bull via the gametes, to determine its genotype.

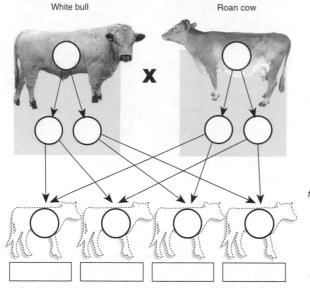

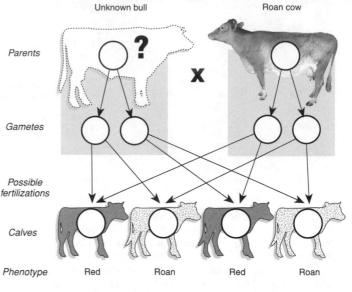

5. A white bull is mated with a roan cow (above, left).

   (a) Fill in the spaces on the diagram (above, left) to show the genotype and phenotype for parents and calves.

   (b) State the phenotype ratio for this cross: _____

   (c) Suggest how the farmer who owns these cattle could control the breeding so that the herd ultimately consisted of red colored cattle only:

   _____

   _____

   _____

6. A unknown bull is mated with a roan cow (above, right). A farmer has only roan shorthorn cows on his farm. He suspects that one of the bulls from his next door neighbors may have jumped the fence to mate with his cows earlier in the year. Half of the calves born were red and half were roan. One neighbor has a red bull, the other has a roan bull.

   (a) Fill in the spaces on the diagram (above, right) to show the genotype and phenotype for parents and calves.

   (b) State which of the neighbor's bulls must have mated with the cows: **red** or **roan** (*delete one*)

7. A plant breeder crossed two plants of the plant variety known as Japanese four o'clock. This plant is known to have its flower color controlled by a gene which possesses incomplete dominant alleles. Pollen from a pink flowered plant was placed on the stigma of a red flowered plant.

   (a) Fill in the spaces on the diagram on the right to show the genotype and phenotype for parents and offspring.

   (b) State the phenotype ratio:

   _____

   _____

   _____

   _____

Multiple Alleles in Blood Groups

The four common blood groups of the human 'ABO blood group system' are determined by three alleles: **A**, **B**, and **O** (also represented in some textbooks as: I^A, I^B, and I^O or just **i**). This is an example of a **multiple allele** system for a gene. The ABO antigens consist of sugars attached to the surface of red blood cells. The alleles code for enzymes (proteins) that join together these sugars. The allele **O** produces a non-functioning enzyme that is unable to make any changes to the basic antigen (sugar) molecule. The other two alleles *(A, B)* are **codominant** and are expressed equally. They each produce a different functional enzyme that adds a different, specific sugar to the basic sugar molecule. The blood group A and B antigens are able to react with antibodies present in the blood from other people and must be matched for transfusion.

Recessive allele: **O** produces a non-functioning protein
Dominant allele: **A** produces an enzyme which forms **A antigen**
Dominant allele: **B** produces an enzyme which forms **B antigen**

Blood group (phenotype)	Possible genotypes	Frequency*		
		White	Black	Native American
O	*OO*	45%	49%	79%
A	*AA AO*	40%	27%	16%
B		11%	20%	4%
AB		4%	4%	1%

* Frequency is based on North American population
Source: www.kcom.edu/faculty/chamberlain/Website/MSTUART/Lect13.htm

If a person has the **AO** allele combination then their blood group will be group **A**. The presence of the recessive allele has no effect on the blood group in the presence of a dominant allele. Another possible allele combination that can create the same blood group is **AA**.

1. Use the information above to complete the table for the possible genotypes for blood group B and group AB.

2. Below are six crosses possible between couples of various blood group types. The first example has been completed for you. Complete the genotype and phenotype for the other five crosses shown:

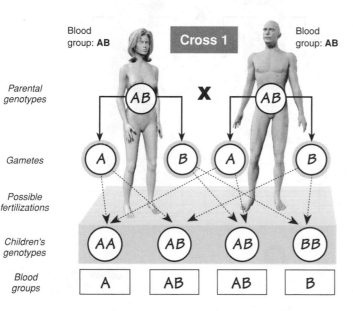

Blood group: **AB** — Cross 1 — Blood group: **AB**

Parental genotypes: AB X AB
Gametes: A, B, A, B
Children's genotypes: AA, AB, AB, BB
Blood groups: A, AB, AB, B

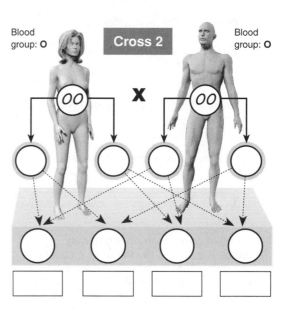

Blood group: **O** — Cross 2 — Blood group: **O**

Parental genotypes: OO X OO

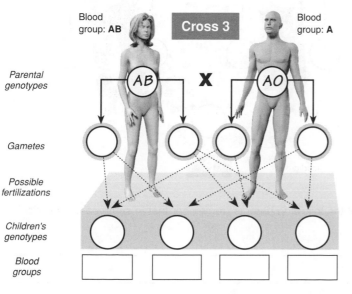

Blood group: **AB** — Cross 3 — Blood group: **A**

Parental genotypes: AB X AO

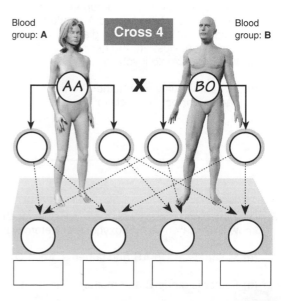

Blood group: **A** — Cross 4 — Blood group: **B**

Parental genotypes: AA X BO

Parental genotypes / Gametes / Possible fertilizations / Children's genotypes / Blood groups

Heredity

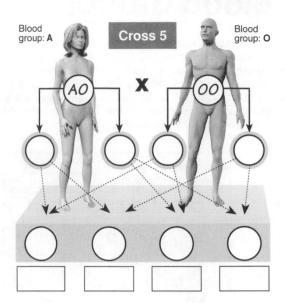

Blood group: A — Cross 5 — Blood group: O

AO X OO

| Parental genotypes |
| Gametes |
| Possible fertilizations |
| Children's genotypes |
| Blood groups |

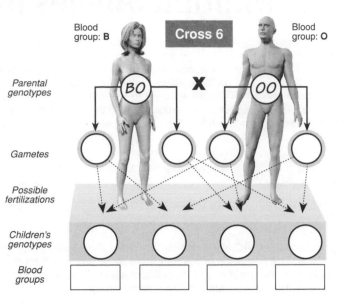

Blood group: B — Cross 6 — Blood group: O

BO X OO

3. A wife is heterozygous for blood group **A** and the husband has blood group **O**.

(a) Give the genotypes of each parent (fill in spaces on the diagram on the right).

Determine the probability of:

(b) One child having blood group **O**:

_____

(c) One child having blood group **A**:

_____

(d) One child having blood group **AB**:

_____

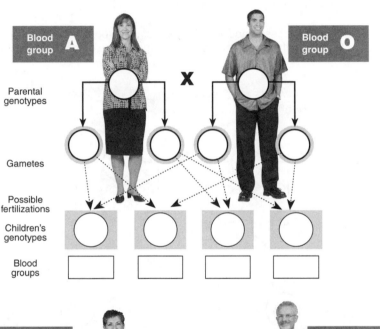

Blood group **A** — X — Blood group **O**

Parental genotypes

Gametes

Possible fertilizations

Children's genotypes

Blood groups

4. In a court case involving a paternity dispute (i.e. who is the father of a child) a man claims that a male child (blood group **B**) born to a woman is his son and wants custody. The woman claims that he is not the father.

(a) If the man has a blood group **O** and the woman has a blood group **A**, could the child be his son? Use the diagram on the right to illustrate the genotypes of the three people involved.

(b) State with reasons whether the man can be correct in his claim:

_____

_____

_____

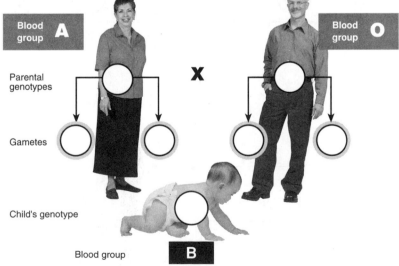

Blood group **A** — X — Blood group **O**

Parental genotypes

Gametes

Child's genotype

Blood group **B**

5. Give the blood groups which are possible for children of the following parents (remember that in some cases you don't know if the parent is homozygous or heterozygous).

(a) Mother is group **AB** and father is group **O**: _____

(b) Father is group **B** and mother is group **A**: _____

Dihybrid Cross

A cross (or mating) between two organisms where the inheritance patterns of **two genes** are studied is called a **dihybrid cross** (compared with the study of one gene in a monohybrid cross). There are a greater number of gamete types (four) produced when two genes are considered. Remember that the genes described are being carried by separate chromosomes and are sorted independently of each other during meiosis (that is why you get four kinds of gamete). The two genes below control two unrelated characteristics **hair color** and **coat length**. Black and short are dominant.

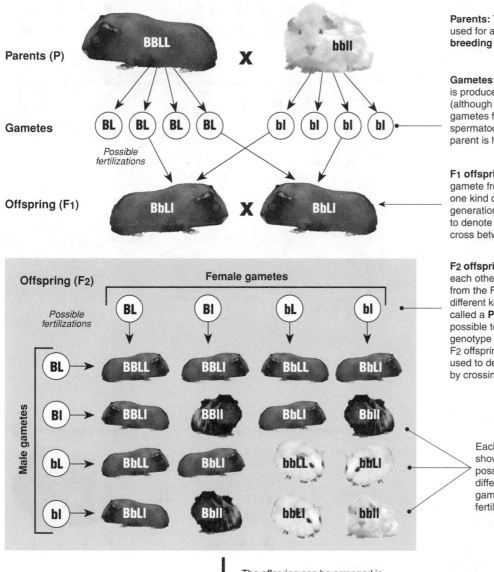

Homozygous black, short hair   **Homozygous white, long hair**

Parents (P)   BBLL   X   bbll

Gametes   BL BL BL BL   bl bl bl bl

Possible fertilizations

Offspring (F1)   BbLl   X   BbLl

Parents: The notation **P**, is only used for a cross between **true breeding** (homozygous) parents.

Gametes: Only one type of gamete is produced from each parent (although they will produce four gametes from each oocyte or spermatocyte). This is because each parent is homozygous for both traits.

F1 offspring: There is only one **kind** of gamete from each parent, therefore only one kind of offspring produced in the first generation. The notation **F1** is only used to denote the heterozygous offspring of a cross between two true breeding parents.

Offspring (F2)

Female gametes

Possible fertilizations

	BL	Bl	bL	bl
BL	BBLL	BBLl	BbLL	BbLl
Bl	BBLl	BBll	BbLl	Bbll
bL	BbLL	BbLl	bbLL	bbLl
bl	BbLl	Bbll	bbLl	bbll

Male gametes

F2 offspring: The F1 were mated with each other (**selfed**). Each individual from the F1 is able to produce four different kinds of gamete. Using a grid called a **Punnett square** (left), it is possible to determine the expected genotype and phenotype ratios in the F2 offspring. The notation **F2** is only used to denote the offspring produced by crossing F1 heterozygotes.

Each of the 16 animals shown here represents the possible zygotes formed by different combinations of gametes coming together at fertilization.

The offspring can be arranged in groups with similar phenotypes:

Genotype   **Phenotype**

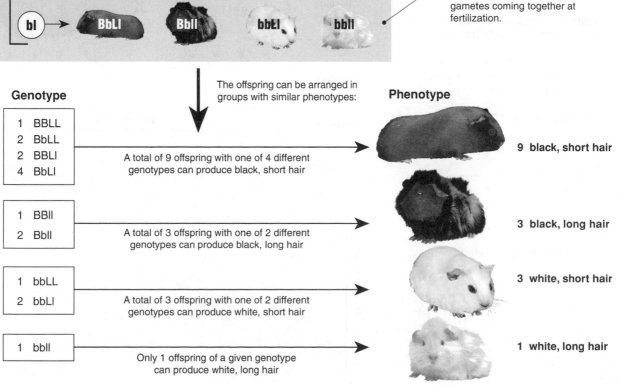

| 1 BBLL |
| 2 BbLL |
| 2 BBLl |
| 4 BbLl |

A total of 9 offspring with one of 4 different genotypes can produce black, short hair

9 black, short hair

| 1 BBll |
| 2 Bbll |

A total of 3 offspring with one of 2 different genotypes can produce black, long hair

3 black, long hair

| 1 bbLL |
| 2 bbLl |

A total of 3 offspring with one of 2 different genotypes can produce white, short hair

3 white, short hair

| 1 bbll |

Only 1 offspring of a given genotype can produce white, long hair

1 white, long hair

Related activities: Basic Genetic Crosses, Dihybrid Cross with Linkage
Web links: Drag and Drop Genetics, Using Chi-Squared in Genetics

A 2

Heredity

Cross Nº 1

The dihybrid cross on the right has been partly worked out for you. You must determine:

1. The genotype and phenotype for each animal (write your answers in its dotted outline).

2. Genotype **ratio** of the offspring:

   _____
   _____
   _____
   _____
   _____
   _____

3. Phenotype **ratio** of the offspring:

   _____
   _____
   _____
   _____

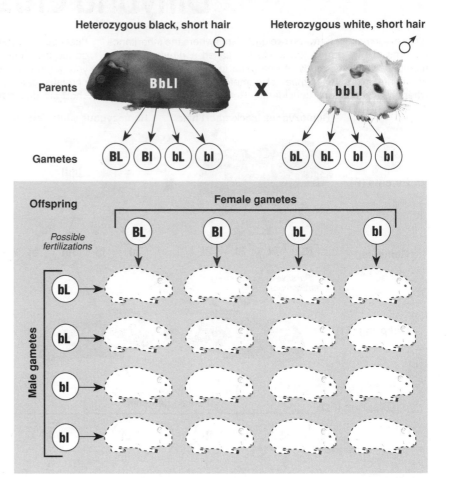

Cross Nº 2

For the dihybrid cross on the right, determine:

1. Gametes produced by each parent (write these in the circles).

2. The genotype and phenotype for each animal (write your answers in its dotted outline).

3. Genotype **ratio** of the offspring:

   _____
   _____
   _____
   _____
   _____

4. Phenotype **ratio** of the offspring:

   _____
   _____
   _____
   _____

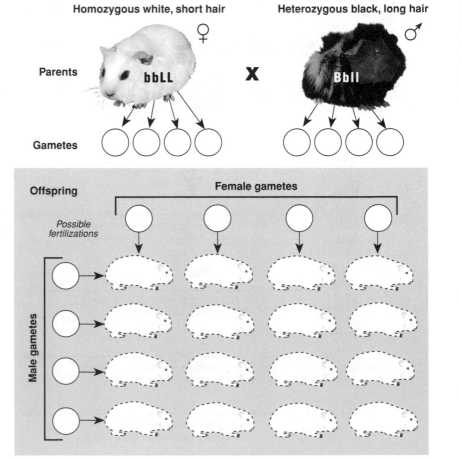

Dihybrid Cross with Linkage

In dihybrid inheritance involving independent assortment of alleles, a cross between two heterozygotes produces the expected 9:3:3:1 ratio in the offspring. In cases of dihybrid inheritance involving linkage, the offspring of a cross between two heterozygotes produces a 3:1 ratio of the parental types with no recombinants. However, because total linkage is uncommon, this 3:1 ratio is rarely achieved. Most dihybrid crosses involving linkage produce equal numbers of parental types and a much smaller number of recombinants. The examples below show the inheritance of body color and wing shape in *Drosophila*. The genes for these two characters are linked and do not assort independently. The example on the left shows the expected phenotype ratios from a mating between heterozygotes without crossing over. The example on the right shows the results of a test cross involving recombination of alleles. A test cross reveals the frequency of recombination for the gene involved. It is expressed as a **crossover value** (%) and calculated by the number of recombinant types / total number of offspring X100.

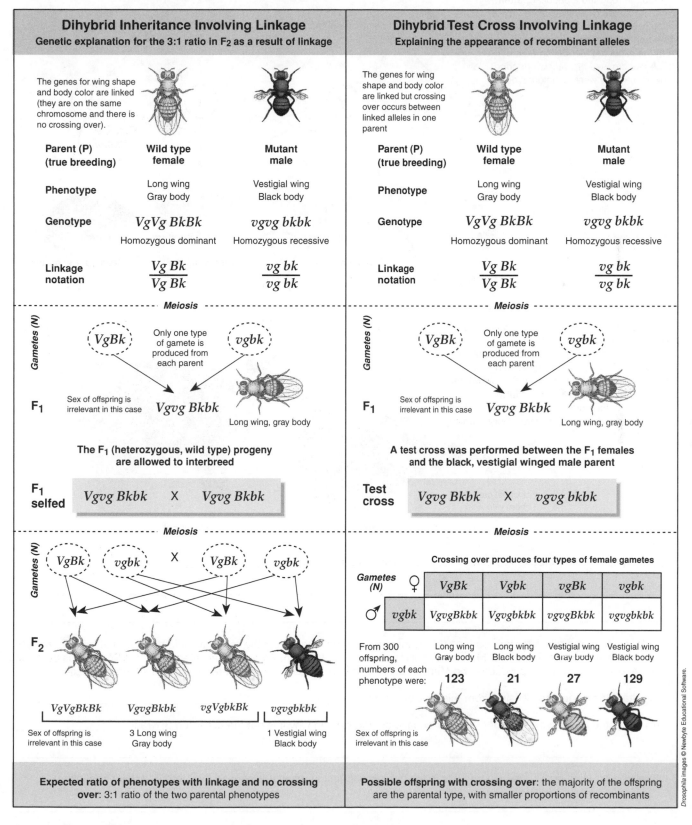

Dihybrid Inheritance Involving Linkage
Genetic explanation for the 3:1 ratio in F₂ as a result of linkage

The genes for wing shape and body color are linked (they are on the same chromosome and there is no crossing over).

Parent (P) (true breeding): Wild type female / Mutant male

Phenotype: Long wing Gray body / Vestigial wing Black body

Genotype: *VgVg BkBk* (Homozygous dominant) / *vgvg bkbk* (Homozygous recessive)

Linkage notation: $\frac{Vg\ Bk}{Vg\ Bk}$ / $\frac{vg\ bk}{vg\ bk}$

Meiosis

Gametes (N): *VgBk* / *vgbk* — Only one type of gamete is produced from each parent

F₁: Sex of offspring is irrelevant in this case — *Vgvg Bkbk* — Long wing, gray body

The F₁ (heterozygous, wild type) progeny are allowed to interbreed

F₁ selfed: *Vgvg Bkbk* X *Vgvg Bkbk*

Meiosis

Gametes (N): *VgBk* / *vgbk* X *VgBk* / *vgbk*

F₂: *VgVgBkBk* / *VgvgBkbk* / *vgVgbkBk* / *vgvgbkbk*

Sex of offspring is irrelevant in this case / 3 Long wing Gray body / 1 Vestigial wing Black body

Expected ratio of phenotypes with linkage and no crossing over: 3:1 ratio of the two parental phenotypes

Dihybrid Test Cross Involving Linkage
Explaining the appearance of recombinant alleles

The genes for wing shape and body color are linked but crossing over occurs between linked alleles in one parent

Parent (P) (true breeding): Wild type female / Mutant male

Phenotype: Long wing Gray body / Vestigial wing Black body

Genotype: *VgVg BkBk* (Homozygous dominant) / *vgvg bkbk* (Homozygous recessive)

Linkage notation: $\frac{Vg\ Bk}{Vg\ Bk}$ / $\frac{vg\ bk}{vg\ bk}$

Meiosis

Gametes (N): *VgBk* / *vgbk* — Only one type of gamete is produced from each parent

F₁: Sex of offspring is irrelevant in this case — *Vgvg Bkbk* — Long wing, gray body

A test cross was performed between the F₁ females and the black, vestigial winged male parent

Test cross: *Vgvg Bkbk* X *vgvg bkbk*

Meiosis

Crossing over produces four types of female gametes

Gametes (N) ♀	*VgBk*	*Vgbk*	*vgBk*	*vgbk*
♂ *vgbk*	*VgvgBkbk*	*Vgvgbkbk*	*vgvgBkbk*	*vgvgbkbk*

From 300 offspring, numbers of each phenotype were:

	Long wing Gray body	Long wing Black body	Vestigial wing Gray body	Vestigial wing Black body
	123	**21**	**27**	**129**

Sex of offspring is irrelevant in this case

Possible offspring with crossing over: the majority of the offspring are the parental type, with smaller proportions of recombinants

Drosophila images © Newbyte Educational Software.

Heredity

1. Calculate the crossover (value) for the offspring of the test cross, above: _____

Related activities: Dihybrid Cross, Recombination
Web links: Chromosome Mapping

ERDA 2

Lethal Alleles

Lethal alleles are mutations of a gene that produce a gene product, which is not only nonfunctional, but may affect the organism's survival. Some lethal alleles are fully dominant and kill in one dose in the heterozygote. Others, such as in the **Manx** cat and yellow mice (below), produce viable offspring with a recognizable phenotype in the heterozygote. In some lethal alleles, the lethality is fully recessive and the alleles confer no detectable effect in the heterozygote at all. Furthermore, lethal alleles may take effect at different stages in development (e.g. in juveniles or, as in **Huntington disease**, in adults).

When **Lucien Cuenot** investigated inheritance of coat color in yellow mice in 1905, he reported a peculiar pattern. When he mated two yellow mice, about ²/₃ of their offspring were yellow, and ¹/₃ were non-yellow. This was a departure from the expected Mendelian ratio of 3:1. A test cross of the yellow offspring showed that they were all heterozygous.

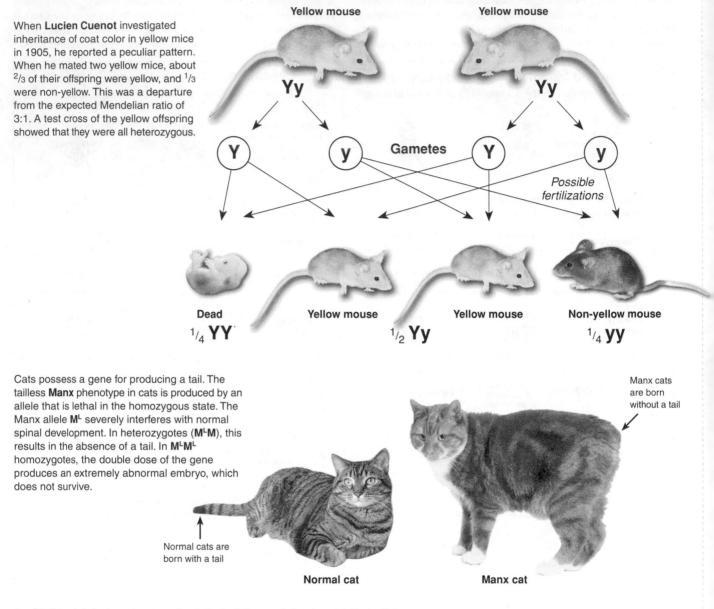

Cats possess a gene for producing a tail. The tailless **Manx** phenotype in cats is produced by an allele that is lethal in the homozygous state. The Manx allele M^L severely interferes with normal spinal development. In heterozygotes (M^LM), this results in the absence of a tail. In M^LM^L homozygotes, the double dose of the gene produces an extremely abnormal embryo, which does not survive.

Manx cats are born without a tail

Normal cats are born with a tail

Normal cat   **Manx cat**

1. Distinguish between recessive lethal alleles and dominant lethal alleles: _____

_____

_____

2. In Manx cats, the allele for taillessness (M^L) is incompletely dominant over the recessive allele for normal tail (M). Tailless Manx cats are heterozygous (M^LM) and carry a recessive allele for normal tail. Normal tailed cats are MM. A cross between two Manx (tailless) cats, produces two Manx to every one normal tailed cat (not a regular 3 to 1 Mendelian ratio).

   (a) State the genotypes arising from this type of cross: _____

   (b) State the phenotype ratio of Manx to normal cats and explain why it is not the expected 3:1 ratio: _____

   _____

3. Explain why Huntington disease persists in the human population when it is caused by a lethal, dominant allele:

   _____

   _____

Related activities: Dominance of Alleles, A Gene That Can Tell Your Future, Genetic Counselling

Problems in Mendelian Genetics

The following problems involve Mendelian crosses through to the F₂ generation. The alleles involved are associated with various phenotypic traits in domestic breeds. See *Basic Genetic Crosses* if you need to review test crosses and back crosses.

1. The Himalayan color-pointed, long-haired cat is a breed developed by crossing a pedigree (true-breeding), uniform-colored, long-haired Persian with a pedigree color-pointed (darker face, ears, paws, and tail) short-haired Siamese.
The genes controlling hair coloring and length are on separate chromosomes: uniform color **U**, color pointed **u**, short hair **S**, long hair **s**.

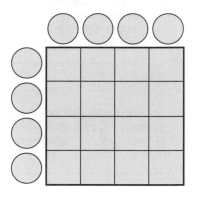

Persian   Siamese   Himalayan

   (a) Using the symbols above, indicate the genotype of each breed below its photograph (above, right). _____ _____ _____

   (b) State the genotype of the **F₁** (Siamese X Persian): _____

   (c) State the phenotype of the **F₁**: _____

   (d) Use the Punnett square to show the outcome of a cross between the F₁ (the F₂):

   (e) State the ratio of the F₂ that would be Himalayan: _____

   (f) State whether the Himalayan would be true breeding: _____

   (g) State the ratio of the F₂ that would be color-point, short-haired cats: _____

   (h) Explain how two F₂ cats of the same phenotype could have different genotypes:

   _____

   _____

   (i) Explain how you could determine which of the F₂ color-point, short-hairs were true breeding for these characters:

   _____

   _____

2. In rabbits, spotted coat **S** is dominant to solid color **s**, while for coat color: black **B** is dominant to brown **b**. A brown spotted rabbit is mated with a solid black one and all the offspring are black spotted (the genes are not linked).

   (a) State the genotypes: Male parent: _____ Female parent: _____ Offspring: _____

   (b) Use the Punnett square to show the outcome of a cross between the F₁ (the F₂):

   (c) Using ratios, state the phenotypes of the F₂ generation: _____

   _____

   (d) State the name given to this type of cross: _____

3. In guinea pigs, rough coat **R** is dominant over smooth coat **r** and black coat **B** is dominant over white **b**. The genes for coat texture and color are not linked. In a cross of a homozygous rough black animal with a homozygous smooth white:

   (a) State the genotype of the **F₁**: _____

   (b) State the phenotype of the **F₁**: _____

   (c) Use the Punnett square to show the outcome of a cross between the F₁ (the F₂):

   (d) Using ratios, state the phenotypes of the F₂ generation:

   _____

   _____

   _____

   _____

Heredity

Related activities: Basic Genetic Crosses, Dominance of Alleles, Lethal Alleles
Web links: Drag and Drop Genetics

RA 3

(e) Use the Punnett square (right) to show the outcome of a cross between the offspring of a **back cross** of the **F₁** to the rough, black parent:

(f) Using ratios, state the phenotype of the F₂ generation:

_____

(g) Use the Punnett square to show the outcome of a cross between the offspring of a **test cross** of the **F₁** to the smooth, white parent:

(h) Using ratios, state the phenotypes of the F₂ generation:

_____

_____

_____

(i) A rough black guinea pig was crossed with a rough white one produced the following offspring: 28 rough black, 31 rough white, 11 smooth black, and 10 smooth white. Determine the genotypes of the parents:

_____

_____

4. Chickens with shortened wings and legs are called creepers. When creepers are mated to normal birds, they produce creepers and normals with equal frequency. When creepers are mated to creepers they produce two creepers to one normal. Crosses between normal birds produce only normal progeny. Explain these results:

_____

_____

5. Black wool of sheep is due to a recessive allele (**b**), and white wool to its dominant allele (**B**). A white ram is crossed to a white ewe. Both animals carry the allele for black (b). They produce a white ram lamb, which is then back crossed to the female parent. Determine the probability of the **back cross** offspring being black:

_____

6. Mallard ducks have their plumage color controlled by a gene with three alleles: **M^R** restricted mallard pattern, **M** mallard pattern, and **m** dusky mallard pattern. The dominance hierarchy is: **M^R > M > m** (i.e. **M^R** is more dominant than **M**, which is more dominant than **m**). Determine the genotypic and phenotypic ratios expected in the **F₁** of the following crosses:

(a) **M^{R}M^R X M^{R}M**: Genotypes: _____

Phenotypes: _____

(b) **M^{R}M^R X M^{R}m**: Genotypes: _____

Phenotypes: _____

(c) **M^{R}M X M^{R}m**: Genotypes: _____

Phenotypes: _____

(d) **M^{R}m X Mm**: Genotypes: _____

Phenotypes: _____

(e) **Mm X mm**: Genotypes: _____

Phenotypes: _____

7. A dominant gene (**W**) produces wire-haired texture in dogs; its recessive allele (**w**) produces smooth hair. A group of heterozygous wire-haired individuals are crossed and their F₁ progeny are then test-crossed. Determine the expected genotypic and phenotypic ratios among the **test cross** progeny:

_____

_____

Human Genotypes

An estimated 25 000 genes determine all human **traits**. While most traits are determined by more than one gene, a number are determined by a single gene system, with dominant/recessive, codominant, or multiple allele inheritance. Single gene traits (below) show **discontinuous variation** in a population, with individuals showing only one of a limited number of phenotypes (usually two or three). Single gene traits may, however, show variable **penetrance**. Penetrance describes the extent to which the properties controlled by a gene will be expressed. Highly penetrant genes will be expressed regardless of the effects of the environment, whereas a gene with low penetrance will only sometimes produce the trait with which it is associated.

Trait: Handedness

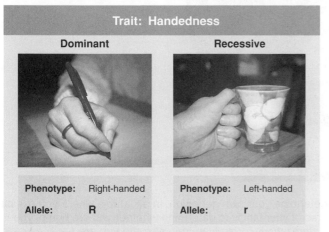

Dominant	Recessive
Phenotype: Right-handed	**Phenotype:** Left-handed
Allele: R	**Allele:** r

The trait of left or right handedness is genetically determined. Right-handed people have the dominant allele, while left handedness is recessive. People that consider themselves ambidextrous can assume they have the dominant allele for this trait.

Trait: Hand clasp

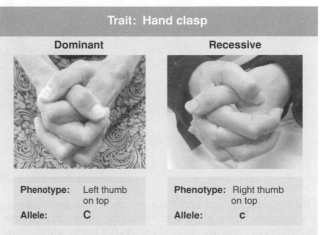

Dominant	Recessive
Phenotype: Left thumb on top	**Phenotype:** Right thumb on top
Allele: C	**Allele:** c

Like handedness, hand clasping shows dominance/recessiveness. When the hands are clasped together, either the left or the right thumb will naturally come to rest on top. The left thumb on top is the dominant trait (C), while the right thumb on top is recessive (c).

Trait: Dimpled chin

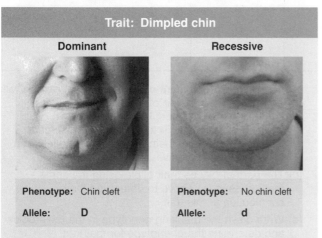

Dominant	Recessive
Phenotype: Chin cleft	**Phenotype:** No chin cleft
Allele: D	**Allele:** d

A cleft or dimple on the chin is inherited. A cleft is dominant (D), while the absence of a cleft is recessive (d), although this gene shows **variable penetrance**, probably as a result of modifier genes.

Trait: Middle digit hair

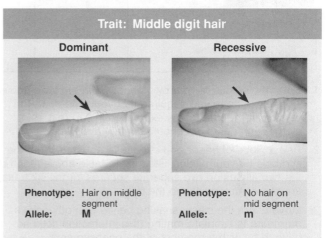

Dominant	Recessive
Phenotype: Hair on middle segment	**Phenotype:** No hair on mid segment
Allele: M	**Allele:** m

Some people have a dominant allele that causes hair to grow on the middle segment of their fingers. It may not be present on all fingers, and in some cases may be very fine and hard to see.

Trait: Ear lobe shape

Dominant	Recessive
Phenotype: Lobes free	**Phenotype:** Lobes attached
Allele: F	**Allele:** f

In people with only the recessive allele (homozygous recessive), ear lobes are attached to the side of the face. The presence of a dominant allele causes the ear lobe to hang freely.

Trait: Thumb hyperextension

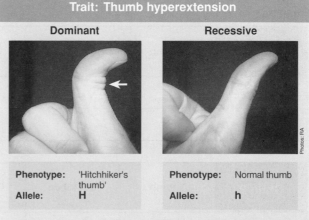

Dominant	Recessive
Phenotype: 'Hitchhiker's thumb'	**Phenotype:** Normal thumb
Allele: H	**Allele:** h

There is a gene that controls the trait known as 'hitchhiker's thumb' which is technically termed distal hyperextensibility. People with the dominant phenotype are able to curve their thumb backwards without assistance, so that it forms an arc shape.

Photos: RA

Heredity

Related activities: Polygenes

P 1

Your Genotype Profile

Use the descriptions and the symbols on the previous page to determine your own genotype. In situations where you exhibit the dominant form of the trait, it may be helpful to study the features of your family to determine whether you are homozygous dominant or heterozygous. If you do not know whether you are heterozygous for a given trait, assume you are.

Your traits:	Thumb	Ear lobes	Chin cleft	Middle digit hair	Handedness	Hand clasp
Phenotype:						
Genotype:						

1. Enter the details of your own genotype in the table above. The row: 'Phenotype' requires that you write down the version of the trait that is expressed in you (e.g. chin cleft). Each genotype should contain two alleles.

2. Use a piece of paper and cut out 12 squares. Write the symbols for your alleles listed in the table above (each of the two alleles on two separate squares for the six traits) and write your initials on the back.

3. Move about the class, shaking hands with other class members to simulate mating (this interaction does not have to be with a member of the opposite sex).

4. Proceed to determine the possible genotypes and phenotypes for your offspring with this other person by:

   (a) Selecting each of the six characters in turn

   (b) Where a genotype for a person is known to be

homozygous (dominant or recessive) that person will simply place down one of the pieces of paper with their allele for that gene. If they are heterozygous for this trait, toss a coin to determine which gets 'donated' with heads being the dominant allele and tails being the recessive.

(c) The partner places their allele using the same method as in (b) above to determine their contribution to this trait.

(d) Write down the resulting genotype in the table below and determine the phenotype for that trait.

(e) Proceed on to the next trait.

5. Try another mating with a different partner or the same partner and see if you end up with a child of the same phenotype.

Child 1	Thumb	Ear lobes	Chin cleft	Middle digit hair	Handedness	Hand clasp
Phenotype:						
Genotype:						

Child 2	Thumb	Ear lobes	Chin cleft	Middle digit hair	Handedness	Hand clasp
Phenotype:						
Genotype:						

Sex Determination

The determination of the sex (gender) of an organism is controlled in most cases by the sex chromosomes provided by each parent. These have evolved to regulate the ratios of males and females produced and preserve the genetic differences between the sexes. In humans, males are referred to as the **heterogametic sex** because each somatic cell has one X chromosome and one Y chromosome. The determination of sex is based on the presence or absence of the Y chromosome. Without the Y chromosome, an individual will develop into a **homogametic** female (each somatic cell with two X chromosomes). In mammals, the male is always the heterogametic sex, but this is not necessarily the case in other taxa. In birds and butterflies, the female is the heterogametic sex, and in some insects the male is simply X whereas the female is XX.

Sex Determination in Humans

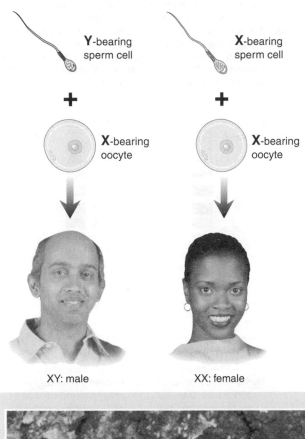

XY: male   XX: female

The Sex Determining Region of the Y Chromosome

Scientists have known since 1959 that the Y chromosome is associated with being male. However, it was not until 1990 that a group of researchers, working for the Medical Research Council in London, discovered the gene on the Y chromosome that determines maleness. It was named **SRY**, for **Sex Determining Region of the Y**. The SRY gene produces a type of protein called a **transcription factor**. This transcription factor switches on the genes that direct the development of male structures in the embryo.

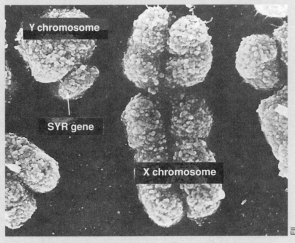

Y chromosome

SYR gene

X chromosome

Smaller and darker female half

Larger and lighter male half

Gynandromorphism occurs when an animal is a genetic mosaic and possesses both male and female characteristics (i.e. some of its cells are genetically male and others are female). This phenomenon is found particularly in insects, but also appears in birds and mammals. Gynandromorphism occurs due to the loss of an X chromosome in a stem cell of a female (XX), so that all the tissues derived from that cell are phenotypically male.

In the pill woodlouse, *Armadillium vulgare*, sex determination is characterized by female heterogamety (ZW) and male homogamety (ZZ). However, in several wild populations this system is overridden by an infectious bacterium. This bacterium causes genetically male woodlice to change into females. The bacteria are transmitted through the egg cytoplasm of the woodlouse. Therefore the conversion of males to females increases the propagation of the bacterium.

1. Explain what determines the sex of the offspring at the moment of conception in humans: _____

_____

2. Explain why human males are called the heterogametic sex: _____

_____

Heredity

Periodicals:
The Y chromosome: it's a man thing

Related activities: Karyotypes
Web links: Sex Determination in Humans

A 1

Genomic Imprinting

The phenotypic effects of some mammalian genes depend on whether they were inherited from the mother or the father. This phenomenon, called **genomic imprinting** (or parental imprinting), is part of **epigenetics**, the study of the heritable changes in gene function that occur without involving changes in the DNA sequence. Just as cells inherit genes, they also inherit the instructions that communicate to the genes when to become active, in which tissue, and to what extent. Epigenetic phenomena are important because they regulate when and at what level genes are expressed.

Genomic Imprinting

Genomic imprinting describes how a small subset of the genes in the genome are expressed according to their parent of origin. 'Imprints' can act as silencers or activators for imprinted genes. A mammal inherits two sets of chromosomes, one from the mother and one from the father. In this way the imprinted gene expression is balanced; a prerequisite for a viable offspring in mammals.

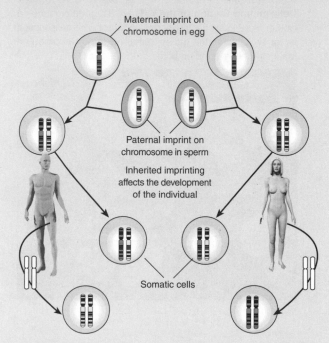

Maternal imprint on chromosome in egg

Paternal imprint on chromosome in sperm

Inherited imprinting affects the development of the individual

Somatic cells

Maternal and paternal chromosomes are differently imprinted. Chromosomes are newly imprinted (reprogrammed) each generation.

Imprinted Genes Are Different

Some imprinted genes are expressed from a maternally inherited chromosome and **silenced** on the paternal chromosome, while other imprinted genes show the opposite expression pattern and are only expressed from a paternally inherited chromosome. Evidence of this is seen in two human genetic disorders. Both are caused by the same mutation; a specific deletion on chromosome 15. The disorder expressed depends on whether the mutation is inherited from the father or the mother.

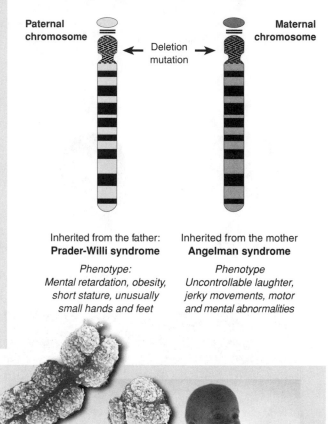

Paternal chromosome

← Deletion mutation →

Maternal chromosome

Inherited from the father:
Prader-Willi syndrome

Phenotype:
Mental retardation, obesity, short stature, unusually small hands and feet

Inherited from the mother
Angelman syndrome

Phenotype
Uncontrollable laughter, jerky movements, motor and mental abnormalities

How Are Genes Silenced?

- In many instances, **gene silencing** achieved through **methylation** of the DNA of genes or regulatory sequences, which results in the gene not being expressed.

- **Methylation** turns off gene expression by adding a methyl group to cytosines in the DNA. This changes the state of the chromatin so that the expression of any genes in the methylated region is inhibited. Methylation is also important in X-inactivation.

- In other instances, phosphorylation or other chemical modification of histone proteins appears to lead to silencing.

Which genes did you inherit from your mother and which from your father? For some genes, imprinting will affect phenotypic expression.

1. (a) Explain what is meant by genomic imprinting: _____

_____

(b) Describe one of the mechanisms by which imprinting is achieved: _____

_____

2. Explain the significance of imprinting to the inheritance of genes: _____

_____

_____

Sex Linkage

Sex linkage is a special case of linkage occurring when a gene is located on a sex chromosome (usually the X). The result of this is that the character encoded by the gene is usually seen only in one sex (the heterogametic sex) and occurs rarely in the homogametic sex. In humans, recessive sex linked genes are responsible for a number of heritable disorders in males, e.g. hemophilia. Women who have the recessive allele are said to be **carriers**. One of the gene loci controlling coat color in cats is sex-linked. The two alleles, red (for orange color) and non-red (for black color), are found only on the X-chromosome.

Allele types

X_o = Non-red (=black)
X_O = Red

Genotypes   **Phenotypes**

X_oX_o, X_oY = Black coated female, male
X_OX_O, X_OY = Orange coated female, male
X_OX_o = Tortoiseshell (intermingled black and orange in fur) in female cats only

1. An owner of a cat is thinking of mating her black female cat with an orange male cat. Before she does this, she would like to know what possible coat colors could result from such a cross. Use the symbols above to fill in the diagram on the right. Summarize the possible genotypes and phenotypes of the kittens in the tables below.

	Genotypes	Phenotypes
Male kittens		

	Genotypes	Phenotypes
Female kittens		

2. A female tortoiseshell cat mated with an unknown male cat in the neighborhood and has given birth to a litter of six kittens. The owner of this female cat wants to know what the appearance and the genotype of the father was of these kittens. Use the symbols above to fill in the diagram on the right. Also show the possible fertilizations by placing appropriate arrows.

Describe the father cat's:

(a) Genotype: _____

(b) Phenotype: _____

3. The owner of another cat, a black female, also wants to know which cat fathered her two tortoiseshell female and two black male kittens. Use the symbols above to fill in the diagram on the right. Show the possible fertilizations by placing appropriate arrows.

Describe the father cat's:

(a) Genotype: _____

(b) Phenotype: _____

(c) Was it the same male cat that fathered both this litter and the one above?
YES / NO   (delete one)

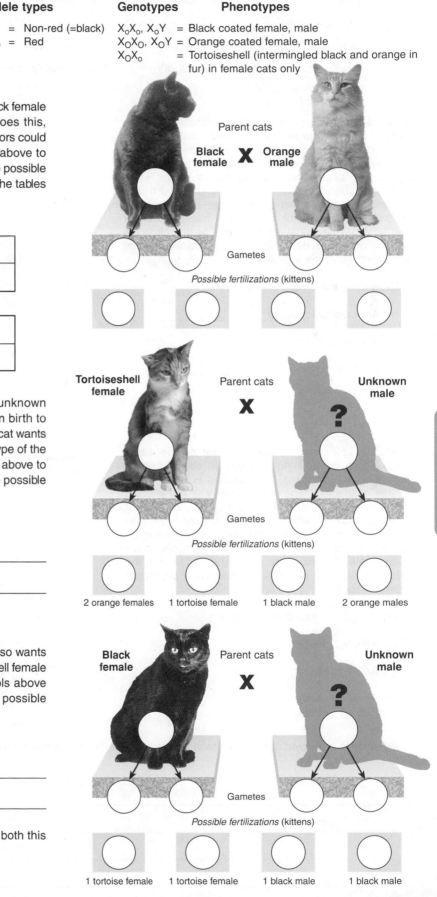

Parent cats
Black female **X** **Orange male**
Gametes
Possible fertilizations (kittens)

Tortoiseshell female **X** Parent cats **Unknown male** **?**
Gametes
Possible fertilizations (kittens)
2 orange females   1 tortoise female   1 black male   2 orange males

Black female **X** Parent cats **Unknown male** **?**
Gametes
Possible fertilizations (kittens)
1 tortoise female   1 tortoise female   1 black male   1 black male

Heredity

Related activities: Dominance of Alleles, Genetic Counseling
Web links: X-linked inheritance

RA 3

Dominant allele in humans

A rare form of rickets in humans is determined by a **dominant** allele of a gene on the **X chromosome** (it is not found on the Y chromosome). This condition is not successfully treated with vitamin D therapy. The allele types, genotypes, and phenotypes are as follows:

Allele types	Genotypes	Phenotypes
X_R = affected by rickets	$X_R X_R, X_R X$ =	Affected female
X = normal	$X_R Y$ =	Affected male
	XX, XY =	Normal female, male

As a genetic counselor you are presented with a married couple where one of them has a family history of this disease. The husband is affected by this disease and the wife is normal. The couple, who are thinking of starting a family, would like to know what their chances are of having a child born with this condition. They would also like to know what the probabilities are of having an affected boy or affected girl. Use the symbols above to complete the diagram right and determine the probabilities stated below (expressed as a proportion or percentage).

4. Determine the probability of having:

   (a) Affected children: _____

   (b) An affected girl: _____

   (c) An affected boy: _____

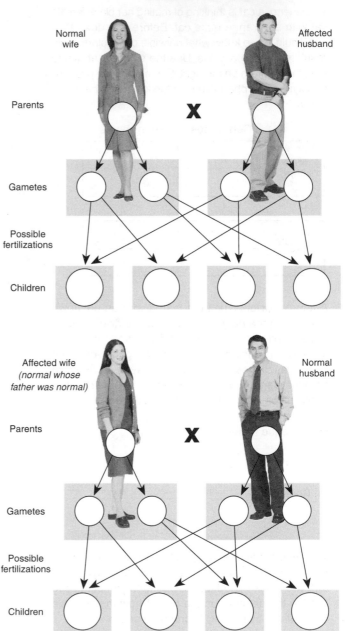

Another couple with a family history of the same disease also come in to see you to obtain genetic counseling. In this case the husband is normal and the wife is affected. The wife's father was not affected by this disease. Determine what their chances are of having a child born with this condition. They would also like to know what the probabilities are of having an affected boy or affected girl. Use the symbols above to complete the diagram right and determine the probabilities stated below (expressed as a proportion or percentage).

5. Determine the probability of having:

   (a) Affected children: _____

   (b) An affected girl: _____

   (c) An affected boy: _____

6. Describing examples other than those above, discuss the role of **sex linkage** in the inheritance of genetic disorders:

_____

_____

_____

_____

_____

Inheritance Patterns

Complete the following monohybrid crosses for different types of inheritance patterns in humans: autosomal recessive, autosomal dominant, sex linked recessive, and sex linked dominant inheritance.

1. **Inheritance of autosomal recessive traits**
   Example: *Albinism*

   Albinism (lack of pigment in hair, eyes and skin) is inherited as an autosomal recessive allele (not sex-linked).

   Using the codes: **PP** (normal)
   Pp (carrier)
   pp (albino)

   (a) Enter the parent phenotypes and complete the Punnett square for a cross between two carrier genotypes.

   (b) Give the ratios for the phenotypes from this cross.

   Phenotype ratios: _____

   _____

2. **Inheritance of autosomal dominant traits**
   Example: *Woolly hair*

   Woolly hair is inherited as an autosomal dominant allele. Each affected individual will have at least one affected parent.

   Using the codes: **WW** (woolly hair)
   Ww (woolly hair, heterozygous)
   ww (normal hair)

   (a) Enter the parent phenotypes and complete the Punnett square for a cross between two heterozygous individuals.

   (b) Give the ratios for the phenotypes from this cross.

   Phenotype ratios: _____

   _____

3. **Inheritance of sex linked recessive traits**
   Example: *Hemophilia*

   Inheritance of hemophilia is sex linked. Males with the recessive (hemophilia) allele, are affected. Females can be carriers.

   Using the codes: **XX** (normal female)
   XX$_h$ (carrier female)
   X$_h$X$_h$ (hemophiliac female)
   XY (normal male)
   X$_h$Y (hemophiliac male)

   (a) Enter the parent phenotypes and complete the Punnett square for a cross between a normal male and a carrier female.

   (b) Give the ratios for the phenotypes from this cross:

   Phenotype ratios: _____

   _____

4. **Inheritance of sex linked dominant traits**
   Example: *Sex linked form of rickets*

   A rare form of rickets is inherited on the X chromosome.

   Using the codes: **XX** (normal female); **XY** (normal male)
   X$_R$X (affected heterozygote female)
   X$_R$X$_R$ (affected female)
   X$_R$Y (affected male)

   (a) Enter the parent phenotypes and complete the Punnett square for a cross between an affected male and heterozygous female.

   (b) Give the ratios for the phenotypes from this cross.

   Phenotype ratios: _____

   _____

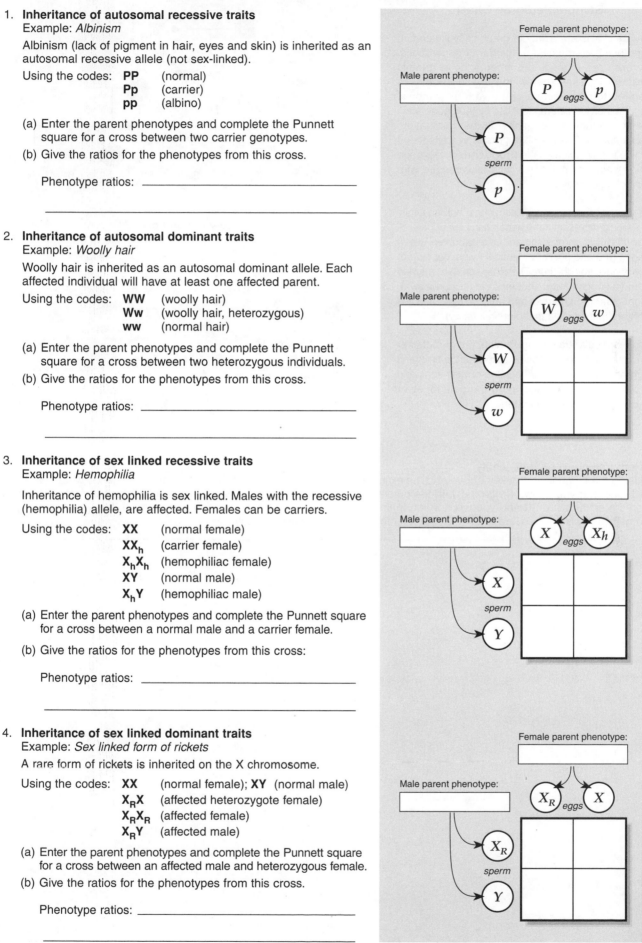

Heredity

Related activities: Monohybrid Cross, Sex Linkage

A 1

Pedigree Analysis

Sample Pedigree Chart

Pedigree charts are a way of graphically illustrating inheritance patterns over a number of generations. They are used to study the inheritance of genetic disorders. The key (below the chart) should be consulted to make sense of the various symbols. Particular individuals are identified by their generation number and their order number in that generation. For example, **II-6** is the sixth person in the second row. The arrow indicates the **propositus**; the person through whom the pedigree was discovered (i.e. who reported the condition).

If the chart on the right were illustrating a human family tree, it would represent three generations: grandparents (I-1 and I-2) with three sons and one daughter. Two of the sons (II-3 and II-4) are identical twins, but did not marry or have any children. The other son (II-1) married and had a daughter and another child (sex unknown). The daughter (II-5) married and had two sons and two daughters (plus a child that died in infancy).

For the particular trait being studied, the grandfather was expressing the phenotype (showing the trait) and the grandmother was a carrier. One of their sons and one of their daughters also show the trait, together with one of their granddaughters.

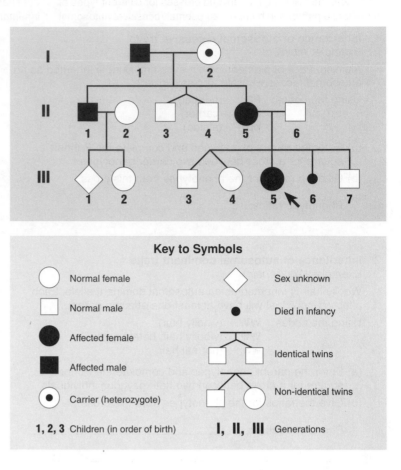

Key to Symbols

- ◯ Normal female
- ▢ Normal male
- ⬤ Affected female
- ◼ Affected male
- ◉ Carrier (heterozygote)
- **1, 2, 3** Children (in order of birth)
- ◇ Sex unknown
- ● Died in infancy
- Identical twins
- Non-identical twins
- **I, II, III** Generations

1. **Pedigree chart of your family**
   Using the symbols in the key above and the example illustrated as a guide, construct a pedigree chart of your own family (or one that you know of) starting with the parents of your mother and/or father on the first line. Your parents will appear on the second line (II) and you will appear on the third line (III). There may be a fourth generation line (IV) if one of your brothers or sisters has had a child. Use a ruler to draw up the chart carefully.

Related activities: Sex Linkage, Inheritance Patterns
Web links: Patterns of Inheritance

Periodicals:
Secrets of the gene

© Biozone International 2001-2010
Photocopying Prohibited

2. Autosomal recessive traits

Albinos lack pigment in the hair, skin and eyes. This trait is inherited as an autosomal recessive allele (i.e. it is not carried on the sex chromosome).

(a) Write the genotype for each of the individuals on the chart using the following letter codes: **PP** normal skin color; **P-** normal, but unknown if homozygous; **Pp** carrier; **pp** albino.

(b) Explain why the parents (II-3) and (II-4) must be **carriers** of a **recessive** allele:

_____

_____

_____

_____

Albinism in humans

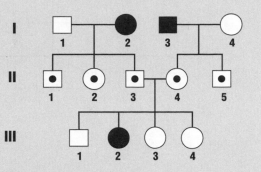

3. Sex linked recessive traits

Haemophilia is a disease where blood clotting is affected. A person can die from a simple bruise (which is internal bleeding). The clotting factor gene is carried on the X chromosome.

(a) Write the genotype for each of the individuals on the chart using the codes: **XY** normal male; X_hY affected male; **XX** normal female; X_hX female carrier; X_hX_h affected female:

(b) Explain why males can never be carriers:

_____

_____

_____

_____

Haemophilia in humans

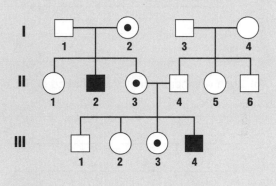

4. Autosomal dominant traits

An unusual trait found in some humans is woolly hair (not to be confused with curly hair). Each affected individual will have at least one affected parent.

(a) Write the genotype for each of the individuals on the chart using the following letter codes:
WW woolly hair; **Ww** woolly hair (heterozygous); **W-** woolly hair, but unknown if homozygous; **ww** normal hair

(b) Describe a feature of this inheritance pattern that suggests the trait is the result of a **dominant** allele:

_____

_____

Woolly hair in humans

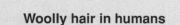

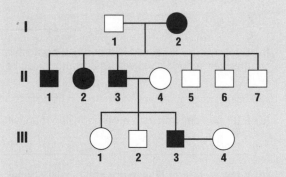

5. Sex linked dominant traits

A rare form of rickets is inherited on the X chromosome. All daughters of affected males will be affected. More females than males will show the trait.

(a) Write the genotype for each of the individuals on the chart using the following letter codes:
XY normal male; X_RY affected male; **XX** normal female; X_{R-} female (unknown if homozygous); X_RX_R affected female.

(b) Explain why more females than males will be affected:

_____

_____

_____

A rare form of rickets in humans

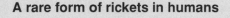

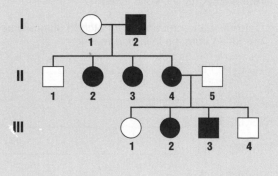

Heredity

6. The pedigree chart below illustrates the inheritance of a trait (darker symbols) in two families joined in marriage.

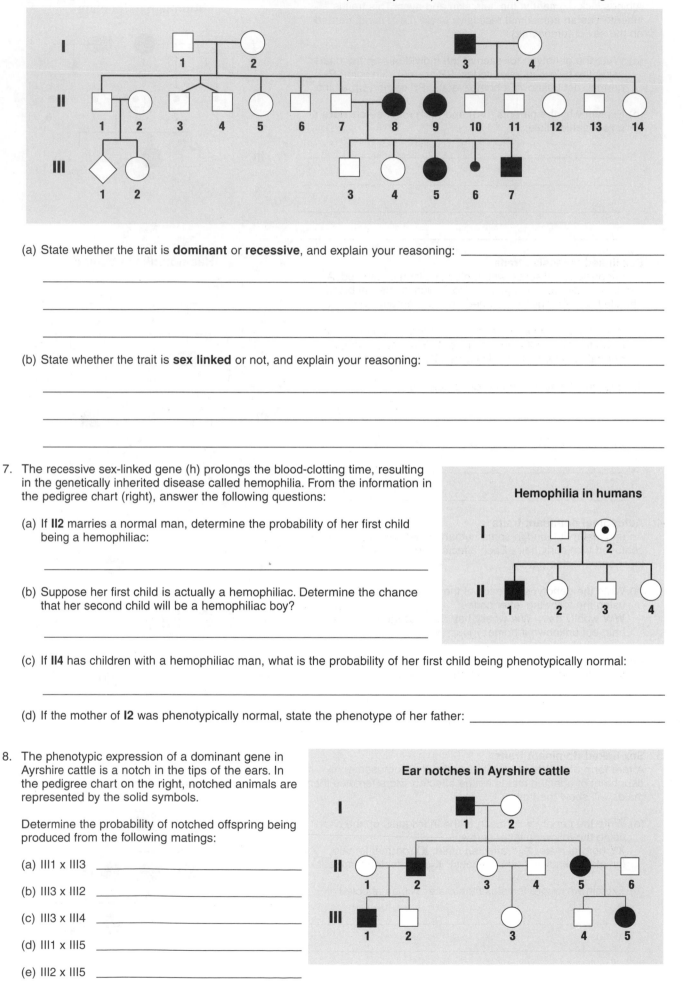

(a) State whether the trait is **dominant** or **recessive**, and explain your reasoning: _____

_____

_____

_____

(b) State whether the trait is **sex linked** or not, and explain your reasoning: _____

_____

_____

_____

7. The recessive sex-linked gene (h) prolongs the blood-clotting time, resulting in the genetically inherited disease called hemophilia. From the information in the pedigree chart (right), answer the following questions:

Hemophilia in humans

(a) If **II2** marries a normal man, determine the probability of her first child being a hemophiliac:

_____

(b) Suppose her first child is actually a hemophiliac. Determine the chance that her second child will be a hemophiliac boy?

_____

(c) If **II4** has children with a hemophiliac man, what is the probability of her first child being phenotypically normal:

_____

(d) If the mother of **I2** was phenotypically normal, state the phenotype of her father: _____

8. The phenotypic expression of a dominant gene in Ayrshire cattle is a notch in the tips of the ears. In the pedigree chart on the right, notched animals are represented by the solid symbols.

Ear notches in Ayrshire cattle

Determine the probability of notched offspring being produced from the following matings:

(a) III1 x III3 _____

(b) III3 x III2 _____

(c) III3 x III4 _____

(d) III1 x III5 _____

(e) III2 x III5 _____

Genetic Counseling

Genetic counselling is an analysis of the risk of producing offspring with known gene defects within a family. Counsellors identify families at risk, investigate the problem present in the family, interpret information about the disorder, analyze inheritance patterns and risks of recurrence, and review available options with the family. Increasingly, there are DNA tests for the identification of specific defective genes. People usually consider genetic counselling if they have a family history of a genetic disorder, or if a routine prenatal screening test yields an unexpected result. While screening for many genetic disorders is now recommended, the use of presymptomatic tests for adult-onset disorders, such as Alzheimer's, is still controversial.

Autosomal Recessive Conditions

Common inherited disorders caused by recessive alleles on autosomes. Recessive conditions are evident only in homozygous recessive genotypes.

Cystic fibrosis: Malfunction of the pancreas and other glands; thick mucus leads to pneumonia and emphysema. Death usually occurs in childhood. CF is the most frequent lethal genetic disorder in childhood (about 1 case in 3700 live births).

Maple syrup urine disease: Mental and physical retardation produced by a block in amino acid metabolism. Isoleucine in the urine produces the characteristic odor.

Tay-Sachs disease: A lipid storage disease which causes progressive developmental paralysis, mental deterioration, and blindness. Death usually occurs by three years of age.

Autosomal Dominant Conditions

Inherited disorders caused by dominant alleles on autosomes. Dominant conditions are evident both in heterozygotes and in homozygous dominant individuals.

Huntington disease: Involuntary movements of the face and limbs with later general mental deterioration. The beginning of symptoms is highly variable, but occurs usually between 30 to 40 years of age.

Genetic testing may involve biochemical tests for gene products such as enzymes and other proteins, microscopic examination of stained or fluorescent chromosomes, or examination of the DNA molecule itself. Various types of genetic tests are performed for various reasons, including:

Carrier Screening
Identifying unaffected individuals who carry one copy of a gene for a disease that requires two copies for the disease to be expressed.

Preimplantation Genetic Diagnosis
Screens for genetic flaws in embryos used for *in vitro* fertilization. The results of the analysis are used to select mutation-free embryos.

Prenatal Diagnostic Testing
Tests for chromosomal abnormalities such as Down syndrome.

Newborn Screening
Newborn babies are screened for a variety of enzyme-based disorders.

Presymptomatic Testing
Testing before symptoms are apparent is important for estimating the risk of developing adult-onset disorders, including Huntington's, cancers, and Alzheimer's disease.

About half of the cases of childhood deafness are the result of an autosomal recessive disorder. Early identification of the problem prepares families and allows early appropriate treatment.

Genetic counseling provides information to families who have members with birth defects or genetic disorders, and to families who may be at risk for a variety of inherited conditions.

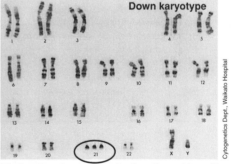

Most pregnant women in developed countries will have a prenatal test to detect chromosomal abnormalities such as Down syndrome and developmental anomalies such as neural defects.

1. Outline the benefits of **carrier screening** to a couple with a family history of a genetic disorder:

_____

_____

2. (a) Suggest why Huntington disease persists in the human population when it is caused by a lethal, dominant allele:

_____

_____

(b) Explain how presymptomatic genetic testing could change this: _____

_____

_____

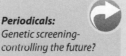

Periodicals:
Genetic screening-controlling the future?

***Related activities**: Examples of Gene Mutations, Aneuploidy in Humans, Down Syndrome, Lethal Alleles* ***Web links**: Patterns of Inheritance*

Heredity

A 3

Interactions Between Genes

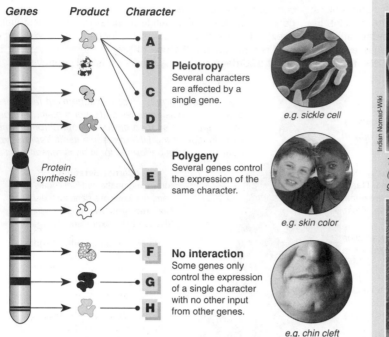

Genes   **Product**   **Character**

A
B

Pleiotropy
Several characters are affected by a single gene.

C
D

e.g. sickle cell

Protein synthesis

E

Polygeny
Several genes control the expression of the same character.

e.g. skin color

F

No interaction
Some genes only control the expression of a single character with no other input from other genes.

G

H

e.g. chin cleft

The widow's peak in the hairline, which is a dominant trait (as seen on Jude Law, left) is not evident if the epistatic gene for male pattern baldness is present (right).

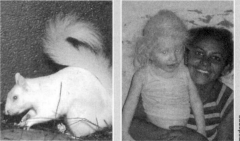

The homozygous condition for the albino allele overrides any other genes present for hair color.

Polygeny and Pleiotropy

Genes do not operate in isolation, and the interactions between genes can result in phenotypes different to those expected on the basis of genotype alone. In some cases a single phenotypic characteristic (e.g. skin color) is controlled by more than one gene. This phenomenon known as **polygeny** and leads to continuous phenotypic variation in the population.

Conversely, **pleiotropy** describes the genetic effect of a single gene on multiple traits. As a consequence of pleiotropy, a mutation in a gene may have a phenotypic effect on some or all traits simultaneously. The human disease, PKU (phenylketonuria) is one example of pleiotropy. This disease can cause mental retardation and reduced hair and skin pigmentation, and can be caused by any of a number of mutations in a single gene that codes for the conversion of phenylalanine to tyrosine, an intermediate in the metabolic pathway leading to melanin production.

Epistasis

A further possible type of gene interaction, called **epistasis**, involves two non-allelic genes (at different loci), where the action of one gene masks or otherwise alters the expression of other genes.

A common example is **albinism**, which appears in rodents (such as the squirrel, left) that are homozygous recessive for the 'albino' allele (a mutation) even if they have the alleles for agouti or black fur. Albinism is widespread in animal phyla. In every case, the presence two albino alleles (cc) overrides pigment expression. Another epistatic gene is that encoding male pattern baldness (top).

1. Explain the differences between **polygeny** and **pleiotropy**, giving examples to illustrate your answer:

_____

_____

_____

2. Explain how epistasis differs from pleiotropy. Describe examples to illustrate your answer:

_____

_____

3. Explain why the genes present for hair color are irrelevant if an animal is homozygous recessive for the albino gene:

_____

_____

4. Using an example (e.g. PKU), suggest how a pleiotropic gene could exert its multiple phenotypic effects: _____

_____

_____

_____

Related activities: Sickle Cell Mutation, Polygenes, Epistasis
Web links: Summary of Gene Interactions

Collaboration

There are genes that may influence the same trait, but produce a phenotype that could not result from the action of either gene independently. These are termed collaborative genes (they show **collaboration**). There are typically four possible phenotypes for this condition. An example of this type of interaction can be found in the comb shape of domestic hens.

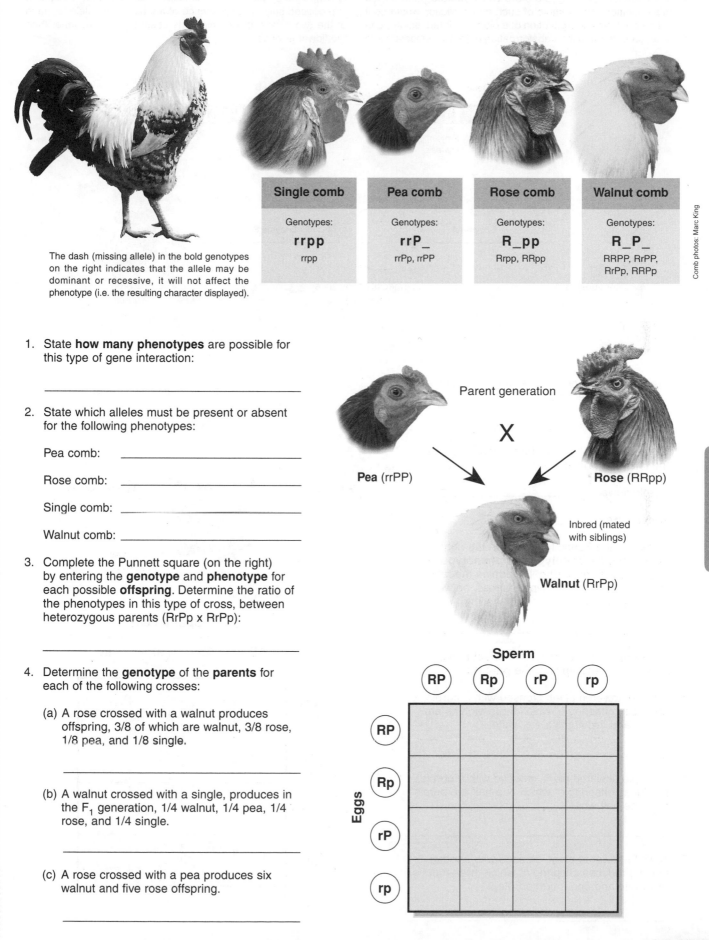

Single comb	Pea comb	Rose comb	Walnut comb
Genotypes:	Genotypes:	Genotypes:	Genotypes:
rrpp	**rrP_**	**R_pp**	**R_P_**
rrpp	rrPp, rrPP	Rrpp, RRpp	RRPP, RrPP, RrPp, RRPp

Comb photos: Marc King

The dash (missing allele) in the bold genotypes on the right indicates that the allele may be dominant or recessive, it will not affect the phenotype (i.e. the resulting character displayed).

1. State **how many phenotypes** are possible for this type of gene interaction:

   _____

2. State which alleles must be present or absent for the following phenotypes:

   Pea comb: _____

   Rose comb: _____

   Single comb: _____

   Walnut comb: _____

3. Complete the Punnett square (on the right) by entering the **genotype** and **phenotype** for each possible **offspring**. Determine the ratio of the phenotypes in this type of cross, between heterozygous parents (RrPp x RrPp):

   _____

4. Determine the **genotype** of the **parents** for each of the following crosses:

   (a) A rose crossed with a walnut produces offspring, 3/8 of which are walnut, 3/8 rose, 1/8 pea, and 1/8 single.

   _____

   (b) A walnut crossed with a single, produces in the F$_1$ generation, 1/4 walnut, 1/4 pea, 1/4 rose, and 1/4 single.

   _____

   (c) A rose crossed with a pea produces six walnut and five rose offspring.

   _____

Parent generation

X

Pea (rrPP)   **Rose** (RRpp)

Inbred (mated with siblings)

Walnut (RrPp)

Sperm

	RP	Rp	rP	rp
RP				
Rp				
rP				
rp				

Eggs

Heredity

Related activities: Dihybrid Cross

Web links: Summary of Gene Interactions

A 3

Complementary Genes

Some genes can only be expressed in the presence of other genes: they are **complementary**. Both genes have to have a dominant allele present for the final end product of the phenotype to be expressed. Typically, there are **two possible phenotypes** for this condition. An example of such an interaction would be if one gene controls the production of a pigment (intermediate) and another gene controls the transformation of that intermediate into

the pigment (by producing a controlling enzyme). Such genes have been found to control some flower colors. The diagram below right illustrates how one kind of flower color in sweet peas is controlled by two complementary genes. The purple pigment is produced only in the presence of the dominant allele for each of the two genes. If a dominant is absent for either gene, then the flower is white.

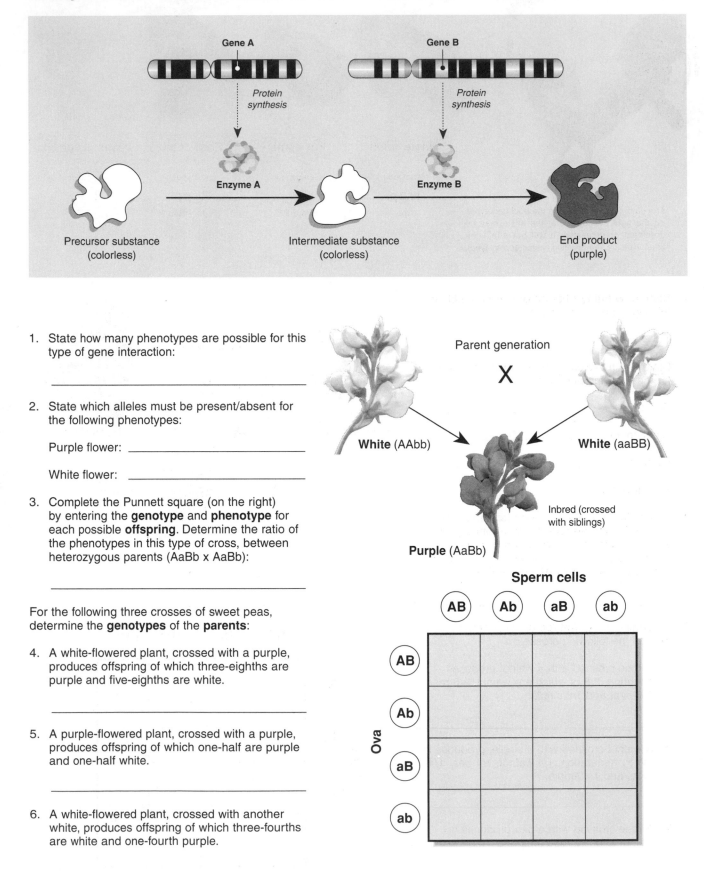

1. State how many phenotypes are possible for this type of gene interaction:

   _____

2. State which alleles must be present/absent for the following phenotypes:

   Purple flower: _____

   White flower: _____

3. Complete the Punnett square (on the right) by entering the **genotype** and **phenotype** for each possible **offspring**. Determine the ratio of the phenotypes in this type of cross, between heterozygous parents (AaBb x AaBb):

   _____

For the following three crosses of sweet peas, determine the **genotypes** of the **parents**:

4. A white-flowered plant, crossed with a purple, produces offspring of which three-eighths are purple and five-eighths are white.

   _____

5. A purple-flowered plant, crossed with a purple, produces offspring of which one-half are purple and one-half white.

   _____

6. A white-flowered plant, crossed with another white, produces offspring of which three-fourths are white and one-fourth purple.

   _____

© Biozone International 2001-2010
Photocopying Prohibited

Related activities: Control of Metabolic Pathways, Dihybrid Cross
Web links: Summary of Gene Interactions

Polygenes

Some phenotypes (e.g. kernel color in maize and skin color in humans) are determined by more than one gene and show **continuous variation** in a population. The production of the skin pigment melanin in humans is controlled by at least three genes. The amount of melanin produced is directly proportional to the number of dominant alleles for either gene (from 0 to 6).

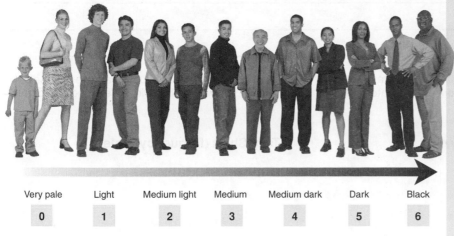

A light-skinned person

A dark-skinned person

Very pale	Light	Medium light	Medium	Medium dark	Dark	Black
0	1	2	3	4	5	6

Number of dark alleles

There are seven shades skin color ranging from very dark to very pale, with most individual being somewhat intermediate in skin color. No dominant allele results in a lack of dark pigment (aabbcc). Full pigmentation (black) requires six dominant alleles (AABBCC).

1. Complete the Punnett square for the F_2 generation (below) by entering the genotypes and the number of dark alleles resulting from a cross between two individuals of intermediate skin color. Color-code the offspring appropriately for easy reference.

   (a) State how many of the 64 possible offspring of this cross will have darker skin than their parents:

   _____

   (b) State how many genotypes are possible for this type of gene interaction:

   _____

2. Explain why in reality we see many more than seven shades of skin color:

   _____

   _____

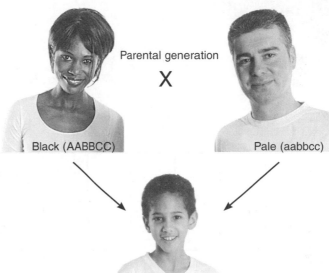

Parental generation

X

Black (AABBCC)

Pale (aabbcc)

Medium (AaBbCc)

F2 generation (AaBbCc X AaBbCc)

GAMETES	ABC	ABc	AbC	Abc	aBC	aBc	abC	abc
ABC								
ABc								
AbC								
Abc								
aBC								
aBc								
abC								
abc								

Heredity

© Biozone International 2001-2010
Photocopying Prohibited

Periodicals:
The color code

Related activities: Descriptive Statistics
Web links: Summary of Gene Interactions

RDA 3

3. Discuss the differences between **continuous** and **discontinuous** variation, giving examples to illustrate your answer:

_____

_____

_____

_____

_____

4 From a sample of no less than 30 adults, collect data (by request or measurement) for one continuous variable (e.g. height, weight, shoe size, or hand span). Record and tabulate your results in the space below, and then plot a frequency histogram of the data on the grid below:

Raw data	Tally Chart (frequency table)

Variable: _____

Frequency

(a) Calculate each of the following for your data. See *Descriptive Statistics* if you need help and attach your working:

Mean: _____ **Mode**: _____ **Median**: _____

Standard deviation: _____

(b) Describe the pattern of distribution shown by the graph, giving a reason for your answer: _____

_____

(c) Explain the genetic basis of this distribution: _____

_____

(d) Explain the importance of a large sample size when gathering data relating to a continuous variable:

_____

Epistasis

In its narrowest definition, **epistatic genes** are those that mask the effect of other genes. Typically there are **three possible phenotypes** for a dihybrid cross involving this type of gene interaction. One well studied example of epistasis occurs between the genes controlling coat color in rodents and other mammals. Skin and hair color is the result of melanin, a pigment which may be either black/brown (eumelanin) or reddish/yellow (phaeomelanin). Melanin itself is made up through several biochemical steps from the amino acid tyrosine. The control of coat color and patterning in mammals is complex and involves at least five major interacting genes. One of these genes (gene C), controls the production of the pigment melanin, while another gene (gene B), is responsible for whether the color is black or brown. The interaction between these genes in determining coat color in mice is illustrated below. Epistasis literally means "standing upon". In albinism, the homozygous recessive condition, cc, "stands upon" the other coat color genes, blocking their expression.

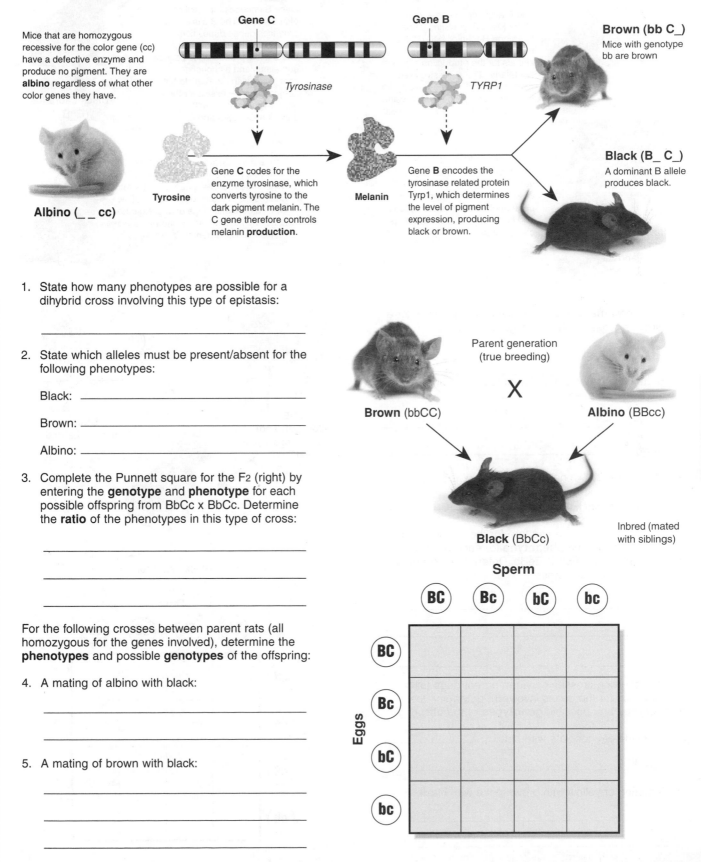

Gene C

Gene B

Brown (bb C_)
Mice with genotype bb are brown

Mice that are homozygous recessive for the color gene (cc) have a defective enzyme and produce no pigment. They are **albino** regardless of what other color genes they have.

Tyrosinase

TYRP1

Albino (_ _ cc)

Tyrosine

Gene **C** codes for the enzyme tyrosinase, which converts tyrosine to the dark pigment melanin. The C gene therefore controls melanin **production**.

Melanin

Gene **B** encodes the tyrosinase related protein Tyrp1, which determines the level of pigment expression, producing black or brown.

Black (B_ C_)
A dominant B allele produces black.

1. State how many phenotypes are possible for a dihybrid cross involving this type of epistasis:

   _____

2. State which alleles must be present/absent for the following phenotypes:

   Black: _____

   Brown: _____

   Albino: _____

3. Complete the Punnett square for the F2 (right) by entering the **genotype** and **phenotype** for each possible offspring from BbCc x BbCc. Determine the **ratio** of the phenotypes in this type of cross:

   _____

   _____

   _____

For the following crosses between parent rats (all homozygous for the genes involved), determine the **phenotypes** and possible **genotypes** of the offspring:

4. A mating of albino with black:

   _____

   _____

5. A mating of brown with black:

   _____

   _____

   _____

Parent generation
(true breeding)

X

Brown (bbCC)

Albino (BBcc)

Black (BbCc)

Inbred (mated with siblings)

Sperm

	BC	Bc	bC	bc
BC				
Bc				
bC				
bc				

Eggs

Heredity

Related activities: Dihybrid Cross
Web links: Summary of Gene Interactions

A 3

Epistatic interactions also regulate coat color in Labrador retrievers. The basic coat colors (yellow, chocolate, and black) are controlled by two genes (E and B). However, coat color variation depends on the allele combinations and the influence of the C gene. In dogs, the C gene always produces phaeomelanin no matter what the E or B genes are producing (phaeomelanin does not show in a dark coat). As a result, four main coat color variations are possible in Labradors, with a further three variations possible in the yellow coated dog.

Chocolate Labrador (E_bb)

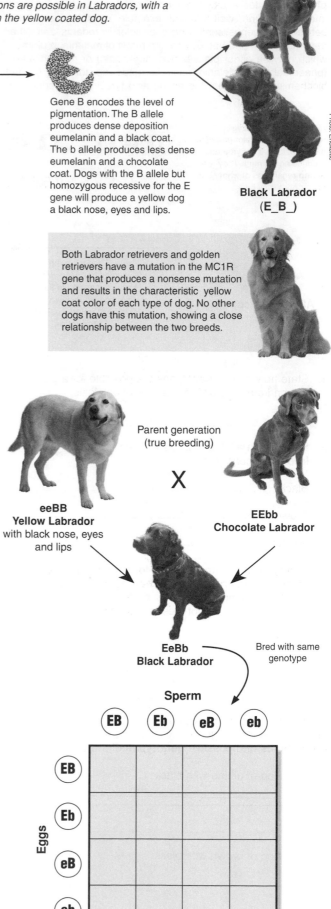

Yellow Labrador (eebb) with brown nose, eyes and lips

The E allele produces a form of the Melanocyte Stimulating Hormone Receptor (MC1R) that is extremely active and constantly signals for the production of eumelanin. The recessive allele does not signal as energetically for the eumelanin production, causing the Labrador to have a yellow coat.

Gene B encodes the level of pigmentation. The B allele produces dense deposition eumelanin and a black coat. The b allele produces less dense eumelanin and a chocolate coat. Dogs with the B allele but homozygous recessive for the E gene will produce a yellow dog a black nose, eyes and lips.

Black Labrador (E_B_)

The C gene alters the color of the yellow Labrador. A homozygous dominant dog will appear dark yellow due to large amounts of phaeomelanin with heterozygous and homozygous recessive dogs being progressively lighter.

Both Labrador retrievers and golden retrievers have a mutation in the MC1R gene that produces a nonsense mutation and results in the characteristic yellow coat color of each type of dog. No other dogs have this mutation, showing a close relationship between the two breeds.

Yellow Labrador (eeB_) with black nose, eyes and lips.

6. State how many main phenotypes are possible for a dihybrid cross involving the genes E and B:

_____

7. State which alleles must be present and absent for the following phenotypes:

Black: _____

Brown: _____

Yellow with brown nose: _____

Yellow with black nose: _____

8. Complete the Punnett square for the F2 by entering the **genotype** and **phenotype** for each possible offspring from EeBb X EeBb. Determine the **ratio** of phenotypes in this cross:

_____

_____

_____

For the following crosses between parent dogs (all homozygous for the genes involved), determine the **phenotypes** and possible **genotypes** of the offspring:

9. A mating of chocolate with black: _____

_____

10. A mating of yellow with brown nose with black:

_____

_____

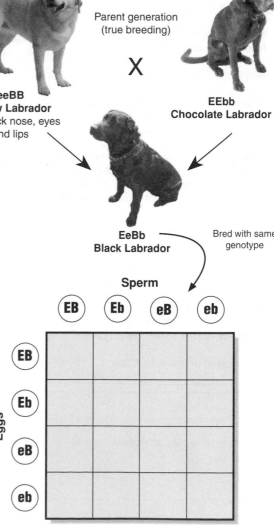

Parent generation (true breeding)

X

eeBB
Yellow Labrador
with black nose, eyes and lips

EEbb
Chocolate Labrador

EeBb
Black Labrador

Bred with same genotype

Sperm

	EB	Eb	eB	eb
EB				
Eb				
eB				
eb				

Eggs

Inheritance in Domestic Cats

Cats have been domesticated for thousands of years. During this time, certain traits or characteristics have been considered fashionable or desirable in a cat by people in different parts of the world. In the domestic cat, the 'wild type' is the short-haired tabby. All the other coat colors found in cats are modifications of this ancestral tabby pattern. Inheritance of coat characteristics and a few other features in cats is interesting because they exhibit the most common genetic phenomena. Some selected traits for domestic cats are identified below, together with a list of the kinds of genetic phenomena easily demonstrated in cats.

Inheritance Patterns in Domestic Cats

Dominance	The polydactylism gene with the dominant allele (Pd) produces a paw with extra digits.
Recessiveness	The dilution gene with the recessive allele (d) produces a diluted black to produce gray, or orange to cream.
Epistasis	The dominant agouti gene (A) must be present for the tabby gene (T) to be expressed.
Multiple alleles	The albino series (C) produces a range of phenotypes from full pigment intensity to true albino.
Incomplete dominance	The spotting gene (S) has three phenotypes ranging from extensive spotting to no spotting at all.
Lethal genes	The Manx gene (M) that produces a stubby or no tail is lethal when in the homozygous dominant condition (MM causes death in the womb).
Pleiotropy	The white gene (W) also affects eye color and can cause congenital deafness (one gene with three effects).
Sex linkage	The orange gene is sex (X) linked and can convert black pigment to orange. Since female cats have two X chromosomes they have three possible phenotypes (black, orange and tortoiseshell) whereas males can normally only exhibit two phenotypes (black and orange).
Environmental effects	The dark color pointing in Siamese and Burmese cats where the gene (cs) is only active in the cooler extremities such as the paws, tail and face.

(NOTE: Some of these genetic phenomena are covered elsewhere)

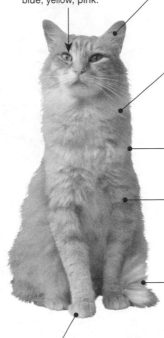

Eyes
May have a range of coloring for the irises: blue, yellow, pink.

Ears
May be normal pointed ears, or the ears may be folded.

Coat color
A wide range of coat colors are available, controlled by a variety of genes. Basic colors include black, white, orange and agouti. Color patterns can range from solid, patched, spotted or tabby.

Coat length
Hair is usually either long or short. There is a breed with extremely short hair; so much so that it looks hairless (sphynx).

Coat texture
Smooth hair is the common phenotype, but there is an allele that causes curly hair.

Tail
Most cats have a long tail. An allele for short, stubby tails is almost completely restricted to the bobcat and Manx breeds.

Paws
Most cats have five digits on the front paw and four on the rear. The occurrence of polydactyly with as many as six or seven digits affects as many as one out of five cats (in some parts of the world it is even higher than this).

Genes controlling inherited traits in domestic cats

Wild forms		Mutant forms		Wild forms		Mutant forms	
Allele	*Phenotype*	*Allele*	*Phenotype*	*Allele*	*Phenotype*	*Allele*	*Phenotype*
A	Agouti	**a**	Black (non-agouti)	**m**	Normal tail	**M**	Manx tail, shorter than normal (stubby)
B	Black pigment	**b**	Brown pigment	**o**	Normal colors (no red, usually black)	**O**	Orange (sex linked)
C	Unicolored	**cch** Silver **cs** Siamese (pointing: dark at extremities) **ca** Albino with blue eyes **c** Albino with pink eyes	**pd**	Normal number of toes	**Pd**	Polydactylism; has extra toes
D	Dense pigment	**d**	Dilute pigment	**R**	Normal, smooth hair	**R**	Rex hair, curly
fd	Normal, pointed ears	**Fd**	Folded ears	**s**	Normal coat color without white spots	**S**	Color interspersed with white patches or spots (piebald white spotting)
Hr	Normal, full coat	**h**	Hairlessness	**T**	Tabby pattern (mackerel striped)	**Ta** **tb**	Abyssinian tabby Blotched tabby, classic pattern of patches or stripes
i	Fur colored all over	**I**	Inhibitor: part of the hair is not colored (silver)	**w**	Normal coat color, not all white	**W** **Wh**	All white coat color (dominant white) Wirehair
L	Short hair	**l**	Long hair, longer than normal				

Variation in Coat Color in Domestic Cats

Non-agouti
A completely jet black cat has no markings on it whatsoever. It would have the genotype: **aaB–D–** since no dominant agouti allele must be present, and the black pigment is not diluted.

Siamese
The color pointing of Siamese cats is caused by warm temperature deactivation of a gene that produces melanin pigment. Cooler parts of the body are not affected and appear dark.

Tortoiseshell
Because this is a sex linked trait, it is normally found only in female cats (**XO, Xo**). The coat is a mixture of orange and black fur irregularly blended together.

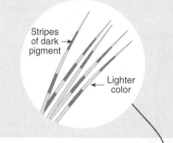

Agouti hair
Enlarged view of agouti hair. Note that the number of darkly pigmented stripes can vary on the same animal.

Stripes of dark pigment

Lighter color

Sex linked orange
The orange (**XO, XO**) cat has an orange coat with little or no patterns such as tabby showing.

Blotched tabby
Lacks stripes but has broad, irregular bands arranged in whorls (**tb**).

Wild type
Mackerel (striped) tabby (**A–B–T–**) with evenly spaced, well-defined, vertical stripes. The background color is **agouti** with the stripes being areas of completely black hairs.

Orange

Black   White

Calico
Similar to a tortoiseshell, but with substantial amounts of white fur present as well. Black, orange and white fur.

Golden yellow coat

Deeper colour stripes

Marmalade
The orange color (**XO, XO**) is expressed, along with the alleles for the tabby pattern. The allele for orange color shows epistatic dominance and overrides the expression of the agouti color so that the tabby pattern appears dark orange.

Other Inherited Features in Domestic Cats

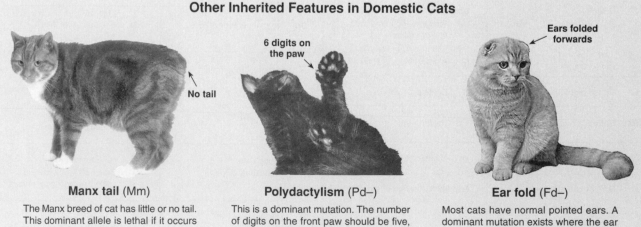

No tail

Manx tail (Mm)
The Manx breed of cat has little or no tail. This dominant allele is lethal if it occurs in the homozygous condition.

6 digits on the paw

Polydactylism (Pd–)
This is a dominant mutation. The number of digits on the front paw should be five, with four digits on the rear paw.

Ears folded forwards

Ear fold (Fd–)
Most cats have normal pointed ears. A dominant mutation exists where the ear is permanently folded forwards.

What Genotype Has That Cat?

Consult the table of genes listed on the previous pages and enter the allele symbols associated with each of the phenotypes in the column headed 'Allele'. For this exercise, study the appearance of real cats around your home or look at color photographs of different cats. For each cat, complete the checklist of traits listed below by simply placing a tick in the appropriate spaces. These traits are listed in the same order as the genes for **wild forms** and **mutant forms** on page 236. On a piece of paper, write each of the cat's genotypes. Use a dash (-) for the second allele for characteristics that could be either heterozygous or homozygous dominant (see the sample at the bottom of the page).

NOTES:
1. *Agouti* fur coloring is used to describe black **hairs** with a light band of pigment close to its tip.
2. Patches of silver fur (called chinchilla) produces the silver tabby phenotype in agouti cats. Can also produce "smoke" phenotype in Persian long-haired cats, causing reduced intensity of the black.
3. Describes the dark extremities (face, tail and paws) with lighter body (e.g. Siamese).
4. The recessive allele makes black cats blue-gray in color and yellow cats cream.
5. Spottiness involving less than half the surface area is likely to be heterozygous.

Phenotype Record Sheet for Domestic Cats

Gene	Phenotype	Allele	Sample	Cat 1	Cat 2	Cat 3	Cat 4
Agouti color	Agouti[1]						
	Non-agouti		✔				
Pigment color	Black		✔				
	Brown						
Color present	Uncolored		✔				
	Silver patches[2]						
	Pointed[3]						
	Albino with blue eyes						
	Albino with pink eyes						
Pigment density	Dense pigment						
	Dilute pigment[4]		✔				
Ear shape	Pointed ears		✔				
	Folded ears						
Hairiness	Normal, full coat		✔				
	Hairlessness						
Hair length	Short hair		✔				
	Long hair						
Tail length	Normal tail (long)		✔				
	Stubby tail or no tail at all						
Orange color	Normal colors (non-orange)		✔				
	Orange						
Number of digits	Normal number of toes		✔				
	Polydactylism (extra toes)						
Hair curliness	Normal, smooth hair		✔				
	Curly hair (rex)						
Spottiness	No white spots						
	White spots (less than half)[5]		✔				
	White spots (more than half)						
Stripes	Mackerel striped (tabby)						
	Blotched stripes						
White coat	Not all white		✔				
	All white coat color						

Sample cat: (see ticks in chart above)

To give you an idea of how to read the chart you have created, here is an example genotype of the author's cat with the following features: *A smoky gray uniform-colored cat, with short smooth hair, normal tail and ears, with 5 digits on the front paws and 4 on the rear paws, small patches of white on the feet and chest.* (Note that the stripe genotype is completely unknown since there is no agouti allele present).

GENOTYPE: aa B– C– dd fdfd Hr– ii L– mm oo pdpd R– Ss ww

Related activities: Inheritance in Domestic Cats
Web links: Cat Color and Fur Length Genetics

A 3

Heredity

KEY TERMS: Crossword

Complete the crossword below, which will test your understanding of key terms in this chapter and their meanings

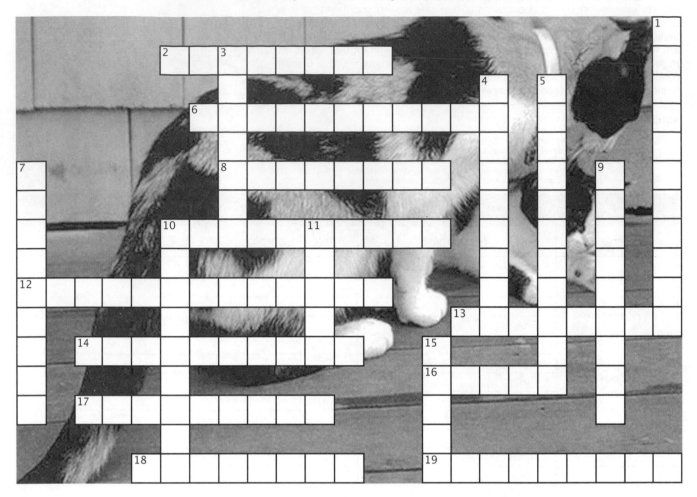

Clues Across

2. Alleles for a gene where there are more than two alleles are called __ __ __ __ __ __ __ alleles.

6. The acquisition of characteristics by the transfer of genetic information from one generation to the next.

8. A chromosome that is not a sex chromosome.

10. The phenomenon of multiple effects by a single gene.

12. Genes that can only be expressed in the presence of other genes. Both genes require the dominant characteristic to be expressed.

13. The family history or inheritance of an organism is often referred to as its __ __ __ __ __ __ __ __.

14. Dominance in which the action of one allele does not completely mask the action of the other.

16. The position of a gene on a chromosome.

17. Characteristics that are a product of multiple genes being expressed are called __ __ __ __ __ __ __ __ __ __ .

18. The specific allele combination of an organism.

19. A gene interaction in which one gene masks the effect of another gene.

Clues Down

1. An inheritance pattern in which both alleles in the heterozygous condition contribute to the phenotype.

3. The term to describe genes that are located on the same chromosome.

4. Allele that expresses its characteristics only when in the homozygous condition.

5. The study of the heritable changes that occur without involving changes in the DNA sequence.

7. A diagnostic cross involving breeding with a recessive with known genotype. (2 words: 4, 5)

9. Any cross between an offspring and one of its parents. (2 words: 4, 5)

10. The physical appearance of the genotype.

11. A particular phenotypic character. Refers to the physical appearance as opposed to the mode of appearance e.g. blue eyes rather than eye color.

15. A gradation in phenotype within the same species, usually along an environmental gradient.

Molecular Genetics

KEY CONCEPTS

▶ DNA controls the behavior of cells.

▶ DNA is a self-replicating molecule constructed according to strict base-pairing rules.

▶ The genetic code, through transcription and translation, contains the information to construct proteins.

▶ The operon model of gene regulation in prokaryotes is useful but not universally applicable.

▶ Viruses can command a cell's metabolic machinery to replicate themselves.

KEY TERMS

amino acids
anticodon
bacteriophage
base-pairing rule
coding strand
codon
DNA
DNA ligase
DNA polymerase
DNA replication
end product inhibition
exons
gene expression
gene induction
genetic code
helicase
hydrogen bonding
introns
lac operon
lagging strand
leading strand
lysogenic cycle
lytic cycle
nucleic acids
nucleotides
Okazaki fragments
promoter
protein
replication fork
reverse transcriptase
RNA (mRNA, rRNA, tRNA)
start codon / stop codon
template strand
transcription
transcription factors
translation
triplet
virus

OBJECTIVES

☐ 1. Use the **KEY TERMS** to help you understand and complete these objectives.

Nucleic Acids
pages 55-56, 202-211

☐ 2. Recall the structure of **nucleotides**. Describe the Watson-Crick double-helix model of **DNA** structure, including reference to the **anti-parallel** nature of DNA, the **base-pairing rule**, and **hydrogen bonding**.

☐ 3. Describe the structure and function of **mRNA**, **tRNA**, and **rRNA**. Contrast the structure and function of RNA and DNA.

☐ 4. Describe the **semi-conservative replication** of DNA. Demonstrate your understanding of the **base-pairing rule** for creating a complementary strand from a single strand of DNA.

☐ 5. Describe and explain the main features of the **genetic code**.

Gene Expression
pages 212-223

☐ 6. Describe the evidence that led to the one gene-one polypeptide hypothesis. Explain how this hypothesis has been modified in light of recent evidence.

☐ 7. Identify the two stages of **gene expression** as transcription and translation.

☐ 8. Describe **transcription**, including the significance of the **coding (sense) strand** and **template (antisense) strand**. Explain the significance of **introns** with the respect to the production of a functional mRNA molecule.

☐ 9. Recall the structure of **amino acids** and how they form the primary structure of proteins (polypeptides). Describe and explain **translation**, including the role of **tRNA molecules**, **ribosomes**, **start codons**, and **stop codons**.

☐ 10. Describe the structure and function of a prokaryote **operon**. Discuss the extent to which the operon model is universally applicable.

☐ 11. Explain how simple metabolic pathways are regulated in prokaryotes, as illustrated by **gene induction** in the *lac* **operon** in *E. coli*.

☐ 12. EXTENSION: Using an example, e.g. control of tryptophan synthesis, explain **end-product** inhibition of a metabolic pathway.

☐ 13. Describe gene regulation through transcriptional control in eukaryotes.

Replication in Viruses
pages 224-227

☐ 14. Describe the structure of viruses. Show how they fulfil their role as specialized intracellular parasites by explaining viral replication in a chosen example.

Periodicals:
listings for this chapter are on page 380

Weblinks:
www.thebiozone.com/
weblink/SB1-2597.html

Teacher Resource CD-ROM:
The Meselson–Stahl Experiment

Does DNA Really Carry the Code?

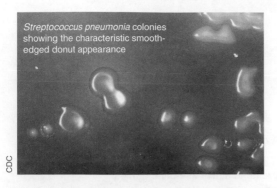

Streptococcus pneumonia colonies showing the characteristic smooth-edged donut appearance

CDC

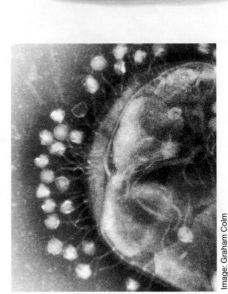

Image: Graham Colm

Scientists had known about DNA since the end of the 19th century, but its role in storing information remained unknown until the 1940s, and its structure remained a mystery for another decade after that. In 1928, experiments by British scientist Fredrick Griffith gave the first indications that DNA was responsible for passing on information. Griffith had been working with two strains of the bacteria *Streptococcus pneumoniae*. Only one strain (the pathogenic strain) caused pneumonia and it was easily identified because it formed colonies with smooth edges. The other, benign strain formed colonies with rough edges. When mice were injected with the pathogenic strain they developed pneumonia and died. The mice injected with the benign strain did not. Mice injected with the heat-killed pathogenic strain did not develop pneumonia either. This showed that the disease was not caused by a chemical associated with the bacteria, or a response by the body to the bacteria, it was the bacterial cells themselves. In a second experiment, Griffith mixed the benign strain with the heat-killed pathogenic strain and injected it into healthy mice. To his surprise, the mice developed pneumonia. When bacteria from the mice were recovered and cultured they produced colonies identical to the pathogenic strain. Somehow the harmless bacteria had acquired information from the dead pathogenic strain. Griffith called this process **transformation**.

In 1944, American scientists, led by Oswald Avery, continued with Griffith's experiments. They made an extract from the heat-killed pathogenic strain and treated it with chemicals to destroy any lipids, carbohydrates, or proteins. This was mixed with the benign strain and transformation still occurred. This established that no proteins, lipids, or carbohydrates were responsible for the transformation. When another identical extract was treated with chemicals that break down DNA, the transformation did not take place - the benign strain failed to acquire the information required to cause pneumonia. From this it was deduced that DNA was the unit that was carrying the information from one bacteria to another.

Another experiment in 1952 by Alfred Hershey, confirmed what the other two experiments had shown. Hershey worked with viruses, which were known to have DNA and to transfer information to their host. However, there was debate over whether the information was transferred by the DNA or by the protein coat of the virus. Hershey use radioactive sulfur and radioactive phosphorus to mark different parts of the virus. The sulfur was incorporated into the protein coat while the phosphorus was incorporated in to the viral DNA. The viruses were then mixed with bacteria and the infected bacteria analysed. The bacteria were found to contain radioactive phosphorus but not radioactive sulfur, showing that the virus had indeed passed information to its host by injecting its own DNA.

1. Explain how Griffith confirmed that it was the bacteria causing the pneumonia and not something else:

    _____

    _____

    _____

2. Explain why sulfur and phosphorus were used in Hershey's experiment: _____

    _____

    _____

    _____

3. Explain why conducting two different experiments (Avery's and Hershey's) is important when confirming an idea:

    _____

    _____

    _____

    _____

    _____

Related activities: Prokaryotic Cells, DNA Molecules

DNA Molecules

Even the smallest DNA molecules are extremely long. The DNA from the small *Polyoma* virus, for example, is 1.7 µm long; about three times longer than the longest proteins. The DNA comprising a bacterial chromosome is 1000 times longer than the cell into which it has to fit. The amount of DNA present in the nucleus of the cells of eukaryotic organisms varies widely from one species to another. In vertebrate sex cells, the quantity of DNA ranges from 40 000 **kb** to 80 000 000 **kb**, with humans about in the middle of the range. The traditional focus of DNA research has been on those DNA sequences that code for proteins, yet protein-coding DNA accounts for less than 2% of the DNA in human chromosomes. The rest of the DNA, once dismissed as non-coding 'evolutionary junk', is now recognized as giving rise to functional RNA molecules, many of which have already been identified as having important regulatory functions. While there is no clear correspondence between the complexity of an organism and the number of protein-coding genes in its genome, this is not the case for non-protein-coding DNA. The genomes of more complex organisms contain much more of this so-called "non-coding" DNA. These RNA-only 'hidden' genes tend to be short and difficult to identify, but the sequences are highly conserved and clearly have a role in inheritance, development, and health.

Total length of DNA in viruses, bacteria, and eukayotes

Taxon	Organism	Base pairs (in 1000s, or kb)	Length
Viruses	Polyoma or SV40	5.1	1.7 µm
	Lambda phage	48.6	17 µm
	T2 phage	166	56 µm
	Vaccinia	190	65 µm
Bacteria	Mycoplasma	760	260 µm
	E. coli (from human gut)	4600	1.56 mm
Eukaryotes	Yeast	13 500	4.6 mm
	Drosophila (fruit fly)	165 000	5.6 cm
	Human	2 900 000	99 cm

Kilobase (kb)

A kilobase (kb) is a unit of length equal to 1000 base pairs of a double-stranded nucleic acid molecule (or 1000 bases of a single-stranded molecule). One kb of double stranded DNA has a length of 0.34 µm (1 µm = 1/1000 mm).

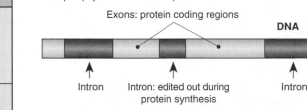

Exons: protein coding regions

DNA

Intron   Intron: edited out during protein synthesis   Intron

Most protein-coding genes in eukaryotic DNA are not continuous and may be interrupted by 'intrusions' of other pieces of DNA. Protein-coding regions (**exons**) are interrupted by non-protein-coding regions called **introns**. Introns range in frequency from 1 to over 30 in a single 'gene' and also in size (100 to more than 10,000 bases). Introns are edited out of the protein-coding sequence during protein synthesis, but probably, after processing, go on to serve a regulatory function.

Giant lampbrush chromosomes

Lampbrush chromosomes are large chromosomes found in amphibian eggs, with lateral loops of DNA that produce a brushlike appearance under the microscope. The two scanning electron micrographs (below and right) show minute strands of DNA giving a fuzzy appearance in the high power view.

Loops of DNA

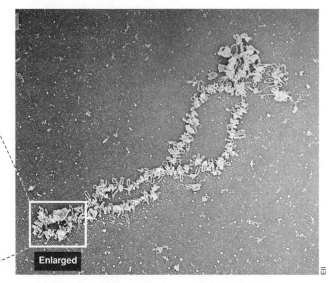

Enlarged

1. Consult the table above and make the following comparisons. Determine how much more DNA is present in:

   (a) The bacterium *E. coli* compared to the Lambda Phage virus: _____

   (b) Human cells compared to the bacteria *E. coli:* _____

2. State what proportion of DNA in a eukaryotic cell is used to code for proteins or structural RNA: _____

3. Describe two reasons why geneticists have reevaluated their traditional view that one gene codes for one polypeptide:

   (a) _____

   _____

   (b) _____

   _____

Periodicals:
DNA: 50 years of the double helix

Related activities: Genomes, The Simplest Case: Genes to Proteins, Gene Expression

Molecular Genetics

DA 1

The Genetic Code

The genetic information that codes for the assembly of amino acids is stored as three-letter codes, called **codons**. Each codon represents one of 20 amino acids used in the construction of polypeptide chains. The **mRNA-amino acid table** (below) can be used to identify the amino acid encoded by each of the mRNA codons. Note that the code is **degenerate** in that for each amino acid, there may be more than one codon. Most of this degeneracy involves the third nucleotide of a codon. The genetic code is **universal**; all living organisms on Earth, from viruses and bacteria, to plants and humans, share the same genetic code book (with a few minor exceptions representing mutations that have occurred over the long history of evolution).

Amino acid		Codons that code for this amino acid	No.	Amino acid		Codons that code for this amino acid	No.
Ala	Alanine	GCU, GCC, GCA, GCG	4	**Leu**	Leucine		
Arg	Arginine			**Lys**	Lysine		
Asn	Asparagine			**Met**	Methionine		
Asp	Aspartic acid			**Phe**	Phenylalanine		
Cys	Cysteine			**Pro**	Proline		
Gln	Glutamine			**Ser**	Serine		
Glu	Glutamic acid			**Thr**	Threonine		
Gly	Glycine			**Try**	Tryptophan		
His	Histidine			**Tyr**	Tyrosine		
Iso	Isoleucine			**Val**	Valine		

1. Use the **mRNA-amino acid table** (below) to list in the table above all the **codons** that code for each of the amino acids and the number of different codons that can code for each amino acid (the first amino acid has been done for you).

2. (a) State how many amino acids could be coded for if a codon consisted of just two bases: _____

   (b) Explain why this number of bases is inadequate to code for the 20 amino acids required to make proteins:

   _____

   _____

3. Describe the consequence of the degeneracy of the genetic code to the likely effect of **point mutations**:

   _____

   _____

mRNA-Amino Acid Table

How to read the table: The table on the right is used to 'decode' the genetic code as a sequence of amino acids in a polypeptide chain, from a given mRNA sequence. To work out which amino acid is coded for by a codon (triplet of bases) look for the first letter of the codon in the row label on the left hand side. Then look for the column that intersects the same row from above that matches the second base. Finally, locate the third base in the codon by looking along the row from the right hand end that matches your codon.

Example: Determine **CAG**

   C on the left row, A on the top column, G on the right row
   CAG is Gln (**glutamine**)

Read second letter here
Read first letter here
Read third letter here

		Second Letter				
First Letter		**U**	**C**	**A**	**G**	Third Letter
U		UUU Phe UUC Phe UUA Leu UUG Leu	UCU Ser UCC Ser UCA Ser UCG Ser	UAU Tyr UAC Tyr UAA STOP UAG STOP	UGU Cys UGC Cys UGA STOP UGG Try	U C A G
C		CUU Leu CUC Leu CUA Leu CUG Leu	CCU Pro CCC Pro CCA Pro CCG Pro	CAU His CAC His CAA Gln CAG Gln	CGU Arg CGC Arg CGA Arg CGG Arg	U C A G
A		AUU Iso AUC Iso AUA Iso AUG Met	ACU Thr ACC Thr ACA Thr ACG Thr	AAU Asn AAC Asn AAA Lys AAG Lys	AGU Ser AGC Ser AGA Arg AGG Arg	U C A G
G		GUU Val GUC Val GUA Val GUG Val	GCU Ala GCC Ala GCA Ala GCG Ala	GAU Asp GAC Asp GAA Glu GAG Glu	GGU Gly GGC Gly GGA Gly GGG Gly	U C A G

Related activities: Genomes, The Simplest Case: Genes to Proteins, Gene Expression, Gene Mutations

Creating a DNA Model

Although DNA molecules can be enormous in terms of their molecular size, they are made up of simple repeating units called **nucleotides**. A number of factors control the way in which these nucleotide building blocks are linked together. These factors cause the nucleotides to join together in a predictable way. This is referred to as the **base pairing rule** and can be used to construct a complementary DNA strand from a template strand, as illustrated in the exercise below:

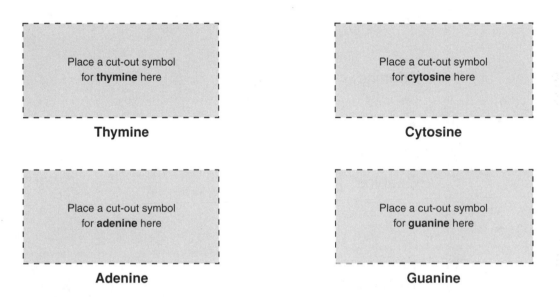

DNA Base Pairing Rule			
Adenine	is always attracted to	**Thymine**	A ⟷ T
Thymine	is always attracted to	**Adenine**	T ⟷ A
Cytosine	is always attracted to	**Guanine**	C ⟷ G
Guanine	is always attracted to	**Cytosine**	G ⟷ C

1. Cut out around the nucleotides on page 207 and separate each of the 24 nucleotides by cutting along the columns and rows (see arrows indicating these cutting points). Although drawn as geometric shapes, these symbols represent chemical structures.

2. Place one of each of the four kinds of nucleotide on their correct spaces below:

> Place a cut-out symbol
> for **thymine** here

Thymine

> Place a cut-out symbol
> for **cytosine** here

Cytosine

> Place a cut-out symbol
> for **adenine** here

Adenine

> Place a cut-out symbol
> for **guanine** here

Guanine

3. Identify and **label** each of the following features on the adenine nucleotide immediately above:
   phosphate, **sugar**, **base**, **hydrogen bonds**

4. Create one strand of the DNA molecule by placing the 9 correct 'cut out' nucleotides in the labeled spaces on the following page (DNA Molecule). Make sure these are the right way up (with the **P** on the left) and are aligned with the left hand edge of each box. Begin with thymine and end with guanine.

5. Create the complementary strand of DNA by using the base pairing rule above. Note that the nucleotides have to be arranged upside down.

6. Under normal circumstances, it is not possible for adenine to pair up with guanine or cytosine, nor for any other mismatches to occur. Describe the two factors that prevent a mismatch from occurring:

   (a) Factor 1: _____

   _____

   (b) Factor 2: _____

   _____

7. Once you have checked that the arrangement is correct, you may glue, paste or tape these nucleotides in place.

> **NOTE:** There may be some value in keeping these pieces loose in order to practise the base pairing rule. For this purpose, *removable tape* would be best.

Related activities: Nucleic Acids, DNA Molecules

PA 2

Molecular Genetics

DNA Molecule

Put the named nucleotides on the left hand side to create the template strand

Thymine | T | A | S

Thymine

Put the matching **complementary** nucleotides opposite the template strand

Cytosine

Adenine

Adenine

Guanine

Thymine

Thymine

Cytosine

Guanine

Nucleotides

Tear out this page along the perforation and separate each of the 24 nucleotides by cutting along the columns and rows (see arrows indicating the cutting points).

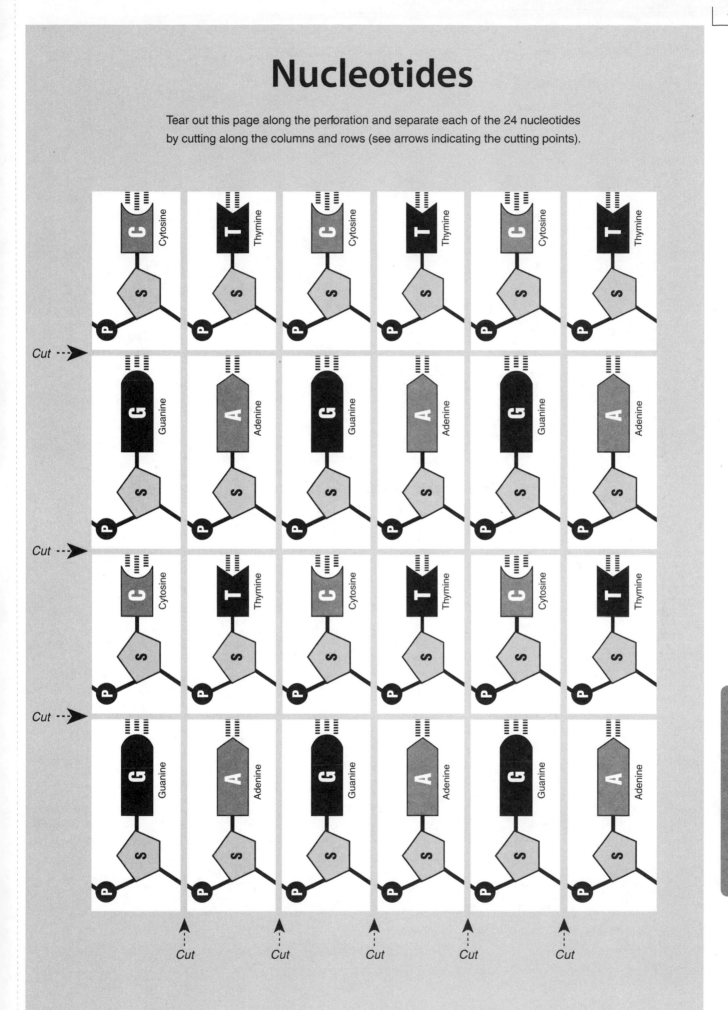

Molecular Genetics

This page is deliberately left blank

DNA Replication

The replication of DNA is a necessary preliminary step for cell division (both mitosis and meiosis). This process creates the **two chromatids** that are found in chromosomes that are preparing to divide. By this process, the whole chromosome is essentially duplicated, but is still held together by a common centromere. Enzymes are responsible for all of the key events. The diagram below shows the essential steps in the process. The diagram on the next page shows how enzymes are involved at each stage.

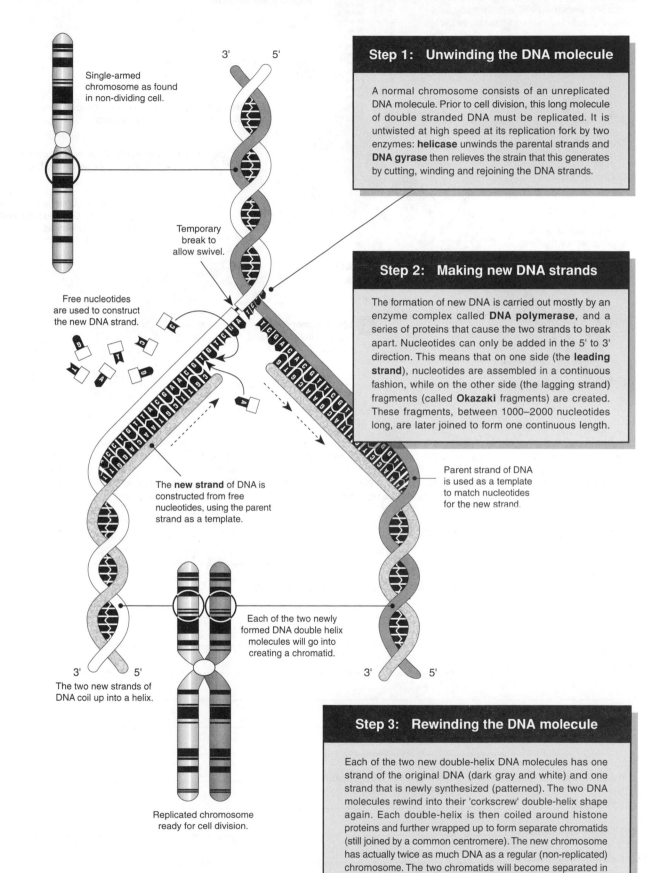

Single-armed chromosome as found in non-dividing cell.

Temporary break to allow swivel.

Free nucleotides are used to construct the new DNA strand.

The **new strand** of DNA is constructed from free nucleotides, using the parent strand as a template.

Parent strand of DNA is used as a template to match nucleotides for the new strand.

The two new strands of DNA coil up into a helix.

Each of the two newly formed DNA double helix molecules will go into creating a chromatid.

Replicated chromosome ready for cell division.

Step 1: Unwinding the DNA molecule

A normal chromosome consists of an unreplicated DNA molecule. Prior to cell division, this long molecule of double stranded DNA must be replicated. It is untwisted at high speed at its replication fork by two enzymes: **helicase** unwinds the parental strands and **DNA gyrase** then relieves the strain that this generates by cutting, winding and rejoining the DNA strands.

Step 2: Making new DNA strands

The formation of new DNA is carried out mostly by an enzyme complex called **DNA polymerase**, and a series of proteins that cause the two strands to break apart. Nucleotides can only be added in the 5' to 3' direction. This means that on one side (the **leading strand**), nucleotides are assembled in a continuous fashion, while on the other side (the lagging strand) fragments (called **Okazaki** fragments) are created. These fragments, between 1000–2000 nucleotides long, are later joined to form one continuous length.

Step 3: Rewinding the DNA molecule

Each of the two new double-helix DNA molecules has one strand of the original DNA (dark gray and white) and one strand that is newly synthesized (patterned). The two DNA molecules rewind into their 'corkscrew' double-helix shape again. Each double-helix is then coiled around histone proteins and further wrapped up to form separate chromatids (still joined by a common centromere). The new chromosome has actually twice as much DNA as a regular (non-replicated) chromosome. The two chromatids will become separated in the cell division process to form two separate chromosomes.

Molecular Genetics

Related activities: Mitosis and the Cell Cycle, Polymerase Chain Reaction
Web links: DNA Replication, The Meselson–Stahl Experiment

DA 3

Enzyme Control of DNA Replication

DNA replication occurs at an astounding rate. As many as 4000 nucleotides per second are replicated. This explains how, under ideal conditions, bacterial cells with as many as 4 million nucleotides, can complete a cell cycle in about 20 minutes. See the activity on **polymerase chain reaction** for a useful application of this process.

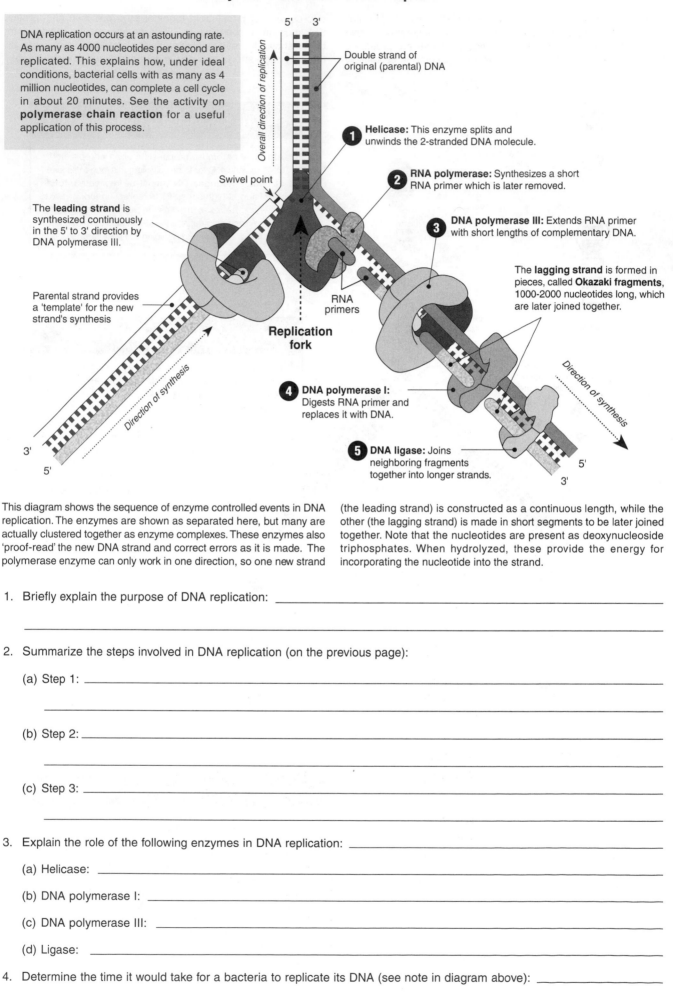

5'   3'

Overall direction of replication

Double strand of original (parental) DNA

1 **Helicase:** This enzyme splits and unwinds the 2-stranded DNA molecule.

2 **RNA polymerase:** Synthesizes a short RNA primer which is later removed.

3 **DNA polymerase III:** Extends RNA primer with short lengths of complementary DNA.

The **lagging strand** is formed in pieces, called **Okazaki fragments**, 1000-2000 nucleotides long, which are later joined together.

Swivel point

The **leading strand** is synthesized continuously in the 5' to 3' direction by DNA polymerase III.

Parental strand provides a 'template' for the new strand's synthesis

RNA primers

Replication fork

Direction of synthesis

Direction of synthesis

3'

5'

4 **DNA polymerase I:** Digests RNA primer and replaces it with DNA.

5 **DNA ligase:** Joins neighboring fragments together into longer strands.

5'

3'

This diagram shows the sequence of enzyme controlled events in DNA replication. The enzymes are shown as separated here, but many are actually clustered together as enzyme complexes. These enzymes also 'proof-read' the new DNA strand and correct errors as it is made. The polymerase enzyme can only work in one direction, so one new strand (the leading strand) is constructed as a continuous length, while the other (the lagging strand) is made in short segments to be later joined together. Note that the nucleotides are present as deoxynucleoside triphosphates. When hydrolyzed, these provide the energy for incorporating the nucleotide into the strand.

1. Briefly explain the purpose of DNA replication: _____

_____

2. Summarize the steps involved in DNA replication (on the previous page):

(a) Step 1: _____

_____

(b) Step 2: _____

_____

(c) Step 3: _____

_____

3. Explain the role of the following enzymes in DNA replication: _____

(a) Helicase: _____

(b) DNA polymerase I: _____

(c) DNA polymerase III: _____

(d) Ligase: _____

4. Determine the time it would take for a bacteria to replicate its DNA (see note in diagram above): _____

Review of DNA Replication

The diagram below summarizes the main steps in DNA replication. You should use this activity to test your understanding of the main features of DNA replication, using your acquired knowledge of DNA replication to fill in the missing information. You should attempt this from what you have learned, but refer to the previous activity if you require help.

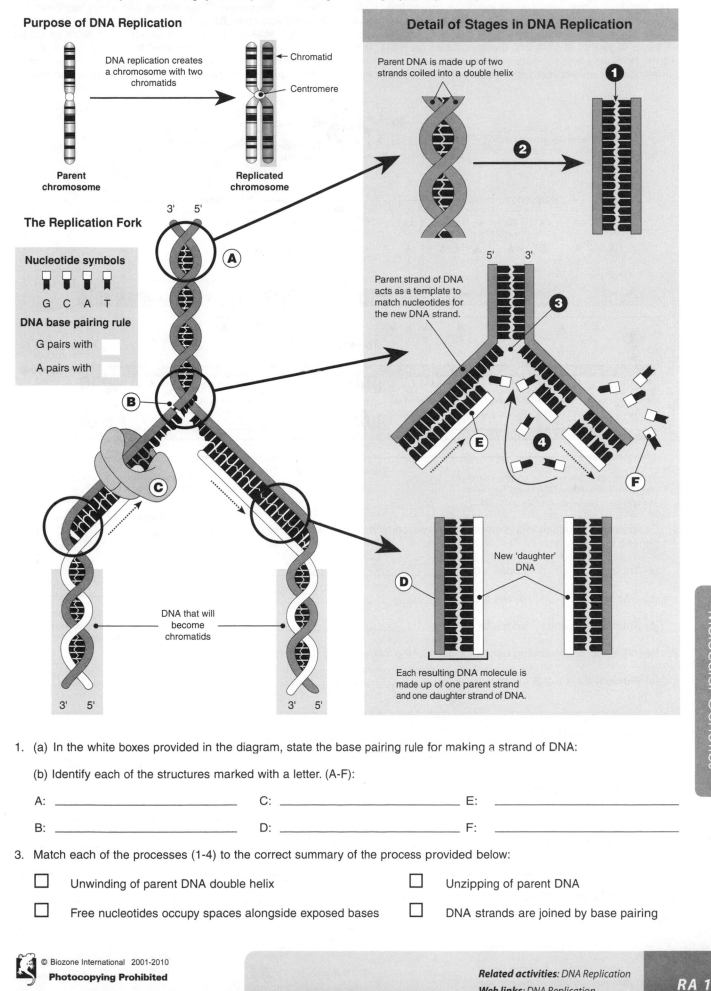

Purpose of DNA Replication

DNA replication creates a chromosome with two chromatids

Chromatid

Centromere

Parent chromosome

Replicated chromosome

The Replication Fork

Nucleotide symbols

G C A T

DNA base pairing rule

G pairs with [ ]

A pairs with [ ]

3' 5'

A

B

C

D

DNA that will become chromatids

3' 5'   3' 5'

Detail of Stages in DNA Replication

Parent DNA is made up of two strands coiled into a double helix

1

2

Parent strand of DNA acts as a template to match nucleotides for the new DNA strand.

5' 3'

3

E

4

F

New 'daughter' DNA

D

Each resulting DNA molecule is made up of one parent strand and one daughter strand of DNA.

Molecular Genetics

1. (a) In the white boxes provided in the diagram, state the base pairing rule for making a strand of DNA:

(b) Identify each of the structures marked with a letter. (A-F):

A: _____   C: _____   E: _____

B: _____   D: _____   F: _____

3. Match each of the processes (1-4) to the correct summary of the process provided below:

[ ] Unwinding of parent DNA double helix

[ ] Free nucleotides occupy spaces alongside exposed bases

[ ] Unzipping of parent DNA

[ ] DNA strands are joined by base pairing

Related activities: DNA Replication
Web links: DNA Replication

RA 1

The Simplest Case: Genes to Proteins

The traditionally held view of genes was as sections of DNA coding only for protein. This view has been revised in recent years with the discovery that much of the nonprotein-coding DNA encodes functional RNAs; it is not all non-coding "junk" DNA as was previously assumed. In fact, our concept of what constitutes a gene is changing rapidly and now encompasses all those segments of DNA that are transcribed (to RNA). This activity considers only the simplest scenario: one in which the gene codes for a functional protein. **Nucleotides**, the basic unit of genetic information, are read in groups of three (**triplets**). Some triplets have a special controlling function in the making of a polypeptide chain. The equivalent of the triplet on the mRNA molecule is the **codon**. Three codons can signify termination of the amino acid chain (UAG, UAA and UGA in the mRNA code). The codon AUG is found at the beginning of every gene (on mRNA) and marks the starting point for reading the gene. The genes required to form a functional end-product (in this case, a functional protein) are collectively called a **transcription unit**.

This polypeptide chain forms one part of the functional protein.

This polypeptide chain forms the other part of the functional protein.

Functional protein

Polypeptide chain

Polypeptide chain

A triplet codes for one amino acid

← Amino acids

Translation

5' AUGCCGUGGAUAUUUCUUUUAUAUUAG 3' 5' AUGCAGCCAGGUAAAGUUCCGUGA 3' ← mRNA

Transcription

DNA: ← **Template** strand

3' TACGGCACCTATAAAGAAAATATAATCTACGTCGGTCCATTTCAAGGCACT 5'

START Triplet Triplet Triplet Triplet Triplet Triplet Triplet STOP START Triplet Triplet Triplet Triplet Triplet Triplet STOP

5' ATGCCGTGGATATTTCTTTTATATTAGATGCAGCCAGGTAAAGTTCCGTGA 3'

DNA: **Coding** strand

|— Gene —| |— Gene —|

Transcription unit

Note: This start code is for the **coding strand** of the DNA. The template DNA strand from which the mRNA is made has the sequence: **TAC**.

Three **nucleotides** make up a **triplet**

Nucleotide

G

In models of nucleic acids, nucleotides are denoted by their base letter. (In this case: **G** is for guanine)

1. Describe the structure in a protein that corresponds to each of the following levels of genetic information:

   (a) Triplet   codes for: _____

   (b) Gene   codes for: _____

   (c) Transcription unit   codes for: _____

2. Describe the basic building blocks for each of the following levels of genetic information:

   (a) **Nucleotide** is made up of: _____

   _____

   (b) **Triplet** is made up of: _____

   _____

   (c) **Gene** is made up of: _____

   _____

   (d) **Transcription unit** is made up of: _____

   _____

3. Describe the steps involved in forming a functional protein: _____

   _____

Related activities: Gene Expression

Periodicals: What is a gene?

Gene Expression

The process of transferring the information encoded in a gene to its functional gene product is called **gene expression**. The central dogma of molecular biology for the past 50 years or so has stated that genetic information, encoded in DNA, is transcribed as molecules of RNA, which are then translated into the amino acid sequences that make up proteins. The established opinion was often stated as "one gene-one protein" and proteins were assumed to be the main regulatory agents for the cell (including its gene expression). The one gene-one protein model is supported by studies of prokaryotic genomes, where the DNA consists almost entirely of protein-coding genes and their regulatory sequences.

Genes and Gene Expression in Prokaryotes

DNA gene for a protein

Transcription

RNA transcript

Translation

Protein

Gene regulation

Because prokaryotic cells lack a nucleus, transcription and translation occur together; RNA is translated into protein almost as fast as it is transcribed from DNA. This feature probably accounts for the lack of large amounts of intronic DNA in prokaryotic genomes. In prokaryotic gene expression, there is insufficient time to remove introns from protein coding sequences and introns would be likely to disable the gene.

Adapted from: *The Hidden Genetic Program of Complex Organisms,* Scientific American, October 2004

Structural and regulatory functions

1. Compare and contrast gene expression in prokaryotes and eukaryotes: _____

_____

_____

_____

_____

_____

_____

2. Study the table (right) summarizing the traditional (old) and revised (new) views of gene expression in eukaryotes. Describe how the two models differ:

_____

_____

_____

_____

_____

_____

_____

_____

Gene Expression in Eukaryotes	
The Old View	**The New View**
• Introns are spliced out of a primary RNA transcript.	• Introns are spliced out of a primary RNA transcript.
• All of the exon RNA (mRNA) is translated into proteins.	• Not all of the exon RNA (mRNA) is translated into proteins. Nonprotein-coding exonic RNA may contribute to microRNAs or has a function on its own.
• Introns are "junk DNA" with no assigned function; they are degraded and recycled.	• Many introns are processed into microRNAs which appear to be involved in regulating development.

Molecular Genetics

Periodicals:
The alternative genome

Related activities: The Simplest Case: Genes to Proteins, Preparing a Gene for Cloning

A 3

In contrast to prokaryotes, eukaryotic genomes contain a large amount of DNA that does not code for proteins. These DNA sequences, called **introns** or intronic DNA, were termed "junk DNA", and were assumed to have no function. However new evidence, arising as more and more diverse genomes are sequenced, suggests that this DNA may encode a vast number of RNA molecules with regulatory functions. Among the eukaryotes, an increase in complexity is associated with an increase in the proportion of nonprotein-coding DNA. This makes sense if the nonprotein-coding DNA has a role in regulating genomic function. These pages contrast gene expression in prokaryotes, where there is very little nonprotein-coding DNA, with the new view of eukaryotic gene expression, where a high proportion of the genomic DNA does not code directly for proteins.

The New View of Gene Expression in Eukaryotes

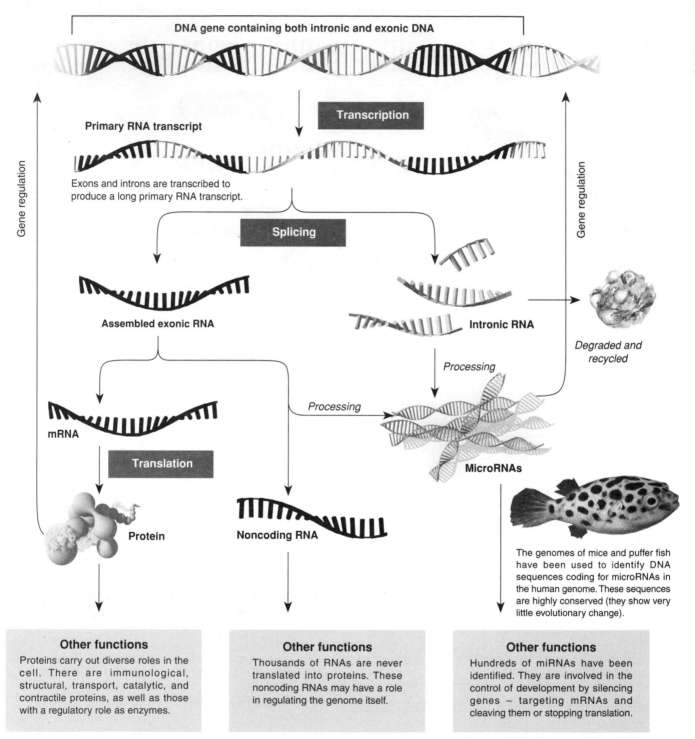

DNA gene containing both intronic and exonic DNA

Gene regulation

Primary RNA transcript

Transcription

Exons and introns are transcribed to produce a long primary RNA transcript.

Splicing

Gene regulation

Assembled exonic RNA

Intronic RNA

Degraded and recycled

Processing

mRNA

Processing

MicroRNAs

Translation

Protein

Noncoding RNA

The genomes of mice and puffer fish have been used to identify DNA sequences coding for microRNAs in the human genome. These sequences are highly conserved (they show very little evolutionary change).

Other functions

Proteins carry out diverse roles in the cell. There are immunological, structural, transport, catalytic, and contractile proteins, as well as those with a regulatory role as enzymes.

Other functions

Thousands of RNAs are never translated into proteins. These noncoding RNAs may have a role in regulating the genome itself.

Other functions

Hundreds of miRNAs have been identified. They are involved in the control of development by silencing genes – targeting mRNAs and cleaving them or stopping translation.

3. The one gene-one protein model does not seem to adequately explain gene expression in eukaryotes, but it is probably still appropriate for prokaryotes. Suggest why:

_____

_____

_____

Analyzing a DNA Sample

The nucleotide (base sequence) of a section of DNA can be determined using DNA sequencing techniques (see the topic *Nucleic Acid Technology* later in this workbook for a description of this technology). The base sequence determines the amino acid sequence of the resultant protein therefore the DNA tells us what type of protein that gene encodes. This exercise reviews the areas of DNA replication, transcription, and translation using an analysis of a gel electrophoresis column. **Attempt it after you have completed the rest of this topic**. Remember that the gel pattern represents the sequence in the synthesized strand.

1. Determine the amino acid sequence of a protein from the nucleotide sequence of its DNA, with the following steps:

   (a) Determine the sequence of **synthesized DNA** in the gel
   (b) Convert it to the complementary sequence of the **sample DNA**
   (c) Complete the **mRNA** sequence
   (d) Determine the **amino acid** sequence by using the *mRNA - amino acid table* in this workbook.

   NOTE: The nucleotides in the gel are read from bottom to top and the sequence is written in the spaces provided from left to right (the first 4 have been done for you).

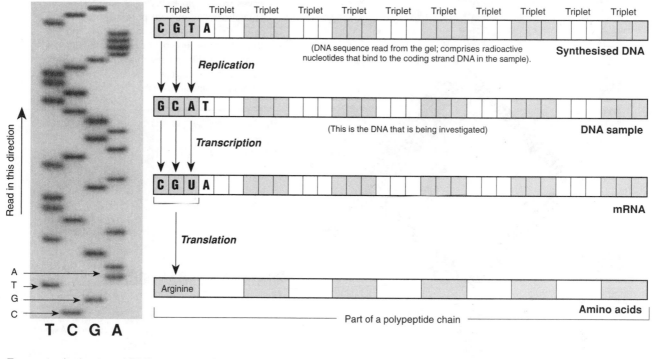

2. For each single strand DNA sequence below, write the base sequence for the **complementary DNA** strand:

   (a) DNA: T A C   T A G   C C G   C G A   T T T   A C A   A T T

   DNA: _____

   (b) DNA: T A C   G C C   T T A   A A G   G G C   C G A   A T C

   DNA: _____

   (c) Identify the cell process that this exercise represents: _____

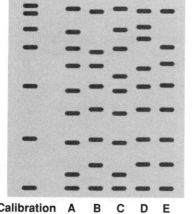

Calibration   A   B   C   D   E

3. Determine the relatedness of each individual (A-E) using each banding pattern on the set of DNA profiles (left). When you have done this, complete the dendrogram by adding the letter of each individual.

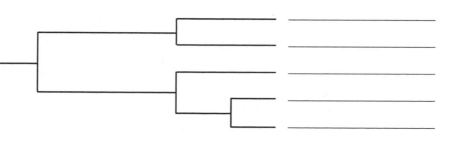

Related activities: Automated DNA Sequencing, The Genetic Code, Gel Electrophoresis

Molecular Genetics

RA 2

Transcription

Transcription is the process by which the code contained in the DNA molecule is transcribed (rewritten) into a **mRNA** molecule. Transcription is under the control of the cell's metabolic processes which must activate a gene before this process can begin. The enzyme that directly controls the process is RNA polymerase, which makes a strand of mRNA using the single strand of DNA (the **template strand**) as the template (hence the term).

The enzyme transcribes only a gene length of DNA at a time and therefore recognizes start and stop signals (codes) at the beginning and end of the gene. Only RNA polymerase is involved in mRNA synthesis; it causes the unwinding of the DNA as well. It is common to find several RNA polymerase enzyme molecules on the same gene at any one time, allowing a high rate of mRNA synthesis to occur.

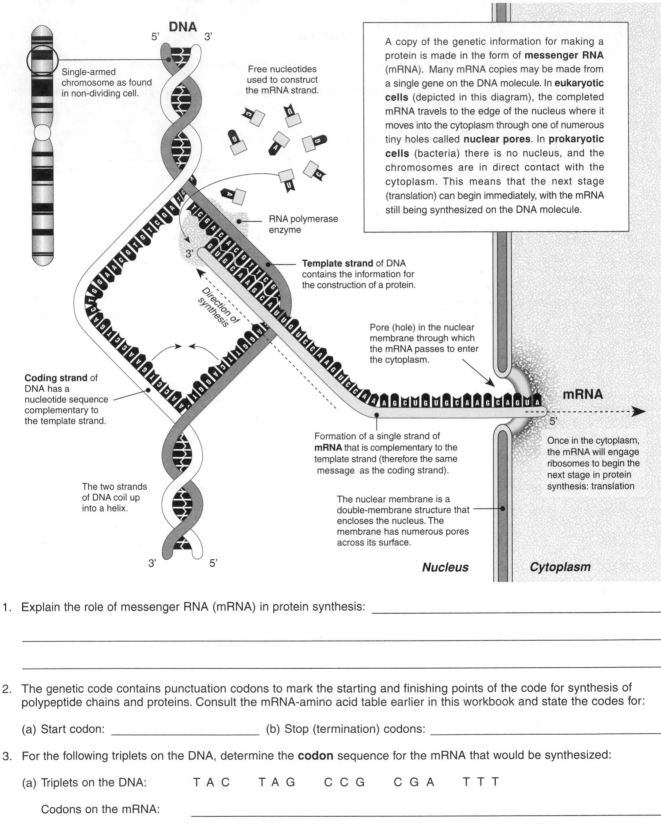

DNA

Single-armed chromosome as found in non-dividing cell.

Free nucleotides used to construct the mRNA strand.

A copy of the genetic information for making a protein is made in the form of **messenger RNA (mRNA)**. Many mRNA copies may be made from a single gene on the DNA molecule. In **eukaryotic cells** (depicted in this diagram), the completed mRNA travels to the edge of the nucleus where it moves into the cytoplasm through one of numerous tiny holes called **nuclear pores**. In **prokaryotic cells** (bacteria) there is no nucleus, and the chromosomes are in direct contact with the cytoplasm. This means that the next stage (translation) can begin immediately, with the mRNA still being synthesized on the DNA molecule.

RNA polymerase enzyme

Direction of Synthesis

Template strand of DNA contains the information for the construction of a protein.

Pore (hole) in the nuclear membrane through which the mRNA passes to enter the cytoplasm.

Coding strand of DNA has a nucleotide sequence complementary to the template strand.

Formation of a single strand of **mRNA** that is complementary to the template strand (therefore the same message as the coding strand).

mRNA

Once in the cytoplasm, the mRNA will engage ribosomes to begin the next stage in protein synthesis: translation

The two strands of DNA coil up into a helix.

The nuclear membrane is a double-membrane structure that encloses the nucleus. The membrane has numerous pores across its surface.

Nucleus   *Cytoplasm*

1. Explain the role of messenger RNA (mRNA) in protein synthesis: _____

   _____

   _____

2. The genetic code contains punctuation codons to mark the starting and finishing points of the code for synthesis of polypeptide chains and proteins. Consult the mRNA-amino acid table earlier in this workbook and state the codes for:

   (a) Start codon: _____   (b) Stop (termination) codons: _____

3. For the following triplets on the DNA, determine the **codon** sequence for the mRNA that would be synthesized:

   (a) Triplets on the DNA: T A C   T A G   C C G   C G A   T T T

   Codons on the mRNA: _____

   (b) Triplets on the DNA: T A C   A A G   C C T   A T A   A A A

   Codons on the mRNA: _____

Related activities: The Genetic Code, Gene Expression
Web links: Transcription in Prokaryotes, Animation of Transcription

Periodicals:
Gene structure and expression

© Biozone International 2001-2010
Photocopying Prohibited

Translation

The diagram below shows the translation phase of protein synthesis. The scene shows how a single mRNA molecule can be 'serviced' by many ribosomes at the same time. The ribosome to the right (lower diagram) is in a more advanced stage of constructing a polypeptide chain because it has 'translated'

more of the mRNA than the ribosome to the left. The anti-codon at the base of each tRNA must make a perfect complementary match with the codon on the mRNA before the amino acid is released. Once released, the amino acid is added to the growing polypeptide chain by enzymes.

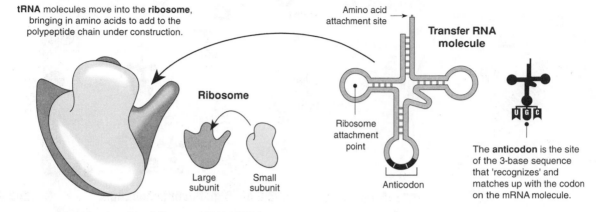

tRNA molecules move into the **ribosome**, bringing in amino acids to add to the polypeptide chain under construction.

Ribosome

Large subunit

Small subunit

Amino acid attachment site

Transfer RNA molecule

Ribosome attachment point

Anticodon

The **anticodon** is the site of the 3-base sequence that 'recognizes' and matches up with the codon on the mRNA molecule.

Ribosomes are made up of a complex of ribosomal RNA (rRNA) and proteins. They exist as two separate sub-units (above) until they are attracted to a binding site on the mRNA molecule, when they join together. Ribosomes have binding sites that attract transfer RNA (**tRNA**) molecules loaded with amino acids. The tRNA molecules are

about 80 nucleotides in length and are made under the direction of genes in the chromosomes. There is a different tRNA molecule for each of the different possible anticodons (see the diagram below) and, because of the degeneracy of the genetic code, there may be up to six different tRNAs carrying the same amino acid.

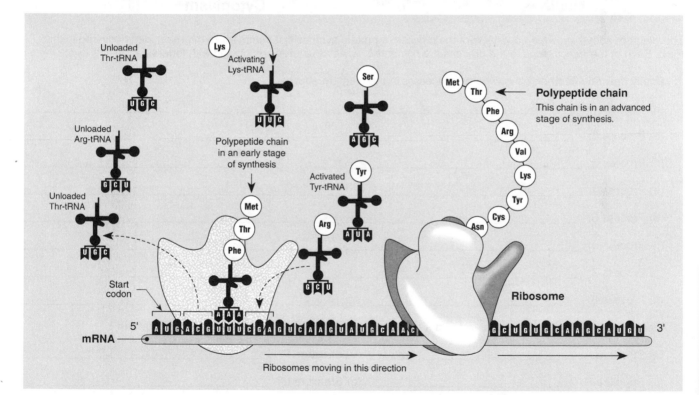

Unloaded Thr-tRNA

Lys
Activating Lys-tRNA

Unloaded Arg-tRNA

Unloaded Thr-tRNA

Polypeptide chain in an early stage of synthesis

Ser

Activated Tyr-tRNA

Tyr

Met Thr Phe Arg Val Lys Tyr Cys Asn

Polypeptide chain
This chain is in an advanced stage of synthesis.

Start codon

mRNA

5' AUGACGUUUCGAGUCAAGUAUGCAAC   GCUGUGCAAGCAUGU 3'

Ribosome

Ribosomes moving in this direction

1. For the following codons on the mRNA, determine the **anti-codons** for each tRNA that would deliver the amino acids:

   Codons on the mRNA: U A C   U A G   C C G   C G A   U U U

   Anti-codons on the tRNAs: _____

2. There are many different types of tRNA molecules, each with a different anti-codon (see the mRNA-amino acid table).

   (a) State how many different tRNA types there are, each with a unique anticodon: _____

   (b) Explain your answer: _____

   _____

   _____

Molecular Genetics

Periodicals:
Transfer RNA

Related activities: The Genetic Code
Web links: Polyribosomes

Protein Synthesis Review

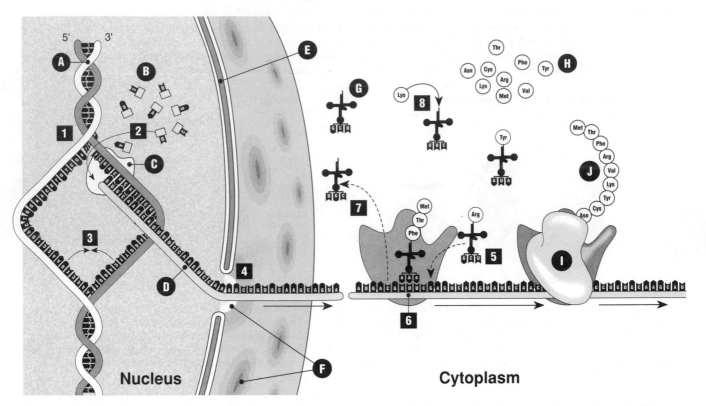

Nucleus

Cytoplasm

The diagram above shows an overview of the process of protein synthesis. It summarizes the main ideas covered in this topic. Each of the major steps in the process are numbered, while structures are labeled with letters.

1. Briefly describe each of the numbered processes in the diagram above:

   (a) Process 1: _____

   (b) Process 2: _____

   (c) Process 3: _____

   (d) Process 4: _____

   (e) Process 5: _____

   (f) Process 6: _____

   (g) Process 7: _____

   (h) Process 8: _____

2. Identify each of the structures marked with a letter and write their names below in the spaces provided:

   (a) Structure A: _____   (f) Structure F: _____

   (b) Structure B: _____   (g) Structure G: _____

   (c) Structure C: _____   (h) Structure H: _____

   (d) Structure D: _____   (i) Structure I: _____

   (e) Structure E: _____   (j) Structure J: _____

3. Describe two factors that would determine whether or not a particular protein is produced in the cell:

   (a) _____

   _____

   (b) _____

   _____

Control of Metabolic Pathways

Metabolism is all the chemical activities of life. The myriad enzyme-controlled **metabolic pathways** that are described as metabolism form a tremendously complex network that is necessary in order to 'maintain' the organism. Errors in the step-wise regulation of enzyme-controlled pathways can result in metabolic disorders that in some cases can be easily identified. An example of a well studied metabolic pathway, the metabolism of **phenylalanine**, is described below.

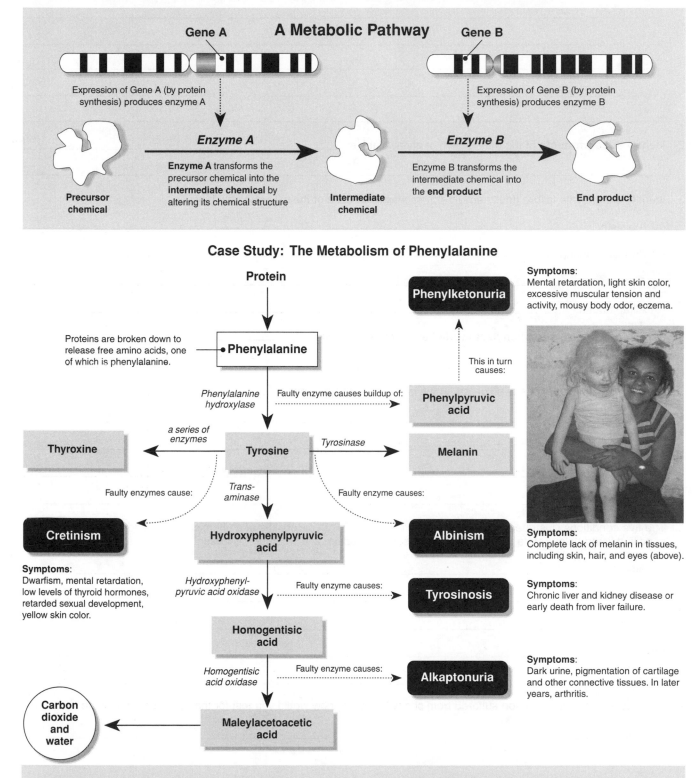

A Metabolic Pathway

Gene A — Expression of Gene A (by protein synthesis) produces enzyme A

Enzyme A transforms the precursor chemical into the **intermediate chemical** by altering its chemical structure

Precursor chemical

Intermediate chemical

Gene B — Expression of Gene B (by protein synthesis) produces enzyme B

Enzyme B transforms the intermediate chemical into the **end product**

End product

Case Study: The Metabolism of Phenylalanine

Protein

Proteins are broken down to release free amino acids, one of which is phenylalanine.

Phenylalanine

Phenylalanine hydroxylase — Faulty enzyme causes buildup of: → **Phenylpyruvic acid**

This in turn causes: → **Phenylketonuria**

Symptoms: Mental retardation, light skin color, excessive muscular tension and activity, mousy body odor, eczema.

a series of enzymes → **Thyroxine**

Tyrosine — *Tyrosinase* → **Melanin**

Faulty enzymes cause: → **Cretinism**

Trans-aminase

Symptoms: Dwarfism, mental retardation, low levels of thyroid hormones, retarded sexual development, yellow skin color.

Hydroxyphenylpyruvic acid

Faulty enzyme causes: → **Albinism**

Symptoms: Complete lack of melanin in tissues, including skin, hair, and eyes (above).

Hydroxyphenyl-pyruvic acid oxidase — Faulty enzyme causes: → **Tyrosinosis**

Symptoms: Chronic liver and kidney disease or early death from liver failure.

Homogentisic acid

Homogentisic acid oxidase — Faulty enzyme causes: → **Alkaptonuria**

Symptoms: Dark urine, pigmentation of cartilage and other connective tissues. In later years, arthritis.

Carbon dioxide and water ← **Maleylacetoacetic acid**

The metabolism of the essential amino acid **phenylalanine** is a well studied metabolic pathway. The first step is carried out by a liver enzyme called phenylalanine hydroxylase, which converts phenylalanine to the amino acid **tyrosine**. Tyrosine, in turn, through a series of intermediate steps, is converted into the skin pigment **melanin** and other substances. If phenylalanine hydroxylase is absent, phenylalanine is converted (in part) into phenylpyruvic acid, which accumulates, together with phenylalanine, in the bloodstream. Phenylpyruvic acid and phenylalanine are central nervous system toxins and produce some of the symptoms of the genetic disease **phenylketonuria**. Other defects in the tyrosine pathway are also known. As indicated above, absence of enzymes operating between tyrosine and melanin, is a cause of **albinism**. **Tyrosinosis** is a rare defect that causes hydroxyphenylpyruvic acid to accumulate in the urine. **Alkaptonuria** causes pigmentation to appear in the cartilage, and produces symptoms of arthritis. A different block in another pathway from tyrosine produces thyroid deficiency leading to goiterous **cretinism** (due to lack of thyroxine).

Molecular Genetics

1. Using the metabolism of phenyalanine as an example, discuss the role of enzymes in **metabolic pathways**:

_____

_____

_____

_____

_____

_____

_____

2. Identify three **products** of the metabolism of phenylalanine: _____

_____

3. Identify the enzyme failure (faulty enzyme) responsible for each of the following conditions:

   (a) Albinism: _____

   (b) Phenylketonuria: _____

   (c) Tyrosinosis: _____

   (d) Alkaptonuria: _____

4. Explain why people with **phenylketonuria** have light skin coloring: _____

_____

5. Discuss the consequences of disorders in the metabolism of **tyrosine**: _____

_____

_____

_____

_____

6. The five conditions illustrated in the diagram are due to too much or too little of a chemical in the body. For each condition listed below, state which chemical causes the problem and whether it is absent or present in excess:

   (a) Albinism: _____

   (b) Phenylketonuria: _____

   (c) Cretinism: _____

   (d) Tyrosinosis: _____

   (e) Alkaptonuria: _____

7. If you suspected that a person suffered from phenylketonuria, how would you test for the condition if you were a doctor:

_____

_____

_____

8. The diagram at the top of the previous page represents the normal condition for a simple metabolic pathway. A starting chemical, called the **precursor**, is progressively changed into a final chemical called the **end product**.

   Consider the effect on this pathway if **gene A** underwent a mutation and the resulting **enzyme A** did not function:

   (a) Identify the chemicals that would be present in **excess**: _____

   (b) Identify the chemicals that would be **absent**: _____

Gene Control in Prokaryotes

The **operon** mechanism was proposed by **Jacob and Monod** to account for the regulation of gene activity in response to the needs of the cell. Their work was carried out with the bacterium *Escherichia coli* and the model is not applicable to eukaryotic cells where the genes are not found as operons. An operon consists of a group of closely linked genes that act together and code for the enzymes that control a particular **metabolic pathway**. These may be for the metabolism of an energy source (e.g. lactose) or the synthesis of a molecule such as an amino acid. The structural genes contain the information for the production of the enzymes themselves and they are transcribed as a single **transcription unit**. These structural genes are controlled by a **promoter**, which initiates the formation of the mRNA, and a region of the DNA in front of the structural genes called the **operator**. A gene outside the operon, called the **regulator gene**, produces a **repressor** molecule that can bind to the operator, and block the transcription of the structural genes. It is the repressor that switches the structural genes on or off and controls the metabolic pathway. Two mechanisms operate in the operon model: gene induction and gene repression. **Gene induction** occurs when genes are switched on by an inducer binding to the repressor molecule and deactivating it. In the *Lac* **operon model** based on *E.coli*, lactose acts as the **inducer**, binding to the repressor and permitting transcription of the structural genes for the utilization of lactose (an infrequently encountered substrate). **Gene repression** occurs when genes that are normally switched on (e.g. genes for synthesis of an amino acid) are switched off by activation of the repressor.

Control of Gene Expression Through Induction: the *Lac* Operon

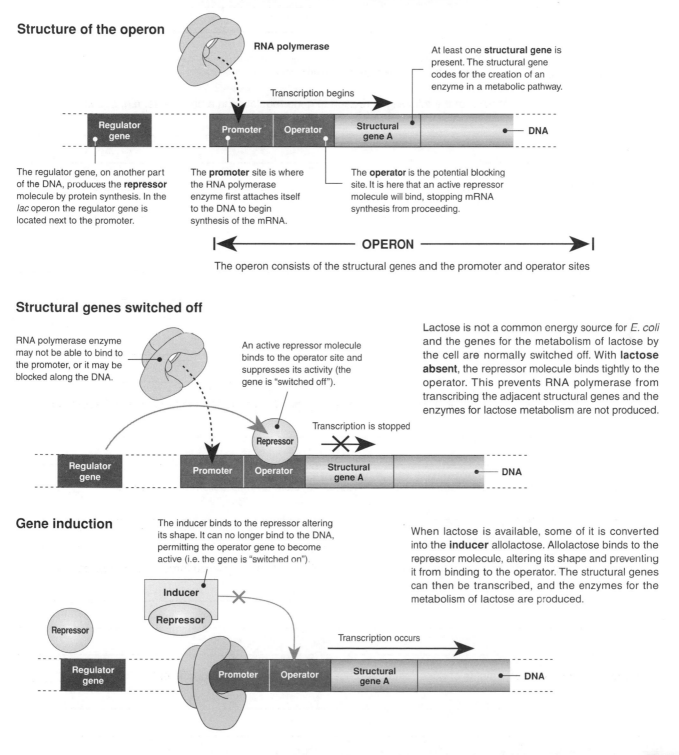

Structure of the operon

RNA polymerase

Transcription begins

At least one **structural gene** is present. The structural gene codes for the creation of an enzyme in a metabolic pathway.

| Regulator gene | Promoter | Operator | Structural gene A | | DNA |

The regulator gene, on another part of the DNA, produces the **repressor** molecule by protein synthesis. In the *lac* operon the regulator gene is located next to the promoter.

The **promoter** site is where the RNA polymerase enzyme first attaches itself to the DNA to begin synthesis of the mRNA.

The **operator** is the potential blocking site. It is here that an active repressor molecule will bind, stopping mRNA synthesis from proceeding.

◄———————— OPERON ————————►

The operon consists of the structural genes and the promoter and operator sites

Structural genes switched off

RNA polymerase enzyme may not be able to bind to the promoter, or it may be blocked along the DNA.

An active repressor molecule binds to the operator site and suppresses its activity (the gene is "switched off").

Lactose is not a common energy source for *E. coli* and the genes for the metabolism of lactose by the cell are normally switched off. With **lactose absent**, the repressor molecule binds tightly to the operator. This prevents RNA polymerase from transcribing the adjacent structural genes and the enzymes for lactose metabolism are not produced.

Repressor

Transcription is stopped

| Regulator gene | Promoter | Operator | Structural gene A | | DNA |

Gene induction

The inducer binds to the repressor altering its shape. It can no longer bind to the DNA, permitting the operator gene to become active (i.e. the gene is "switched on").

When lactose is available, some of it is converted into the **inducer** allolactose. Allolactose binds to the repressor molecule, altering its shape and preventing it from binding to the operator. The structural genes can then be transcribed, and the enzymes for the metabolism of lactose are produced.

Inducer

Repressor

Repressor

Transcription occurs

| Regulator gene | Promoter | Operator | Structural gene A | | DNA |

Molecular Genetics

Related activities: Gene Control in Eukaryotes
Web links: Induction of the Lac Operon

A 3

Control of Gene Expression Through Repression

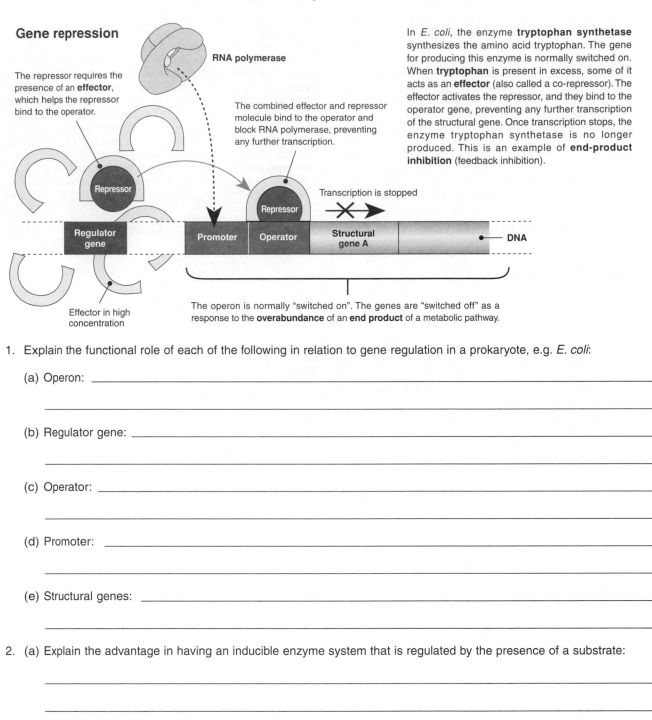

Gene repression

The repressor requires the presence of an **effector**, which helps the repressor bind to the operator.

RNA polymerase

The combined effector and repressor molecule bind to the operator and block RNA polymerase, preventing any further transcription.

In *E. coli*, the enzyme **tryptophan synthetase** synthesizes the amino acid tryptophan. The gene for producing this enzyme is normally switched on. When **tryptophan** is present in excess, some of it acts as an **effector** (also called a co-repressor). The effector activates the repressor, and they bind to the operator gene, preventing any further transcription of the structural gene. Once transcription stops, the enzyme tryptophan synthetase is no longer produced. This is an example of **end-product inhibition** (feedback inhibition).

Repressor

Transcription is stopped

Repressor

Regulator gene

Promoter   Operator   Structural gene A   DNA

Effector in high concentration

The operon is normally "switched on". The genes are "switched off" as a response to the **overabundance** of an **end product** of a metabolic pathway.

1. Explain the functional role of each of the following in relation to gene regulation in a prokaryote, e.g. *E. coli*:

    (a) Operon: _____

    _____

    (b) Regulator gene: _____

    _____

    (c) Operator: _____

    _____

    (d) Promoter: _____

    _____

    (e) Structural genes: _____

    _____

2. (a) Explain the advantage in having an inducible enzyme system that is regulated by the presence of a substrate:

    _____

    _____

    (b) Suggest when it would not be adaptive to have an inducible system for metabolism of a substrate: _____

    _____

    _____

    (c) Giving an example, outline how gene control in a non-inducible system is achieved through **gene repression**:

    _____

    _____

    _____

3. Describe how the two mechanisms of gene control described here are fundamentally different: _____

    _____

    _____

Gene Control in Eukaryotes

All the cells in your body contain identical copies of your genetic instructions. Yet these cells appear very different (e.g. muscle, nerve, and epithelial cells have little in common). These morphological differences reflect profound differences in the expression of genes during the cell's development. For example, muscle cells express the genes for the proteins that make up the contractile elements of the muscle fiber. This wide variety of cell structure and function reflects the precise control over the time, location, and extent of expression of a huge variety of genes. The physical state of the DNA in or near a gene is important in helping to control whether the gene is even available for transcription.

When the **heterochromatin** is condensed, the transcription proteins cannot reach the DNA and the gene is not expressed. To be transcribed, a gene must first be unpacked from its condensed state. Once unpacked, control of gene expression involves the interaction of **transcription factors** with DNA sequences that control the specific gene. Initiation of transcription is the most important and universally used control point in gene expression. A simplified summary of this process is outlined below. Note the differences between this model and the operon model, described earlier, which is not applicable to eukaryotes because eukaryotic genes are not found as operons.

1

RNA polymerase

Transcription factors that bind to RNA polymerase.

Transcription factors (activators) that bind to enhancer.

Promoter region of DNA

Enhancer sequence of DNA

Transcription factors and RNA polymerase bind

Coding region of gene

2

Promoter

Enhancer sequence

Transcription begins and will continue until a terminator is encountered.

Control of Gene Expression in Eukaryotes

- Eukaryotic genes are very different from prokaryotic genes: they have introns (which you recall are removed after the primary transcript is made) and a relatively large number of **control elements** (non-coding DNA that help regulate transcription by binding proteins called transcription factors).

- Each functional eukaryotic gene has a **promoter region** at the upstream end of the gene; a DNA sequence where RNA polymerase binds and starts transcription.

- Eukaryotic RNA polymerase alone cannot initiate the transcription of a gene; it is dependent on **transcription factors** in order to recognize and bind to the **promoter** (step 1).

- Transcription is activated when a hairpin loop in the DNA brings the transcription factors (activators) attached to the **enhancer sequence** in contact with the transcription factors bound to RNA polymerase at the promoter (step 2).

- Protein-protein interactions are crucial to eukaryotic transcription. Only when the complete initiation complex is assembled can the polymerase move along the DNA template strand and produce the complementary strand of RNA.

- Transcription is deactivated when a terminator sequence is encountered. Terminators are nucleotide sequences that function to stop transcription. *Do not confuse these with terminator codons, which are the stop signals for translation.*

- A range of transcription factors and enhancer sequences throughout the genome may selectively activate the expression of specific genes at appropriate stages during cell development.

1. Explain the functional role of each of the following in relation to gene regulation in a eukaryote:

   (a) Promoter: _____

   _____

   (b) Transcription factors: _____

   _____

   (c) Enhancer sequence: _____

   _____

   (d) RNA polymerase: _____

   _____

   (e) Terminator sequence: _____

   _____

2. Identify one difference between the mechanisms of gene control in eukaryotes and prokaryotes:

   _____

   _____

Molecular Genetics

Related activities: Gene Control in Prokaryotes, Control of Metabolic Pathways
Web links: Control of Gene Expression in Eukaryotes

RA 3

The Structure of Viruses

Viruses are non-cellular **obligate intracellular parasites**, requiring a living host cell in order to reproduce. The traditional view of viruses is as a minimal particle, containing just enough genetic information to infect a host and hijack the host's machinery into replicating more viral particles. The identification in 2004 of a new family of viruses, called **mimiviruses**, forced a rethink of this conservative view. Mimiviruses overlap with parasitic cellular organisms in terms of both size (400 nm) and genome complexity (over 1000 genes) and their existence suggests a fourth domain

of life. A typical, fully developed viral particle (**virion**) lacks the metabolic machinery of cells, containing just a single type of nucleic acid (DNA or RNA) encased in a protein coat or **capsid**. Being non-cellular, they do not conform to the existing criteria upon which a five or six kingdom classification system is based. Viruses can be distinguished by their structure (see below) and by the nature of their genetic material (single or double stranded DNA or RNA). Many are difficult to study because they require living animals, embryos, or cell cultures in order to replicate.

Diversity in Viral Structure

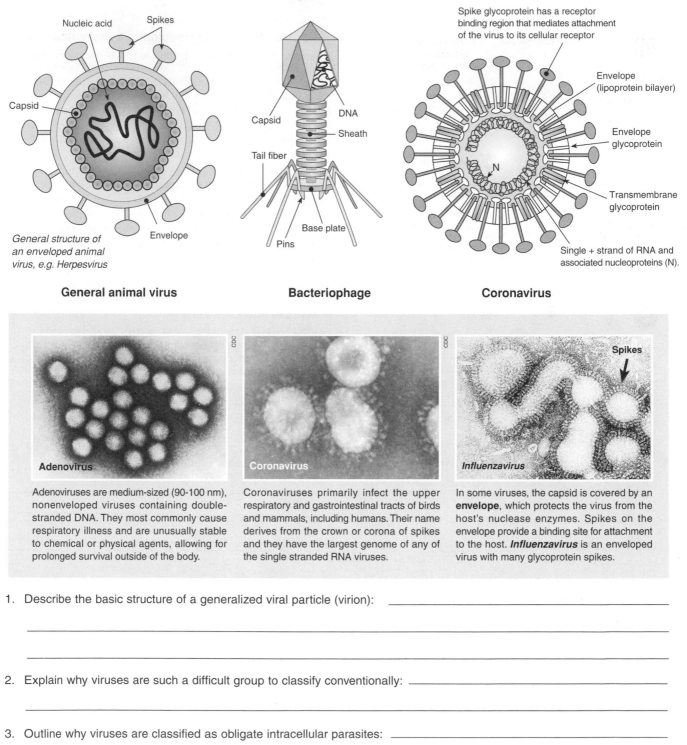

General structure of an enveloped animal virus, e.g. Herpesvirus — Nucleic acid, Spikes, Capsid, Envelope

Bacteriophage — Capsid, DNA, Sheath, Tail fiber, Base plate, Pins

Coronavirus — Spike glycoprotein has a receptor binding region that mediates attachment of the virus to its cellular receptor. Envelope (lipoprotein bilayer). Envelope glycoprotein. Transmembrane glycoprotein. Single + strand of RNA and associated nucleoproteins (N).

General animal virus   **Bacteriophage**   **Coronavirus**

Adenovirus

Adenoviruses are medium-sized (90-100 nm), nonenveloped viruses containing double-stranded DNA. They most commonly cause respiratory illness and are unusually stable to chemical or physical agents, allowing for prolonged survival outside of the body.

Coronavirus

Coronaviruses primarily infect the upper respiratory and gastrointestinal tracts of birds and mammals, including humans. Their name derives from the crown or corona of spikes and they have the largest genome of any of the single stranded RNA viruses.

Influenzavirus — Spikes

In some viruses, the capsid is covered by an **envelope**, which protects the virus from the host's nuclease enzymes. Spikes on the envelope provide a binding site for attachment to the host. **Influenzavirus** is an enveloped virus with many glycoprotein spikes.

1. Describe the basic structure of a generalized viral particle (virion): _____

_____

_____

2. Explain why viruses are such a difficult group to classify conventionally: _____

_____

3. Outline why viruses are classified as obligate intracellular parasites: _____

_____

4. Explain why many viruses are difficult to culture: _____

_____

Related activities: Replication in Animal Viruses, Replication in Bacteriophages

Periodicals: Are viruses alive?

© Biozone International 2001-2010
Photocopying Prohibited

Replication in Animal Viruses

Animal viruses are more complex and varied in structure than the viruses that infect bacteria. Likewise, animal host cells are more diverse in structure and metabolism than bacterial cells. Consequently, animal viruses exhibit a number of different mechanisms for **replicating**, i.e. entering a host cell and producing and releasing new virions. Enveloped viruses bud out from the host cell, whereas those without an envelope are released by rupture of the cell membrane. Three processes (attachment, penetration, and uncoating) are shared by both DNA- and RNA-containing animal viruses but the methods of biosynthesis vary between these two major groups. Generally, **DNA viruses** replicate their DNA in the nucleus of the host cell using viral enzymes, and synthesize their capsid and other proteins in the cytoplasm using the host cell's enzymes. This is outlined below for a typical enveloped DNA virus. **RNA viruses** are more variable in their methods of biosynthesis. The example on the next page describes replication in the retrovirus HIV, where the virus uses its own reverse transcriptase to synthesize viral DNA and produce either **latent proviruses** or active, mature retroviruses.

Entry of an Enveloped Virus into a Cell

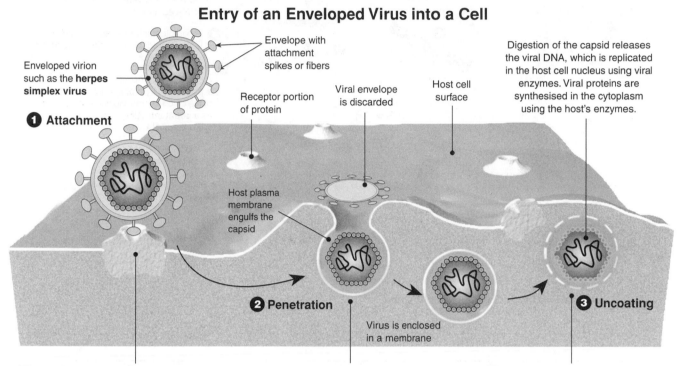

Enveloped virion such as the **herpes simplex virus**

Envelope with attachment spikes or fibers

Receptor portion of protein

Viral envelope is discarded

Host cell surface

Digestion of the capsid releases the viral DNA, which is replicated in the host cell nucleus using viral enzymes. Viral proteins are synthesised in the cytoplasm using the host's enzymes.

1 Attachment

Host plasma membrane engulfs the capsid

2 Penetration

Virus is enclosed in a membrane

3 Uncoating

When a viral particle encounters the cell surface, it attaches to the **receptor sites** of proteins on the cell's plasma membrane.

Once the viral particle is attached, the host cell begins to engulf the virus by **endocytosis**. This is the cell's usual response to foreign particles.

The nucleic acid core is uncoated and the **biosynthesis** of new viruses begins. Mature virions are released by budding from the host cell.

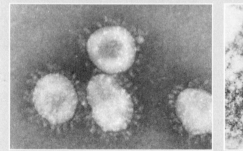

Coronaviruses are irregularly shaped viruses associated with upper respiratory infections and SARS. The envelope bears distinctive projections.

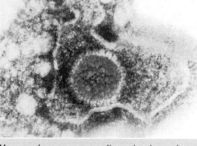

Herpesviruses are medium-sized enveloped viruses that cause various diseases including fever blisters, chickenpox, shingles, and herpes.

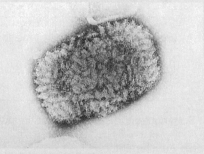

This *Vaccinia* virus belongs to the family of pox viruses; large (200-350 nm), enveloped DNA viruses that cause diseases such as smallpox.

All photos courtesy of CDC

1. Describe the purpose of the glycoprotein spikes found on some enveloped viruses: _____

_____

2. (a) Explain the significance of endocytosis to the entry of an enveloped virus into an animal cell: _____

_____

_____

(b) State where an enveloped virus replicates its viral DNA: _____

(c) State where an enveloped virus synthesizes its proteins: _____

Related activities: The Structure of Viruses, Replication in Bacteriophages
Web links: HSV Infection and Replication, HIV Life Cycle, Retrovirus Life Cycle

RA 2

Molecular Genetics

How HIV Infects a Helper T Cell

HIV, the infectious agent that causes AIDS, is a retrovirus (RNA not DNA). It is able to splice its genes into the host cell's chromosome.

1 HIV particle is attracted to CD4 receptors on a helper T cell.

CD4 receptors

2 HIV particle fuses with the plasma membrane of the T cell and the capsid is removed by enzymes.

3 Reverse transcriptase causes the formation of viral DNA (using the viral RNA as a template).

4 A complementary strand of DNA is formed, producing double stranded DNA.

5 The DNA is integrated into the host's chromosome. The viral DNA is now called a **provirus**. A prophage never comes out of the chromosome. However, it may remain as a **latent infection**, replicating along with the host's DNA.

6 The viral genes are transcribed into mRNA molecules.

7 Viral mRNA is translated into HIV proteins. Some mRNA also provides the genome for the next generation of viruses.

8 Assembly of the capsids around the viral genomes.

9 Budding of the new viruses from the host cell.

Spikes

TEM

Mature **HIV-1** virions budding from a lymphocyte. Note their glycoprotein spikes.

New virion

RNA

Nucleus

DNA

3. (a) Describe how an HIV particle enters a host cell: _____

_____

_____

   (b) Explain the role of the reverse transcriptase in the life cycle of a retrovirus: _____

_____

_____

   (c) Explain the significance of the formation of a provirus: _____

_____

_____

4. Summarize the steps involved in invasion of a host cell by an enveloped viral particle such as *Influenzavirus*:

   (a) Attachment: _____

_____

   (b) Penetration: _____

_____

   (c) Uncoating: _____

_____

   (d) Biosynthesis: _____

_____

   (e) Release: _____

_____

Replication in Bacteriophages

Viruses infect living cells, commanding the metabolism of the host cell and producing new viral particles. In viruses that use bacterial cells as a host (**bacteriophages**), this process may not immediately follow infection. Instead, the virus may integrate its nucleic acid into the host cell's DNA, forming a provirus or **prophage**. This type of cycle, called **lysogenic** or **temperate**, does not kill the host cell outright. Instead, the host cell is occupied by the virus and used to replicate the viral genes. During this time, the viral infection is said to be **latent**. The virus may be **transduced** into becoming active again, entering the **lytic cycle**

and utilizing the host's cellular mechanisms to produce new virions. The lytic cycle results in death of the host cell through **cell lysis**. Although the multiplication of animal viruses follows a similar pattern to that of bacteriophage multiplication there are notable differences. Animal viruses differ in their mechanisms for entering cells (see the previous activity) and, once inside the cell, the production of new virions is different. This is partly because of differences in host cell structure and metabolism and partly because the structure of animal viruses themselves is very variable.

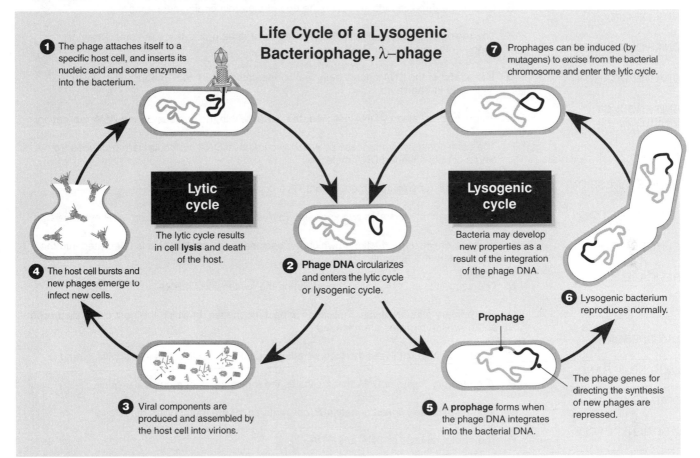

Life Cycle of a Lysogenic Bacteriophage, λ–phage

1. The phage attaches itself to a specific host cell, and inserts its nucleic acid and some enzymes into the bacterium.

7. Prophages can be induced (by mutagens) to excise from the bacterial chromosome and enter the lytic cycle.

Lytic cycle

The lytic cycle results in cell **lysis** and death of the host.

Lysogenic cycle

Bacteria may develop new properties as a result of the integration of the phage DNA.

4. The host cell bursts and new phages emerge to infect new cells.

2. **Phage DNA** circularizes and enters the lytic cycle or lysogenic cycle.

6. Lysogenic bacterium reproduces normally.

Prophage

3. Viral components are produced and assembled by the host cell into virions.

5. A **prophage** forms when the phage DNA integrates into the bacterial DNA.

The phage genes for directing the synthesis of new phages are repressed.

1. Discuss the main stages of replication of a typical bacteriophage (e.g. λ) as seen in:

   (a) The **lytic cycle**: _____

   _____

   _____

   (b) The **lysogenic cycle**: _____

   _____

   _____

2. Explain the purpose of the tail region on a bacteriophage: _____

   _____

3. (a) Identify the cycle during which a bacterium may acquire new properties from a virus: _____

   (b) Explain why this can occur: _____

   (c) Describe the implications of this ability for human health and disease: _____

   _____

   _____

Related activities: The Structure of Viruses
Web links: Lytic Cycle, Lysogeny

RA 2

Molecular Genetics

KEY TERMS: Mix and Match

INSTRUCTIONS: Test your vocab by matching each term to its correct definition, as identified by its preceding letter code.

ANTICODON

BASE-PAIRING RULE

CODING STRAND

CODON

DNA LIGASE

DNA POLYMERASE

DNA REPLICATION

END-PRODUCT INHIBITION

EXONS

GENE EXPRESSION

GENE INDUCTION

GENETIC CODE

HELICASE

INTRONS

LAC OPERON

LAGGING STRAND

LEADING STRAND

LYSOGENIC CYCLE

LYTIC CYCLE

MESSENGER RNA

NUCLEOTIDES

OKAZAKI FRAGMENTS

PROMOTER

REPLICATION FORK

REVERSE TRANSCRIPTASE

STOP CODON

TEMPLATE STRAND

TRANSCRIPTION

TRANSLATION

TRANSFER RNA

A A molecule of RNA encoding a chemical blueprint for a protein product.

B The switching on of genes by deactivation of a repressor molecule.

C The process by which information from a gene is used to produce a functional gene product.

D A set of rules by which information encoded in DNA or mRNA is translated into proteins.

E One of the three codons signalling the end of a protein-coding sequence.

F The region of a transfer RNA with a sequence of three bases that are complementary to a codon in the messenger RNA.

G The strand of the DNA double helix that is orientated in a 3' to 5' manner and which is replicated in fragments.

H A structure, created by DNA helicase, that forms within the nucleus during DNA replication.

I The semi-conservative process by which two identical DNA molecules are produced from a single double-stranded DNA molecule.

J The process of creating an equivalent RNA copy of a sequence of DNA.

K An enzyme that separates two annealed DNA strands using energy from ATP hydrolysis.

L A phase of viral reproduction in which the bacteriophage nucleic acid is integrated into the host bacterium's genome.

M The sequence of DNA that is copied during the synthesis of mRNA.

N The primary method of viral replication in bacteriophages, in which the host cell is destroyed and new viral particles are released.

O An enzyme that catalyzes the incorporation of deoxyribonucleotides into a DNA strand.

P A regulatory region of DNA that facilitates the transcription of a particular gene.

Q A group of closely linked genes in *E. coli* required for the metabolism of lactose.

R The structural units of DNA and RNA.

S A DNA polymerase enzyme that transcribes single-stranded RNA into double-stranded DNA.

T A small RNA molecule that shuttles specific amino acids to a growing polypeptide chain at the ribosomal site of protein synthesis during translation.

U A negative feedback mechanism used to regulate the production of a given molecule.

V A sequence of three adjacent nucleotides constituting the genetic code.

W The stage of gene expression in which mRNA is decoded to produce a specific polypeptide.

X An enzyme that links together two DNA strands that have double-strand break.

Y Nucleic acid sequences that are represented in the mature form of an RNA molecule.

Z A relatively short piece of DNA created on the lagging strand during DNA replication.

AA DNA regions within a gene that are not translated into protein.

BB The rule governing the pairing of complementary bases in DNA.

CC The DNA strand with the same base sequence as the RNA transcript produced (although with thymine replaced by uracil in mRNA).

DD The strand of the DNA double helix that is oriented in a 5' to 3' manner and is replicated in one continuous piece.

Mutation

KEY CONCEPTS

▶ Mutations are the ultimate source of new alleles.

▶ Gene mutations offer the greatest evolutionary potential.

▶ Mutations may be harmful, beneficial, or neutral.

▶ Chromosomal mutations and aneuploidies produce more widespread effects.

▶ Polyploidies have been important in the evolution of plant varieties.

Periodicals:
listings for this
chapter are on page 381

Weblinks:
www.thebiozone.com/
weblink/SB1-2597.html

**Teacher Resource
CD-ROM:**
*The Appearance of
a New Word*

OBJECTIVES

☐ 1. Use the **KEY TERMS** to help you understand and complete these objectives.

The Nature of Mutation
pages 230-231, 236-238

☐ 2. Explain the significance of **mutations** and the species-specific naturally occurring **spontaneous (background) mutation rate**.

☐ 3. Describe and explain the effects of **mutagens** on DNA and mutation rates. Explain the role of **carcinogens** and **oncogenes** in **cancer**.

☐ 4. Explain how the location of a mutation (i.e. **somatic** or **gametic**) can have a different significance in terms of heritable change.

☐ 5. Describe and explain examples of **beneficial** and **harmful** mutations. Explain the evolutionary importance of **neutral mutations**.

Types of Mutation
pages 233-235, 239-243

☐ 6. Describe and explain **gene mutations**, and their significance in terms of evolutionary potential. Describe the cause and effect of gene mutations as illustrated by: **base substitution**, **base deletion**, and **base insertion**.

☐ 7. Describe and explain the effect of a base substitution mutation, as illustrated by the **sickle cell mutation**.

☐ 8. Describe other examples of disorders in humans that arise as a result of gene mutations, e.g. **cystic fibrosis** and **Huntington disease**.

☐ 9. Describe and explain **chromosome** (block) **mutations**. Describe the nature and genetic consequences of the following chromosome mutations: **translocations**, **inversions**, **duplications**, **deletions**.

☐ 10. Describe and explain **aneuploidy** arising as a result of **non-disjunction** during meiosis. Describe **nullisomy**, **monosomy**, **disomy**, and **trisomy**.

☐ 11. Explain the significance of **Barr bodies** to the expression of aneuploidies.

☐ 12. Describe and explain examples of aneuploidy in human sex chromosomes, e.g. Turner and Klinefelter **syndrome**.

☐ 13. Describe and explain an example of aneuploidy in human autosomes, e.g **trisomy** 21 or Down syndrome. Explain the **maternal age effect**.

Polyploidy
page 361

☐ 14. Distinguish between **aneuploidy** and **polyploidy**, and between **monoploid**, **triploid**, and **tetraploid**. Describe and explain the role of polyploidy in the development of crops (e.g. wheat).

Causes of Mutations

Mutations occur in all organisms spontaneously. The natural rate at which a gene will undergo a change is normally very low, but this rate can be increased by environmental factors such as ionizing radiation and mutagenic chemicals (e.g. benzene). Only those mutations taking place in cells producing gametes will be inherited. If they occur in a body cell after the organism has begun to develop beyond the zygote (fertilized egg cell) stage, they are called **somatic mutations**. Such mutations usually give rise to **chimeras** (mixture of cell types, some with the mutation and some without, in the same organism). In some cases a mutation may trigger the onset of **cancer**, if the normal controls over gene regulation and expression are disrupted.

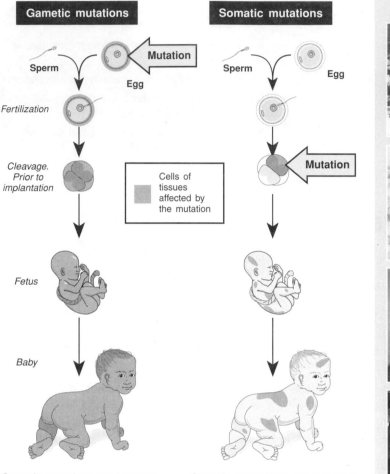

Gametic mutations are inherited and occur in the testes of males and in the ovaries of females.

Somatic mutations occur in body cells. They are not inherited but may affect the person during their lifetime.

Mutagens and their Effects

Ionizing radiation
Nuclear and ultraviolet radiation, gamma rays, and X-rays (e.g. from medical treatment) are forms of high energy, ionizing radiation associated with the development of some cancers (e.g. thyroid cancer, leukemia, and skin cancer).

Viruses & microorganisms
Some viruses, e.g. the herpes virus (left), integrate into the human chromosome, disrupting genes and triggering cancers. Aflatoxins, produced by the fungus *Aspergillus flavus,* can contaminate grain, and are potent inducers of liver cancer.

Environmental poisons
Many chemicals are mutagenic. Synthetic and natural examples include organic solvents such as carbon tetrachloride (used in dry cleaning fluid) and formaldehyde, tobacco and coal tars, benzene, asbestos, vinyl chlorides, some dyes, and nitrites.

Alcohol and diet
High alcohol intake increases the risk of some cancers. Diets high in fat, especially those containing burned meat or meats preserved with nitrates, slow gut passage time, allowing mutagenic irritants to form in the bowel.

1. List examples of environmental factors that induce mutations under the following headings:

   (a) Radiation: _____

   (b) Chemical agents: _____

2. Discuss the role of **mutagens** in causing **mutations**: _____

   _____

   _____

   _____

   _____

3. (a) Distinguish between gametic and somatic mutations: _____

   _____

   (b) Describe the importance of gametic mutations in an evolutionary sense: _____

   _____

   _____

Related activities: The Effects of Mutations
Web links: Radiation and DNA

Periodicals:
What is a mutation?
Radiation and risk

The Effects of Mutations

It is not correct to assume that all mutations are harmful. There are many documented cases where mutations conferring a survival advantage have arisen in a population. Such **beneficial mutations** occur most often among viruses and bacteria but occur in multicellular organisms also (e.g. insects). Sometimes, a mutation may be **neutral** and have no immediate effect. If there is no selective pressure against it, a mutation may be carried in the population and be of benefit (or harm) at some future time.

Harmful Mutations

There are many well-documented examples of mutations that cause harmful effects. Examples are the mutations giving rise to **cystic fibrosis** (CF) and **sickle cell disease**. The sickle cell mutation involves a change to only one base in the DNA sequence, whereas the CF mutation involves the loss of three nucleotides. The malformed proteins that result from these mutations cannot carry out their normal biological functions.

Albinism is caused by a mutation in the gene that produces an enzyme in the metabolic pathway to produce melanin. It occurs in a large number of animals (photos, right). Albinos are not common in the wild because they tend to be more vulnerable to predation.

Albinism is widespread in the animal kingdom. A mutation in the metabolic pathway to melanin results in a lack of pigment.

Silent Mutations

Some mutations are termed silent because a change in the gene sequence does not result in a change to the amino acid sequence. Such mutations have been seen as neutral (no effect) at the time they occur*, but may be important in an evolutionary sense if they later become subject to selection pressure. The example shows how a change to the DNA sequence (normal vs. mutant) can be silenced if there is no change to the amino acid sequence. The redundancy in the code means that both the original (GAA) and the mutated (GAG) triplet still code for glutamic acid. *Note that recent evidence indicates that silent mutations may not necessarily be neutral if they cause mRNA instability.*

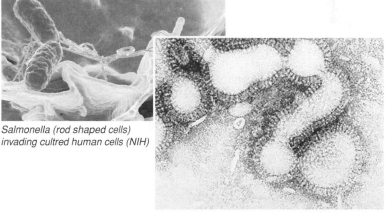

Normal DNA / mRNA / Mutant DNA / mRNA

Phe / Tyr / Glu / Glu / Val

Resulting amino acid chain

Beneficial Mutations

Bacteria reproduce asexually by binary fission. They are susceptible to antibiotics (substances that harm them or inhibit their growth) but are well-known for acquiring **antibiotic resistance** through mutation. The genes for bacterial resistance can be transferred within or even between bacterial species. New, multi-resistant bacterial superbugs have arisen in this way.

Viruses, including HIV and *Influenzavirus*, have membrane envelopes coated with glycoproteins. These are used by the host to identify the virus so that it can be destroyed. The genes coding for these glycoproteins are constantly mutating. The result is that each new viral 'strain' goes undetected by the immune system until well after the infection is established.

Salmonella (rod shaped cells) invading cultred human cells (NIH)

Influenzavirus showing glycoprotein spikes (Photo: DS)

1. Explain the evolutionary significance of **neutral mutations**:

_____

_____

2. Giving examples, explain the difference between **harmful** and **beneficial mutations**:

_____

_____

_____

Periodicals: How do mutations lead to evolution?

Related activities: Causes of Mutations Cancer: Cells Out of Control
Web links: Mutation Exercise

Mutation

A 1

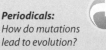

Gene Mutations

Gene mutations are small, localized changes in the structure of a DNA strand. These changes may be induced by a **mutagen** or arise spontaneously as a result of errors during DNA replication. The changes may involve a single nucleotide (often called **point mutations**), or changes to a triplet (e.g. triplet deletion or triplet repeat). The diagrams below show how point mutations can occur by substitution, insertion, or deletion. These alterations in the DNA are at the **nucleotide** level where individual **codons** are affected. Alteration of the precise nucleotide sequence of a coded gene in turn alters the mRNA transcribed from the mutated DNA and may affect the polypeptide chain that it creates. Note that mutations do not always result in altered proteins, because more than one codon may code for the same amino acid. Most of this **degeneracy** in the code occurs at the third base of a codon.

Normal DNA

| A | A | A | A | T | G | C | T | T | C | T | C | C | A | A |

mRNA

| U | U | U | U | A | C | G | A | A | G | A | G | G | U | U |

Amino acids

Phe — Tyr — Glu — Glu — Val

Amino acid sequence forms a normal polypeptide chain

Mutation: Substitute **T** instead of **C**

Mutant DNA

| A | A | A | A | T | G | T | T | T | C | T | C | C | A | A |

mRNA

| U | U | U | U | A | C | A | A | A | G | A | G | G | U | U |

Amino acids

Phe — Tyr — Lys — Glu — Val

Substitution produces a polypeptide chain with the wrong amino acid

Mutation: Insertion of **C**

Mutant DNA

| A | A | A | A | T | G | C | C | T | T | C | T | C | C | A |

mRNA

| U | U | U | U | A | C | G | G | A | A | G | A | G | G | U |

Amino acids

Phe — Tyr — Gly — Arg — Gly

The insertion creates a large scale **frame shift** resulting in a completely new sequence of amino acids. The resulting protein is unlikely to have any biological activity.

Mutation: Deletion of **C**

Mutant DNA

| A | A | A | A | T | G | T | T | C | T | C | C | A | A | G |

mRNA

| U | U | U | U | A | C | A | A | G | A | G | G | U | U | C |

Amino acids

Phe — Tyr — Lys — Arg — Phe

Large scale frame shift resulting in a completely new sequence of amino acids. The resulting protein is unlikely to have any biological activity.

1. Explain what is meant by a **frame shift mutation**:_____

_____

2. Some gene mutations are more disruptive to an organism than others.

   (a) Identify which type of gene mutations are the most damaging to an organism: _____

   (b) Explain why they are the most disruptive: _____

   _____

   (c) Describe what type of gene mutation is least likely to cause a change in protein structure and explain your answer:

   _____

   _____

3. In the following DNA sequence, replace the **G** of the second codon with an A to create a new mutant DNA, then determine the new mRNA sequence, and the amino acid sequence. Refer to the mRNA-amino acid table to identify the amino acids coded in each case.

   (a) Original DNA: **AAA   ATG   TTT   CTC   CAA   GAT** _____

      Mutated DNA: _____

      mRNA: _____

      Amino acids: _____

   (b) Identify the amino acid coded by codon 2 (ATG) in the original DNA: _____

   (c) Explain the effect of the mutation: _____

Related activities: The Genetic Code, Control of Metabolic Pathways, Cystic Fibrosis Mutation, Sickle Cell Mutation   **Web links:** Point Mutation Problems

© Biozone International 2001-2010
Photocopying Prohibited

Examples of Gene Mutations

Humans have more than 6000 physiological diseases attributed to mutations in single genes and over one hundred syndromes known to be caused by chromosomal abnormality. The number of genetic disorders identified increases every year. The work of the Human Genome Project is enabling the identification of the genetic basis of these disorders. This will facilitate the development of new drug therapies and gene therapies. Four genetic disorders are summarized below.

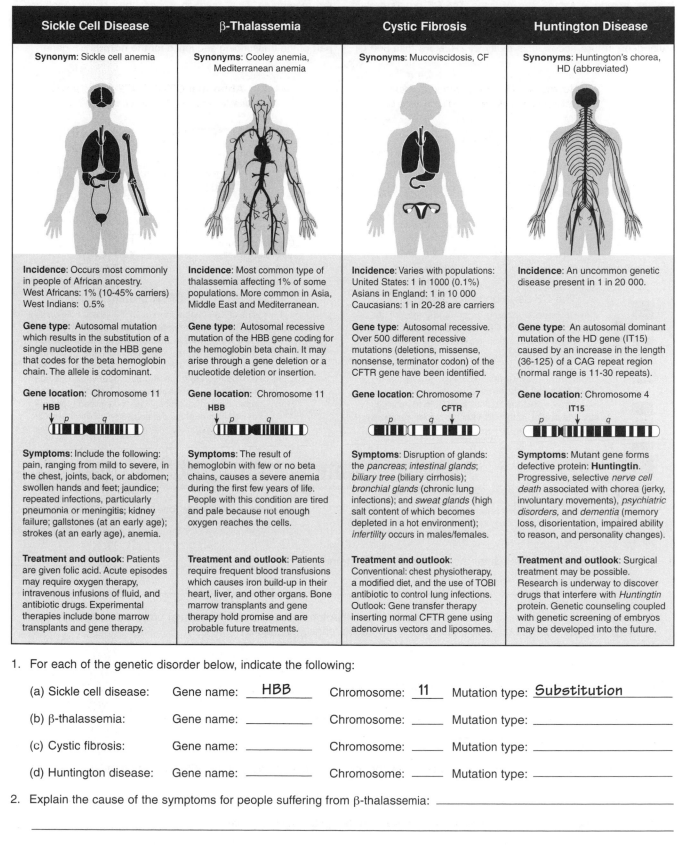

Sickle Cell Disease	β-Thalassemia	Cystic Fibrosis	Huntington Disease
Synonym: Sickle cell anemia	**Synonyms:** Cooley anemia, Mediterranean anemia	**Synonyms:** Mucoviscidosis, CF	**Synonyms:** Huntington's chorea, HD (abbreviated)
Incidence: Occurs most commonly in people of African ancestry. West Africans: 1% (10-45% carriers) West Indians: 0.5%	**Incidence:** Most common type of thalassemia affecting 1% of some populations. More common in Asia, Middle East and Mediterranean.	**Incidence:** Varies with populations: United States: 1 in 1000 (0.1%) Asians in England: 1 in 10 000 Caucasians: 1 in 20-28 are carriers	**Incidence:** An uncommon genetic disease present in 1 in 20 000.
Gene type: Autosomal mutation which results in the substitution of a single nucleotide in the HBB gene that codes for the beta hemoglobin chain. The allele is codominant.	**Gene type:** Autosomal recessive mutation of the HBB gene coding for the hemoglobin beta chain. It may arise through a gene deletion or a nucleotide deletion or insertion.	**Gene type:** Autosomal recessive. Over 500 different recessive mutations (deletions, missense, nonsense, terminator codon) of the CFTR gene have been identified.	**Gene type:** An autosomal dominant mutation of the HD gene (IT15) caused by an increase in the length (36-125) of a CAG repeat region (normal range is 11-30 repeats).
Gene location: Chromosome 11 HBB	**Gene location:** Chromosome 11 HBB	**Gene location:** Chromosome 7 CFTR	**Gene location:** Chromosome 4 IT15
Symptoms: Include the following: pain, ranging from mild to severe, in the chest, joints, back, or abdomen; swollen hands and feet; jaundice; repeated infections, particularly pneumonia or meningitis; kidney failure; gallstones (at an early age); strokes (at an early age), anemia.	**Symptoms:** The result of hemoglobin with few or no beta chains, causes a severe anemia during the first few years of life. People with this condition are tired and pale because not enough oxygen reaches the cells.	**Symptoms:** Disruption of glands: the *pancreas*; *intestinal glands*; *biliary tree* (biliary cirrhosis); *bronchial glands* (chronic lung infections); and *sweat glands* (high salt content of which becomes depleted in a hot environment); *infertility* occurs in males/females.	**Symptoms:** Mutant gene forms defective protein: **Huntingtin**. Progressive, selective *nerve cell death* associated with chorea (jerky, involuntary movements), *psychiatric disorders,* and *dementia* (memory loss, disorientation, impaired ability to reason, and personality changes).
Treatment and outlook: Patients are given folic acid. Acute episodes may require oxygen therapy, intravenous infusions of fluid, and antibiotic drugs. Experimental therapies include bone marrow transplants and gene therapy.	**Treatment and outlook:** Patients require frequent blood transfusions which causes iron build-up in their heart, liver, and other organs. Bone marrow transplants and gene therapy hold promise and are probable future treatments.	**Treatment and outlook:** Conventional: chest physiotherapy, a modified diet, and the use of TOBI antibiotic to control lung infections. Outlook: Gene transfer therapy inserting normal CFTR gene using adenovirus vectors and liposomes.	**Treatment and outlook:** Surgical treatment may be possible. Research is underway to discover drugs that interfere with *Huntingtin* protein. Genetic counseling coupled with genetic screening of embryos may be developed into the future.

1. For each of the genetic disorder below, indicate the following:

   (a) Sickle cell disease: Gene name: __HBB__   Chromosome: __11__   Mutation type: _Substitution_

   (b) β-thalassemia: Gene name: _____   Chromosome: _____   Mutation type: _____

   (c) Cystic fibrosis: Gene name: _____   Chromosome: _____   Mutation type: _____

   (d) Huntington disease: Gene name: _____   Chromosome: _____   Mutation type: _____

2. Explain the cause of the symptoms for people suffering from β-thalassemia: _____
   _____

3. Suggest a reason for the differences in the country-specific incidence rates for some genetic disorders:
   _____
   _____

Mutation

Related activities: The Human Genome Project, Sickle Cell Mutation, Cystic Fibrosis Mutation

A 2

Cystic Fibrosis Mutation

Cystic fibrosis an inherited disorder caused by a mutation of the **CF gene**. It is one of the most common lethal autosomal recessive conditions affecting caucasians, with an incidence in the US of 1 in 1000 live births and a **carrier frequency** of 4%. It is uncommon in Asians and Africans. The CF gene's protein product, **CFTR**, is a membrane-based protein with a function in regulating transport of chloride across the membrane. A faulty gene in turn codes for a faulty CFTR. More than 500 CF mutations have been described, giving rise to disease symptoms of varying severity. One common mutation accounts for more than 70% of all defective CF genes. This mutation, called δ(delta) F508, leads to the absence of CFTR from its proper position in the membrane. This mutation is described below. Another CF mutation, R117H, which is also relatively common, produces a partially functional CFTR protein. The DNA sequence below is part of the transcribing sequence for the **normal** CF gene.

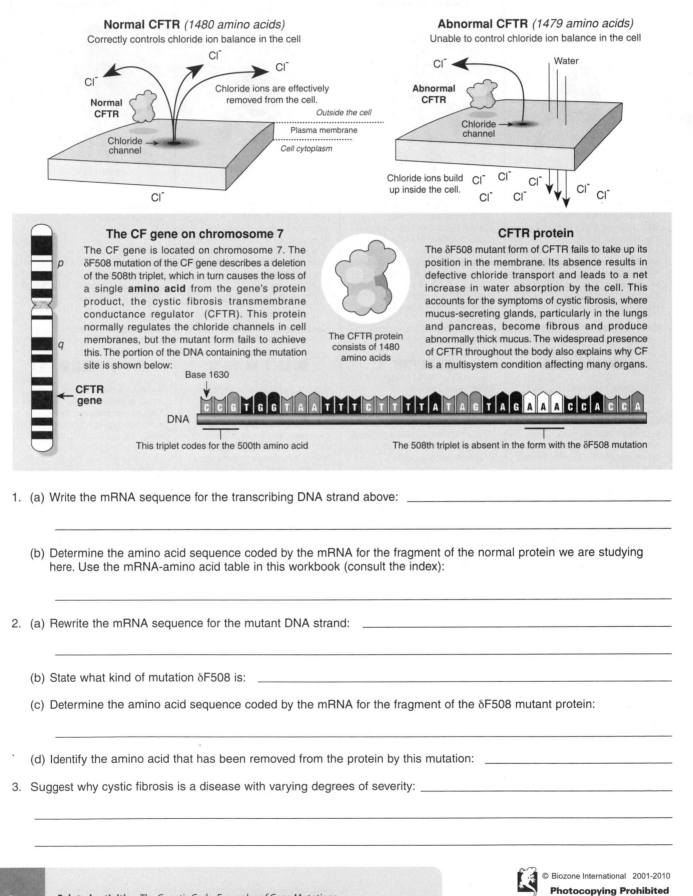

Normal CFTR *(1480 amino acids)*
Correctly controls chloride ion balance in the cell

Cl⁻ Cl⁻ Cl⁻
Chloride ions are effectively removed from the cell.
Normal CFTR
Outside the cell
Plasma membrane
Chloride channel
Cell cytoplasm
Cl⁻

Abnormal CFTR *(1479 amino acids)*
Unable to control chloride ion balance in the cell

Cl⁻ Water
Abnormal CFTR
Chloride channel
Chloride ions build up inside the cell. Cl⁻ Cl⁻ Cl⁻ Cl⁻ Cl⁻ Cl⁻ Cl⁻

The CF gene on chromosome 7

The CF gene is located on chromosome 7. The δF508 mutation of the CF gene describes a deletion of the 508th triplet, which in turn causes the loss of a single **amino acid** from the gene's protein product, the cystic fibrosis transmembrane conductance regulator (CFTR). This protein normally regulates the chloride channels in cell membranes, but the mutant form fails to achieve this. The portion of the DNA containing the mutation site is shown below:

The CFTR protein consists of 1480 amino acids

CFTR protein

The δF508 mutant form of CFTR fails to take up its position in the membrane. Its absence results in defective chloride transport and leads to a net increase in water absorption by the cell. This accounts for the symptoms of cystic fibrosis, where mucus-secreting glands, particularly in the lungs and pancreas, become fibrous and produce abnormally thick mucus. The widespread presence of CFTR throughout the body also explains why CF is a multisystem condition affecting many organs.

p
q
CFTR gene

Base 1630

DNA CCGTGGTAATTTCTTTTTATAGTAGAAACCACCA

This triplet codes for the 500th amino acid

The 508th triplet is absent in the form with the δF508 mutation

1. (a) Write the mRNA sequence for the transcribing DNA strand above: _____

_____

(b) Determine the amino acid sequence coded by the mRNA for the fragment of the normal protein we are studying here. Use the mRNA-amino acid table in this workbook (consult the index):

_____

2. (a) Rewrite the mRNA sequence for the mutant DNA strand: _____

_____

(b) State what kind of mutation δF508 is: _____

(c) Determine the amino acid sequence coded by the mRNA for the fragment of the δF508 mutant protein:

_____

(d) Identify the amino acid that has been removed from the protein by this mutation: _____

3. Suggest why cystic fibrosis is a disease with varying degrees of severity: _____

_____

_____

Related activities: The Genetic Code, Examples of Gene Mutations

Sickle Cell Mutation

Sickle cell disease (formerly called sickle cell anemia) is an inherited disorder caused by a gene mutation which codes for a faulty beta (β) chain hemoglobin (Hb) protein. This in turn causes the red blood cells to deform causing a whole range of medical problems. The DNA sequence below is the beginning of the transcribing sequence for the **normal** β-chain Hb molecule.

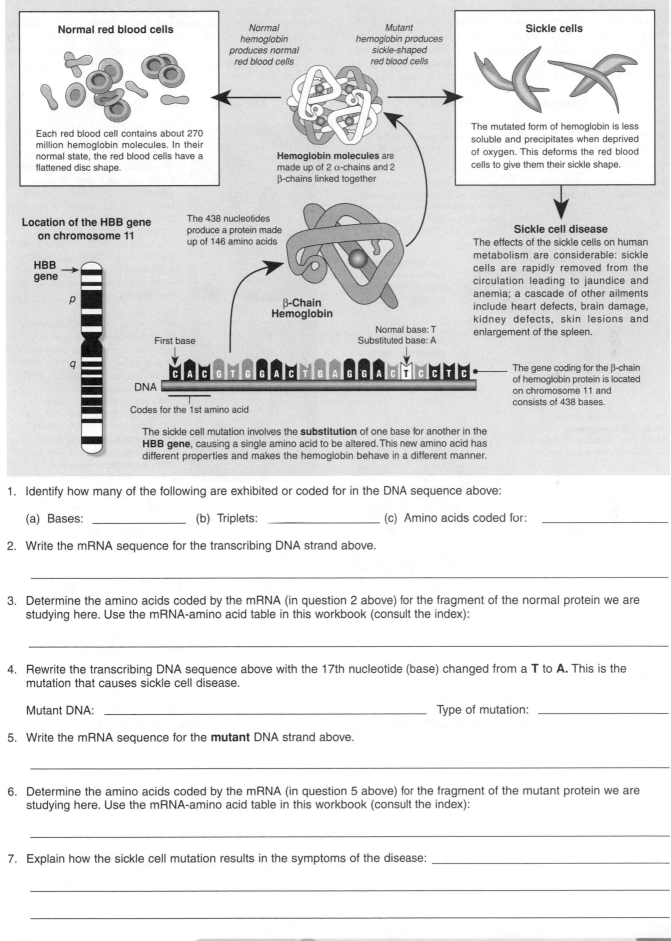

Normal red blood cells

Each red blood cell contains about 270 million hemoglobin molecules. In their normal state, the red blood cells have a flattened disc shape.

Normal hemoglobin produces normal red blood cells

Hemoglobin molecules are made up of 2 α-chains and 2 β-chains linked together

Mutant hemoglobin produces sickle-shaped red blood cells

Sickle cells

The mutated form of hemoglobin is less soluble and precipitates when deprived of oxygen. This deforms the red blood cells to give them their sickle shape.

Location of the HBB gene on chromosome 11

HBB gene

p

q

The 438 nucleotides produce a protein made up of 146 amino acids

β-Chain Hemoglobin

First base

Normal base: T
Substituted base: A

DNA   C A C G T G G A C T G A G G A C T C C T C

Codes for the 1st amino acid

Sickle cell disease
The effects of the sickle cells on human metabolism are considerable: sickle cells are rapidly removed from the circulation leading to jaundice and anemia; a cascade of other ailments include heart defects, brain damage, kidney defects, skin lesions and enlargement of the spleen.

The gene coding for the β-chain of hemoglobin protein is located on chromosome 11 and consists of 438 bases.

The sickle cell mutation involves the **substitution** of one base for another in the **HBB gene**, causing a single amino acid to be altered. This new amino acid has different properties and makes the hemoglobin behave in a different manner.

1. Identify how many of the following are exhibited or coded for in the DNA sequence above:

   (a) Bases: _____ (b) Triplets: _____ (c) Amino acids coded for: _____

2. Write the mRNA sequence for the transcribing DNA strand above.

   _____

3. Determine the amino acids coded by the mRNA (in question 2 above) for the fragment of the normal protein we are studying here. Use the mRNA-amino acid table in this workbook (consult the index):

   _____

4. Rewrite the transcribing DNA sequence above with the 17th nucleotide (base) changed from a **T** to **A**. This is the mutation that causes sickle cell disease.

   Mutant DNA: _____ Type of mutation: _____

5. Write the mRNA sequence for the **mutant** DNA strand above.

   _____

6. Determine the amino acids coded by the mRNA (in question 5 above) for the fragment of the mutant protein we are studying here. Use the mRNA-amino acid table in this workbook (consult the index):

   _____

7. Explain how the sickle cell mutation results in the symptoms of the disease: _____

   _____

   _____

Periodicals: Genetics of sickle cell anemia

Related activities: The Genetic Code, Examples of Gene Mutations

RA 3

Mutation

Antigenic Variability in Pathogens

Influenza (flu) is a disease of the upper respiratory tract caused by the viral genus *Influenzavirus*. Globally, up to 500 000 people die from influenza every year. It is estimated that 5-20% of Americans are affected by the flu every year, and a small number of deaths occur as a result. Three types of *Influenzavirus* affect humans. They are simply named *Influenzavirus* A, B, and C. The most common and most virulent of these strains is *Influenzavirus* A, which is discussed in more detail below. Influenza viruses are constantly undergoing genetic changes. **Antigenic drifts** are small changes in the virus which happen continually over time.

Such changes mean that the influenza vaccine must be adjusted each year to include the most recently circulating influenza viruses. **Antigenic shift** occurs when two or more different viral strains (or different viruses) combine to form a new subtype. The changes are large and sudden and most people lack immunity to the new subtype. New influenza viruses arising from antigenic shift have caused influenza pandemics that have killed millions people over the last century. *Influenzavirus* A is considered the most dangerous to human health because it is capable of antigenic shift.

Structure of *Influenzavirus*

Viral strains are identified by the variation in their H and N surface antigens. Viruses are able to combine and readily rearrange their RNA segments, which alters the protein composition of their H and N glycoprotein spikes.

The *influenzavirus* is surrounded by an **envelope** containing protein and lipids.

The genetic material is actually closely surrounded by protein capsomeres (these have been omitted here and below right in order to illustrate the changes in the RNA more clearly).

The **neuraminidase (N) spikes** help the virus to detach from the cell after infection.

Hemagglutinin (H) spikes allow the virus to recognize and attach to cells before attacking them.

The viral genome is contained on **eight RNA segments**, which enables the exchange of genes between different viral strains.

Photo right: *Electron micrograph of Influenzavirus showing the glycoprotein spikes projecting from the viral envelope*

Spikes

Antigenic Shift in *Influenzavirus*

Influenza vaccination is the primary method for preventing influenza and is 75% effective. The ability of the virus to recombine its RNA enables it to change each year, so that different strains dominate in any one season. The 'flu' vaccination is updated annually to incorporate the antigenic properties of currently circulating strains. Three strains are chosen for each year's vaccination. Selection is based on estimates of which strains will be predominant in the following year.

H1N1, H1N2, and H3N2 (below) are the known *Influenza A* viral subtypes currently circulating among humans. Although the body will have acquired antibodies from previous flu strains, the new combination of N and H spikes is sufficiently different to enable new viral strains to avoid detection by the immune system. The World Health Organization coordinates strain selection for each year's influenza vaccine.

H1N1   **H1N2**   **H3N2**

1. The *Influenzavirus* is able to mutate readily and alter the composition of H and N spikes on its surface.

   (a) Explain why this is the case: _____

   _____

   _____

   (b) Explain how this affects the ability of the immune system to recognise the virus and launch an attack:

   _____

   _____

   _____

2. Discuss why a virus capable of antigenic shift is more dangerous to humans than a virus undergoing antigenic drift:

   _____

   _____

   _____

Resistance in Pathogens

Although many pathogens are controlled effectively with drugs and vaccines, the spread of drug resistance amongst microorganisms is increasingly undermining the ability to treat and control diseases such as tuberculosis and malaria. Methicillin resistant strains of the common bacterium *Staphylococcus aureus* (MRSA) have acquired genes that confer antibiotic resistance to all penicillins, including **methicillin** and other narrow-spectrum pencillin-type drugs. Such strains, called "superbugs", were discovered in the UK in 1961 and are now widespread, and the infections they cause are exceedingly difficult to treat. Genes for drug resistance arise through mutation, and the high mutation rates and short generation times of viral, bacterial, and protozoan pathogens have contributed to the rapid spread of drug resistance through populations. This is well documented for malaria, TB, and HIV/ AIDS. Rapid evolution in pathogens is exacerbated too by the strong selection pressure created by the wide use and misuse of antimicrobial drugs, the poor quality of available drugs, and poor patient compliance. The most successful treatment for several diseases now appears to be a multi-pronged attack using a cocktail of drugs to target the pathogen at many stages.

Global Spread of Chloroquine Resistance

1980s
1970s
1960s
1957
1960s
1980s
1959
1978
1970s
1960s
1980s

Areas of chloroquine resistance in *P. falciparum*.

Malaria in humans is caused by various species of *Plasmodium*, a protozoan parasite transmitted by *Anopheles* mosquitoes. The inexpensive antimalarial drug **chloroquine** was used successfully to treat malaria for many years, but its effectiveness has declined since resistance to the drug was first recorded in the 1960s. Chloroquine resistance has spread steadily (above) and now two of the four *Plasmodium* species, *P. falciparum* and *P. vivax* are chloroquine-resistant. *P. falciparum* alone accounts for 80% of all human malarial infections and 90% of the deaths, so this rise in resistance is of global concern. New anti-malarial drugs have been developed, but are expensive and often have undesirable side effects. Resistance to even these newer drugs is already evident, especially in *P. falciparum*, although this species is currently still susceptible to artemisinin, a derivative of the medicinal herb *Artemisia annua*.

Drug Resistance in HIV

Strains of drug-resistant HIV arise when the virus mutates during replication. Resistance may develop as a result of a single mutation, or through a step-wise accumulation of specific mutations. These mutations may alter drug binding capacity or increase viral fitness, or they may be naturally occurring polymorphisms (which occur in untreated patients). Drug resistance is likely to develop in patients who do not follow their treatment schedule closely, as the virus has an opportunity to adapt more readily to a "non-lethal" drug dose. The best practice for managing the HIV virus is to treat it with a cocktail of anti-retroviral drugs with different actions to minimise the number of viruses in the body. This minimises the replication rate, and also the chance of a drug resistant mutation being produced.

Anti-HIV drug

Drug stops replication of susceptible variants

Resistant variant replicates and comes to predominate

HIV variants susceptible to drug

HIV variant resistant to drug

1. Describe how genes for drug resistance arise in a microbial population:

_____

_____

2. Briefly describe three mechanisms by which bacteria achieve drug resistance:

(a) _____

(b) _____

(c) _____

3. Discuss the implications (to humans) of bacteria acquiring several different mechanisms of resistance: _____

_____

_____

_____

Mutation

Related activities: The Evolution of Drug Resistance in Bacteria

RA 2

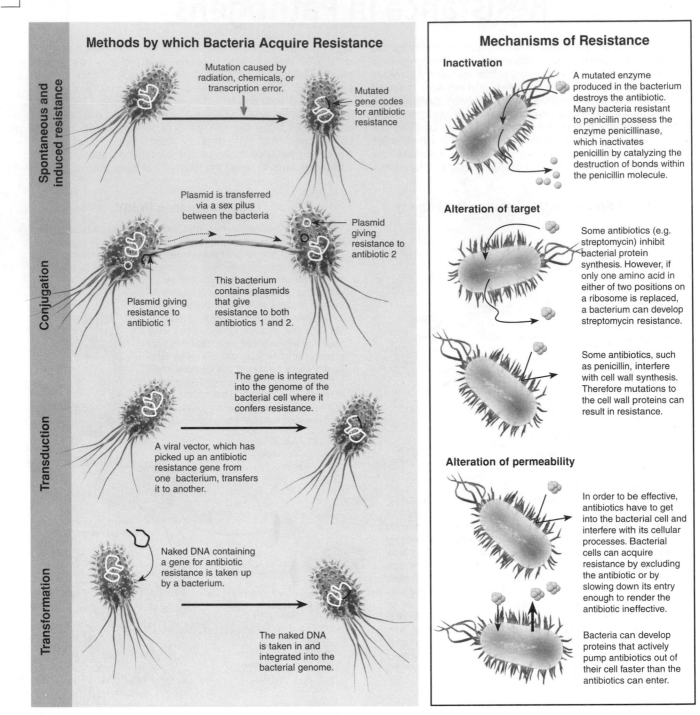

Methods by which Bacteria Acquire Resistance

Spontaneous and induced resistance

Mutation caused by radiation, chemicals, or transcription error.

Mutated gene codes for antibiotic resistance

Conjugation

Plasmid is transferred via a sex pilus between the bacteria

Plasmid giving resistance to antibiotic 2

Plasmid giving resistance to antibiotic 1

This bacterium contains plasmids that give resistance to both antibiotics 1 and 2.

Transduction

The gene is integrated into the genome of the bacterial cell where it confers resistance.

A viral vector, which has picked up an antibiotic resistance gene from one bacterium, transfers it to another.

Transformation

Naked DNA containing a gene for antibiotic resistance is taken up by a bacterium.

The naked DNA is taken in and integrated into the bacterial genome.

Mechanisms of Resistance

Inactivation

A mutated enzyme produced in the bacterium destroys the antibiotic. Many bacteria resistant to penicillin possess the enzyme penicillinase, which inactivates penicillin by catalyzing the destruction of bonds within the penicillin molecule.

Alteration of target

Some antibiotics (e.g. streptomycin) inhibit bacterial protein synthesis. However, if only one amino acid in either of two positions on a ribosome is replaced, a bacterium can develop streptomycin resistance.

Some antibiotics, such as penicillin, interfere with cell wall synthesis. Therefore mutations to the cell wall proteins can result in resistance.

Alteration of permeability

In order to be effective, antibiotics have to get into the bacterial cell and interfere with its cellular processes. Bacterial cells can acquire resistance by excluding the antibiotic or by slowing down its entry enough to render the antibiotic ineffective.

Bacteria can develop proteins that actively pump antibiotics out of their cell faster than the antibiotics can enter.

4. Discuss the factors contributing to the rapid spread of drug resistance in pathogens: _____

_____

_____

_____

_____

_____

_____

_____

_____

_____

_____

_____

Chromosome Mutations

The diagrams below show the different types of **chromosome mutation** that can occur only during **meiosis**. These mutations (sometimes also called **block mutations**) involve the rearrangement of whole blocks of genes, rather than individual bases within a gene. Each type of mutation results in an alteration in the number and/or sequence of whole sets of genes (represented by letters) on the chromosome. In humans, **translocations** occur with varying frequency (several

rare types of Down syndrome occur in this way). Individuals with a **balanced translocation** have the correct amount of genetic material and appear phenotypically normal but have an increased chance of producing faulty gametes. Translocation may sometimes involve the fusion of whole chromosomes, thereby reducing the chromosome number of an organism. This is thought to be an important mechanism by which **instant speciation** can occur.

Deletion

A break may occur at two points on the chromosome and the middle piece of the chromosome falls out. The two ends then rejoin to form a chromosome deficient in some genes. Alternatively, the end of a chromosome may break off and is lost.

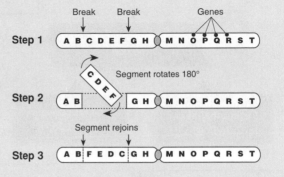

Inversion

The middle piece of the chromosome falls out and rotates through 180° and then rejoins. There is no loss of genetic material. The genes will be in a reverse order for this segment of the chromosome.

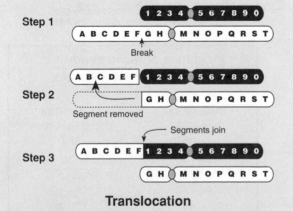

Translocation

Translocation involves the movement of a group of genes between different chromosomes. The large chromosome (white) and the small chromosome (black) are not homologous. A piece of one chromosome breaks off and joins onto another chromosome. This will cause major problems when the chromosomes are passed to gametes. Some will receive extra genes, while some will be deficient.

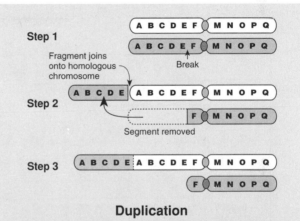

Duplication

A segment is lost from one chromosome and is added to its homologue. In this diagram, the darker chromosome on the bottom is the 'donor' of the duplicated piece of chromosome. The chromosome with the segment removed is deficient in genes. Some gametes will receive double the genes while others will have no genes for the affected segment.

1. For each of the chromosome (block) mutations illustrated above, write the original gene sequence and the new gene sequence after the mutation has occurred (the first one has been done for you):

	Original sequence(s)	Mutated sequence(s)
(a) Deletion:	A B C D E F G H M N O P Q R S T	A B G H M N O P Q R S T
(b) Inversion:	_____	_____
(c) Translocation:	_____	_____
	_____	_____
(d) Duplication:	_____	_____
	_____	_____

2. Identify which type of block mutation is likely to be the least damaging to the organism, explaining your answer:

_____

A 2

Mutation

The Fate of Conceptions

A significant number of conceptions do not end in live births. A large proportion of miscarriages, which are spontaneous natural abortions, are caused by **chromosome disorders** such as trisomy and polyploidy. Some of these disorders are less severe than others and those affected survive into childhood or beyond.

There is a good correlation between the age of the mother and the incidence in chromosome abnormalities, called the **maternal age effect**. Prospective mothers older than 35-40 years are encouraged to have a prenatal test (e.g. **amniocentesis** or **CVS**) to determine the **karyotype** of the fetus.

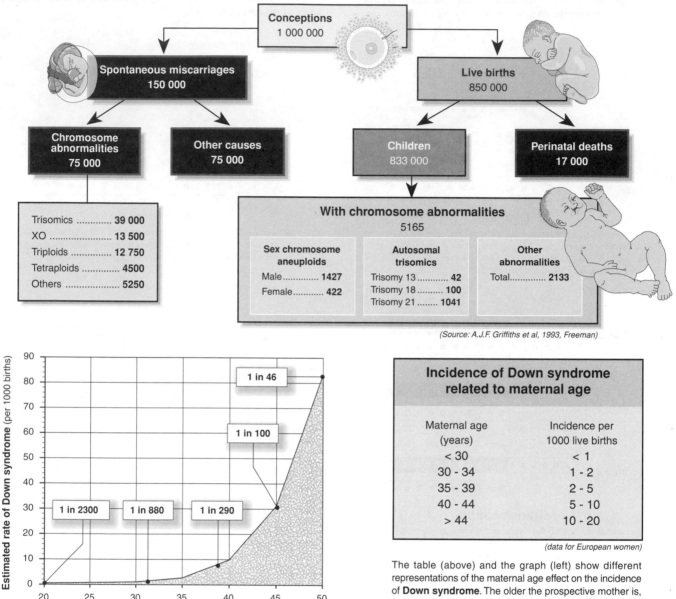

Conceptions
1 000 000

Spontaneous miscarriages
150 000

Live births
850 000

Chromosome abnormalities
75 000

Other causes
75 000

Children
833 000

Perinatal deaths
17 000

Trisomics **39 000**
XO **13 500**
Triploids **12 750**
Tetraploids **4500**
Others **5250**

With chromosome abnormalities
5165

Sex chromosome aneuploids	Autosomal trisomics	Other abnormalities
Male............. 1427	Trisomy 13 42	Total............. 2133
Female........... 422	Trisomy 18 100	
	Trisomy 21 1041	

(Source: A.J.F. Griffiths et al, 1993, Freeman)

Graph: Estimated rate of Down syndrome (per 1000 births) vs Maternal age (years)
1 in 46
1 in 100
1 in 2300
1 in 880
1 in 290

Incidence of Down syndrome related to maternal age

Maternal age (years)	Incidence per 1000 live births
< 30	< 1
30 - 34	1 - 2
35 - 39	2 - 5
40 - 44	5 - 10
> 44	10 - 20

(data for European women)

The table (above) and the graph (left) show different representations of the maternal age effect on the incidence of **Down syndrome**. The older the prospective mother is, the more likely it is that she will have an affected child.

1. Discuss the role of the **maternal age effect** in the incidence rate of Down syndrome and other trisomic syndromes:

_____

_____

_____

2. Explain the role of **amniocentesis** in detecting trisomic disorders: _____

_____

_____

3. Explain why, in recent times, most Down syndrome babies are born to younger mothers: _____

_____

_____

Related activities: Aneuploidy in Humans, Down Syndrome
Web links: Prenatal Diagnosis

Aneuploidy in Humans

Euploidy is the condition of having an exact multiple of the haploid number of chromosomes. Normal euploid humans have 46 chromosomes (2N). **Aneuploidy** is the condition where the chromosome number is not an exact multiple of the normal haploid set for the species (the number may be more, e.g. 2N+2, or less, e.g. 2N–1). **Polysomy** is aneuploidy involving reduplication of some of the chromosomes beyond the normal diploid number (e.g. 2N+1). Aneuploidy usually results from the **non-disjunction** (failure to separate) of homologous chromosomes during meiosis. The two most common forms are monosomy (e.g. Turner syndrome) and trisomy (e.g. Down and Klinefelter syndrome) as outlined on the following pages.

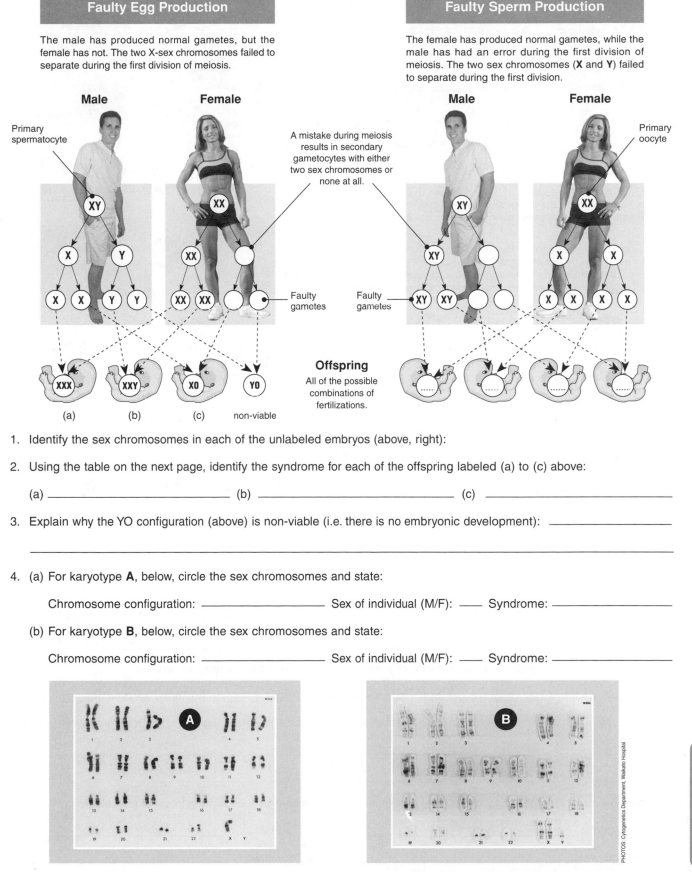

1. Identify the sex chromosomes in each of the unlabeled embryos (above, right):

2. Using the table on the next page, identify the syndrome for each of the offspring labeled (a) to (c) above:

   (a) _____ (b) _____ (c) _____

3. Explain why the YO configuration (above) is non-viable (i.e. there is no embryonic development): _____
_____

4. (a) For karyotype **A**, below, circle the sex chromosomes and state:

   Chromosome configuration: _____ Sex of individual (M/F): ___ Syndrome: _____

   (b) For karyotype **B**, below, circle the sex chromosomes and state:

   Chromosome configuration: _____ Sex of individual (M/F): ___ Syndrome: _____

Related activities: Down Syndrome

Mutation

Examples of Aneuploidy in Human Sex Chromosomes

Sex chromosomes and chromosome condition	Apparent sex	Phenotype
XO, monosomic	Female	Turner syndrome
XX, disomic	Female	**Normal female**
XXX, trisomic	Female	Metafemale. Most appear normal; they have a greater tendency to criminality
XXXX, tetrasomic	Female	Rather like Down syndrome, low fertility and intelligence
XY, disomic	Male	**Normal male**
XYY, trisomic	Male	Jacob syndrome, apparently normal male, tall, aggressive
XXY, trisomic	Male	Klinefelter syndrome (infertile). Incidence rate 1 in 1000 live male births, with a maternal age effect.
XXXY, tetrasomic	Male	Extreme Klinefelter, mentally retarded

Above: Features of selected aneuploidies in humans. Note that this list represents only a small sample of the possible sex chromosome aneuploidies in humans.

Right: Symbolic representation of Barr body occurrence in various human karyotypes. The chromosome number is given first, and the inactive X chromosomes are framed by a black box. Note that in aneuploid syndromes, such as those described here, all but one of the X chromosomes are inactivated, regardless of the number present.

Barr Bodies

In the nucleus of any non-dividing somatic cell, one of the X chromosomes condenses to form a visible piece of chromatin, called a **Barr body**. This chromosome is inactivated, so that only one X chromosome in a cell ever has its genes expressed. The inactivation is random, and the inactive X may be either the maternal homologue (from the mother) or the paternal homologue (from the father).

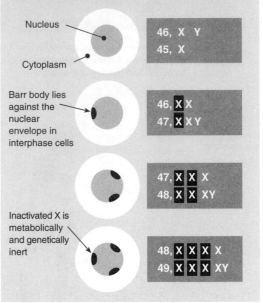

46, X Y
45, X

Barr body lies against the nuclear envelope in interphase cells

46, X X
47, X X Y

47, X X X
48, X X XY

Inactivated X is metabolically and genetically inert

48, X X X X
49, X X X XY

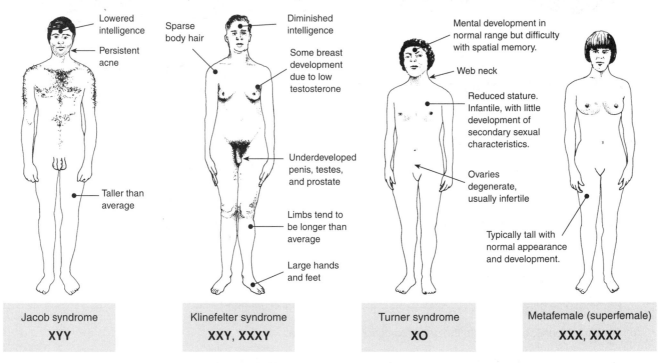

Jacob syndrome: Lowered intelligence, Persistent acne, Taller than average.
XYY

Klinefelter syndrome: Sparse body hair, Diminished intelligence, Some breast development due to low testosterone, Underdeveloped penis, testes, and prostate, Limbs tend to be longer than average, Large hands and feet.
XXY, XXXY

Turner syndrome: Mental development in normal range but difficulty with spatial memory, Web neck, Reduced stature. Infantile, with little development of secondary sexual characteristics, Ovaries degenerate, usually infertile.
XO

Metafemale (superfemale): Typically tall with normal appearance and development.
XXX, XXXX

5. State how many Barr bodies are present in each somatic cell for each of the following syndromes:

(a) Jacob syndrome: _____ (b) Klinefelter syndrome: _____ (c) Turner syndrome: _____

6. Explain the consequence of X-chromosome inactivation in terms of the proteins encoded by the X chromosome genes:

_____

_____

7. State how many chromosomes for each set of homologues are present for the following forms of aneuploidy:

(a) Nullisomy: _____ (c) Trisomy: _____

(b) Monosomy: _____ (d) Polysomy: _____

Trisomy in Human Autosomes

Trisomy is a form of **polysomy** where the nucleus of the cells have one chromosome pair represented by three chromosomes (2N+1). The extra chromosome disturbs the overall chromosomal balance causing abnormalities or death. In humans, about 50% of all spontaneous abortions result from chromosomal abnormalities, and trisomies are responsible for about half of these (25% of all spontaneous abortions). About 6% of live births involve children with chromosomal abnormalities. Autosomal trisomies make up only 0.1% of all pregnancies. Of the three autosomal trisomies surviving to birth, trisomy 21 (**Down** syndrome) is the most common. The other two, **Edward** and **Patau**, show severe physical and mental abnormalities. Trisomies in other autosomes are rare.

Down Syndrome (Trisomy 21)

Down syndrome is the most common of the human aneuploidies. The incidence rate in humans is about 1 in 800 births for women aged 30 to 31 years, with a maternal age effect (the rate increases rapidly with maternal age). The most common form of this condition arises when meiosis fails to separate the pair of chromosome number 21s in the eggs that are forming in the woman's ovaries (it is apparently rare for males to be the cause of this condition). In addition to growth failure and mental retardation, there are a number of well known phenotypic traits (see diagram right).

Down syndrome may arise from several causes:

Non-disjunction: Nearly all cases (approximately 95%) result from **non-disjunction** of chromosome 21 during **meiosis**. When this happens, a gamete (usually the oocyte) ends up with 24 rather than 23 chromosomes, and fertilization produces a trisomic offspring (see the karyotype photo, above right).

Translocation: One in twenty cases of Down syndrome (fewer than 3-4%) arise from a **translocation** mutation where one parent is a translocation carrier (chromosome 21 is fused to another chromosome, usually number 14).

Mitotic errors: A very small proportion of cases (fewer than 3%) arise from the failure of the pair of chromosomes 21 to separate during **mitosis** at an early embryonic stage. The resulting individual is a **mosaic** in which two cell lines exist, one of which is trisomic. If the mitotic abnormality occurs very early in development, a large number of cells are affected and the full Down syndrome is expressed. If only a few cells are affected, there are only mild expressions of the syndrome.

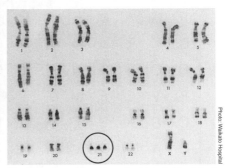

A child showing features of the Down syndrome phenotype.

The karyotype of a trisomic 21 individual that produces the phenotype known as Down syndrome.

Photo: Waikato Hospital

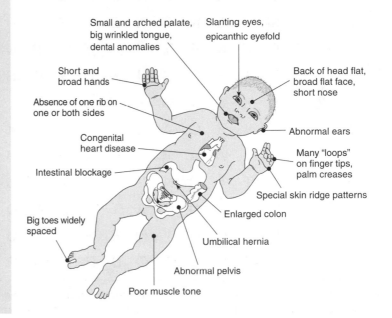

- Small and arched palate, big wrinkled tongue, dental anomalies
- Slanting eyes, epicanthic eyefold
- Short and broad hands
- Back of head flat, broad flat face, short nose
- Absence of one rib on one or both sides
- Abnormal ears
- Congenital heart disease
- Many "loops" on finger tips, palm creases
- Intestinal blockage
- Special skin ridge patterns
- Enlarged colon
- Big toes widely spaced
- Umbilical hernia
- Abnormal pelvis
- Poor muscle tone

---

1. Distinguish between an autosomal aneuploidy and one involving the sex chromosomes: _____

_____

2. (a) Suggest a possible reason why the presence of an extra chromosome causes such a profound effect on the development of a person's phenotype:

_____

(b) With reference to Down syndrome, explain what you understand by the term **syndrome**: _____

_____

3. (a) Describe the main cause of Down syndrome: _____

_____

(b) Describe the features of the Down phenotype: _____

_____

4. Describe one other cause of Down syndrome: _____

_____

5. State how many chromosomes would be present in the somatic cells of an individual with Down syndrome: _____

Periodicals:
The biological aspects of Down syndrome

Related activities: *Aneuploidy in Humans*
Web links: *Down Syndrome*

Mutation

A 2

KEY TERMS Word Find

Use the clues below to find the relevant key terms in the WORD FIND grid

```
M  I  G  A  M  E  T  I  C  P  T  V  M  N  P  O  L  Y  S  O  M  Y  R  A
X  T  K  E  V  A  J  X  O  R  G  W  A  P  I  N  E  U  T  R  A  L  N  T
F  R  A  M  E  S  H  I  F  T  C  W  S  Y  N  D  R  O  M  E  H  D  A  J
C  X  I  V  D  N  P  R  A  G  Y  L  O  E  F  J  E  I  E  B  U  E  Z  E
B  Q  F  A  N  T  I  G  E  N  I  C  D  R  I  F  T  B  A  A  E  H  B  F
F  V  C  F  M  T  T  I  Q  Y  R  U  R  E  F  G  F  Q  F  S  A  B  E  A
C  S  X  S  Z  F  J  P  O  L  Y  P  L  O  I  D  Y  W  S  V  N  I  J  D
D  A  I  R  V  A  N  T  I  G  E  N  I  C  S  H  I  F  T  F  E  U  W  K
Q  O  R  N  N  D  K  Z  I  N  S  E  R  T  I  O  N  K  F  M  U  G  I  P
E  R  G  C  V  Q  V  K  N  C  B  O  Y  I  S  D  G  X  W  Y  P  L  V  X
M  Y  T  M  I  E  N  O  N  D  I  S  J  U  N  C  T  I  O  N  L  U  K  J
N  H  B  R  E  N  R  Y  U  S  Y  L  P  N  P  W  S  K  K  Q  O  W  V  P
N  C  M  T  I  K  O  S  J  K  P  Y  O  S  R  S  O  M  A  T  I  C  M  A
C  U  F  E  X  S  N  G  I  M  L  H  X  I  B  L  H  P  T  X  D  E  U  R
Z  G  K  V  K  Z  O  A  E  O  F  Z  F  X  I  S  F  H  C  V  Y  L  T  V
O  D  Y  D  R  W  A  M  U  N  N  B  E  B  A  R  R  B  O  D  Y  B  A  N
M  U  T  A  T  I  O  N  Y  A  I  V  Y  O  C  C  I  T  F  E  J  V  G  L
T  R  A  N  S  L  O  C  A  T  I  O  N  D  U  D  A  C  W  R  Q  P  E  U
Q  F  J  N  O  C  H  I  M  E  R  A  K  U  G  K  Y  H  V  P  X  N  N  H
K  X  F  G  D  Q  S  W  C  D  R  C  A  R  R  I  E  R  C  I  R  Z  Z  X
```

An inactivated and condensed X chromosome found in a non-dividing somatic cell (2 words).

A group of symptoms that occur together. Usually identified with a disease or abnormality.

Deletion or insertion of a base is likely to cause this error (2 words).

A type of aneuploidy where the chromosome number is more than the normal 2N.

Any chemical, influence, or object that is able to increase the mutation rate of an organism.

Event occurring during meiosis in which chromosome pairs may not split evenly into different cells.

Condition or individual in which a portion of the cells contain the normal number of chromosomes while the rest do not.

The movement of part of one chromosome onto another non-homologous chromosome.

A spontaneous or induced change to the DNA sequence of an organism.

A mutation where a new base (or sequence of bases) is placed into the DNA sequence.

Mutation in which the a block of DNA is removed and re-inserted in the reverse direction.

The condition of having a chromosome number that is not an exact multiple of the normal 2N.

The condition of having a chromosome complement of more than 2N (e.g. 3N).

A type of aneuploidy where the chromosome number for the affected chromosome is 2n+1 (3).

Heritable mutations affect the germline cells and are referred to as this.

An individual heterozygous for a recessive characteristic, who therefore does not express that characteristic but may pass it on.

Mutation in which a change in the base sequence of a codon has no effect on the functionality of the protein produced.

Any chemical, influence or object that is able to increase the likelihood of cancer.

The combination of two or more viral strains to form a new subtype (2 words).

Mutations to the body's cells that are not inherited but may affect an individual during their lifetime are called this.

Small changes in viral structure that occur continually over time.

Nucleic Acid
Technology

KEY CONCEPTS

- ► Biotechnology often makes use of GMOs.
- ► Genetic modification relies on a few basic techniques, widely applied.
- ► Genetic modification potentially offers huge benefits, but presents ethical concerns for many.
- ► Cloning technology has many applications, including in the rapid production of transgenics.
- ► Stem cell technology offers a way to provide immune-compatible tissues for medicine.

KEY TERMS

annealing
biotechnology
blunt end
cloning
DNA (gene) probes
DNA (genetic) profiling
DNA chip (microarray)
DNA ligase
DNA ligation
DNA polymerase
DNA sequencing
forensics
gel electrophoresis
gene technology
gene therapy
genetic modification
GMO
marker gene
microsatellite
molecular clone
plasmid
polymerase chain reaction
primer
recognition site
recombinant DNA
    technology
restriction digestion
restriction enzyme
reverse transcription
stem cell
sticky end
Taq polymerase
transgenesis
vector

OBJECTIVES

☐ 1. Use the **KEY TERMS** to help you understand and complete these objectives.

Basic Principles
pages 246-262

☐ 2. Describe the scope of biotechnology, distinguishing it from gene technology, but recognising the connections between the two.

☐ 3. Recognize that the same, relatively few, basic techniques (#4-8) are used in a range of different processes and applications.

☐ 4. Explain the use of **restriction enzymes** in **recombinant DNA technology**.

☐ 5. Explain the technique and purpose of **DNA ligation** and **annealing**, including the role of **DNA ligase**.

☐ 6. Explain the role of **gel electrophoresis** (of DNA) in gene technology. Explain how the DNA fragments on a gel are made visible and the role of **DNA markers** in identifying fragments of different size.

☐ 7. Describe and explain the role of **polymerase chain reaction** (PCR) in **DNA amplification**, including the role of **primers** and **DNA polymerase**.

☐ 8. Describe the construction of a **DNA chip** (**microarray**), identifying the principles by which the chip operates.

Processes and Applications
pages 117-118, 263-284

☐ 9. SEQUENCING: Describe and explain the use of **PCR**, **radioactive labelling**, and **gel electrophoresis** in **DNA sequencing**. Explain the role of automated sequencing in the feasibility of large scale **genome analyses** and discuss current and future applications of this technology.

☐ 10. PROFILING: Describe and explain **DNA profiling** using PCR. Discuss the applications of DNA profiling as a **forensic** and **diagnostic tool**.

☐ 11. *IN-VIVO* GENE CLONING: Describe and explain gene cloning using **plasmids**. Discuss one or more of the applications of gene cloning.

☐ 12. TRANSGENESIS: Describe and explain **transgenesis**, including the role of **vectors** in integrating foreign DNA into another genome. Describe the applications of transgenic organisms, and any ethical concerns with their use.

☐ 13. CASE STUDIES IN BIOTECHNOLOGY: Describe and explain how gene technology is used to meet human needs or demands in the areas of medicine, forensics, agriculture, or phylogenetics. Examples could include enzyme production, **gene therapy**, crop improvement, or protein production.

☐ 14. MEDICAL BIOTECHNOLOGY: Not all medical applications of biotechnology involve genetic modification. Describe and explain the use of **cloning**, **monoclonal antibodies**, **tissue engineering**, organ transplants, and **stem cell technology**. Discuss any ethical concerns with these applications.

Periodicals:
listings for this
chapter are on page 381

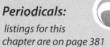

Weblinks:
www.thebiozone.com/
weblink/SB1-2597.html

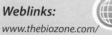

Teacher Resource CD-ROM:
Engineering Solutions

Amazing Organisms, Amazing Enzymes

Before the 1980s scientists knew of only a few organisms that could survive in extreme conditions. Indeed, many scientists believed that life in highly saline or high temperature and pressure environments was impossible. That view changed with the discovery of bacteria inhabiting the deep sea hydrothermal vents. They tolerate temperatures over 110°C and pressures of over 200 atmospheres. Bacteria were also found in volcanic hot pools on land, some surviving at temperatures in excess of 80°C. Most enzymes are denatured at temperatures above 40°C, but these **thermophilic** bacteria have enzymes that are fully functional at high temperatures. This discovery led to the development of one of most important techniques in biotechnology, the **polymerase chain reaction** (PCR).

PCR is a technique, first described in the 1970s, that allows scientists to copy and multiply a piece of DNA millions of times. The DNA is heated to 98°C so that it separates into single strands and polymerase enzyme is added to synthesize new DNA strands from supplied free nucleotides. This earlier technique was labour intensive and expensive because the polymerase denatured at the high temperatures and had to be replaced every cycle. In 1985, a thermophilic polymerase (*Taq* polymerase) was isolated from the bacterium *Thermophilus aquaticus,* which inhabited the hot springs of Yellowstone National Park. Isolating this enzyme enabled automation of the PCR process, because the polymerase was stable throughout multiple cycles of synthesis. This led to an rapid growth in biotechnology, and gene technology in particular, because DNA samples could be easily copied for sequencing.

Searching for novel compounds in organisms from extreme environments is important in the development of new biotechnologies. Organisms must have compounds that can work in their specific environment, and the identification and extraction of these may allow them to be adapted for human use. For example, the Antarctic sea sponge *Kirkpatrickia variolosa* produces an alkaloid excreted as a toxic defence to prevent other organisms growing nearby. Tests indicate that this same chemical may have biological activity against cancer cells. Compounds from other sponge species are currently being assessed to treat a range of diseases including cancer, AIDS, tuberculosis and other bacterial infections, and cystic fibrosis.

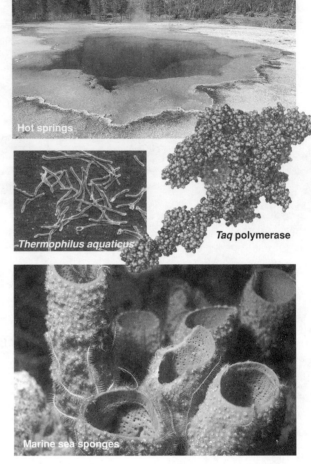

Hot springs

Thermophilus aquaticus

Taq polymerase

Marine sea sponges

1. Explain why PCR was not a viable technique until the mid 1980s: _____

_____

_____

_____

2. Explain why *Taq* polymerase was so important in the development of PCR: _____

_____

_____

_____

3. Explain how investigating the lifestyles of other organisms can lead to advances in unrelated areas of science:

_____

_____

_____

_____

_____

_____

_____

_____

Related activities: Polymerase Chain Reaction

What is Genetic Modification?

The genetic modification of organisms is a vast industry, and the applications of the technology are exciting and far reaching. It brings new hope for medical cures, promises to increase yields in agriculture, and has the potential to help solve the world's pollution and resource crises. Organisms with artificially altered DNA are referred to as **genetically modified organisms** or **GMOs**. They may be modified in one of three ways (outlined below). Some of the current and proposed applications of gene technology raise complex ethical and safety issues, where the benefits of their use must be carefully weighed against the risks to human health, as well as the health and well-being of other organisms and the environment as a whole.

Producing Genetically Modified Organisms (GMOs)

Foreign gene is inserted into host DNA

Host DNA

Existing gene is altered

Host DNA

Gene is deleted or deactivated

Host DNA

Add a foreign gene

A novel (foreign) gene is inserted from another species. This will enable the GMO to express the trait coded by the new gene. Organisms genetically altered in this way are referred to as **transgenic**.

Alter an existing gene

An existing gene may be altered to make it express at a higher level (e.g. growth hormone) or in a different way (in tissue that would not normally express it). This method is also used for gene therapy.

Delete or 'turn off' a gene

An existing gene may be deleted or deactivated (switched off) to prevent the expression of a trait (e.g. the deactivation of the ripening gene in tomatoes produced the Flavr-Savr tomato).

Human insulin, used to treat diabetic patients, is now produced using transgenic bacteria.

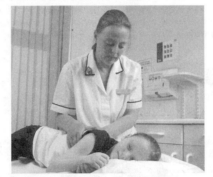

Gene therapy could be used treat genetic disrorders, such as cystic fibrosis.

Manipulating gene action is one way in which to control processes such as ripening in fruit.

1. Using examples, discuss the ways in which an organism may be genetically modified (to produce a GMO):

_____

_____

_____

_____

_____

_____

2. Explain how human needs or desires have provided a stimulus for the development of the following biotechnologies:

(a) Gene therapy: _____

_____

_____

(b) The production and use of transgenic organisms: _____

_____

_____

(c) Plant micropropagation (tissue culture): _____

_____

_____

Related activities: Applications of GMOs
Web links: Biotechnology Tmeline

RA 2

Applications of GMOs

Techniques for genetic manipulation are now widely applied throughout modern biotechnology: in food and enzyme technology, in industry and medicine, and in agriculture and horticulture. Microorganisms are among the most widely used GMOs, with applications ranging from pharmaceutical production and vaccine development to environmental clean-up. Crop plants are also popular candidates for genetic modification although their use, as with much of genetic engineering of higher organisms, is controversial and sometimes problematic.

Applications of GMOs

Extending shelf life

Some fresh produce (e.g. tomatoes) have been engineered to have an extended keeping quality. In the case of tomatoes, the gene for ripening has been switched off, delaying the natural process of softening in the fruit.

Pest or herbicide resistance

Plants can be engineered to produce their own insecticide and become pest resistant. Genetically engineered herbicide resistance is also common. In this case, chemical weed killers can be used freely without crop damage.

Crop improvement

Gene technology is now an integral part of the development of new crop varieties. Crops can be engineered to produce higher protein levels or to grow in inhospitable conditions (e.g. salty or arid conditions).

Environmental clean-up

Some bacteria have been engineered to thrive on waste products, such as liquefied newspaper pulp or oil. As well as degrading pollutants and wastes, the bacteria may be harvested as a commercial protein source.

Biofactories

Transgenic bacteria are widely used to produce desirable products: often hormones or proteins. Large quantities of a product can be produced using bioreactors (above). Examples: insulin production by recombinant yeast, production of bovine growth hormone.

Vaccine development

The potential exists for multipurpose vaccines to be made using gene technology. Genes coding for vaccine components (e.g. viral protein coat) are inserted into an unrelated live vaccine (e.g. polio vaccine), and deliver proteins to stimulate an immune response.

Livestock improvement using transgenic animals

Transgenic sheep have been used to enhance wool production in flocks (above, left). The keratin protein of wool is largely made of a single amino acid, cysteine. Injecting developing sheep with the genes for the enzymes that generate cysteine produces woollier transgenic sheep. In some cases, transgenic animals have been used as biofactories. Transgenic sheep carrying the human gene for a protein, α-1-antitrypsin produce the protein in their milk. The antitrypsin is extracted from the milk and used to treat hereditary emphysema.

1. In a short account discuss one of the applications of GMOs described above: _____

_____

_____

_____

_____

_____

_____

_____

_____

_____

_____

_____

_____

Related activities: Golden Rice, Chymosin Production, Production of Insulin

Restriction Enzymes

One of the essential tools of genetic engineering is a group of special **restriction enzymes** (also known as restriction endonucleases). These have the ability to cut DNA molecules at very precise sequences of 4 to 8 base pairs called **recognition sites**. These enzymes are the "molecular scalpels" that allow genetic engineers to cut up DNA in a controlled way. Although first isolated in 1970, these enzymes were discovered earlier in many bacteria (see panel on the next page). The purified forms of these bacterial restriction enzymes are used today as tools to

cut DNA (see table on the next page for examples). Enzymes are named according to the bacterial species from which they were first isolated. By using a 'tool kit' of over 400 restriction enzymes recognizing about 100 recognition sites, genetic engineers can isolate, sequence, and manipulate individual genes derived from any type of organism. The sites at which the fragments of DNA are cut may result in overhanging "sticky ends" or non-overhanging "blunt ends". Pieces may later be joined together using an enzyme called **DNA ligase** in a process called **ligation**.

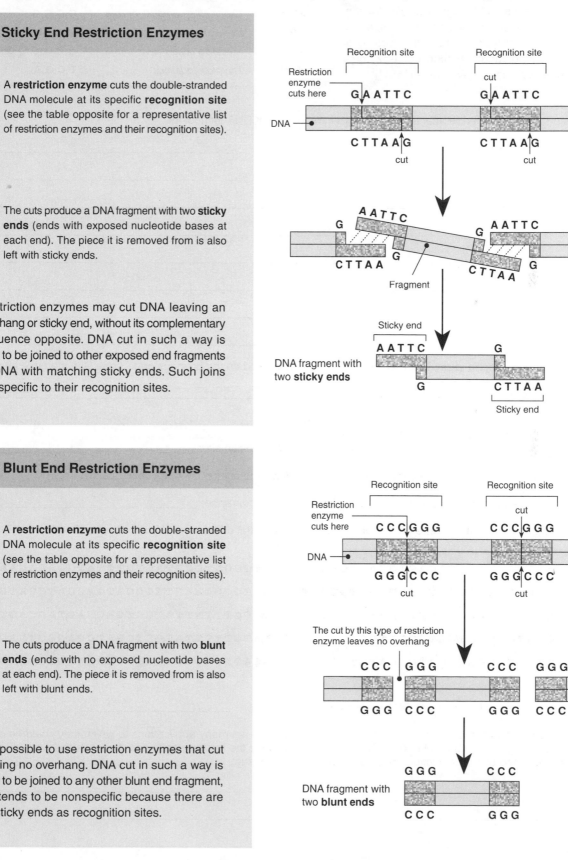

Sticky End Restriction Enzymes

1 A **restriction enzyme** cuts the double-stranded DNA molecule at its specific **recognition site** (see the table opposite for a representative list of restriction enzymes and their recognition sites).

2 The cuts produce a DNA fragment with two **sticky ends** (ends with exposed nucleotide bases at each end). The piece it is removed from is also left with sticky ends.

Restriction enzymes may cut DNA leaving an overhang or sticky end, without its complementary sequence opposite. DNA cut in such a way is able to be joined to other exposed end fragments of DNA with matching sticky ends. Such joins are specific to their recognition sites.

Blunt End Restriction Enzymes

1 A **restriction enzyme** cuts the double-stranded DNA molecule at its specific **recognition site** (see the table opposite for a representative list of restriction enzymes and their recognition sites).

2 The cuts produce a DNA fragment with two **blunt ends** (ends with no exposed nucleotide bases at each end). The piece it is removed from is also left with blunt ends.

It is possible to use restriction enzymes that cut leaving no overhang. DNA cut in such a way is able to be joined to any other blunt end fragment, but tends to be nonspecific because there are no sticky ends as recognition sites.

Related activities: Ligation
Web links: DNA Interactive: Cut and Paste

A 3

Origin of Restriction Enzymes

Restriction enzymes have been isolated from many bacteria. It was observed that certain *bacteriophages* (viruses that infect bacteria) could not infect bacteria other than their usual hosts. The reason was found to be that other potential hosts could destroy almost all of the phage DNA using *restriction enzymes* present naturally in their cells; a defense mechanism against the entry of foreign DNA. Restriction enzymes are named according to the species they were first isolated from, followed by a number to distinguish different enzymes isolated from the same organism.

Recognition sites for selected restriction enzymes

Enzyme	Source	Recognition Sites
EcoRI	*Escherichia coli* RY13	G A A T T C
BamHI	*Bacillus amyloliquefaciens* H	G G A T C C
HaeIII	*Haemophilus aegyptius*	G G C C
HindIII	*Haemophilus influenzae* Rd	A A G C T T
HpaI	*Haemophilus parainfluenzae*	G T T A A C
HpaII	*Haemophilus parainfluenzae*	C C G G
MboI	*Moraxella bovis*	G A T C
NotI	*Norcardia otitidis-caviarum*	G C G G C C G C
TaqI	*Thermus aquaticus*	T C G A

1. Explain the following terms, identifying their role in recombinant DNA technology:

   (a) Restriction enzyme: _____

   _____

   (b) Recognition site: _____

   _____

   (c) Sticky end: _____

   _____

   (d) Blunt end: _____

   _____

2. The action of a specific sticky end restriction enzyme is illustrated on the previous page (top). Use the table above to:

   (a) Name the **restriction enzyme** used: _____

   (b) Name the organism from which it was first isolated: _____

   (c) State the **base sequence** for this restriction enzyme's recognition site: _____

3. A genetic engineer wants to use the restriction enzyme ***Bam*HI** to cut the DNA sequence below:

   (a) Consult the table above and state the recognition site for this enzyme: _____

   (b) Circle every **recognition site** on the DNA sequence below that could be cut by the enzyme ***Bam*HI**:

```
          10              20              30              40              50              60
|AATGGGTACG|CACAGTGGAT|CCACGTAGTA|TGCGATGCGT|AGTGTTTATG|GAGAGAAGAA|
          70              80              90             100             110             120
|AACGCGTCGC|CTTTTATCGA|TGCTGTACGG|ATGCGGAAGT|GGCGATGAGG|ATCCATGCAA|
         130             140             150             160             170             180
|TCGCGGCCGA|TCGXGTAATA|TATCGTGGCT|GCGTTTATTA|TCGTGACTAG|TAGCAGTATG|
         190             200             210             220             230             240
|CGATGTGACT|GATGCTATGC|TGACTATGCT|ATGTTTTTAT|GCTGGATCCA|GCGTAAGCAT|
         250             260             270             280             290             300
|TTCGCTGCGT|GGATCCCATA|TCCTTATATG|CATATATTCT|TATACGGATC|GCGCACGTTT|
```

   (c) State how many fragments of DNA were created by this action: _____

4. When restriction enzymes were first isolated in 1970 there were not many applications to which they could be put to use. They are now an important tool in genetic engineering. Describe the human needs and demands that have driven the development and use of restriction enzymes in genetic engineering:

   _____

   _____

   _____

Ligation

DNA fragments produced using restriction enzymes may be reassembled by a process called **ligation**. Pieces are joined together using an enzyme called **DNA ligase**. DNA of different origins produced in this way is called **recombinant DNA** (because it is DNA that has been recombined from different sources). The combined techniques of using restriction enzymes and ligation are the basic tools of genetic engineering (also known as recombinant DNA technology).

Creating a Recombinant DNA Plasmid

1 If two pieces of DNA are cut by the same restriction enzyme, they will produce fragments with matching **sticky ends** (ends with exposed nucleotide bases at each end).

2 When two such matching sticky ends come together, they can join by base-pairing. This process is called **annealing**. This can allow DNA fragments from a different source, perhaps a **plasmid**, to be joined to the DNA fragment.

3 The joined fragments will usually form either a linear molecule or a circular one, as shown here for a **plasmid**. However, other combinations of fragments can occur.

4 The fragments of DNA are joined together by the enzyme **DNA ligase**, producing a molecule of **recombinant DNA**.

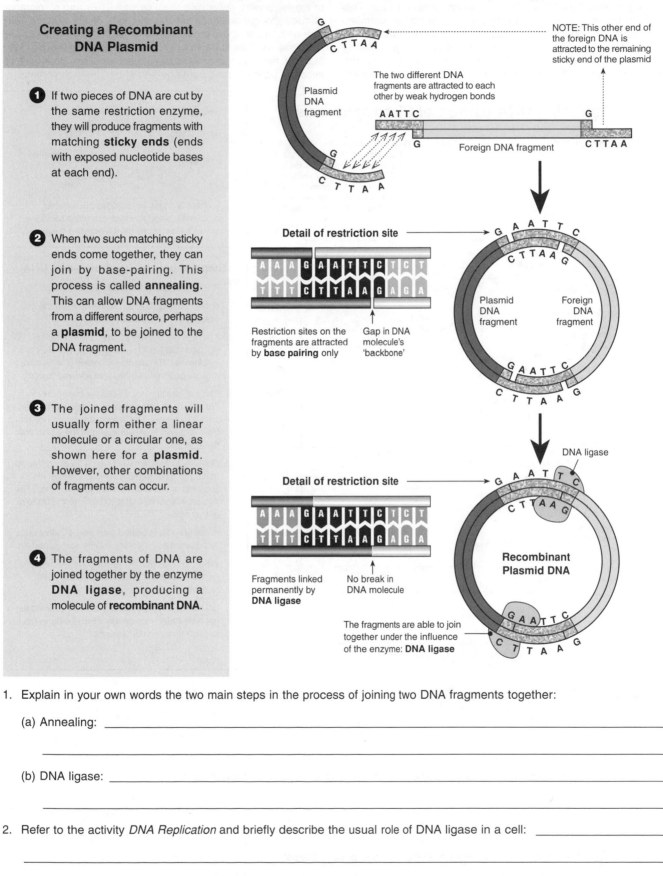

1. Explain in your own words the two main steps in the process of joining two DNA fragments together:

   (a) Annealing: _____

   (b) DNA ligase: _____

2. Refer to the activity *DNA Replication* and briefly describe the usual role of DNA ligase in a cell: _____

3. Explain why ligation can be considered the *reverse* of the restriction enzyme process: _____

Gel Electrophoresis

Gel electrophoresis is a method that separates large molecules (including nucleic acids or proteins) on the basis of size, electric charge, and other physical properties. Such molecules possess a slight electric charge (see DNA below). To prepare DNA for gel electrophoresis the DNA is often cut up into smaller pieces. This is done by mixing DNA with restriction enzymes in controlled conditions for about an hour. Called **restriction digestion**, it produces a range of DNA fragments of different lengths. During electrophoresis, molecules are forced to move through the pores of a **gel** (a jelly-like material), when the electrical current is applied. Active electrodes at each end of the gel provide the driving force. The electrical current from one electrode repels the molecules while the other electrode simultaneously attracts the molecules. The frictional force of the gel material resists the flow of the molecules, separating them by size. Their rate of migration through the gel depends on the strength of the electric field, size and shape of the molecules, and on the ionic strength and temperature of the buffer in which the molecules are moving. After staining, the separated molecules in each lane can be seen as a series of bands spread from one end of the gel to the other.

Analyzing DNA using Gel Electrophoresis

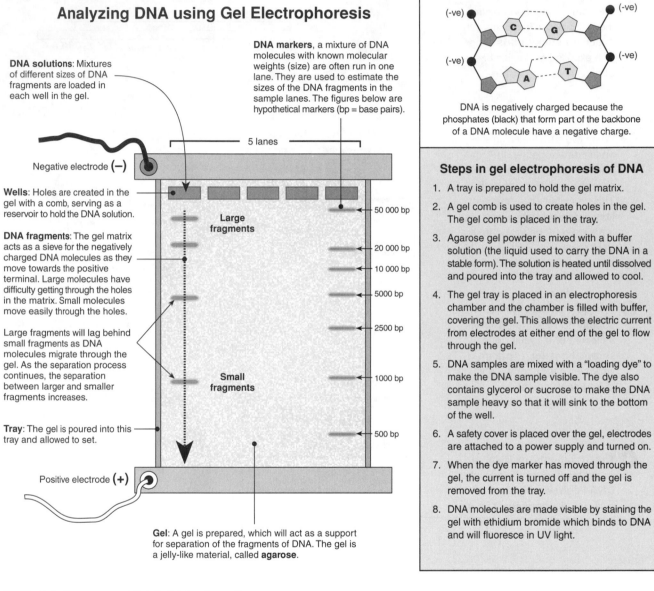

DNA solutions: Mixtures of different sizes of DNA fragments are loaded in each well in the gel.

DNA markers, a mixture of DNA molecules with known molecular weights (size) are often run in one lane. They are used to estimate the sizes of the DNA fragments in the sample lanes. The figures below are hypothetical markers (bp = base pairs).

DNA is negatively charged because the phosphates (black) that form part of the backbone of a DNA molecule have a negative charge.

Negative electrode **(−)**

Wells: Holes are created in the gel with a comb, serving as a reservoir to hold the DNA solution.

DNA fragments: The gel matrix acts as a sieve for the negatively charged DNA molecules as they move towards the positive terminal. Large molecules have difficulty getting through the holes in the matrix. Small molecules move easily through the holes.

Large fragments will lag behind small fragments as DNA molecules migrate through the gel. As the separation process continues, the separation between larger and smaller fragments increases.

Tray: The gel is poured into this tray and allowed to set.

Positive electrode **(+)**

5 lanes

Large fragments

Small fragments

- 50 000 bp
- 20 000 bp
- 10 000 bp
- 5000 bp
- 2500 bp
- 1000 bp
- 500 bp

Gel: A gel is prepared, which will act as a support for separation of the fragments of DNA. The gel is a jelly-like material, called **agarose**.

Steps in gel electrophoresis of DNA

1. A tray is prepared to hold the gel matrix.
2. A gel comb is used to create holes in the gel. The gel comb is placed in the tray.
3. Agarose gel powder is mixed with a buffer solution (the liquid used to carry the DNA in a stable form). The solution is heated until dissolved and poured into the tray and allowed to cool.
4. The gel tray is placed in an electrophoresis chamber and the chamber is filled with buffer, covering the gel. This allows the electric current from electrodes at either end of the gel to flow through the gel.
5. DNA samples are mixed with a "loading dye" to make the DNA sample visible. The dye also contains glycerol or sucrose to make the DNA sample heavy so that it will sink to the bottom of the well.
6. A safety cover is placed over the gel, electrodes are attached to a power supply and turned on.
7. When the dye marker has moved through the gel, the current is turned off and the gel is removed from the tray.
8. DNA molecules are made visible by staining the gel with ethidium bromide which binds to DNA and will fluoresce in UV light.

1. Explain the purpose of gel electrophoresis: _____

_____

_____

2. Describe the two forces that control the speed at which fragments pass through the gel:

(a) _____

(b) _____

3. Explain why the smallest fragments travel through the gel the fastest: _____

_____

_____

Related activities: Nucleic Acids, DNA Profiling, Automated DNA Sequencing
Web links: DNA Extraction, Gel Electrophoresis

Polymerase Chain Reaction

Many procedures in DNA technology (such as DNA sequencing and DNA profiling) require substantial amounts of DNA to work with. Some samples, such as those from a crime scene or fragments of DNA from a long extinct organism, may be difficult to get in any quantity. The diagram below describes the laboratory technique called **polymerase chain reaction** (**PCR**). Using this technique, vast quantities of DNA identical to trace samples can be created. This process is often termed **DNA amplification**. Although only one cycle of replication is shown below, following cycles replicate DNA at an exponential rate. PCR can be used to make literally billions of copies in only a few hours. **Linear PCR** differs from regular PCR in that the same original DNA templates are used repeatedly. It is used to make many radio-labeled DNA fragments for DNA sequencing.

A Single Cycle of the Polymerase Chain Reaction

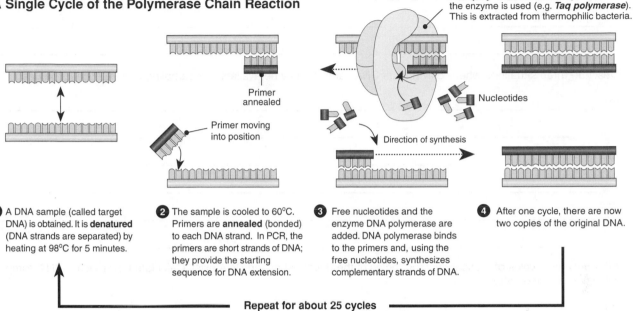

DNA polymerase: A thermally stable form of the enzyme is used (e.g. *Taq polymerase*). This is extracted from thermophilic bacteria.

Primer annealed

Primer moving into position

Nucleotides

Direction of synthesis

1 A DNA sample (called target DNA) is obtained. It is **denatured** (DNA strands are separated) by heating at 98°C for 5 minutes.

2 The sample is cooled to 60°C. Primers are **annealed** (bonded) to each DNA strand. In PCR, the primers are short strands of DNA; they provide the starting sequence for DNA extension.

3 Free nucleotides and the enzyme DNA polymerase are added. DNA polymerase binds to the primers and, using the free nucleotides, synthesizes complementary strands of DNA.

4 After one cycle, there are now two copies of the original DNA.

Repeat for about 25 cycles

Repeat cycle of heating and cooling until enough copies of the target DNA have been produced

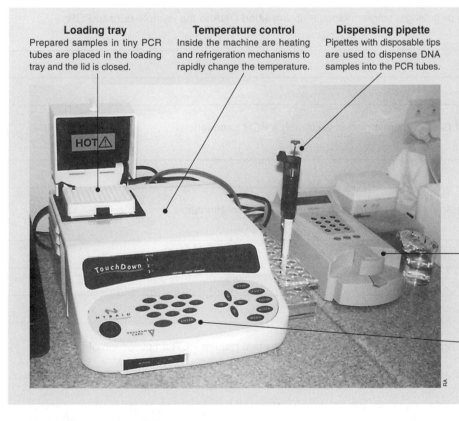

Loading tray
Prepared samples in tiny PCR tubes are placed in the loading tray and the lid is closed.

Temperature control
Inside the machine are heating and refrigeration mechanisms to rapidly change the temperature.

Dispensing pipette
Pipettes with disposable tips are used to dispense DNA samples into the PCR tubes.

Thermal Cycler

Amplification of DNA can be carried out with simple-to-use machines called **thermal cyclers**. Once a DNA sample has been prepared, in just a few hours the amount of DNA can be increased billions of times. Thermal cyclers are in common use in the biology departments of universities, as well as other kinds of research and analytical laboratories. The one pictured on the left is typical of this modern piece of equipment.

DNA quantitation
The amount of DNA in a sample can be determined by placing a known volume in this quantitation machine. For many genetic engineering processes, a minimum amount of DNA is required.

Controls
The control panel allows a number of different PCR programs to be stored in the machine's memory. Carrying out a PCR run usually just involves starting one of the stored programs.

1. Explain the purpose of PCR: _____

_____

_____

_____

Periodicals:
The polymerase chain reaction

Related activities: DNA Profiling Using PCR, In Vivo Gene Cloning
Web links: PCR Animation

2. Briefly describe how the **polymerase chain reaction** (PCR) works: _____

_____

_____

_____

_____

_____

_____

3. Describe three situations where only minute DNA samples may be available for sampling and PCR could be used:

(a) _____

_____

(b) _____

_____

(c) _____

_____

4. After only two cycles of replication, four copies of the double-stranded DNA exist. Calculate how much a DNA sample will have increased after:

(a) 10 cycles: _____ (b) 25 cycles: _____

5. The risk of contamination in the preparation for PCR is considerable.

(a) Explain what the effect would be of having a single molecule of unwanted DNA in the sample prior to PCR:

_____

_____

_____

(b) Describe two possible sources of DNA contamination in preparing a PCR sample:

Source 1: _____

Source 2: _____

(c) Describe two precautions that could be taken to reduce the risk of DNA contamination:

Precaution 1: _____

_____

Precaution 2: _____

_____

6. Describe two other genetic engineering/genetic manipulation procedures that require PCR amplification of DNA:

(a) _____

_____

_____

(b) _____

_____

_____

DNA Profiling Using PCR

In chromosomes, some of the DNA contains simple, repetitive sequences. These *noncoding* nucleotide sequences repeat themselves over and over again and are found scattered throughout the genome. Some repeating sequences are short (2-6 base pairs) called **microsatellites** or **short tandem repeats** (STRs) and can repeat up to 100 times. The human genome has numerous different microsatellites. Equivalent sequences in different people vary considerably in the numbers of the repeating unit. This phenomenon has been used to develop **DNA profiling**, which identifies the natural variations found in every person's DNA. Identifying such differences in the DNA of individuals is a useful tool for forensic investigations.

In 1998, the FBI's Combined Offender DNA Index System (CODIS) was established, providing a national database of DNA samples from convicted criminals, suspects, and crime scenes. In the USA, there are many laboratories approved for forensic DNA testing. Increasingly, these are targeting the 13 core STR loci recommended by the FBI; enough to guarantee that the odds of someone else sharing the same result are extremely unlikely (less than one in a thousand million). The CODIS may be used to solve previously unsolved crimes and to assist in current or future investigations. DNA profiling can also be used to establish genetic relatedness (e.g. in paternity or pedigree disputes), or when searching for a specific gene (e.g. screening for disease).

Microsatellites (Short Tandem Repeats)

Microsatellites consist of a variable number of tandem repeats of a 2 to 6 base pair sequence. In the example below it is a two base sequence (CA) that is repeated.

Telomeres

Centromeres

Homologous pair of chromosomes

Microsatellites are found throughout the genome: within genes (introns) and between genes, and particularly near **centromeres** and **telomeres**.

The human genome contains about 100 000 separate blocks of tandem repeats of the dinucleotide: **CA**. One such block at a known location on a chromosome is shown below:

DNA

DNA

Flanking regions to which PCR primers can be attached

The tandem repeat may exist in two versions (alleles) in an individual; one on each homologous chromosome. Each of the strands shown left is a double stranded DNA, but only the CA repeat is illustrated.

How short tandem repeats are used in DNA profiling

This diagram shows how three people can have quite different microsatellite arrangements at the same point (locus) in their DNA. Each will produce a different DNA profile using gel electrophoresis:

❶ Extract DNA from sample

A sample collected from the tissue of a living or dead organism is treated with chemicals and enzymes to extract the DNA, which is separated and purified.

❷ Amplify microsatellite using PCR

Specific primers (arrowed) that attach to the flanking regions (light gray) either side of the microsatellite are used to make large quantities of the microsatellite and flanking regions sequence only (no other part of the DNA is amplified/replicated).

❸ Visualize fragments on a gel

The fragments are separated by length, using **gel electrophoresis**. DNA, which is negatively charged, moves toward the positive terminal. The smaller fragments travel faster than larger ones.

DNA from individual 'A':

DNA from individual 'B':

DNA from individual 'C':

Microsatellite

Microsatellite from individual 'A':

Microsatellite from individual 'B':

Microsatellite from individual 'C':

Primers   Flanking region   STR   DNA

The results of PCR are many fragments

A   B   C

The products of PCR amplification (making many copies) are fragments of different sizes that can be directly visualized using gel electrophoresis.

Largest fragments

Smallest fragments

Related activities: Gel Electrophoresis, Forensic Applications of DNA Profiling, Polymerase Chain Reaction   *Web links*: DNA Profiling Using Probes

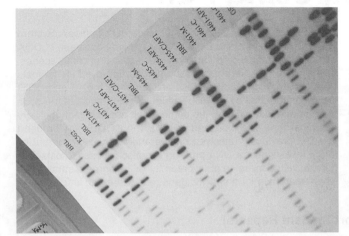

The photo above shows a film output from a DNA profiling procedure. Those lanes with many regular bands are used for calibration; they contain DNA fragment sizes of known length. These calibration lanes can be used to determine the length of fragments in the unknown samples.

DNA profiling can be automated in the same way as DNA sequencing. Computer software is able to display the results of many samples run at the same time. In the photo above, the sample in lane 4 has been selected. It displays fragments of different length on the left of the screen.

1. Describe the properties of **short tandem repeats** that are important to the application of **DNA profiling** technology:

_____

_____

_____

_____

_____

2. Explain the role of each of the following techniques in the process of DNA profiling:

(a) Gel electrophoresis: _____

_____

_____

(b) PCR: _____

_____

_____

3. Describe the three main steps in DNA profiling using PCR:

(a) _____

_____

_____

(b) _____

_____

_____

(c) _____

_____

_____

_____

4. Explain why as many as 10 STR sites are used to gain a DNA profile for forensic evidence: _____

_____

Forensic Applications of DNA Profiling

The use of DNA as a tool for solving crimes such as homicide is well known, but it can also be used to as a solution to many other problems. DNA evidence has been used to identify body parts, solve cases of industrial sabotage and contamination, for paternity testing, and even in identifying animal products illegally made from endangered species.

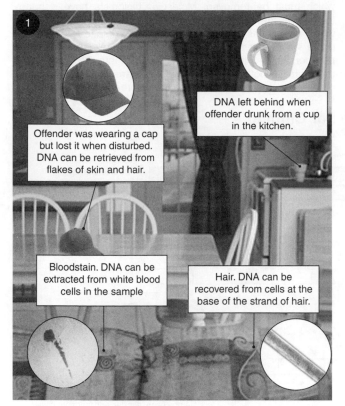

1

Offender was wearing a cap but lost it when disturbed. DNA can be retrieved from flakes of skin and hair.

DNA left behind when offender drunk from a cup in the kitchen.

Bloodstain. DNA can be extracted from white blood cells in the sample

Hair. DNA can be recovered from cells at the base of the strand of hair.

During the initial investigation, samples of material that may contain DNA are taken for analysis. At a crime scene, this may include blood and body fluids as well as samples of clothing or objects that the offender might have touched. Samples from the victim are also taken to eliminate them as a possible source of contamination.

2 DNA is isolated and profiles are made from all samples and compared to known DNA profiles such as that of the victim.

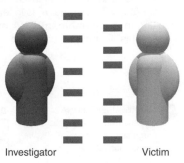

Profiles from collected DNA   Investigator   Victim

A B C D

3 Unknown DNA samples are compared to DNA databases of convicted offenders and to the DNA of the alleged offender.

Alleged offender

Profiles from DNA database

A E F G

4 Although it does not make a complete case, DNA profiling, in conjunction with other evidence, is one of the most powerful tools in identifying offenders or unknown tissues.

1. In the above case two sets of DNA profiles are shown. Describe the purpose of lane A in each set of profiles:

_____

_____

2. Explain why DNA profiles are obtained for both the victim and investigator:

_____

_____

3. Use the evidence to decide if the alleged offender is innocent or guilty and explain your decision:

_____

_____

_____

4. Explain how DNA profiling could be used to refute official claims of the number of whales being captured and sold in fish markets:

_____

_____

_____

Whale DNA: Tracking Illegal Slaughter

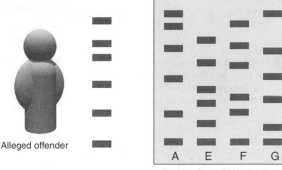

Under International Whaling Commission regulations, some species of whales can be captured for scientific research and their meat sold legally. Most, including humpback and blue whales, are fully protected and to capture or kill them for any purpose is illegal. Between 1999 and 2003 Scott Baker and associates from Oregon State University's Marine Mammal Institute investigated whale meat sold in markets in Japan and South Korea. Using DNA profiling techniques, they found around 10% of the samples tested were from fully protected whales including western gray whales and humpbacks. They also found that many more whales were being killed than were being officially reported.

Preparing a Gene for Cloning

Gene cloning is the process of making large quantities of a piece of DNA once it has been isolated. Its purpose is to yield large quantities of an individual gene or its protein product when the gene is expressed. Genes can be cloned *in vitro* in an automated process called PCR, which amplifies the DNA. Genes can also be cloned when they are part of an organism, in a technique called *in vivo* cloning. A gene of interest (e.g. a human gene) is inserted into the DNA of a vector, resulting in a **recombinant DNA molecule** called a **molecular clone**. This technique uses the self-replicating properties of the vector to make copies of the gene. The genes of interest are rarely ready for cloning in their native form because they include pieces of non-protein coding DNA, called **introns**, which must be removed. Molecular biologists have a handy tool in the form of an enzyme, called **reverse transcriptase**, which makes this possible. Reverse transcriptase is a common name for an enzyme that functions as a RNA-dependent DNA polymerase and it is used to copy RNA into DNA. This task is integral to both *in vitro* and *in vivo* gene cloning because it produces a reconstructed gene that is ready for amplification.

Preparing a Gene For Cloning

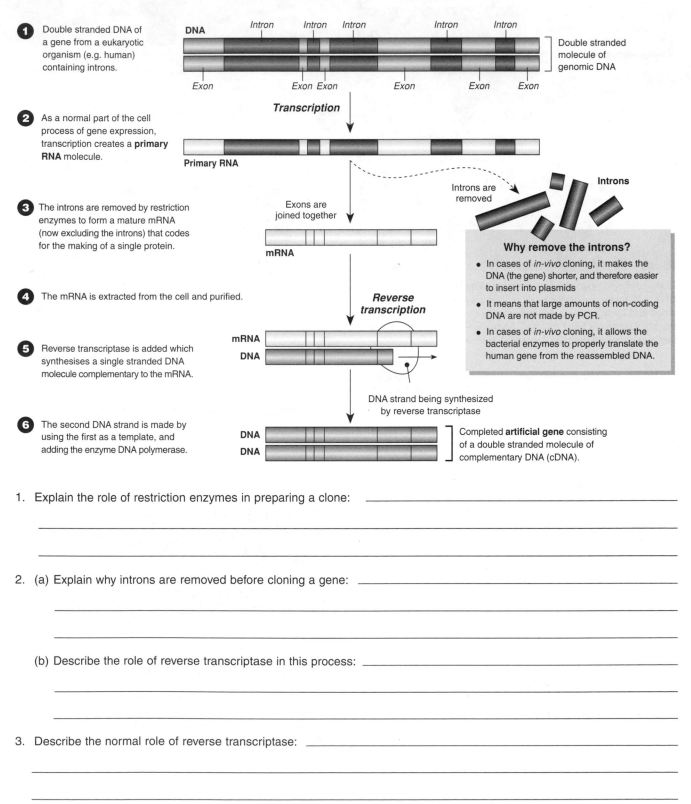

1. Double stranded DNA of a gene from a eukaryotic organism (e.g. human) containing introns.

2. As a normal part of the cell process of gene expression, transcription creates a **primary RNA** molecule.

3. The introns are removed by restriction enzymes to form a mature mRNA (now excluding the introns) that codes for the making of a single protein.

4. The mRNA is extracted from the cell and purified.

5. Reverse transcriptase is added which synthesises a single stranded DNA molecule complementary to the mRNA.

6. The second DNA strand is made by using the first as a template, and adding the enzyme DNA polymerase.

Why remove the introns?

- In cases of *in-vivo* cloning, it makes the DNA (the gene) shorter, and therefore easier to insert into plasmids
- It means that large amounts of non-coding DNA are not made by PCR.
- In cases of *in-vivo* cloning, it allows the bacterial enzymes to properly translate the human gene from the reassembled DNA.

1. Explain the role of restriction enzymes in preparing a clone: _____

_____

_____

2. (a) Explain why introns are removed before cloning a gene: _____

_____

_____

(b) Describe the role of reverse transcriptase in this process: _____

_____

_____

3. Describe the normal role of reverse transcriptase: _____

_____

_____

In Vivo Gene Cloning

It is possible to use the internal replication machinery of a cell to clone a gene, or even many genes, at once. By using cells to copy desired genes, it is also possible to produce any protein product the genes may code for. Recombinant DNA techniques (restriction digestion and ligation) are used to insert a gene of interest into the DNA of a vector (e.g. plasmid or viral DNA). This produces a **recombinant DNA molecule** that can used to transmit the gene of interest to another organism. To be useful, all vectors must be able to replicate inside their host organism, they must have one or more sites at which a restriction enzyme

can cut, and they must have some kind of **genetic marker** that allows them to be easily identified. Viruses, and organisms such as bacteria and yeasts have DNA that behaves in this way. Bacterial plasmids are commonly used because they are easy to manipulate, their restriction sites are well known, and they are readily taken up by cells in culture. Once the recombinant plasmid vector (containing the desired gene) has been taken up by bacterial cells, and those cells are identified, the gene can be replicated many times as the bacteria grow and divide.

Cloning a Human Gene

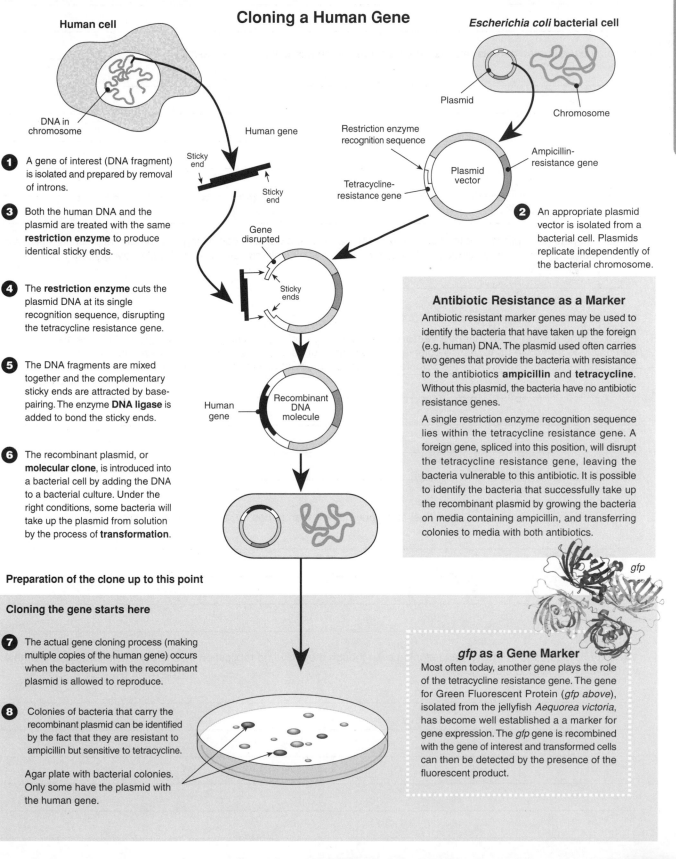

1 A gene of interest (DNA fragment) is isolated and prepared by removal of introns.

3 Both the human DNA and the plasmid are treated with the same **restriction enzyme** to produce identical sticky ends.

4 The **restriction enzyme** cuts the plasmid DNA at its single recognition sequence, disrupting the tetracycline resistance gene.

5 The DNA fragments are mixed together and the complementary sticky ends are attracted by base-pairing. The enzyme **DNA ligase** is added to bond the sticky ends.

6 The recombinant plasmid, or **molecular clone**, is introduced into a bacterial cell by adding the DNA to a bacterial culture. Under the right conditions, some bacteria will take up the plasmid from solution by the process of **transformation**.

2 An appropriate plasmid vector is isolated from a bacterial cell. Plasmids replicate independently of the bacterial chromosome.

Antibiotic Resistance as a Marker

Antibiotic resistant marker genes may be used to identify the bacteria that have taken up the foreign (e.g. human) DNA. The plasmid used often carries two genes that provide the bacteria with resistance to the antibiotics **ampicillin** and **tetracycline**. Without this plasmid, the bacteria have no antibiotic resistance genes.

A single restriction enzyme recognition sequence lies within the tetracycline resistance gene. A foreign gene, spliced into this position, will disrupt the tetracycline resistance gene, leaving the bacteria vulnerable to this antibiotic. It is possible to identify the bacteria that successfully take up the recombinant plasmid by growing the bacteria on media containing ampicillin, and transferring colonies to media with both antibiotics.

Preparation of the clone up to this point

Cloning the gene starts here

7 The actual gene cloning process (making multiple copies of the human gene) occurs when the bacterium with the recombinant plasmid is allowed to reproduce.

8 Colonies of bacteria that carry the recombinant plasmid can be identified by the fact that they are resistant to ampicillin but sensitive to tetracycline.

Agar plate with bacterial colonies. Only some have the plasmid with the human gene.

gfp as a Gene Marker

Most often today, another gene plays the role of the tetracycline resistance gene. The gene for Green Fluorescent Protein (*gfp* above), isolated from the jellyfish *Aequorea victoria*, has become well established a a marker for gene expression. The *gfp* gene is recombined with the gene of interest and transformed cells can then be detected by the presence of the fluorescent product.

Related activities: Restriction Enzymes, Preparing a Gene for Cloning
Web links: Gene Cloning

RA 3

1. Explain why it might be desirable to use *in vivo* methods to clone genes rather than PCR: _____
    _____
    _____
    _____

2. Explain when it may not be desirable to use bacteria to clone genes: _____
    _____
    _____

3. Explain how a human gene is removed from a chromosome and placed into a plasmid. _____
    _____
    _____
    _____
    _____
    _____
    _____

4. A bacterial plasmid replicates at the same rate as the bacteria. If a bacteria containing a recombinant plasmid replicates and divides once every thirty minutes, calculate the number of plasmid copies there will be after twenty four hours:
    _____
    _____

5. When cloning a gene using **plasmid vectors**, the bacterial colonies containing the recombinant plasmids are mixed up with colonies that have none. All the colonies look identical, but some have taken up the plasmids with the human gene, and some have not. Explain how the colonies with the recombinant plasmids are identified:
    _____
    _____
    _____
    _____

6. Explain why the *gfp* marker is a more desirable gene marker than genes for antibiotic resistance:
    _____
    _____
    _____
    _____

7. Viruses are also used in *in vivo* gene cloning even though they have no replication machinery themselves. Explain how viruses can be used to clone genes:
    _____
    _____
    _____
    _____
    _____
    _____

DNA Chips

Microarrays (DNA chips or gene chips) are relatively recent tools in gene research. Their development a decade ago built on earlier DNA probe technology and provided a tool to quickly compare the (known) DNA on a chip with (unknown) DNA to determine which genes were present in a sample or to determine the code of an unsequenced string of DNA. Microarrays have also provided a tool which, increasingly, is being used to investigate the activity level (the expression) of those genes. Microarrays rely on **nucleic acid hybridization**, in which a known DNA fragment is used as a **probe** to find complementary sequences. In a microarray, DNA fragments, corresponding to known genes, are fixed to a solid support in an orderly pattern, usually as a series of dots. The fragments are tested for hybridization with samples of labeled cDNA molecules. Computer analysis then reveals which genes are active in different tissues, in different stages of development, or in tissues in different states of health.

What is a DNA Chip?

A **microarray** (DNA chip) consists of DNA probes fixed to a small solid support such as a glass slide or a nylon filter. Each spot on the microarray has thousands to millions of copies of a different **DNA probe**. The probes are single stranded DNA molecules, each representing a gene.

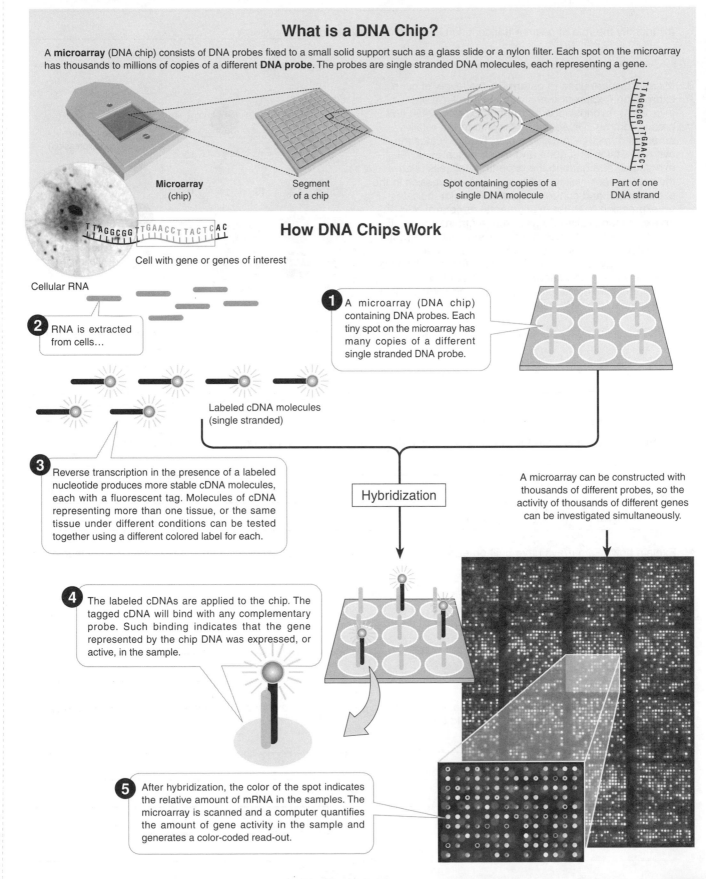

Microarray (chip)

Segment of a chip

Spot containing copies of a single DNA molecule

Part of one DNA strand

Cell with gene or genes of interest

How DNA Chips Work

Cellular RNA

2 RNA is extracted from cells...

1 A microarray (DNA chip) containing DNA probes. Each tiny spot on the microarray has many copies of a different single stranded DNA probe.

Labeled cDNA molecules (single stranded)

3 Reverse transcription in the presence of a labeled nucleotide produces more stable cDNA molecules, each with a fluorescent tag. Molecules of cDNA representing more than one tissue, or the same tissue under different conditions can be tested together using a different colored label for each.

Hybridization

A microarray can be constructed with thousands of different probes, so the activity of thousands of different genes can be investigated simultaneously.

4 The labeled cDNAs are applied to the chip. The tagged cDNA will bind with any complementary probe. Such binding indicates that the gene represented by the chip DNA was expressed, or active, in the sample.

5 After hybridization, the color of the spot indicates the relative amount of mRNA in the samples. The microarray is scanned and a computer quantifies the amount of gene activity in the sample and generates a color-coded read-out.

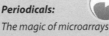

Periodicals:
The magic of microarrays

Related activities: In Vivo Gene Cloning
Web links: Genomics, DNA Microarray

RA 3

1. Describe one purpose of microarrays: _____

_____

_____

2. (a) Identify the basic principle by which microarrays work: _____

_____

_____

(b) Identify the role of reverse transcription in microarray technology: _____

_____

3. Microarrays are used to determine the levels of gene expression (expression analysis). In one type of microarray, hybridization of the red (experimental) and green (control) cDNAs is proportional to the relative amounts of mRNA in the samples. Red indicates the overexpression of a gene and green indicates under-expression of a gene in the experimental cells relative the control cells, yellow indicates equal expression in the experimental and control cells, and no color indicates no expression in either experimental or control cells. In an experiment, cDNA derived from a strain of antibiotic resistant bacteria (experimental cells) was labeled with a red fluorescent tag and cDNA derived from a a non-resistant strain of the same bacterium (control cells) was labeled with a green fluorescent tag. The cDNAs were mixed and hybridized to a chip containing spots of DNA from genes 1-25. The results are shown on the right.

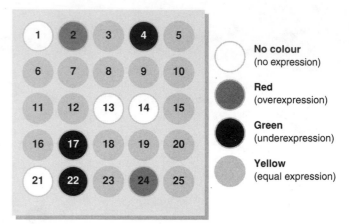

No colour (no expression)

Red (overexpression)

Green (underexpression)

Yellow (equal expression)

(a) Discuss the conclusions you could make about which genes might be implicated in antibiotic resistance in this case:

_____

_____

_____

(b) Suggest how this information could be used to design new antibiotics that are less vulnerable to resistance:

_____

_____

_____

4. Explain how microarrays have built on earlier DNA probe technology and describe the advantages they offer in studies of gene expression:

_____

_____

_____

_____

5. Microarrays are frequently used in diagnostic medicine to compare gene expression in cancerous and non-cancerous tissue. Suggest how this information could be used:

_____

_____

_____

_____

_____

Automated DNA Sequencing

DNA sequencing can be automated using **gel electrophoresis** machines that can sequence up to 600 bases at a time. Automation improves the speed at which samples can be sequenced and has made large scale sequencing projects (such as the **Human Genome Project**) possible. Automated sequencing uses nucleotides labeled with **fluorescent dyes**. With this technique, the entire base sequence for a sample can be determined from a single lane on the gel. Computer software automatically interprets the data from the gel and produces a base sequence.

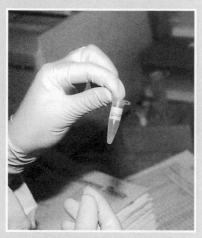

1. DNA sample arrives

Purified DNA samples may contain linear DNA or plasmids. The sample should contain about 1×10^{11} DNA molecules. The sample is checked to ensure that there is enough DNA present in the sample to work with.

2. Primer and reaction mix added

A **DNA primer** is added to the sample which provides a starting sequence for synthesis. Also added is the **sequencing reaction mix** containing the *polymerase enzyme* and free nucleotides, some which are labeled with dye.

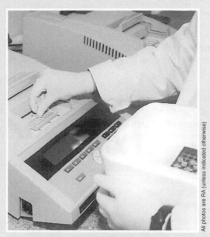

All photos are RA (unless indicated otherwise)

3. Create dye-labeled fragments

A PCR machine creates fragments of DNA complementary to the original template DNA. Each fragment is tagged with a fluorescent dye-labeled nucleotide. Running for 25 cycles, it creates 25×10^{11} single-stranded DNA molecules.

4. Centrifuge to create DNA pellet

The sample is chemically precipitated and centrifuged to settle the DNA fragments as a solid pellet at the bottom of the tube. Unused nucleotides, still in the liquid, are discarded.

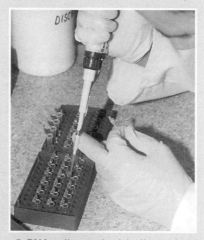

5. DNA pellet washed, buffer added

The pellet is washed with ethanol, dried, and a gel loading buffer is added. All that remains now is single stranded DNA with one dye-labeled nucleotide at the end of each molecule.

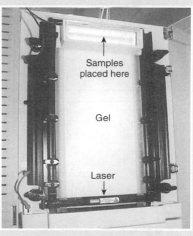

Samples placed here

Gel

Laser

6. Acrylamide gel is loaded

The DNA sequencer is prepared by placing the gel (sandwiched between two sheets of glass) into position. A 36 channel 'comb' for receiving the samples is placed at the top of the gel.

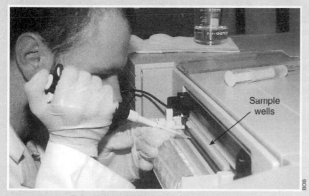

Sample wells

BOB

7. Loading DNA samples onto gel

Different samples can be placed in each of the 36 wells (funnel shaped receptacles) above the gel. A control DNA sample of known sequence is applied to the first lane of the sequencer. If there are problems with the control sequence then results for all other lanes are considered invalid.

8. Running the DNA sequencer

Powerful computer software controls the activity of the DNA sequencer. The gel is left to run for up to 10 hours. During this time an argon laser is constantly scanning across the bottom of the gel to detect the passing of dye-labeled nucleotides attached to DNA fragments.

Related activities: Gel Electrophoresis, The Human Genome Project
Web links: Manual DNA Sequencing

How a DNA Sequencer Operates

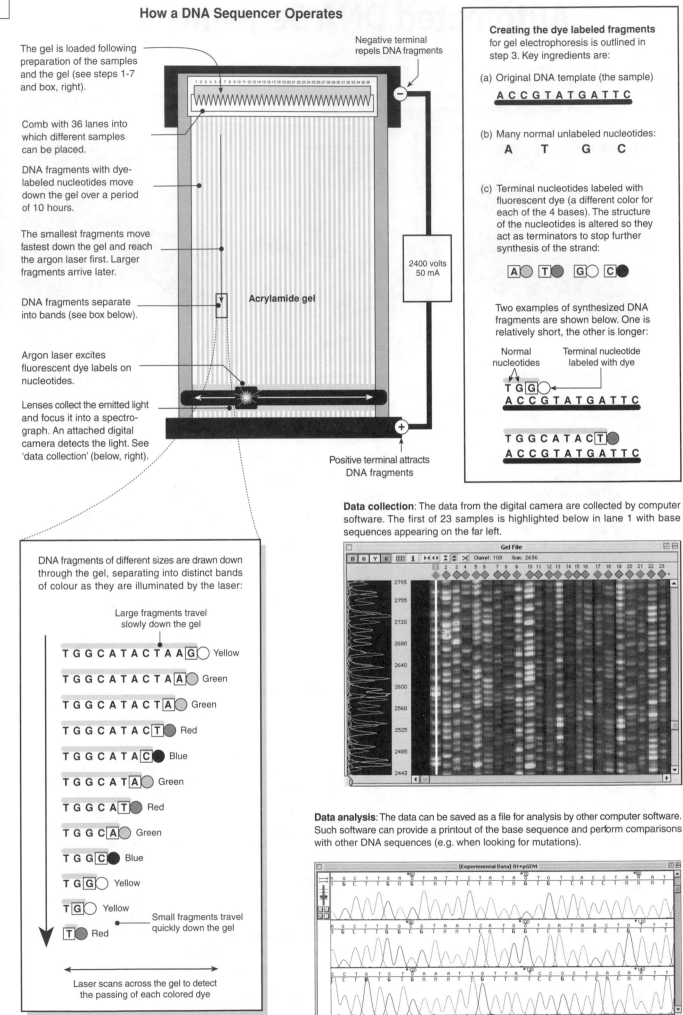

The gel is loaded following preparation of the samples and the gel (see steps 1-7 and box, right).

Comb with 36 lanes into which different samples can be placed.

DNA fragments with dye-labeled nucleotides move down the gel over a period of 10 hours.

The smallest fragments move fastest down the gel and reach the argon laser first. Larger fragments arrive later.

DNA fragments separate into bands (see box below).

Argon laser excites fluorescent dye labels on nucleotides.

Lenses collect the emitted light and focus it into a spectrograph. An attached digital camera detects the light. See 'data collection' (below, right).

Negative terminal repels DNA fragments

2400 volts 50 mA

Acrylamide gel

Positive terminal attracts DNA fragments

Creating the dye labeled fragments for gel electrophoresis is outlined in step 3. Key ingredients are:

(a) Original DNA template (the sample)

A C C G T A T G A T T C

(b) Many normal unlabeled nucleotides:

A   T   G   C

(c) Terminal nucleotides labeled with fluorescent dye (a different color for each of the 4 bases). The structure of the nucleotides is altered so they act as terminators to stop further synthesis of the strand:

A○ T● G○ C●

Two examples of synthesized DNA fragments are shown below. One is relatively short, the other is longer:

Normal nucleotides

Terminal nucleotide labeled with dye

T G G ○
A C C G T A T G A T T C

T G G C A T A C T ●
A C C G T A T G A T T C

Data collection: The data from the digital camera are collected by computer software. The first of 23 samples is highlighted below in lane 1 with base sequences appearing on the far left.

Gel File

B G Y R Channel: 108 Scan: 2636

2793
2755
2720
2680
2640
2600
2560
2525
2485
2443

Data analysis: The data can be saved as a file for analysis by other computer software. Such software can provide a printout of the base sequence and perform comparisons with other DNA sequences (e.g. when looking for mutations).

(Experimental Data) 01•pGEM

DNA fragments of different sizes are drawn down through the gel, separating into distinct bands of colour as they are illuminated by the laser:

Large fragments travel slowly down the gel

T G G C A T A C T A A G ○ Yellow
T G G C A T A C T A A ● Green
T G G C A T A C T A ○ Green
T G G C A T A C T ● Red
T G G C A T A C ● Blue
T G G C A T A ○ Green
T G G C A T ● Red
T G G C A ○ Green
T G G C ● Blue
T G G ○ Yellow
T G ○ Yellow
T ● Red

Small fragments travel quickly down the gel

Laser scans across the gel to detect the passing of each colored dye

Chymosin Production

Nucleic Acid Technology

The Issue

► **Chymosin** (also known as **rennin**) is an enzyme that digests milk proteins. It is the active ingredient in rennet, a substance used by cheesemakers to clot milk into curds.

► Traditionally rennin is extracted from "chyme", i.e. the stomach secretions of suckling calves (hence its name of chymosin).

► By the 1960s, a shortage of chymosin was limiting the volume of cheese produced.

► Enzymes from fungi were used as an alternative but were unsuitable because they caused variations in the cheese flavour.

Concept 1	Concept 2	Concept 3	Concept 4
Enzymes are proteins made up of amino acids. The amino acid sequence of chymosin can be determined and the mRNA coding sequence for its translation identified.	**Reverse transcriptase** can be used to synthesize a DNA strand from the mRNA. This process produces DNA without the introns, which cannot be processed by bacteria.	DNA can be cut at specific sites using **restriction enzymes** and rejoined using **DNA ligase.** New genes can be inserted into self-replicating bacterial **plasmids**.	Under certain conditions, bacteria are able to lose or take up plasmids from their environment. Bacteria are readily grown in vat cultures at little expense.

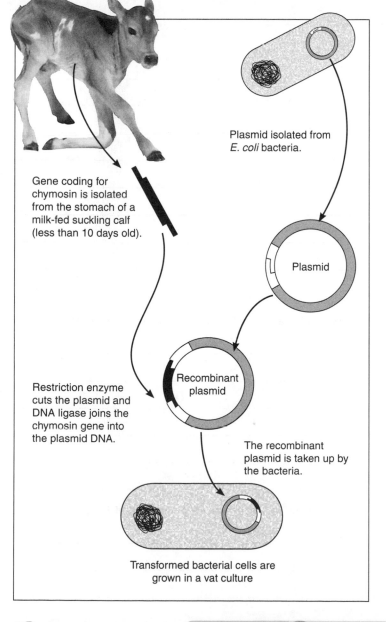

Gene coding for chymosin is isolated from the stomach of a milk-fed suckling calf (less than 10 days old).

Plasmid isolated from *E. coli* bacteria.

Plasmid

Restriction enzyme cuts the plasmid and DNA ligase joins the chymosin gene into the plasmid DNA.

Recombinant plasmid

The recombinant plasmid is taken up by the bacteria.

Transformed bacterial cells are grown in a vat culture

Techniques

The amino acid sequence of chymosin is first determined and the RNA codons for each amino acid identified.

mRNA matching the identified sequence is isolated from the stomach of young calves. **Reverse transcriptase** is used to transcribe mRNA into DNA. The DNA sequence can also be made synthetically once the sequence is determined.

The DNA is amplified using PCR.

Plasmids from *E. coli* bacteria are isolated and cut using **restriction enzymes.** The DNA sequence for chymosin is inserted using **DNA ligase**.

Plasmids are returned to *E. coli* by placing the bacteria under conditions that induce them to take up plasmids.

Outcomes

The transformed bacteria are grown in vat culture. Chymosin is produced by *E. coli* in packets within the cell that are separated during the processing and refining stage.

Recombinant chymosin entered the marketplace in 1990. It established a significant market share because cheesemakers found it to be cost effective, of high quality, and in consistent supply. Most cheese is now produced using recombinant chymosin such as CHY-MAX.

Further Applications

A large amount of processing is required to extract chymosin from *E.coli*. There are now a number of alternative bacteria and fungi that have been engineered to produce the enzyme. Most chymosin is now produced using the fungi *Aspergillus niger* and *Kluyveromyces lactis*. Both are produced in a similar way as that described for *E. coli*.

Examples of Food industry Enzymes Produced by GMOs		
Enzyme	Application	Genetically modified organism
Acetolactate decarboxylase	Acceleration of beer maturation	*Bacillus* spp. *(bacteria)*
Maltogenic alpha amylase	Anti-staling in bread	*Bacillus subtilis (bacteria)*
Xylanase	Improvement of dough, crumb structure and volume in bread	*Aspergillus oryzae (fungi)*
Lipase	Processing of palm oil to produce cocoa butter substitutes	*Aspergillus oryzae (fungi)*
Hemicellulases	Improvement of dough, crumb structure and volume in bread	*Bacillus subtilis (bacteria)*
Cyclomaltodextrin glycosytransferase	Development of flavor and aroma in foodstuffs	*Bacillus spp (bacteria)*

1. Describe the main use of chymosin: _____

_____

2. Describe the traditional source of chymosin: _____

_____

3. Summarize the key concepts that led to the development of the technique for producing chymosin:

   (a) Concept 1: _____

   (b) Concept 2: _____

   (c) Concept 3: _____

   (d) Concept 4: _____

4. Discuss how the gene for chymosin was isolated and how the technique could be applied to isolating other genes:

_____

_____

_____

_____

_____

_____

_____

5. Describe three advantages of using chymosin produced by GE bacteria over chymosin from traditional sources:

   (a) _____

   _____

   (b) _____

   _____

   (c) _____

   _____

6. Explain why the fungus *Aspergillus niger* is now more commonly used to produce chymosin instead of *E. coli*:

_____

_____

_____

Golden Rice

The Issue

► **Beta-carotene** (β-carotene) is a precursor to **vitamin A** which is involved in many functions including vision, immunity, foetal development, and skin health.

► Vitamin A deficiency is common in developing countries where up to 500,000 children suffer from night blindness, and death rates due to infections are high due to a lowered immune response.

► Providing enough food containing useful quantities of β-carotene is difficult and expensive in many countries.

Concept 1

Rice is a staple food in many developing countries. It is grown in large quantities and is available to most of the population, but it lacks many of the essential nutrients required by the human body for healthy development. It is low in β-carotene.

Concept 2

Rice plants produce β-carotene but not in the edible rice **endosperm**. Engineering a new biosynthetic pathway would allow β-carotene to be produced in the endosperm. Genes expressing enzymes for carotene synthesis can be inserted into the rice genome.

Concept 3

The enzyme **carotene desaturase (CRT1)** in the soil bacterium *Erwinia uredovora*, catalyses multiple steps in carotenoid biosynthesis. **Phytoene synthase (PSY)** overexpresses a colourless carotene in the daffodil plant *Narcissus pseudonarcissus.*

Concept 4

DNA can be inserted into an organism's genome using a suitable **vector**. *Agrobacterium tumefaciens* is a tumor-forming bacterial plant pathogen that is commonly used to insert novel DNA into plants.

The Development of Golden Rice

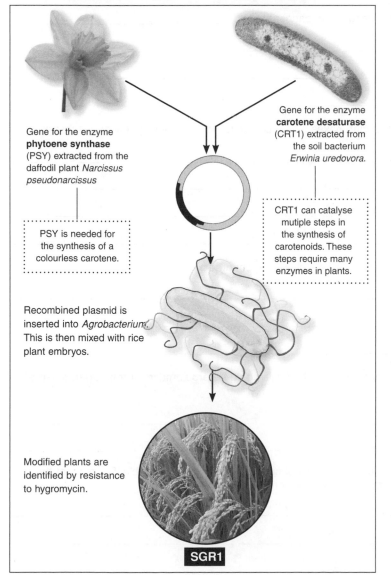

Gene for the enzyme **phytoene synthase** (PSY) extracted from the daffodil plant *Narcissus pseudonarcissus*

Gene for the enzyme **carotene desaturase** (CRT1) extracted from the soil bacterium *Erwinia uredovora*.

PSY is needed for the synthesis of a colourless carotene.

CRT1 can catalyse mutiple steps in the synthesis of carotenoids. These steps require many enzymes in plants.

Recombined plasmid is inserted into *Agrobacterium*. This is then mixed with rice plant embryos.

Modified plants are identified by resistance to hygromycin.

SGR1

Techniques

The **PSY** gene from daffodils and the **CRT1** gene from *Erwinia uredovora* are sequenced.

DNA sequences are synthesized into packages containing the CRT1 or PSY gene, terminator sequences, and **endosperm specific promoters** (these ensure expression of the gene only in the edible portion of the rice).

The *Ti* plasmid from *Agrobacterium* is modified using restriction enzymes and DNA ligase to delete the tumor-forming gene and insert the synthesized DNA packages. A gene for resistance to the antibiotic **hygromycin** is also inserted so that transformed plants can be identified later. The parts of the *Ti* plasmid required for plant transformation are retained.

Modified *Ti* plasmid is inserted into the bacterium.

Agrobacterium is incubated with rice plant embryo. Transformed embryos are identified by their resistance to hygromycin.

Outcomes

The rice produced had endosperm with a distinctive yellow colour. Under greenhouse conditions golden rice (**SGR1**) contained 1.6 µg per g of carotenoids. Levels up to five time higher were produced in the field, probably due to improved growing conditions.

Further Applications

Further research on the action of the PSY gene identified more efficient methods for the production of β-carotene. The second generation of golden rice now contains up to 37 µg per g of carotenoids. Golden rice was the first instance where a complete biosynthetic pathway was engineered. The procedures could be applied to other food plants to increase their nutrient levels.

Periodicals: Rice, risk, and regulations, The engineering of crop plants

Related activities: Restriction Enzymes, Ligation, In Vivo Gene Cloning

All photos USDA

The ability of *Agrobacterium* to transfer genes to plants is exploited for crop improvement. The tumour-inducing *Ti* plasmid is modified to delete the tumour-forming gene and insert a gene coding for a desirable trait. The parts of the *Ti* plasmid required for plant transformation are retained.

Soybeans are one of the many food crops that have been genetically modified for broad spectrum herbicide resistance. The first GM soybeans were planted in the US in 1996. By 2007, nearly 60% of the global soybean crop was genetically modified; the highest of any other crop plant.

GM cotton was produced by inserting the gene for the BT toxin into its genome. The bacterium *Bacillus thuringiensis* naturally produces BT toxin, which is harmful to a range of insects, including the larvae that eat cotton. The BT gene causes cotton to produce this insecticide in its tissues.

1. Describe the basic methodology used to create golden rice: _____

_____

_____

_____

_____

2. Explain how scientists ensured β-carotene was produced in the endosperm: _____

_____

_____

_____

3. Describe the property of *Agrobacterium tumefaciens* that makes it an ideal vector for introducing new genes into plants:

_____

_____

_____

_____

4. (a) Explain how this new variety of rice could reduce disease in developing countries:

_____

_____

_____

(b) Absorption of vitamin A requires sufficient dietary fat. Explain how this could be problematic for the targeted use of golden rice in developing countries:

_____

_____

5. As well as increasing nutrient content as in golden rice, other traits of crop plants are also desirable. For each of the following traits, suggest features that could be desirable in terms of increasing yield:

(a) Grain size or number: _____

(b) Maturation rate: _____

(c) Pest resistance: _____

Production of Insulin

Insulin B chain

Insulin A chain

The Issue

▶ **Type 1 diabetes mellitus** is a metabolic disease caused by a lack of **insulin**. Around 25 people in every 100,000 suffer from type 1 diabetes.

▶ It is treatable only with injections of insulin.

▶ In the past, insulin was taken from the pancreases of cows and pigs and purified for human use. The method was expensive and some patients had severe allergic reactions to the foreign insulin or its contaminants.

Concept 1

DNA can be cut at specific sites using **restriction enzymes** and joined together using **DNA ligase**. New genes can be inserted into self-replicating bacterial **plasmids** at the point where the cuts are made.

Concept 2

Plasmids are circular pieces of DNA found in some bacteria. They usually carry genes useful to the bacterium. In the bacterium *E. coli*, the gene controlling gene transcription is carried on a plasmid.

Concept 3

Under certain conditions, Bacteria are able to lose or pick up plasmids from their environment. Bacteria can be readily grown in vat cultures at little expense.

Concept 4

The DNA sequences coding for the production of the two polypeptide chains (A and B) that form human insulin can be isolated from the human genome.

The nucleotide sequences for each insulin chain are synthesized separately and placed into separate plasmids

The recombinant plasmids are introduced into the bacterial cells

β-galactosidase + chain A

The gene is expressed as separate chains

β-galactosidase + chain B

Insulin A chain

Disulfide bond

Insulin B chain

Techniques

The **gene** is **chemically synthesized** as two nucleotide sequences, one for the **insulin A chain** and one for the **insulin B chain**. The two sequences are small enough to be inserted into a plasmid.

Plasmids are extracted from *Escherichia coli*. The gene for the bacterial enzyme **β-galactosidase**, which controls the transcription of genes, is located on the plasmid. To make the bacteria produce insulin, the insulin gene must be linked to the **β-galactosidase** gene.

Restriction enzymes are used to cut plasmids at the appropriate site and the A and B insulin sequences are inserted. The sequences are joined with the plasmid DNA using **DNA ligase**.

The **recombinant plasmids** are inserted back into the bacteria by placing them together in a culture that favours plasmid uptake by bacteria.

The bacteria are then grown and multiplied in vats under carefully controlled growth conditions.

Outcomes

The product consists partly of β-galactosidase, joined with either the A or B chain of insulin. The chains are extracted, purified, and mixed together. The A and B insulin chains connect via **disulfide cross linkages** to form the functional insulin protein. The insulin can then be made ready for injection in various formulations.

Further Applications

The techniques involved in producing human insulin from genetically modified bacteria can be applied to a range human proteins and hormones. Proteins currently being produced include human growth hormone, interferon, and factor VIII.

Related activities: Restriction Enzymes, Ligation, Chymosin Production

A 2

Insulin production in *Saccharomyces*

Yeast cells are **eukaryotic** and hence are much larger than bacterial cells. This enables them to accommodate much larger plasmids and proteins within them.

The gene for human insulin is inserted into a plasmid. The yeast plasmid is larger than that of *E.coli*, so the entire gene can be inserted in one piece rather than as two separate pieces.

Cleavage site

The **proinsulin** protein that is produced folds into a specific shape and is cleaved by the yeast's own cellular enzymes, producing the completed insulin chain.

By producing insulin this way, the secondary step of combining the separate protein chains is eliminated, making the refining process much simpler.

Cleavage site

1. Describe the three major problems associated with the traditional method of obtaining insulin to treat diabetes:

   (a) _____

   (b) _____

   (c) _____

2. Explain the reasoning behind using *E. coli* to produce insulin and the benefits that GM technology has brought to diabetics:

   _____

   _____

   _____

   _____

3. Explain why, when using *E. coli*, the insulin gene is synthesized as two separate A and B chain nucleotide sequences:

   _____

   _____

4. Explain why the synthetic nucleotide sequences ('genes') are 'tied' to the β-galactosidase gene: _____

   _____

   _____

5. Yeast (*Saccharomyces cerevisiae*) is also used in the production of human insulin. Discuss the differences in the production of insulin using yeast and *E. coli* with respect to:

   (a) Insertion of the gene into the plasmid: _____

   _____

   _____

   (b) Secretion and purification of the protein product: _____

   _____

   _____

   _____

Gene Therapy

Gene therapy refers to the application of gene technology to treat disease by correcting or replacing faulty genes. It was first envisioned as a treatment, or even a cure, for genetic disorders, but it could also be used to treat a wide range of diseases, including those that resist conventional treatments. Although varying in detail, all gene therapies are based around the same technique. Normal (non-faulty) DNA containing the correct gene is inserted into a vector, which is able to transfer the DNA into the patient's cells in a process called **transfection**. The vector may be a virus, liposome, or any one of a variety of other molecular transporters. The vector is introduced into a sample of the patient's cells and

these are cultured to amplify the correct gene. The cultured cells are then transferred back to the patient. The use of altered stem cells instead of mature somatic cells has so far achieved longer lasting results in many patients. The treatment of somatic cells or stem cells may be **therapeutic** but the changes are not inherited. **Germline therapy** would enable genetic changes to be passed on. To date there have been limited successes with gene therapy because transfection of targeted cells is inefficient and side effects can be severe or even fatal. However there have been a small number of successes, including the treatment of one form of SCID, a genetic disease affecting the immune system.

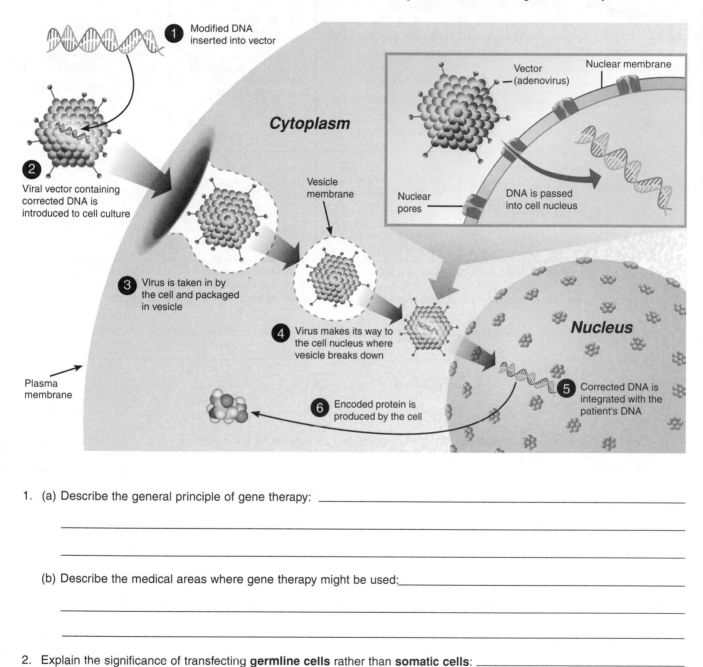

1. (a) Describe the general principle of gene therapy: _____

_____

_____

(b) Describe the medical areas where gene therapy might be used: _____

_____

_____

2. Explain the significance of transfecting **germline cells** rather than **somatic cells**: _____

_____

_____

3. Explain the purpose of **gene amplification** in gene therapy: _____

_____

_____

_____

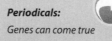
© Biozone International 2001-2010
Photocopying Prohibited

Periodicals:
Genes can come true

Related activities: Gene Delivery Systems
Web links: Gene Therapy, Gene Therapy Primer

RA 2

Vectors That Can Be Used For Gene Therapy

	Retrovirus	Adenovirus	Liposome	Naked DNA
Insert size:	8000 bases	8000 bases	>20 000 bases	>20 000 bases
Integration:	Yes	No	No	No
***In vivo* delivery:**	Poor	High	Variable	Poor
Advantages	• Integrate genes into the chromosomes of the human host cell. • Offers chance for long-term stability.	• Modified for gene therapy, they infect human cells and express the normal gene. • Most do not cause disease. • Have a large capacity to carry foreign genes.	• Liposomes seek out target cells using sugars in their membranes that are recognised by cell receptors. • Have no viral genes that may cause disease.	• Have no viral genes that may cause disease. • Expected to be useful for vaccination.
Disadvantages	• Many infect only cells that are dividing. • Genes integrate randomly into chromosomes, so might disrupt useful genes in the host cell.	• Viruses may have poor survival due to attack by the host's immune system. • Genes may function only sporadically because they are not integrated into host cell's chromosome.	• Less efficient than viruses at transferring genes into cells, but recent work on using sugars to aid targeting have improved success rate.	• Unstable in most tissues of the body. • Inefficient at gene transfer.

In the table above, the following terms are defined as follows: **Naked DNA**: the genes are applied by ballistic injection (firing using a gene gun) or by regular hypodermic injection of plasmid DNA. **Insert size**: size of gene that can be inserted into the vector. **Integration**: whether or not the gene is integrated into the host DNA (chromosomes). ***In vivo* delivery**: ability to transfer a gene directly into a patient.

4. Explain why genetically modified stem cells offer greater potential in gene therapy treatments than GM somatic cells:

_____

_____

_____

5. (a) Describe the features of viruses that make them well suited as **vectors** for gene therapy: _____

_____

_____

(b) Describe two problems with using viral vectors for gene therapy: _____

_____

_____

6. (a) Explain why it may be beneficial for a (therapeutic) gene to integrate into the patient's chromosome:

_____

_____

(b) Explain why this has the potential to cause problems for the patient: _____

_____

7. (a) Explain why naked DNA is likely to be unstable within a patient's tissues: _____

_____

(b) Explain why enclosing the DNA within liposomes might provide greater stability: _____

_____

Gene Delivery Systems

The mapping of the human genome has improved the feasibility of gene therapy as a option for treating an increasingly wide range of diseases, but it remains technically difficult to deliver genes successfully to a patient. Even after a gene has been identified, cloned, and transferred to a patient, it must be expressed normally. To date, the success of gene therapy has been generally poor, and improvements have been short-lived or counteracted by adverse side effects. Inserted genes may reach only about 1% of target cells and those that reach their destination may work inefficiently and produce too little protein, too slowly to be of benefit. In addition, many patients react immunologically to the vectors used in gene transfer. Much of the current research is focussed on improving the efficiency of gene transfer and expression. One of the first gene therapy trials was for **cystic fibrosis** (CF). CF was an obvious candidate for gene therapy because, in most cases, the disease is caused by a single, known gene mutation. However, despite its early promise, gene therapy for this disease has been disappointing (below right). Another candidate for gene therapy, again because the disease is caused by single, known mutations, is Severe Combined Immune Deficiency (SCID) (below left). Gene therapies developing for this disease have so far proved promising.

Nucleic Acid Technology

Treating SCID using Gene Therapy

The most common form of **SCID** (Severe Combined Immune Deficiency) is **X-linked SCID**, which results from mutations to a gene on the X chromosome encoding the **common gamma chain**, a protein forming part of a receptor complex for numerous types of leukocytes. A less common form of the disease, (**ADA-SCID**) is caused by a defective gene that codes for the enzyme adenosine deaminase (ADA).

Both of these types of SCID lead to immune system failure. A common treatment for SCID is bone marrow transplant, but this is not always successful and runs the risks of infection from unscreened viruses. **Gene therapy** appears to hold the best chances of producing a cure for SCID because the mutation affects only one gene whose location is known. DNA containing the corrected gene is placed into a **gutted retrovirus** and introduced to a sample of the patient's **bone marrow.** The treated cells are then returned to the patient.

In some patients with ADA-SCID, treatment was so successful that supplementation with purified ADA was no longer required. The treatment carries risks though. In early trials, two of ten treated patients developed leukemia when the corrected gene was inserted next to a gene regulating cell growth.

Samples of bone marrow being extracted prior to treatment with gene therapy.

Georgetown University Hospital

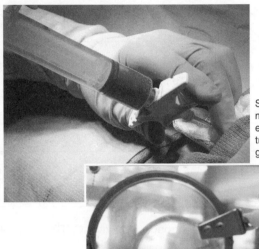

Jacoplane

Detection of SCID is difficult for the first months of an infant's life due to the mother's antibodies being present in the blood. Suspected SCID patients must be kept in sterile conditions at all times to avoid infection.

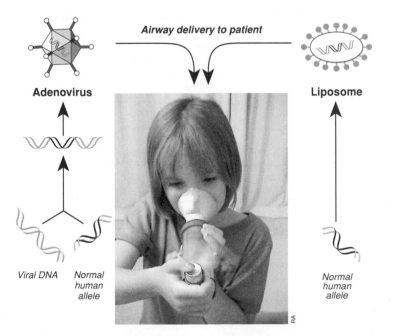

Airway delivery to patient

Adenovirus

Liposome

Viral DNA *Normal human allele*

Normal human allele

RA

An **adenovirus** that normally causes colds is genetically modified to make it safe and to carry the normal (unmutated) CFTR ('cystic fibrosis') gene.

Liposomes are tiny fat globules. Normal CF genes are enclosed in liposomes, which fuse with plasma membranes and deliver the genes into the cells.

Gene Therapy - Potential Treatment for Cystic Fibrosis?

Cystic fibrosis (CF) is caused by a mutation to the gene coding for a chloride ion channel important in creating sweat, digestive juices, and mucus. The dysfunction results in abnormally thick, sticky mucus that accumulates in the lungs and intestines. The identification and isolation of the CF gene in 1989 meant that scientists could look for ways in which to correct the genetic defect rather than just treating the symptoms using traditional therapies.

The main target of CF gene therapy is the lung, because the progressive lung damage associated with the disease is eventually lethal.

In trials, normal genes were isolated and inserted into patients using vectors such as **adenoviruses** and **liposomes**, delivered via the airways (left). The results of trials were disappointing: on average, there was only a 25% correction, the effects were short lived, and the benefits were quickly reversed. Alarmingly, the adenovirus used in one of the trials led to the death of one patient.

Source: Cystic Fibrosis Trust, UK.

Related activities: Gene Therapy, Cystic Fibrosis Mutation

A 3

1. A great deal of current research is being devoted to discovering a gene therapy solution to treat **cystic fibrosis** (CF):

    (a) Describe the symptoms of CF: _____

    _____

    _____

    (b) Explain why this genetic disease has been so eagerly targeted by gene therapy researchers: _____

    _____

    _____

    _____

    _____

    (c) Outline some of the problems so far encountered with gene therapy for CF: _____

    _____

    _____

    _____

    _____

    _____

2. Identify two vectors for introducing healthy CFTR genes into CF patients.

    (a) Vector 1: _____

    (b) Vector 2: _____

3. (a) Describe the difference between X-linked SCID and ADA-SCID: _____

    _____

    (b) Identify the vector used in the treatment of SCID: _____

    _____

4. Briefly outline the differences in the gene therapy treatment of CF and SCID:_____

    _____

    _____

    _____

    _____

5. Changes made to chromosomes as a result of gene therapy involving somatic cells are not inherited. Germ-line gene therapy has the potential to cure disease, but the risks and benefits are still not clear. For each of the points outlined below, evaluate the risk of germ-line gene therapy relative to somatic cell gene therapy and explain your answer:

    (a) Chance of interfering with an essential gene function: _____

    _____

    _____

    _____

    (b) Misuse of the therapy to selectively alter phenotype: _____

    _____

    _____

    _____

Investigating Genetic Diversity

PCR and **DNA sequencing** can be used in assessing **genetic biodiversity**. For conservationists, large amounts of genetic variation within a species may indicate a greater ability to adapt to environmental change (e.g. climate shifts). The amount of variation between populations of a species is of particular interest. Sometimes the genetic variation found between populations is enough to warrant separating them into two or more 'morphologically cryptic' species (containing populations that are identical in appearance, but different genetically). **Springtails** are abundant arthropods, closely related to insects, which live in soil throughout the world. One particular species, *Gomphiocephalus hodgsoni*, is the largest year-round inhabitant of the Antarctic continent. It is being studied in an area of

Antarctica known as the Dry Valleys, particularly in Taylor Valley. This region is largely ice-free, and the springtails survive in moist habitats such as at the edges of lakes and glacial streams. Springtails collected throughout Taylor Valley appear to be morphologically identical. However, after DNA analysis of a gene from springtail **mitochondrial DNA**, significant genetic biodiversity has been found between populations. This may indicate the presence of more than one species. As climate change and the presence of humans affect the habitat of Taylor Valley over time, it is important to understand and monitor the genetic structure of the springtail populations in order to ensure that biodiversity is conserved.

Nucleic Acid Technology

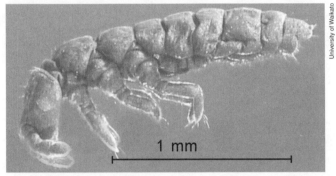

University of Waikato

The springtail *Gomphiocephalus hodgsoni* (above) is a small arthropod, just over 1 mm long. Liam Nolan investigated the genetic relatedness of populations in and around Taylor Valley in Antarctica.

Leo Sanchez

Canada Glacier

Stream

Taylor Valley, one of the Dry Valleys in Antarctica, is clear of snow much of the year. The ephemeral stream is ideal springtail habitat.

The Process of DNA Analysis of Springtails is Illustated Below

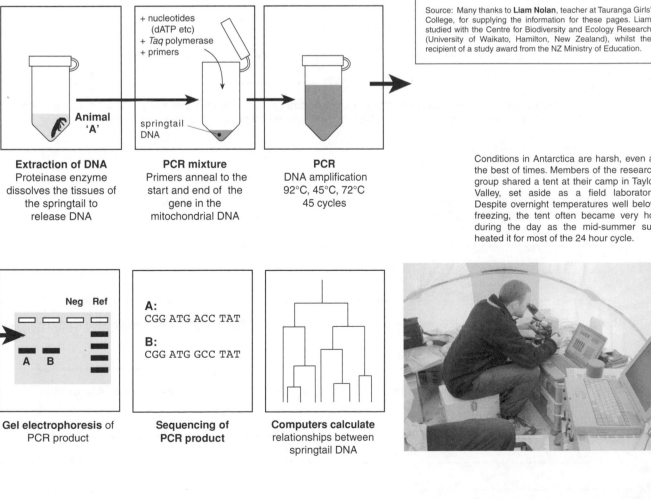

+ nucleotides (dATP etc)
+ *Taq* polymerase
+ primers

Animal 'A'

springtail DNA

Extraction of DNA
Proteinase enzyme dissolves the tissues of the springtail to release DNA

PCR mixture
Primers anneal to the start and end of the gene in the mitochondrial DNA

PCR
DNA amplification 92°C, 45°C, 72°C 45 cycles

Source: Many thanks to **Liam Nolan**, teacher at Tauranga Girls' College, for supplying the information for these pages. Liam studied with the Centre for Biodiversity and Ecology Research (University of Waikato, Hamilton, New Zealand), whilst the recipient of a study award from the NZ Ministry of Education.

Conditions in Antarctica are harsh, even at the best of times. Members of the research group shared a tent at their camp in Taylor Valley, set aside as a field laboratory. Despite overnight temperatures well below freezing, the tent often became very hot during the day as the mid-summer sun heated it for most of the 24 hour cycle.

Neg   Ref

A:
CGG ATG ACC TAT

B:
CGG ATG GCC TAT

A   B

Burkhard Budel

Gel electrophoresis of PCR product

Sequencing of PCR product

Computers calculate relationships between springtail DNA

Related activities: Gel Electrophoresis, Polymerase Chain Reaction, Manual DNA Sequencing

RA 3

Expeditions in Antarctica are fraught with logistical problems. Apart from making sure that scientific equipment functions properly in freezing temperatures, just getting the equipment from Scott Base to the field station is difficult, requiring the use of helicopters.

Scott Base is the center for research scientists from around the orld. They work mostly during the summer months, when there is perpetual daylight. There are facilities for carrying out some lab work, as well as recreational facilities for the expedition members.

1. Explain why a **proteinase** enzyme is helpful in the extraction of DNA from springtails:

_____

_____

2. (a) Describe the function of *Taq* polymerase in **PCR**: _____

_____

(b) Explain why nucleotides are added to the PCR mixture: _____

_____

(c) Explain the effect of different temperatures (used in PCR) on the DNA and primers:

_____

_____

_____

3. (a) The **electrophoresis gel** is also loaded with a known '**negative**'; a substance that will produce a definite negative result, for comparison with samples A and B. Describe what would be put into the negative well:

_____

_____

(b) A **reference**, which contains a mixture of DNA segments of known length, is also loaded onto the gel. Explain why such fragments are added into the reference well:

_____

_____

_____

4. (a) The given DNA sequences (on the previous page) are taken from two different individuals. Describe the kind of mutation observed:

_____

(b) Mutations are most frequently found at the third base of a codon. Discuss the significance of this mutation in the springtails' DNA sequence:

_____

_____

_____

The Human Genome Project

The **Human Genome Project** (HGP) is a publicly funded venture involving many different organizations throughout the world. In 1998, Celera Genomics in the USA began a competing project, as a commercial venture, in a race to be the first to determine the human genome sequence. In 2000, both organizations reached the first draft stage, and the entire genome is now available as a high quality (golden standard) sequence. In addition to determining the order of bases in the human genome, genes are being identified, sequenced, and mapped (their specific chromosomal location identified). The next challenge is to assign functions to the identified genes. By identifying and studying the protein products of genes (a field known as **proteomics**),

scientists can develop a better understanding of genetic disorders. Long term benefits of the HGP are both medical and non-medical (see next page). Many biotechnology companies have taken out patents on gene sequences. This practice is controversial because it restricts the use of the sequence information to the patent holders. Other genome sequencing projects have arisen as a result of the initiative to sequence the human one. In 2002 the International HapMap Project was started with the aim of developing a haplotype map (HapMap) of the human genome. Initially data was gathered from four populations with African, Asian and European ancestry and additional populations may be included as analysis of human genetic variation continues.

Nucleic Acid Technology

Gene Mapping

This process involves determining the precise position of a gene on a chromosome. Once the position is known, it can be shown on a diagram.

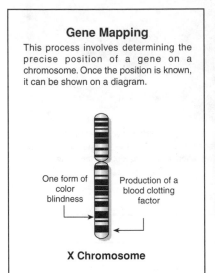

One form of color blindness

Production of a blood clotting factor

X Chromosome

Equipment used for DNA Sequencing

Banks of PCR machines prepare DNA for the sequencing gel stage. The DNA is amplified and chemically tagged (to make the DNA fluoresce and enable visualization on a gel).

Genesis Research and Development Corp, Auckland

Banks of DNA sequencing gels and powerful computers are used to determine the base order in DNA.

HGSI

Count of Mapped Genes

The length and number of mapped genes to date for each chromosome are tabulated below. The entire human genome contains approximately 20 000-25 000 genes.

Chromosome	Length (Mb)	No. of Mapped Genes
1	263	1873
2	255	1113
3	214	965
4	203	614
5	194	782
6	183	1217
7	171	995
8	155	591
9	145	804
10	144	872
11	144	1162
12	143	894
13	114	290
14	109	1013
15	106	510
16	98	658
17	92	1034
18	85	302
19	67	1129
20	72	599
21	50	386
22	56	501
X	164	1021
Y	59	122
Total:		**19 447**

Data to March 2008 from gdb.org (now offline)

Examples of Mapped Genes

The positions of an increasing number of genes have been mapped onto human chromosomes (see below). Sequence variations can cause or contribute to identifiable disorders. Note that chromosome 21 (the smallest human chromosome) has a relatively low gene density, while others are gene rich. This is possibly why trisomy 21 (Down syndrome) is one of the few viable human autosomal trisomies.

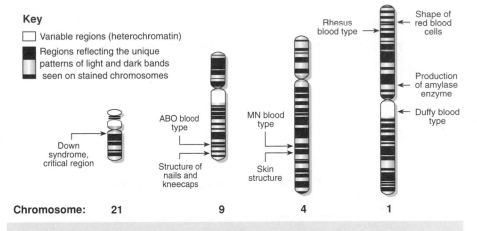

Key

☐ Variable regions (heterochromatin)

▨ Regions reflecting the unique patterns of light and dark bands seen on stained chromosomes

Down syndrome, critical region

ABO blood type

Structure of nails and kneecaps

MN blood type

Skin structure

Rhesus blood type

Shape of red blood cells

Production of amylase enzyme

Duffy blood type

Chromosome: 21   9   4   1

The aim of the HGP was to produce a continuous block of sequence information for each chromosome. Initially the sequence information was obtained to draft quality, with an error rate of 1 in 1000 bases. The **Gold Standard sequence**, with an error rate of <1 per 100 000 bases, was completed in October 2004. Key results of the research are:

- The analysis suggests that there are perhaps only 20 000-25 000 protein-coding genes in our human genome.
- The number of gaps has been reduced 400-fold to only 341
- It covers 99% of the gene containing parts of the genome and is 99.999% accurate.
- The new sequence correctly identifies almost all known genes (99.74%).
- Its accuracy and completeness allows systematic searches for causes of disease.

Periodicals:
Genes, the genome, & disease,
Understanding the HGP

Related activities: Genome Projects

RA 2

Benefits and ethical issues arising from the Human Genome Project

Medical benefits

- Improved **diagnosis** of disease and predisposition to disease by genetic testing.
- Better identification of disease carriers, through genetic testing.
- Better **drugs** can be designed using knowledge of protein structure (from gene sequence information) rather than by trial and error.
- Greater possibility of successfully using **gene therapy** to correct genetic disorders.

Non-medical benefits

- Greater knowledge of **family relationships** through genetic testing, e.g. paternity testing in family courts.
- Advances **forensic science** through analysis of DNA at crime scenes.
- Improved knowledge of the evolutionary relationships between humans and other organisms, which will help to develop better, more accurate classification systems.

Possible ethical issues

- It is unclear whether third parties, e.g. health insurers, have rights to genetic test results.
- If treatment is unavailable for a disease, genetic knowledge about it may have no use.
- Genetic tests are costly, and there is no easy answer as to who should pay for them.
- Genetic information is hereditary so knowledge of an individual's own genome has implications for members of their family.

Couples can already have a limited range of genetic tests to determine the risk of having offspring with some disease-causing mutations.

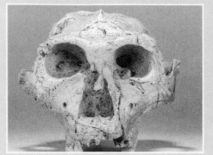

When DNA sequences are available for humans and their ancestors, comparative analysis may provide clues about human evolution.

Legislation is needed to ensure that there is no discrimination on the basis of genetic information, e.g. at work or for health insurance.

1. Briefly describe the objectives of the Human Genome Project (HGP): _____

_____

_____

_____

2. Suggest a reason for developing a HapMap of the human genome: _____

_____

_____

3. Describe two possible **benefits** of Human Genome Project (HGP):

    (a) Medical: _____

    _____

    (b) Non-medical: _____

    _____

4. Explain what is meant by **proteomics** and explain its significance to the HGP and the ongoing benefits arising from it:

_____

_____

_____

_____

5. Suggest two possible points of view for one of the **ethical issues** described in the list above (top right):

    (a) _____

    _____

    (b) _____

    _____

Genome Projects

There are many genome projects underway around the world, including the Human Genome Project. The aim of most genome projects is to determine the DNA sequence of the organism's entire genome. Over one hundred bacterial and viral genomes, as well as a number of larger genomes (including honeybee, nematode worm, African clawed frog, pufferfish, zebra fish, rice, cow, dog, and rat) have already been sequenced. Genomes that are, for a variety of reasons, high priority for DNA sequencing include the sea urchin, kangaroo, pig, cat, baboon, silkworm, rhesus monkey, turkey and even Neanderthals (prehumans). Genome sequencing is costly, so candidates are carefully chosen. Important factors in this choice include the value of the knowledge to practical applications, the degree of technical difficulty involved, and the size of the genome (very large genomes are generally avoided). Genome sizes and the number of genes per genome vary, and are not necessarily correlated with the size and structural complexity of the organism itself. Once completed, genome sequences are analyzed by computer to identify genes.

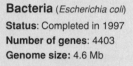

Artist's impression

Yeast (*Saccharomyces cerevisiae*)

Status: Completed in 1996
Number of genes: 6000
Genome size: 13 Mb

The first eukaryotic genome to be completely sequenced. Yeast is used as a model organism to study human cancer.

Bacteria (*Escherichia coli*)

Status: Completed in 1997
Number of genes: 4403
Genome size: 4.6 Mb

E. coli has been used as a laboratory organism for over 70 years. Various strains of *E. coli* are responsible for several human diseases.

Fruit fly (*Drosophila melanogaster*)

Status: Completed in 2000
Number of genes: 14 000
Genome size: 150 Mb

Drosophila has been used extensively for genetic studies for many years. About 50% of all fly proteins show similarities to mammalian proteins.

Mouse (*Mus musculus*)

Status: Completed in 2002
Number of genes: 30 000
Genome size: 2500 Mb

New drugs destined for human use are often tested on mice because more than 90% of their proteins show similarities to human proteins.

Chimpanzee (*Pan troglodytes*)

Status: Draft, Dec. 2003, Completed, Sept. 2005
Genome size: 3000 Mb

Chimp and human genomes differ by <2%. Identifying differences could provide clues to the genetics of diseases such as cancer, to which chimps are less prone.

Banana (*Musa acuminata*)

Status: In progress. Due 2006
Genome size: 500-600 Mb

The first tropical crop to be sequenced. Bananas have high economic importance. Knowledge of the genome will assist in producing disease resistant varieties of banana.

Neanderthal
(*H. neanderthalensis*)

Status: In progress
Genome size: 3000 Mb

This ambitious project is attempting to reconstruct the genome of a Neanderthal. To date, 2% of the genome has been sequenced from fossil remains.

Chicken (*Gallus gallus*)

Status: Completed in Feb. 2004
Genome size: 1200 Mb

Various human viruses were first found in chickens making this species important for the study of human disease and cross-species transfers. It was the first bird genome to be sequenced.

1. Calculate the number of genes per megabase (Mb) of DNA for the organisms above:

   (a) Yeast: _____ (b) *E. coli*: _____ (c) Fruit fly: _____ (d) Mouse: _____

2. Suggest why the number of genes per Mb of DNA varies between organisms (hint: consider relative sizes of introns):

   _____

   _____

3. Suggest why researchers want to sequence the genomes of plants such as wheat, rice, and maize:

   _____

   _____

4. Use a web engine search to find:

   (a) First **multicellular animal genome** to be sequenced: _____ Date: _____

   (b) First **plant genome** to be sequenced: _____ Date: _____

Periodicals:
Genomes for all,
What is genomics?

Related activities: *The Human Genome Project, Genetically Modified Plant*
Web links: *Landmarks in Biotechnology, A Guide to Sequenced Genomes*

RDA 2

Cloning by Embryo Splitting

Livestock breeds frequently produce only one individual per pregnancy and all individuals in a herd will have different traits. Cloning (by embryo splitting or other means) makes it possible to produce high value herds with identical traits more quickly. Developed in the 1980s, and adopted by livestock breeders, embryo splitting, or artificial twinning, is the simplest way in which to create a clone. Embryo splitting simply replicates the natural twinning process. A fertilized egg is grown into eight cells before being split into four individual embryos, each consisting of just two cells. The four genetically identical embryos are then implanted into surrogate mothers. While this technique produces multiple clones, the clones are derived from an embryo whose physical characteristics are not completely known. This represents a serious limitation for practical applications when the purpose of the procedure is to produce high value livestock. In 2000, a rhesus macaque was cloned in this manner, with the goal of producing identical individuals that could be used to perfect new therapies for human disease. Cloning technology can also be used to produce early embryos from which undifferentiated **stem cells** can be isolated for use in tissue and cell engineering.

Livestock are selected for cloning on the basis of desirable qualities such as wool, meat, or milk productivity.

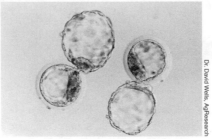

Cloned embryos immediately prior to implantation into a surrogate. These are at the blastocyst stage (a mass of cells that have begun to differentiate).

The individuals produced by embryo splitting have the same characteristics as the parents.

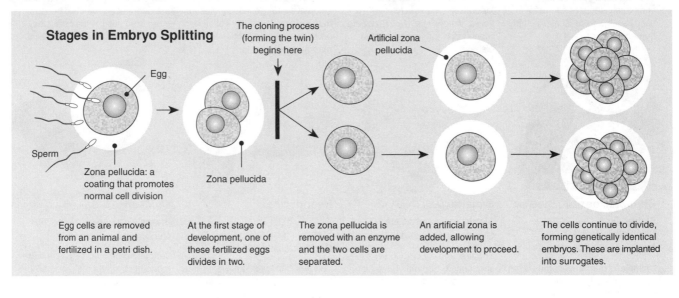

Stages in Embryo Splitting

The cloning process (forming the twin) begins here

Artificial zona pellucida

Egg

Sperm

Zona pellucida: a coating that promotes normal cell division

Zona pellucida

Egg cells are removed from an animal and fertilized in a petri dish.

At the first stage of development, one of these fertilized eggs divides in two.

The zona pellucida is removed with an enzyme and the two cells are separated.

An artificial zona is added, allowing development to proceed.

The cells continue to divide, forming genetically identical embryos. These are implanted into surrogates.

1. Explain how **embryo splitting** differs from adult cloning: _____

_____

_____

2. Briefly describe the possible benefits to be gained from cloning the following:

(a) Stem cells for medical use: _____

_____

_____

(b) High milk yielding cows: _____

_____

_____

3. Suggest one reason why it would be undesirable to produce all livestock using embryo splitting: _____

_____

_____

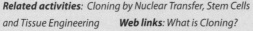
Related activities: Cloning by Nuclear Transfer, Stem Cells and Tissue Engineering **Web links:** *What is Cloning?*

Cloning by Nuclear Transfer

Blackface ewe

Dolly

The Issue

► Individuals vary in characteristics, even within specific breeds of animal, such as sheep.

► Clones remove the variability and produce livestock that would develop in a predictable way or produce a consistent quality of product such as wool or milk.

► Clones produced using traditional embryo-splitting are derived from an embryo whose physical characteristics are not completely known. Scientists wanted to speed up the process and produce clones from a proven phenotype.

Concept 1
Somatic cells can be made to return to a dormant or embryonic state so that their genes will not be expressed.

Concept 2
The nucleus of a cell can be removed and replaced with the nucleus of an unrelated cell. Cells can be made to fuse together.

Concept 3
Fertilized egg cells produce embryos. Egg cells that contain the nucleus of a donor cell will produce embryos with DNA identical to the donor cell.

Concept 4
Embryos can be implanted into surrogate mothers and develop to full term with seemingly no ill effects.

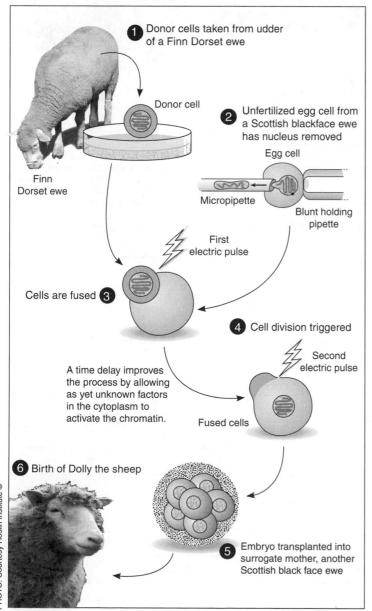

① Donor cells taken from udder of a Finn Dorset ewe

Donor cell

Finn Dorset ewe

② Unfertilized egg cell from a Scottish blackface ewe has nucleus removed

Egg cell

Micropipette

Blunt holding pipette

First electric pulse

Cells are fused ③

④ Cell division triggered

Second electric pulse

A time delay improves the process by allowing as yet unknown factors in the cytoplasm to activate the chromatin.

Fused cells

⑥ Birth of Dolly the sheep

⑤ Embryo transplanted into surrogate mother, another Scottish black face ewe

Techniques

Donor cells from the udder of a Finn Dorset ewe are taken and cultured in a low nutrient media for a week. The nutrient deprived cells stop dividing and become **dormant**.

An **unfertilized egg** from a Scottish blackface ewe has the nucleus removed using a micropipette. The rest of the cell contents are left intact.

The dormant udder cell and the recipient denucleated egg cell are fused using a mild electric pulse.

A second electric pulse triggers cellular activity and cell division, jump starting the cell into development. This can also be triggered by chemical means.

After six days the embryo is transplanted into a surrogate mother, another Scottish blackface ewe.

After a 148 day gestation 'Dolly' is born. DNA profiling shows she is genetically identical to the original Finn Dorset cell donor.

Outcomes

Dolly, a Finn Dorset lamb, was born at the Roslin Institute (near Edinburgh) in July 1996. She was the first mammal to be cloned from **non-embryonic** cells, i.e. cells that had already differentiated into their final form. Dolly's birth showed that the process leading to cell specialization is not irreversible and that cells can be 'reprogrammed' into an embryonic state. Although cloning seems relatively easy there are many problems that occur. Of the hundreds of eggs that were reconstructed only 29 formed embryos and only Dolly survived to birth.

Further Applications

In animal reproductive technology, cloning has facilitated the rapid production of genetically superior stock. These animals may then be dispersed among commercial herds. The **primary focus** of the new cloning technologies is to provide an economically viable way to rapidly produce transgenic animals with very precise genetic modifications.

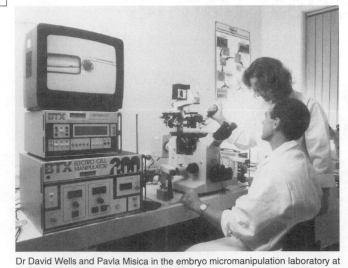

Dr David Wells and Pavla Misica in the embryo micromanipulation laboratory at AgResearch in Hamilton, New Zealand (monitor's image is enlarged on the right).

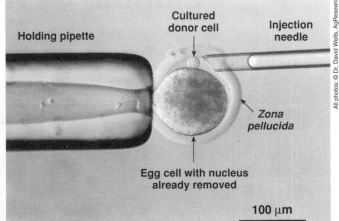

A single cultured cell is injected underneath the *zona pellucida* (the outer membrane) and positioned next to the egg cell (step 3 of diagram on the left).

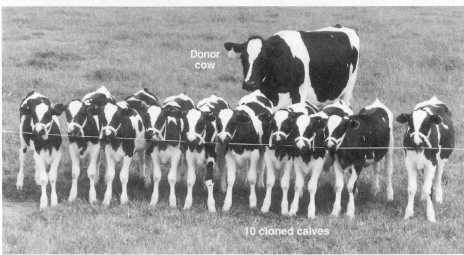

Adult cloning heralds a new chapter in the breeding of livestock. Traditional breeding methods are slow, unpredictable, and suffer from a time delay in waiting to see what the phenotype is like before breeding the next generation. Adult cloning methods now allow a rapid spread of valuable livestock into commercial use among farmers. It will also allow the livestock industry to respond rapidly to market changes in the demand for certain traits in livestock products. In New Zealand, 10 healthy clones were produced from a single cow (the differences in coat colour patterns arise from the random migration of pigment cells in early embryonic development).

Lady is the last surviving cow of the rare Enderby Island (south of N.Z.) cattle breed. Adult cloning was used to produce her genetic duplicate, Elsie (born 31 July 1998). This result represents the first demonstration of the use of adult cloning in animal conservation.

1. Explain what is meant by **adult cloning** (as it relates to **nuclear transfer** techniques involving adult animals):

   _____

   _____

2. Explain how each of the following events is controlled in the **nuclear transfer** process:

   (a) The switching off of all genes in the donor cell: _____

   _____

   (b) The fusion (combining) of donor cell with enucleated egg cell: _____

   _____

   (c) The activation of the cloned cell into producing an embryo: _____

   _____

3. Describe two potential applications of nuclear transfer technology for the cloning of animals:

   (a) _____

   _____

   (b) _____

   _____

Plant Tissue Culture

The Issue

▶ Individuals vary in characteristics, even within specific varieties of the same species.

▶ Plants that are uniform are easier to manage, but are difficult to produce in large numbers from naturally pollinated stock.

▶ It is advantageous for crop management, quality control, and dissemination of transgenics to be able to produce large numbers of genetically uniform plants. The technology can also be applied in the recovery of endangered species.

Concept 1	Concept 2	Concept 3	Concept 4
Plants can be cut into many pieces, almost all of which have the potential to grow into new plants.	Plants propagated from the same tissues will all have the same genetic make up. They will all be clones.	Different **growth hormones** in plants cause the development of different parts of the plant, e.g. the shoots or roots.	If desirable plants are used for the original tissue then all the plants propagated from them will also be desirable.

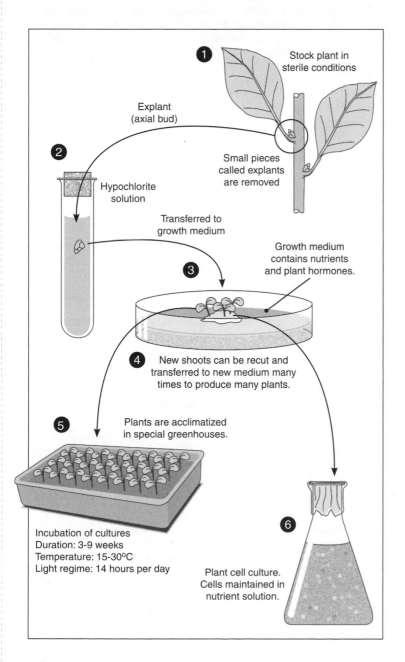

1 Stock plant in sterile conditions

Explant (axial bud)

Small pieces called explants are removed

2 Hypochlorite solution

Transferred to growth medium

Growth medium contains nutrients and plant hormones.

3

4 New shoots can be recut and transferred to new medium many times to produce many plants.

5 Plants are acclimatized in special greenhouses.

Incubation of cultures
Duration: 3-9 weeks
Temperature: 15-30°C
Light regime: 14 hours per day

6 Plant cell culture. Cells maintained in nutrient solution.

Techniques

Pest free stock plants have small pieces cut (**excised**) from them. These pieces, called **explants**, may be stem tissue, nodes, flower buds, leaves or sections of shoot tip meristem.

The surface of the explants are sterilized using a solution of **sodium hypochlorite**.

The explants are transferred to a **culture vessel** under sterile conditions.

Growth hormones are added. By changing the concentrations of different hormones, the explants can be made to grow shoots, roots, stems, or an undifferentiated cell mass (callus).

Plants induced to grow shoots and roots are transferred to growing containers under controlled greenhouse conditions to be acclimatized before planting outside.

Calluses are mechanically broken up into separate cells and placed into a growth medium where they can be maintained indefinitely.

Outcomes

Plants propagated in this way may be genetically unstable or infertile, with chromosomes structurally altered or in unusual numbers. The success of tissue culture is affected by factors such as selection of **explant** material, the composition of the culturing media, plant hormone levels, lighting, and temperature. New genetic stock may be introduced into cloned lines periodically to prevent reduction in genetic diversity.

Further Applications

Plant tissue culture is extensively used in **forestry** to produce trees of uniform height and diameter. It has a number of advantages, including the ability to generate large numbers of plants from one explant and rapid propagation of species that have a long generation times or low seed production.

Micropropagation of the Tasmanian blackwood tree *(Acacia melanoxylon)*

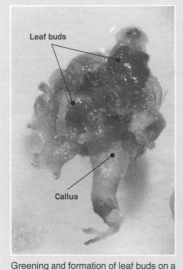

Greening and formation of leaf buds on a callus growing on culturing medium.

Normal shoots with juvenile leaves growing from a callus on media. They appear identical to those produced directly from seeds.

Seedling with juvenile foliage 6 months after transfer to greenhouse.

Micropropagation is increasingly used in conjunction with genetic engineering to propagate transgenic plants. Genetic engineering and micropropagation achieve similar results to conventional selective breeding but more precisely, quickly, and independently of growing season. The **Tasmanian blackwood** (above) is well suited to this type of manipulation. It is a versatile hardwood tree now being extensively trialled in some countries as a replacement for tropical hardwoods. The timber is of high quality, but genetic variations between individual trees lead to differences in timber quality and colour. Tissue culture allows the multiple propagation of trees with desirable traits (e.g. uniform timber colour). Tissue culture could also help to find solutions to problems that cannot be easily solved by forestry management. When combined with genetic engineering (introduction of new genes into the plant) problems of pest and herbicide susceptibility may be resolved. Genetic engineering may also be used to introduce a gene for male sterility, thereby stopping pollen production. This would improve the efficiency of conventional breeding programmes by preventing self-pollination of flowers (the manual removal of stamens is difficult and very labour intensive).

Information courtesy of Raewyn Poole, University of Waikato (Unpublished Msc. thesis).

1. Describe the general purpose of **micropropagation** (plant tissue culture):_____

_____

2. (a) Explain what a **callus** is: _____

_____

(b) Explain how a callus may be stimulated to initiate root and shoot formation: _____

_____

3. Explain a potential problem with micropropagation in terms of long term ability to adapt to environmental changes:

_____

_____

4. Discuss the **advantages** and **disadvantages** of micropropagation compared with traditional methods of plant propagation such as grafting:

_____

_____

_____

_____

_____

_____

_____

_____

The Ethics of GMO Technology

The risks of using **genetically modified organisms** (GMOs) have been the subject of considerable debate in recent times. Most experts agree that, provided GMOs are tested properly, the health risks to individuals should be minimal from plant products, although minor problems will occur. Health risks from animal GMOs are potentially more serious, especially when the animals are for human consumption. The potential benefits to be gained from the use of GMOs creates enormous pressure to apply the existing technology. However, there are many concerns, including the environmental and socio-economic effects, and problems of unregulated use. There is also concern about the environmental and economic costs of possible GMO accidents. GMO research is being driven by heavy investment on the part of biotechnology companies seeking new applications for GMOs. Currently a matter of great concern to consumers is the adequacy of government regulations for the labeling of food products with GMO content. This may have important trade implications for countries exporting and importing GMO produce.

Some important points about GMOs

1. The modified DNA is in every cell of the GMO.

2. The mRNA is only expressed in specific tissues.

3. The foreign protein is only expressed in particular tissues but it may circulate in the blood or lymph or be secreted (e.g. milk).

4. In animals, the transgene is only likely to be transmitted from parent to offspring. However, viral vectors may enable accidental transfer of the transgene between unrelated animals.

5. In plants, transmission of the transgene in GMOs is possible by pollen, cuttings, and seeds (even between species).

6. If we eat the animal or plant proper, we will also be eating DNA. The DNA will remain 'intact' if raw, but "degraded" if cooked.

7. Non-transgenic food products may be processed using genetically modified bacteria or yeast, and cells containing their DNA may be in the food product.

8. A transgenic product (e.g. a protein, polypeptide or a carbohydrate) may be in the GMO, but not in the portions sold to the consumer.

Potential effects of GMOs on the world

1. Increase in food production.

2. Decrease in use of pesticides, herbicides and animal remedies.

3. Improvement in the health of the human population and the medicines used to achieve it.

4. Possible development of transgenic products which may be harmful to some (e.g. new proteins causing allergies).

5. May have little real economic benefit to farmers (and the consumer) when increased production (as little as 10%) is weighed against cost, capital, and competition.

6. Possible (uncontrollable) spread of transgenes into other species: plants, indigenous species, animals, and humans.

7. Concerns that the release of GMOs into the environment may be irreversible.

8. Economic sanctions resulting from a consumer backlash against GMO foods and products.

9. Animal welfare and ethical issues: GM animals may suffer poor health and reduced life span.

10. GMOs may cause the emergence of pest, insect, or microbial resistance to traditional control methods.

11. May create a monopoly and dependence of developing countries on companies who are seeking to control the world's commercial seed supply.

GMO protestors are arrested

Issue: The accidental release of GMOs into the environment.

Problem: Recombinant DNA may be taken up by non-target organisms. Thes then may have the potential to become pests or cause disease.

Solution: Rigorous controls on the production and release of GMOs. GMOs could have specific genes deleted so that their growth requirements are met only in controlled environments.

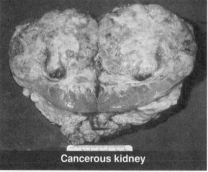
Cancerous kidney

Issue: A new gene or genes may disrupt normal gene function.

Problem: Gene disruption may trigger cancer. Successful expression of the desired gene is frequently very low.

Solution: A combination of genetic engineering, cloning, and genetic screening so that only those cells that have been successfully transformed are used to produce organisms.

Protest against GMOs in the environment

Issue: Targeted use of transgenic organisms in the environment.

Problem: Once their desired function, e.g. environmental clean-up, is completed, they may be undesirable invaders in the ecosystem.

Solution: GMOs can be engineered to contain "suicide genes" or metabolic deficiencies so that they do not survive for long in the new environment after completion of their task.

1. Suggest why genetically modified (GM) plants are thought to pose a greater environmental threat than GM animals:

_____

_____

_____

_____

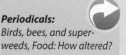
Periodicals:
Birds, bees, and super-
weeds, Food: How altered?

Related activities: Applications of GMOs, Production of Insulin,
Golden Rice, Chymosin Production,

2. Describe an advantage and a problem with the use of genetically engineered herbicide resistant crop plants:

(a) Advantage: _____

_____

(b) Problem: _____

_____

3. Describe an advantage and a problem with using tropical crops genetically engineered to grow in cold regions:

(a) Advantage: _____

_____

(b) Problem: _____

_____

4. Describe an advantage and a problem with using crops that are genetically engineered to grow in marginal habitats (for example, in very saline or poorly aerated soils):

(a) Advantage: _____

_____

(b) Problem: _____

_____

5. Describe two uses of transgenic animals within the livestock industry:

(a) _____

(b) _____

6. Some years ago, Britain banned the import of a GM, pest resistant corn variety containing marker genes for ampicillin antibiotic resistance. Suggest why the use of antibiotic-resistance genes as markers is no longer common practice:

_____

_____

7. Many agricultural applications of DNA technology make use of transgenic bacteria which infect plants and express a foreign gene. Explain one advantage of each of the following applications of genetic engineering to crop biology:

(a) Development of nitrogen-fixing *Rhizobium* bacteria that can colonise non-legumes such as corn and wheat:

_____

(b) Addition of transgenic *Pseudomonas fluorescens* bacteria into seeds (bacterium produces a pathogen-killing toxin):

_____

8. Some of the public's fears and concerns about genetically modified food stem from moral or religious convictions, while others have a biological basis and are related to the potential biological threat posed by GMOs.
(a) Conduct a class discussion or debate to identify these fears and concerns, and list them below:

_____

_____

_____

_____

(b) Identify which of those you have listed above pose a real biological threat: _____

_____

_____

_____

KEY TERMS Mix and Match

INSTRUCTIONS: Test your vocab by matching each term to its correct definition, as identified by its preceding letter code.

AMPLIFICATION

ANNEALING

BLUNT END

DNA CHIP

DNA LIGATION

DNA MARKER

DNA POLYMERASE

DNA PROFILING

GEL ELECTROPHORESIS

GENE TECHNOLOGY

GMO

MARKER GENE

mtDNA

MICROSATELLITE

MODEL ORGANISM

MOLECULAR CLONE

PCR

PLASMID

PRIMER

RECOGNITION SITE

RECOMBINANT DNA

RESTRICTION ENZYME

STICKY END

TRANSFORMATION

VECTOR

A An organism or artificial vehicle that is capable of transferring a DNA sequence to another organism.

B The attachment of a DNA primer to a length of DNA as a starter point for the process of replication by a polymerase enzyme.

C The process of locating regions of a DNA sequence that are variable between individuals in order to distinguish between individuals.

D A short length of DNA used to identify the starting sequence for PCR so that polymerase enzymes can begin amplification.

E An enzyme that is able to cut a length of DNA at a specific sequence or site.

F A sequence of DNA on a chromosome known to be linked with a specific gene.

G A species that is extensively studied in order to understand how certain biological systems work.

H A reaction that is used to amplify fragments of DNA using cycles of heating and cooling (abbreviation).

I A small circular piece of DNA commonly found in bacteria.

J The process of producing more copies of a length of DNA, normally using PCR.

K DNA that has had a new sequence added so that the original sequence has been changed.

L The acquisition of genetic material by the uptake of naked DNA by the recipient.

M A short (normally two base pairs) piece of DNA that repeats a variable number of times between people and so can be used to distinguish between individuals.

N An array of thousands of microscopic spots of DNA oligonucleotides that can be used to measure the level of gene expression or to search for novel genes.

O An enzyme that is able to replicate DNA and commonly used in PCR to amplify a length of DNA.

P An organism that has had part of its DNA sequence altered either by the removal or insertion of a piece of DNA.

Q A process that is used to separate different lengths of DNA by placing them in a gel matrix placed in a buffered solution through which an electric current is passed.

R The manipulation of DNA and gene sequences in order to modify the characteristics of organisms.

S The site or sequence of DNA at which a restriction enzyme attaches and cuts the DNA.

T A cut in a length of DNA by a restriction enzyme that results in two strands of DNA being different lengths with one strand overhanging the other.

U A type of cut in a length of DNA caused by a restriction enzyme that results in both strands of DNA being the same length.

V The repairing or attaching of fragmented DNA by ligase enzymes.

W Circular DNA found in the mitochondria.

X A gene, with an identifiable effect, used yo determine if a piece of DNA has been successfully inserted into the host organism.

Y One of multiple copies of an isolated defined sequence of (often recombinant) DNA.

© Biozone International 2001-2010
Photocopying Prohibited

R 2

The Origin and Evolution of Life

The origin of life	• Conditions on prebiotic Earth • RNA and the first cells • The origin of eukaryotes
The evidence for evolution	• Fossils and stratigraphy • Comparative molecular analyses • Comparative anatomy and evo-devo • Biogeography and continental drift

Speciation

The species concept	• The biological species concept • Hybrids, ring species, & cryptic species
The new synthesis of evolution	• The Darwin-Wallace theory • Neo-Darwinism: updating the picture • Adaptation and fitness
Mechanisms of evolution	• Mutation and gene flow • Natural selection and genetic drift • Changing selection pressures
Evolution in real time	• Drug and insecticide resistance • Finches, flies, and salamanders • Skin color in humans - what is race?
Selective breeding	• Selective breeding in animals • Selective breeding in crop plants • Polyploidy and instant speciation

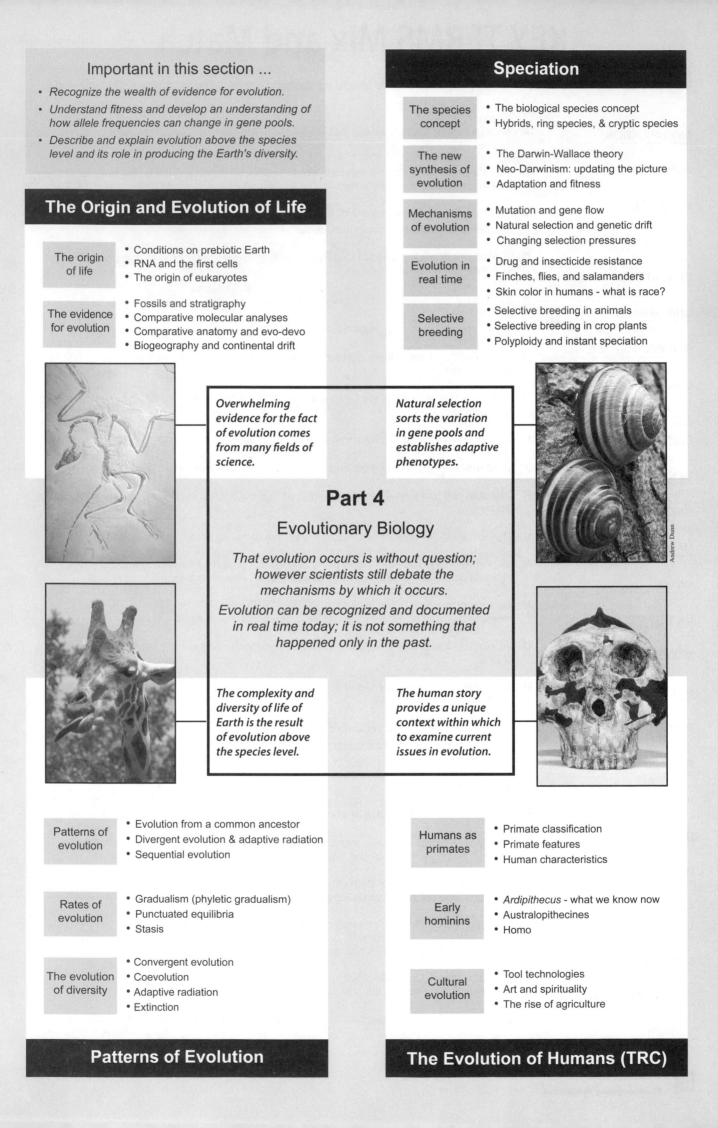

Overwhelming evidence for the fact of evolution comes from many fields of science.

Natural selection sorts the variation in gene pools and establishes adaptive phenotypes.

Andrew Dunn

Part 4

Evolutionary Biology

That evolution occurs is without question; however scientists still debate the mechanisms by which it occurs.

Evolution can be recognized and documented in real time today; it is not something that happened only in the past.

The complexity and diversity of life of Earth is the result of evolution above the species level.

The human story provides a unique context within which to examine current issues in evolution.

Patterns of evolution	• Evolution from a common ancestor • Divergent evolution & adaptive radiation • Sequential evolution
Rates of evolution	• Gradualism (phyletic gradualism) • Punctuated equilibria • Stasis
The evolution of diversity	• Convergent evolution • Coevolution • Adaptive radiation • Extinction

Humans as primates	• Primate classification • Primate features • Human characteristics
Early hominins	• *Ardipithecus* - what we know now • Australopithecines • Homo
Cultural evolution	• Tool technologies • Art and spirituality • The rise of agriculture

Patterns of Evolution

The Evolution of Humans (TRC)

The Origin and Evolution of Life

KEY CONCEPTS

▶ The primordial environment on Earth was important in the origin of the first organic compounds.

▶ Fossilized cyanobacterial colonies provide evidence of life 3.5 billion years ago.

▶ The dating of the fossil record gives a history of life on Earth and a geologic time scale.

▶ Overwhelming evidence for the fact of evolution comes from many fields of science.

KEY TERMS

absolute dating
Archaea
biogeographical evidence
common ancestor
comparative anatomy
endosymbiotic theory
epoch
era
Eubacteria
Eukarya
evo-devo
fossil
fossil record
Hox genes
geologic time scale
Miller-Urey experiments
molecular clock
paleontological evidence
panspermia
period
phylogeny
primordial environment
radiometric dating
relative dating
rock strata
transitional fossil

OBJECTIVES

☐ 1. Use the **KEY TERMS** to help you understand and complete these objectives.

The Origin and History of Life on Earth   pages 290-296

☐ 2. Describe the primordial environment and the likely events that led to the formation of life on Earth.

☐ 3. Recognize major stages in the evolution of life on Earth. Summarize the main ideas related to where life originated: ocean surface, extraterrestrial (panspermia), and deep sea thermal vents.

☐ 4. Outline the experiments of Miller and Urey that attempted to simulate the prebiotic environment on Earth. Describe their importance in our understanding of the probable origin of organic compounds.

☐ 5. Describe the geological and paleontological evidence for when life originated on Earth. Describe the likely role of RNA as the first self-replicating molecule. Outline its role as an enzyme and in the origin of the first self-replicating cells.

☐ 6. Discuss the possible origin of membranes and the first prokaryotic cells.

☐ 7. Discuss the **endosymbiotic** (endosymbiont) theory for the evolution of eukaryotic cells. Summarize the evidence in support of this theory.

The Evidence for Evolution   pages 297-316, 333-334, 375

☐ 8. Explain how fossils are formed and dated. Explain how the dating of fossil-bearing rocks has provided the data for dividing the history of life on Earth into geological periods, which collectively form the **geologic time scale**.

☐ 9. Explain the significance of **transitional fossils**. Offer an explanation for the apparent lack of transitional forms in the **fossil record**.

☐ 10. Explain the biochemical evidence by the universality of DNA, amino acids, and protein structures (e.g. cytochrome C) for the common ancestry of living organisms. Describe how comparisons of specific molecules between species are used as an indication of relatedness or **phylogeny**.

☐ 11. Discuss how biochemical variations can be used as a **molecular clock** to determine probable dates of divergence from a **common ancestor**.

☐ 12. Explain how **comparative anatomy** and physiology have contributed to an understanding of evolutionary relationships.

☐ 13. Explain how **evolutionary developmental biology** (evo-devo) has provided some of the strongest evidence for the mechanisms of evolution, particularly for the evolution of novel forms.

☐ 14. Discuss the significance of vestigial organs as indicators of evolutionary trends in some groups.

☐ 15. Describe and explain the **biogeographical evidence** for evolution.

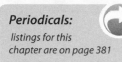

Periodicals:
listings for this chapter are on page 381

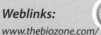

Weblinks:
www.thebiozone.com/
weblink/SB1-2597.html

Teacher Resource CD-ROM:
Dating the Past

Life in the Universe

Life 'as we know it' requires three basic ingredients: a source of energy, carbon, and liquid water. Complex organic molecules (as are found in living things) have been detected beyond Earth in interstellar dust clouds and in meteorites that have landed on Earth. More than 4 billion years ago, one such dust cloud collapsed into a swirling **accretion disk** that gave rise to the sun and planets. Some of the fragile molecules survived the heat of solar system formation by sticking together in comets at the disk's fringe where temperatures were freezing. Later, the comets and other cloud remnants carried the molecules to Earth.

The formation of these organic molecules and their significance to the origin of life on Earth are currently being investigated experimentally (see below). The study of the origin of life on Earth is closely linked to the search for life elsewhere in our solar system. There are further plans to send solar and lunar orbiters to other planets and their moons, and even to land on a comet over the next decade. Their objective will be to look for signs of life (present or past) or its chemical precursors. If detected, such a discovery would suggest that life (at least 'primitive' life) may be widespread in the universe.

Planet Formation

Interstellar dust and gas

Sun

Planets forming

Accretion disc

Galaxy

Nebula

(an artist's impression)

How Organic Molecules Might Form in Space

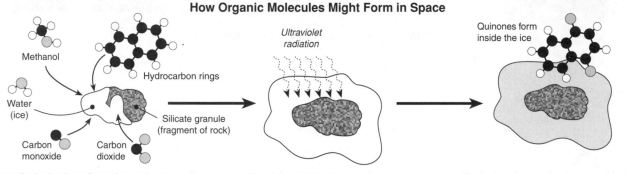

Methanol

Hydrocarbon rings

Water (ice)

Silicate granule (fragment of rock)

Carbon monoxide

Carbon dioxide

Ultraviolet radiation

Quinones form inside the ice

Interstellar ice begins to form when molecules such as methanol, water, and hydrocarbon freeze onto sandlike granules of silicate drifting in dense interstellar clouds.

Ultraviolet radiation from nearby stars cause some of the chemical bonds of the frozen compounds in the ice to break.

The broken down molecules recombine into structures such as quinones, which would never form if the fragments were free to float away.

Two **Mars Exploration rovers** landed on Mars in early 2004. Each rover carried sophisticated instruments, which were used to determine the history of climate and water at two sites where conditions may once have been favorable for life.

Organic Molecules Detected in Space

In a simple cloud-chamber experiment with simulated space ice (frozen water, methanol, and ammonia), complex compounds were yielded, including: ketones, nitriles, ethers, alcohols, and quinones (nearly identical in structure to those that help chlorophyll). These same organic molecules are found in carbon-rich meteorites. A six-carbon molecule (known as HMT) was also created. In warm, acidic water it is known to produce amino acids.

In another investigation into compounds produced in this way, some of the molecules displayed a tendency to form capsule-like droplets in water. These capsules were similar to those produced using extracts of a meteorite from Murchison, Australia in 1989. When organic compounds from the meteorite were mixed with water, they spontaneously assembled into spherical structures similar to cell membranes. These capsules were found to be made up of a host of complex organic molecules.

Source: *Life's far-flung raw materials*, Scientific American, July 1999, pp. 26-33

1. Suggest how sampling the chemical makeup of a comet might assist our understanding of life's origins:

    Compare it to molecule composition

2. Explain the significance of molecules from space that naturally form capsule-like droplets when added to water:

    Similar to cell membranes

3. Explain how scientists are able to know about the existence of complex organic molecules in space:

    meteorites

Related activities: *Prebiotic Experiments*
Web links: *Life in the Universe*

Periodicals:
The ice of life

© Biozone International 2001-2010

Photocopying Prohibited

The Origin of Life on Earth

Recent discoveries of **prebiotic** conditions on other planets and their moons has rekindled interest in the origin of life on primeval Earth. Experiments demonstrate that both peptides and nucleic acids may form polymers naturally in the conditions that are thought to have existed in a primitive terrestrial environment. RNA has also been shown to have enzymatic properties (**ribozymes**) and is capable of self-replication. These discoveries have removed some fundamental obstacles to creating a plausible scientific model for the origin of life from a prebiotic soup. Much research is now underway and space probes have been sent to Mercury, Venus, Mars, Pluto and its moon, Charon. They will search for evidence of prebiotic conditions or primitive microorganisms. The study of life in such regions beyond our planet is called **exobiology**.

Steps Proposed in the Origin of Life

The appearance of life on our planet may be understood as the result of evolutionary processes that involve the following major steps:

1. Formation of the Earth (4600 mya) and its acquisition of volatile organic chemicals by collision with comets and meteorites, which provided the precursors of biochemical molecules.

2. Prebiotic synthesis and accumulation of amino acids, purines, pyrimidines, sugars, lipids, and other organic molecules in the primitive terrestrial environment.

3. Prebiotic condensation reactions involving the synthesis of polymers of peptides (proteins), and nucleic acids (most probably just RNA) with self-replicating and catalytic (enzymatic) abilities.

4. Synthesis of lipids, their self-assembly into double-layered membranes and liposomes, and the 'capturing' of prebiotic (self-replicating and catalytic) molecules within their boundaries.

5. Formation of a **protobiont**; an immediate precursor to the first living systems. Such protobionts would exhibit cooperative interactions between small catalytic peptides, replicative molecules, proto-tRNA, and protoribosomes.

An RNA World

RNA has the ability to act as both genes and enzymes and offers a way around the "chicken-and-egg" problem: genes require enzymes to form; enzymes require genes to form. The first stage of evolution may have proceeded by RNA molecules performing the catalytic activities necessary to assemble themselves from a nucleotide soup. At the next stage, RNA molecules began to synthesize proteins. There is a problem with RNA as a prebiotic molecule because the ribose is unstable. This has led to the idea of a pre-RNA world (PNA).

Photo: Ron Lind

These living **stromatolites** from a beach in Western Australia are created by mats of bacteria. Similar, fossilized stromatolites have been found in rocks dating back to 3500 million years ago.

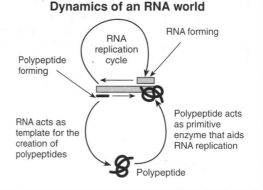

Dynamics of an RNA world

RNA replication cycle

RNA forming

Polypeptide forming

RNA acts as template for the creation of polypeptides

Polypeptide acts as primitive enzyme that aids RNA replication

Polypeptide

Scenarios for the Origin of Life on Earth

The origin of life remains a matter of scientific speculation. Three alternative views of how the key processes occurred are illustrated below:

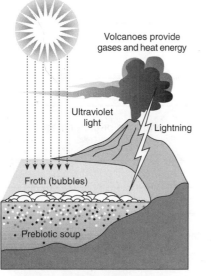

Volcanoes provide gases and heat energy

Ultraviolet light

Lightning

Froth (bubbles)

Prebiotic soup

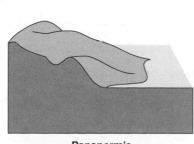

Comet or meteorite from elsewhere in the solar system harboring microorganisms

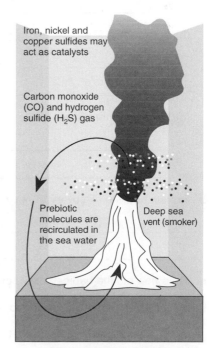

Iron, nickel and copper sulfides may act as catalysts

Carbon monoxide (CO) and hydrogen sulfide (H_2S) gas

Prebiotic molecules are recirculated in the sea water

Deep sea vent (smoker)

Ocean surface (tidal pools)

This popular theory suggests that life arose in a tidepool, pond or on moist clay on the primeval Earth. Gases from volcanoes would have been energized by UV light or electrical discharges to form the prebiotic molecules in froth.

Panspermia

Cosmic ancestry (panspermia) is a serious scientific theory that proposes living organisms were 'seeded' on Earth as 'passengers' aboard comets and meteors. Such incoming organisms would have to survive the heat of re-entry.

Undersea thermal vents

A recently proposed theory suggests that life may have arisen at ancient volcanic vents (called smokers). This environment provides the necessary gases, energy, and a possible source of catalysts (metal sulfides).

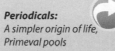

Periodicals:
A simpler origin of life,
Primeval pools

Related activities: *Prebiotic Experiments*
Web links: *4 Billion Years of Evolution*

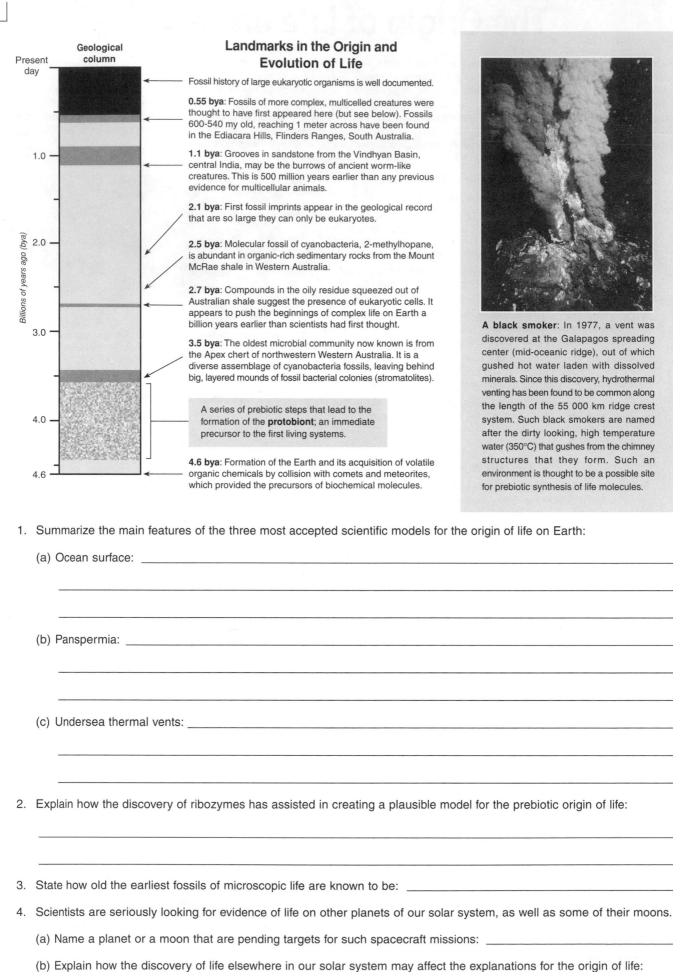

Geological column

Present day

Billions of years ago (bya)

1.0

2.0

3.0

4.0

4.6

Landmarks in the Origin and Evolution of Life

Fossil history of large eukaryotic organisms is well documented.

0.55 bya: Fossils of more complex, multicelled creatures were thought to have first appeared here (but see below). Fossils 600-540 my old, reaching 1 meter across have been found in the Ediacara Hills, Flinders Ranges, South Australia.

1.1 bya: Grooves in sandstone from the Vindhyan Basin, central India, may be the burrows of ancient worm-like creatures. This is 500 million years earlier than any previous evidence for multicellular animals.

2.1 bya: First fossil imprints appear in the geological record that are so large they can only be eukaryotes.

2.5 bya: Molecular fossil of cyanobacteria, 2-methylhopane, is abundant in organic-rich sedimentary rocks from the Mount McRae shale in Western Australia.

2.7 bya: Compounds in the oily residue squeezed out of Australian shale suggest the presence of eukaryotic cells. It appears to push the beginnings of complex life on Earth a billion years earlier than scientists had first thought.

3.5 bya: The oldest microbial community now known is from the Apex chert of northwestern Western Australia. It is a diverse assemblage of cyanobacteria fossils, leaving behind big, layered mounds of fossil bacterial colonies (stromatolites).

A series of prebiotic steps that lead to the formation of the **protobiont**; an immediate precursor to the first living systems.

4.6 bya: Formation of the Earth and its acquisition of volatile organic chemicals by collision with comets and meteorites, which provided the precursors of biochemical molecules.

A black smoker: In 1977, a vent was discovered at the Galapagos spreading center (mid-oceanic ridge), out of which gushed hot water laden with dissolved minerals. Since this discovery, hydrothermal venting has been found to be common along the length of the 55 000 km ridge crest system. Such black smokers are named after the dirty looking, high temperature water (350°C) that gushes from the chimney structures that they form. Such an environment is thought to be a possible site for prebiotic synthesis of life molecules.

1. Summarize the main features of the three most accepted scientific models for the origin of life on Earth:

   (a) Ocean surface: _____

   _____

   _____

   (b) Panspermia: _____

   _____

   _____

   (c) Undersea thermal vents: _____

   _____

   _____

2. Explain how the discovery of ribozymes has assisted in creating a plausible model for the prebiotic origin of life:

   _____

   _____

3. State how old the earliest fossils of microscopic life are known to be: _____

4. Scientists are seriously looking for evidence of life on other planets of our solar system, as well as some of their moons.

   (a) Name a planet or a moon that are pending targets for such spacecraft missions: _____

   (b) Explain how the discovery of life elsewhere in our solar system may affect the explanations for the origin of life:

   _____

   _____

Prebiotic Experiments

In the 1950s, Stanley Miller and Harold Urey used equipment (illustrated below) to attempt to recreate the conditions on the primitive Earth. They hoped that the experiment might give rise to the biological molecules that were forerunners to the development of the first living organisms. Researchers at the time believed that the Earth's early atmosphere was made up of methane, water vapor, ammonia, and hydrogen gas. Many variations on this experiment, using a variety of recipes, have produced similar results. It seems that the building blocks of life are relatively easy to create. Many types of organic molecules have even been detected in deep space.

The Miller-Urey Experiment

The experiment (right) was run for a week after which samples were taken from the collection trap for analysis. Up to 4% of the carbon (from the methane) had been converted to amino acids. In this and subsequent experiments, it has been possible to form all 20 amino acids commonly found in organisms, along with nucleic acids, several sugars, lipids, adenine, and even ATP (if phosphate is added to the flask). Researchers now believe that the early atmosphere may be similar to the vapors given off by modern volcanoes: carbon monoxide (CO), carbon dioxide (CO_2), and nitrogen (N_2). Note the absence of free atmospheric oxygen.

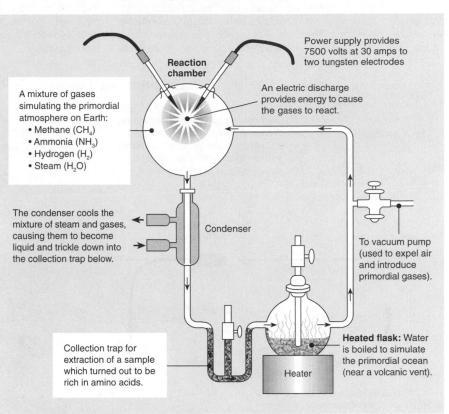

Reaction chamber

A mixture of gases simulating the primordial atmosphere on Earth:
- Methane (CH_4)
- Ammonia (NH_3)
- Hydrogen (H_2)
- Steam (H_2O)

Power supply provides 7500 volts at 30 amps to two tungsten electrodes

An electric discharge provides energy to cause the gases to react.

The condenser cools the mixture of steam and gases, causing them to become liquid and trickle down into the collection trap below.

Condenser

To vacuum pump (used to expel air and introduce primordial gases).

Collection trap for extraction of a sample which turned out to be rich in amino acids.

Heater

Heated flask: Water is boiled to simulate the primordial ocean (near a volcanic vent).

Iron pyrite, or 'fools gold' (above), has been proposed as a possible stabilizing surface for the synthesis of organic compounds in the prebiotic world.

Some scientists envisage a global winter scenario for the formation of life. Organic compounds are more stable in colder temperatures and could combine in a lattice of ice. This frozen world could be thawed later.

Lightning is a natural phenomenon associated with volcanic activity. It may have supplied a source of electrical energy for the formation of new compounds (such as oxides of nitrogen) which were incorporated into organic molecules.

The early Earth was subjected to volcanism everywhere. At volcanic sites such as deep sea hydrothermal vents and geysers (like the one above), gases delivered vital compounds to the surface, where reactions took place.

1. In the Miller-Urey experiment simulating the conditions on primeval Earth, identify parts of the apparatus equivalent to:

   (a) Primeval atmosphere: _____

   (b) Primeval ocean: _____

   (c) Lightning: _____

   (d) Volcanic heat: _____

2. Name the organic molecules that were created by this experiment: _____

   _____

3. (a) Suggest a reason why the Miller-Urey experiment is not an accurate model of what happened on the primeval Earth:

   _____

   (b) Suggest changes to the experiment that could help it to better fit our understanding of the Earth's primordial conditions:

   _____

The Origin and Evolution of Life

Periodicals:
Earth in the beginning

Related activities: The Origin of Life on Earth
Web links: An Interview with Stanley Miller

A 3

The Origin of Eukaryotes

The first firm evidence of eukaryote cells is found in the fossil record at 540-600 mya. It is thought that eukaryote cells evolved from large prokaryote cells that ingested other free-floating prokaryotes. They formed a symbiotic relationship with the cells they engulfed (**endosymbiosis**). The two most important organelles that developed in eukaryote cells were mitochondria, for aerobic respiration, and chloroplasts, for photosynthesis in aerobic conditions. Primitive eukaryotes probably acquired mitochondria by engulfing purple bacteria. Similarly, chloroplasts may have been acquired by engulfing primitive cyanobacteria (which were already capable of photosynthesis). In both instances the organelles produced became dependent on the nucleus of the host cell to direct some of their metabolic processes. The sequence of evolutionary change shown below suggests that the lines leading to animal cells diverged before those leading to plant cells, but the reverse could also be true. Animal cells might have evolved from plant-like cells which subsequently lost their chloroplasts.

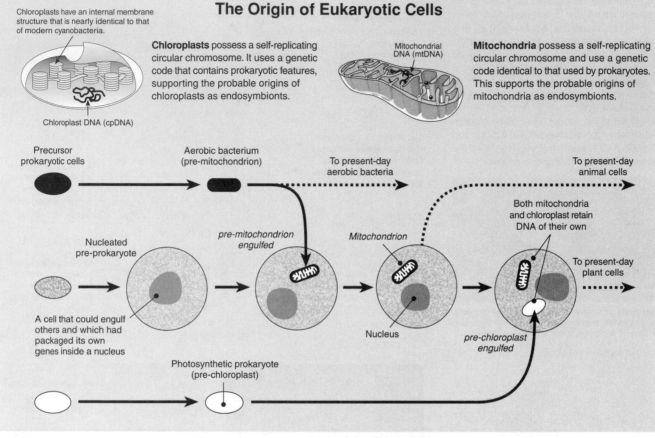

The Origin of Eukaryotic Cells

Chloroplasts have an internal membrane structure that is nearly identical to that of modern cyanobacteria.

Chloroplast DNA (cpDNA)

Chloroplasts possess a self-replicating circular chromosome. It uses a genetic code that contains prokaryotic features, supporting the probable origins of chloroplasts as endosymbionts.

Mitochondrial DNA (mtDNA)

Mitochondria possess a self-replicating circular chromosome and use a genetic code identical to that used by prokaryotes. This supports the probable origins of mitochondria as endosymbionts.

Precursor prokaryotic cells

Aerobic bacterium (pre-mitochondrion)

To present-day aerobic bacteria

To present-day animal cells

Nucleated pre-prokaryote

pre-mitochondrion engulfed

Mitochondrion

Both mitochondria and chloroplast retain DNA of their own

To present-day plant cells

A cell that could engulf others and which had packaged its own genes inside a nucleus

Nucleus

pre-chloroplast engulfed

Photosynthetic prokaryote (pre-chloroplast)

1. Distinguish between the two possible sequences of evolutionary change suggested in the endosymbiosis theory:

_____

_____

_____

2. Explain how the endosymbiosis theory accounts for the origins of the following organelles in eukaryotic cells:

(a) Mitochondria: _____

_____

(b) Chloroplasts: _____

_____

3. Describe the evidence that is found in modern mitochondria and chloroplasts that supports the endosymbiosis theory:

_____

_____

4. Comment on how the fossil evidence of early life supports or contradicts the endosymbiotic theory: _____

_____

_____

RA 2

Related activities: The Origin of Life on Earth
Web links: The Endosymbiotic Theory

Periodicals:
The rise of life on Earth

© Biozone International 2001-2010
Photocopying Prohibited

The History of Life on Earth

The scientific explanation the origin of life on Earth is based soundly on the extensive fossil record, as well as the genetic comparison of modern life forms. Together they clearly indicate that modern life forms arose from ancient ancestors that have long since become extinct. These ancient life forms themselves originally arose from primitive cells living some 3500 million years ago in conditions quite different from those on Earth today. The earliest fossil records of living things show only simple cell types. It is believed that the first cells arose as a result of evolution at the chemical level in a 'primordial soup' (a rich broth of chemicals in a warm pool of water, perhaps near a volcanic vent). Life appears very early in Earth's history, but did not evolve beyond the simple cell stage until much later, (about 600 mya). This would suggest that the development of complex life forms required more difficult evolutionary hurdles to be overcome. The buildup of free oxygen in the atmosphere, released as a by-product from photosynthesizing organisms, was important for the evolutionary development of animal life.

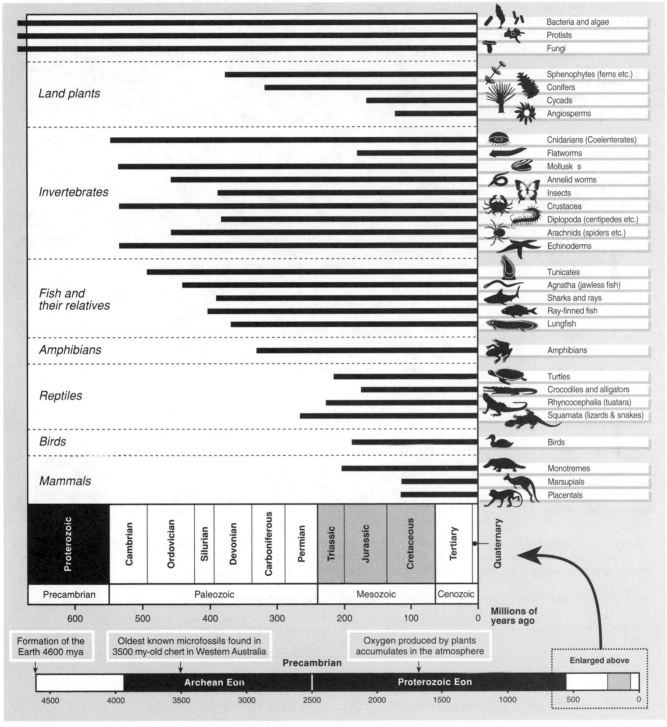

1. Explain the importance of the buildup of free oxygen in the atmosphere for the evolution of animal life:

_____

2. Using the diagram above, determine how many millions of years ago the fossil record shows the first appearance of:

(a) Invertebrates: _____   (b) Fish (ray-finned): _____   (c) Land plants: _____

(d) Reptiles: _____   (e) Birds: _____   (f) Mammals: _____

Periodicals:
Life's long fuse, A cool
early life

Related activities: The Origin of Eukaryotes, Extinction
Web links: Deep Time, A Brief History of Life

A 2

Cenozoic

1.65 mya: Modern humans evolve and their hunting activities, starting at the most recent ice age, cause the most recent mass extinction.

3-5 mya: Early humans arise from ape ancestors

65-1.65 mya: Major shifts in climate. Major adaptive radiations of angiosperms (flowering plants), insects, birds, and mammals.

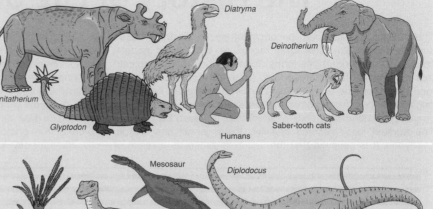

Diatryma

Deinotherium

Unitatherium

Glyptodon

Humans

Saber-tooth cats

Mesozoic

65 mya: Apparent asteroid impact causes mass extinctions of many marine species and all dinosaurs.

135-65 mya: Major radiations of dinosaurs, fishes, and insects. Origin of angiosperms.

181-135 mya: Major radiations of dinosaurs.

240-205 mya: Recoveries, adaptive radiation of marine invertebrates, dinosaurs and fishes. Origin of mammals. Gymnosperms become dominant land plants.

Mesosaur

Diplodocus

Early gymnosperms

Deinonychus

Early mammals

Torosaur

Later Paleozoic

240 mya: Mass extinction of nearly all species on land and in the sea.

435-280 mya: Vast swamps with the first vascular plants. Origin and adaptive radiation of reptiles, insects, and spore bearing plants (including gymnosperms).

500-435 mya: Major adaptive radiations of marine invertebrates and early fishes.

Armored fish

Early insects

Early vascular plants

Trilobite

Ammonite

Early amphibians

Into

Early Paleozoic (Cambrian)

550-500 mya: Origin of animals with hard parts (appear as fossils in rocks). Simple marine communities. A famous Canadian site with a rich collection of early Cambrian fossils is known as the Burgess Shale deposits; examples are shown on the right.

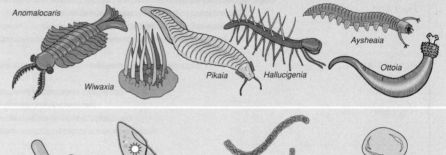

Anomalocaris

Aysheaia

Ottoia

Wiwaxia

Pikaia

Hallucigenia

Precambrian

2500–570 mya: Origin of protists, fungi, algae and animals.

3800–2500 mya: Origin of photosynthetic bacteria.

4600–3800 mya: Chemical and molecular evolution leading to origin of life; protocells to anaerobic bacteria.

4600 mya: Origin of Earth.

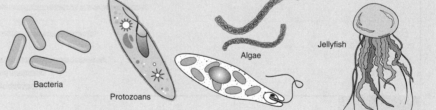

Algae

Jellyfish

Bacteria

Protozoans

3. An important feature of the history of life is that it has not been a steady progression of change. There have been bursts of evolutionary change as newly evolved groups undergo **adaptive radiations** and greatly increase in biodiversity. Such events are often associated with the sudden mass extinction of other, unrelated groups.

(a) Explain the significance of mass extinctions in stimulating new biodiversity: _____

_____

_____

_____

(b) Briefly describe how the biodiversity of the Earth has changed since the origin of life:

_____

_____

_____

_____

_____

Fossil Formation

Fossils are the remains of long-dead organisms that have escaped decay and have, after many years, become part of the Earth's crust. A fossil may be the preserved remains of the organism itself, the impression of it in the sediment (mold), or marks made by it during its lifetime (called trace fossils). For fossilization to occur, rapid burial of the organism is required (usually in water-borne sediment). This is followed by chemical alteration, where minerals are added or removed. Fossilization requires the normal processes of decay to be permanently arrested. This can occur if the remains are isolated from the air or water and decomposing microbes are prevented from breaking them down. Fossils provide a record of the appearance and extinction of organisms, from species to whole taxonomic groups. Once this record is calibrated against a time scale (by using a broad range of dating techniques), it is possible to build up a picture of the evolutionary changes that have taken place.

Modes of preservation

Silicification: Silica from weathered volcanic ash is gradually incorporated into partly decayed wood (also called petrification).

Phosphatization: Bones and teeth are preserved in phosphate deposits.

Pyritization: Iron pyrite replaces hard remains of the dead organism.

Tar pit: Animals fall into and are trapped in mixture of tar and sand.

Trapped in amber: Gum from conifers traps insects and then hardens.

Limestone: Calcium carbonate from the remains of marine plankton is deposited as a sediment that traps the remains of other sea creatures.

Brachiopod (lamp shell), Jurassic (New Zealand)

Mold: This impression of a lamp shell is all that is left after the original shell material was dissolved after fossilization.

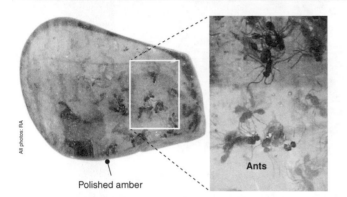
Polished amber — Ants

Insects in amber: The fossilized resin or gum produced by some ancient conifers trapped these insects (including the ants visible in the enlargement) about 25 million years ago (Madagascar).

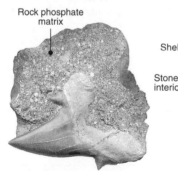
Ray structure — Bark — Growth rings largely destroyed

Petrified wood: A cross-section of a limb from a coniferous tree (Madagascar).

Rock phosphate matrix

Shark tooth: The tooth of a shark *Lamna obliqua* from phosphate beds, Eocene (Khouribga, Morocco).

Shell — Stone interior

Ammonite: This ammonite still has a layer of the original shell covering the stone interior, Jurassic (Madagascar).

Sand and tar matrix — Wing bones

Bird bones: Fossilized bones of a bird that lived about 5 million years ago and became stuck in the tar pits at la Brea, Los Angeles, USA.

Shell and chambers replaced by iron pyrite

Cast: This ammonite has been preserved by a process called pyritization, late Cretaceous (Charmouth, England).

Fossil fern: This compression fossil of a fern leaf shows traces of carbon and wax from the original plant, Carboniferous (USA).

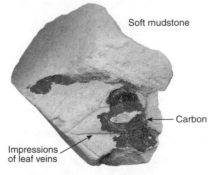
Soft mudstone — Carbon — Impressions of leaf veins

Sub-fossil: Leaf impression in soft mudstone (can be broken easily with fingers) with some of the remains of the leaf still intact (a few thousand years old, New Zealand).

The Origin and Evolution of Life

Periodicals: Meet your ancestor, The quick and the dead

Related activities: The Origin of Life on Earth, Dating a Fossil Site
Web links: Getting into the Fossil Record

A 1

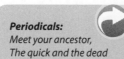

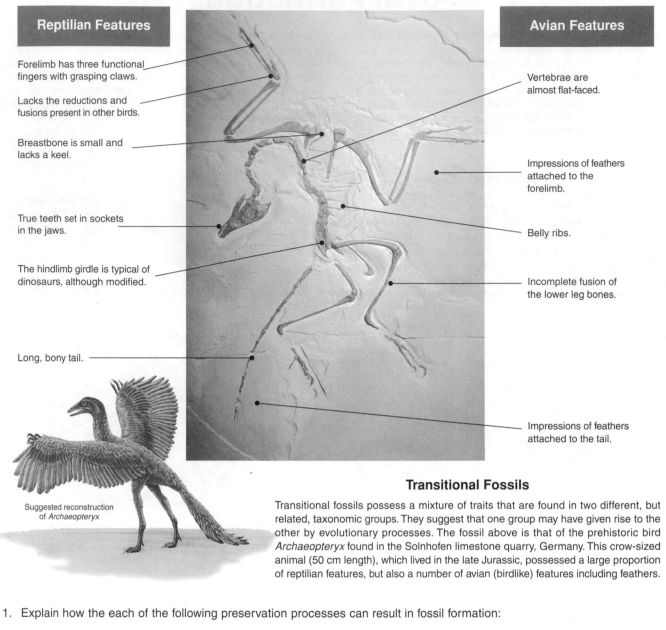

Reptilian Features

Forelimb has three functional fingers with grasping claws.

Lacks the reductions and fusions present in other birds.

Breastbone is small and lacks a keel.

True teeth set in sockets in the jaws.

The hindlimb girdle is typical of dinosaurs, although modified.

Long, bony tail.

Avian Features

Vertebrae are almost flat-faced.

Impressions of feathers attached to the forelimb.

Belly ribs.

Incomplete fusion of the lower leg bones.

Impressions of feathers attached to the tail.

Suggested reconstruction of *Archaeopteryx*

Transitional Fossils

Transitional fossils possess a mixture of traits that are found in two different, but related, taxonomic groups. They suggest that one group may have given rise to the other by evolutionary processes. The fossil above is that of the prehistoric bird *Archaeopteryx* found in the Solnhofen limestone quarry, Germany. This crow-sized animal (50 cm length), which lived in the late Jurassic, possessed a large proportion of reptilian features, but also a number of avian (birdlike) features including feathers.

1. Explain how the each of the following preservation processes can result in fossil formation:

   (a) Pyritization: _____

   _____

   (b) Amber: _____

   _____

   (c) Petrification: _____

   _____

   (d) Phosphatization: _____

   _____

   (e) Tar pit: _____

   _____

2. Name the natural process that must be arrested in order for fossilization to take place: _____

3. Comment on the importance of **transitional fossils** to our understanding of evolutionary change: _____

   _____

   _____

The Fossil Record

The diagram below represents a cutting into the earth revealing the layers of rock. Some of these layers may have been laid down by water (sedimentary rocks) or by volcanic activity (volcanic rocks). Fossils are the actual remains or impressions of plants, animals, or other organisms that become trapped in the sediments after their death. Layers of sedimentary rock are arranged in the order that they were deposited, with the most recent layers near the surface (unless they have been disturbed).

Profile with Sedimentary Rocks Containing Fossils

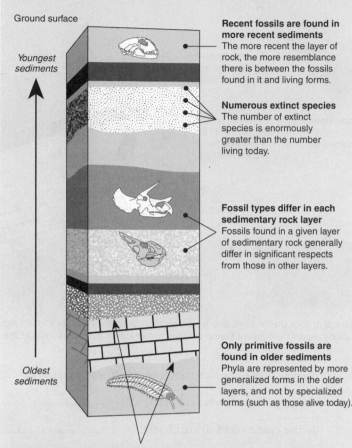

Ground surface

Youngest sediments

Oldest sediments

Recent fossils are found in more recent sediments
The more recent the layer of rock, the more resemblance there is between the fossils found in it and living forms.

Numerous extinct species
The number of extinct species is enormously greater than the number living today.

Fossil types differ in each sedimentary rock layer
Fossils found in a given layer of sedimentary rock generally differ in significant respects from those in other layers.

Only primitive fossils are found in older sediments
Phyla are represented by more generalized forms in the older layers, and not by specialized forms (such as those alive today).

New fossil types mark changes in environment
In the rocks marking the end of one geological period, it is common to find many new fossils that become dominant in the next. Each geological period had an environment very different from those before and after. Their boundaries coincided with drastic environmental changes and the appearance of new niches. These produced new selection pressures resulting in new adaptive features in the surviving species, as they responded to the changes.

The rate of evolution can vary

According to the fossil record, rates of evolutionary change seem to vary. There are bursts of species formation and long periods of relative stability within species (stasis). The occasional rapid evolution of new forms apparent in the fossil record, is probably a response to a changing environment. During periods of stable environmental conditions, evolutionary change may slow down.

The Fossil Record of Proboscidea

African and Indian elephants have descended from a diverse group of animals known as **proboscideans** (named for their long trunks). The first pig-sized, trunkless members of this group lived in Africa 40 million years ago. From Africa, their descendants invaded all continents except Antarctica and Australia. As the group evolved, they became larger; an effective evolutionary response to deter predators. Examples of extinct members of this group are illustrated below:

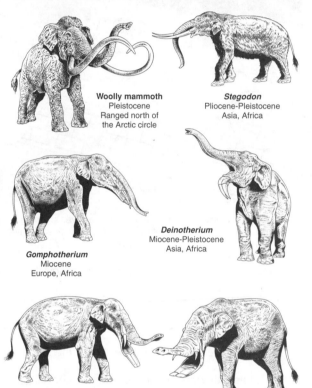

Woolly mammoth
Pleistocene
Ranged north of the Arctic circle

Stegodon
Pliocene-Pleistocene
Asia, Africa

Gomphotherium
Miocene
Europe, Africa

Deinotherium
Miocene-Pleistocene
Asia, Africa

Ambelodon
Middle Miocene
North America

Patybelodon
Middle Miocene
Northern Asia, Europe, Africa

- **Modern day species can be traced:** The evolution of many present-day species can be very well reconstructed. For instance, the evolutionary history of the modern elephants is exceedingly well documented for the last 40 million years. The modern horse also has a well understood fossil record spanning the last 50 million years.

- **Fossil species are similar to but differ from today's species:** Most fossil animals and plants belong to the same major taxonomic groups as organisms living today. However, they do differ from the living species in many features.

The Origin and Evolution of Life

1. Name an animal or plant taxon (e.g. family, genus, or species) that has:

   (a) A good fossil record of evolutionary development: _____

   (b) Appeared to have changed very little over the last 100 million years or so: _____

2. Discuss the importance of **fossils** as a record of evolutionary change over time: _____

   _____

   _____

   _____

Periodicals:
The accidental discovery of a feathered giant dinosaur

Related activities: Dating a Fossil Site

RA 2

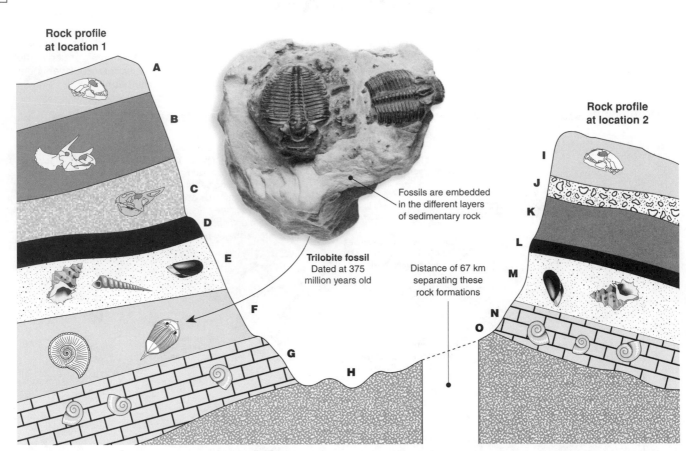

Rock profile at location 1

A B C D E F G H

Fossils are embedded in the different layers of sedimentary rock

Trilobite fossil
Dated at 375 million years old

Distance of 67 km separating these rock formations

Rock profile at location 2

I J K L M N O

The questions below relate to the diagram above, showing a hypothetical rock profile from two locations separated by a distance of 67 km. There are some differences between the rock layers at the two locations. Apart from layers D and L which are volcanic ash deposits, all other layers are comprised of sedimentary rock.

3. Assuming there has been no geological activity (e.g. tilting or folding), state in which rock layer (**A-O**) you would find:

   (a) The youngest rocks at Location 1: _____ (c) The youngest rocks at Location 2: _____

   (b) The oldest rocks at Location 1: _____ (d) The oldest rocks at Location 2: _____

4. (a) State which layer at location 1 is of the same age as layer M at location 2: _____

   (b) Explain the reason for your answer above: _____

   _____

5. The rocks in layer H and O are sedimentary rocks. Explain why there are no visible fossils in layers:

   _____

   _____

6. (a) State which layers present at location 1 are missing at location 2: _____

   (b) State which layers present at location 2 are missing at location 1: _____

7. Describe three methods of dating rocks: _____

   _____

   _____

8. Using radiometric dating, the trilobite fossil was determined to be approximately 375 million years old. The volcanic rock layer (D) was dated at 270 million years old, while rock layer B was dated at 80 million years old. Give the approximate **age range** (i.e. greater than, less than or between given dates) of the rock layers listed below:

   (a) Layer A: _____ (d) Layer G: _____

   (b) Layer C: _____ (e) Layer L: _____

   (c) Layer E: _____ (f) Layer O: _____

Dating a Fossil Site

The diagram below shows a rock shelter typical of those found in the Dordogne Valley of Southwest France. Such shelters have yielded a rich source of Neanderthal and modern human remains. It illustrates the way in which hominin activity is revealed at archaeological excavations. Occupation sites included shallow caves or rocky overhangs of limestone. The floors of these caves accumulated the debris of natural rockfalls, together with the detritus of human occupation at various layers, called **occupation horizons**. A wide array of techniques can be used for dating, some of which show a high degree of reliability (see the table below). The use of several appropriate techniques to date material improves the reliability of the date determined.

Rock shelter used
by early humans

Enlarged
below

Dating method	Dating range (years ago)	Datable materials
Radiocarbon (^{14}C)	1000 - 50 000+	Bone, shell, charcoal
Potassium-argon (K/Ar)	10 000 - 100 million	Volcanic rocks and minerals
Uranium series decay	less than 1 million	Marine carbonate, coral, shell
Thermoluminescence	less than 200 000	Ceramics (burnt clay)
Fission track	1000 - 100 million	Volcanic rock, glass, pottery
Electron spin resonance	2000 - 500 000	Bone, teeth, loess, burnt flint

Limestone cave formations can be dated using **uranium series** decay measurements. This method can be used to date calcite deposits up to the age of 300 000 years.

Rock fall from the roof of the overhanging shelter.

Occupation horizon **A**, with evidence of an ancient hearth in its uppermost layer.

Occupation horizon **B**, with evidence of a human burial.

Zone without any evidence of human occupation.

Charcoal

Pottery
Pottery bowl dated at
7000 ± 350 years old.

Bones
Skull of an early human but unable
to directly determine its age.

Hearth
The remains of an ancient fireplace
was dated at 18 500 ± 1000 years old.

Tooth
A bison's tooth was dated at
45 000 ± 2500 years old.

1. Discuss the significance of **occupation horizons**: _____

_____

_____

_____

2. Determine the approximate date range for the items below: (Hint: take into account layers/artifacts with known dates)

   (a) The skull at point B: _____

   (b) Occupation horizon A: _____

3. Name the dating methods that could have been used to date each of the following, at the site above:

   (a) Pottery bowl: _____   (c) Hearth: _____

   (b) Skull: _____   (d) Tooth: _____

Periodicals:
How old is...?

Related activities: The Fossil Record
Web links: Dating Methods

RDA 2

The Origin and Evolution of Life

Interpreting Fossil Sites

Human skull

Charcoal fragments (possible evidence of fire use and excellent for radiocarbon dating).

Bones from a large mammal with evidence of butchering (cut and scrape marks from stone tools). These provide information on the past ecology and environment of the hominins in question.

Excavation through rock strata (layers). The individual layers can be dated using both chronometric (absolute) and relative dating methods.

Stone tools

Photo: RA

istock

Searching for ancient human remains, including the evidence of culture, is the work of **paleoanthropologists**. Organic materials, such as bones and teeth, are examined and analyzed by physical anthropologists, while cultural materials, such as tools, weapons, shelters, and artworks, are examined by archaeologists. Both these disciplines, **paleoanthropology** and **archeology**, are closely associated with other scientific disciplines, including **geochemistry** (for **chronometric dates**), **geology** (for reconstructions of past physical landscapes), and **paleontology** (for knowledge of the past species assemblages).

The reconstruction of a **dig site**, pictured above, illustrates some of the features that may be present at a site of hominin activity. Naturally, the type of information recovered from a site will depend on several factors, including the original nature of the site and its contents, the past and recent site environment, and earlier disturbance by people or animals. During its period of occupation, a site represents an interplay between additive and subtractive processes; building vs destruction, growth vs decay. Organic matter decays, and other features of the site, such as tools, can be disarranged, weathered, or broken down. The archaeologists goal is to maximize the recovery of information, and recent trends have been to excavate and process artifacts immediately, and sometimes to leave part of the site intact so that future work, perhaps involving better methodologies, is still possible.

4. Explain why paleoanthropologists date and interpret all of the remains at a particular site of interest (e.g. animal bones, pollen, and vegetation, as well as hominin remains):

_____

_____

_____

_____

5. Discuss the importance of involving several scientific disciplines when interpreting a site of hominin activity:

_____

_____

_____

_____

_____

Protein Homologies

Traditionally, phylogenies were based largely on anatomical or behavioral traits and biologists attempted to determine the relationships between organisms based on overall degree of similarity or by tracing the appearance of key characteristics. With the advent of molecular techniques, homologies can now be studied at the molecular level as well and these can be compared to the phylogenies established using other methods. Protein sequencing provides an excellent tool for establishing **homologies** (similarities resulting from shared ancestry). Each protein has a specific number of amino acids arranged in a specific order. Any differences in the sequence reflect changes in the DNA sequence. Commonly studied proteins include blood proteins, such as **hemoglobin** (below), and the respiratory protein **cytochrome *c*** (overleaf). Many of these proteins are highly conserved, meaning they change very little over time, presumably because mutations would be detrimental to basic function. Conservation of protein sequences is indicated by the identical amino acid residues at corresponding parts of proteins.

Amino Acid Differences in Hemoglobin

Human beta chain	0
Chimpanzee	0
Gorilla	1
Gibbon	2
Rhesus monkey	8
Squirrel monkey	9
Dog	15
Horse, cow	25
Mouse	27
Gray kangaroo	38
Chicken	45
Frog	67

When the sequence of the **beta hemoglobin chain** (right), which is 146 amino acids long, is compared between humans, five other primates, and six other vertebrates, the results support the phylogenies established using other methods. The numbers in the table (left) represent the number of amino acid differences between the beta chain of humans and those of other species. In general, the number of amino acid differences between the hemoglobins of different vertebrates is inversely proportional to genetic relatedness.

Shading indicates (from top) primates, non-primate placental mammals, marsupials, and non-mammals.

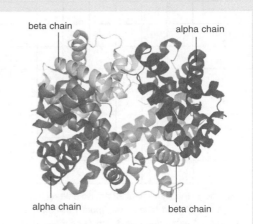

beta chain   alpha chain

alpha chain   beta chain

In most vertebrates, the oxygen-transporting blood protein hemoglobin is composed of four polypeptide chains, two alpha chains and two beta chains. Hemoglobin is derived from myoglobin, and ancestral species had just myoglobin for oxygen transport. When the amino acid sequences of myoglobin, the hemoglobin alpha chain, and the hemoglobin beta chain are compared, there are several amino acids that remain conserved between all three. These amino acid sequences must be essential for function because they have remained unchanged throughout evolution.

Using Immunology to Determine Phylogeny

The immune system of one species will recognize the blood proteins of another species as foreign and form antibodies against them. This property can be used to determine the extent of homology between species. Blood proteins, such as albumins, are used to prepare **antiserum** in rabbits. The antiserum contains antibodies against the test blood proteins (e.g. human) and will react to those proteins in any blood sample they are mixed with. The extent of the reaction indicates how different the proteins are; the greater the reaction, the greater the homology. This principle is illustrated (right) for antiserum produced to human blood and its reaction with the blood of other primates and a rat.

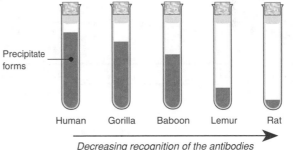

Precipitate forms

Human   Gorilla   Baboon   Lemur   Rat

Decreasing recognition of the antibodies against human blood proteins

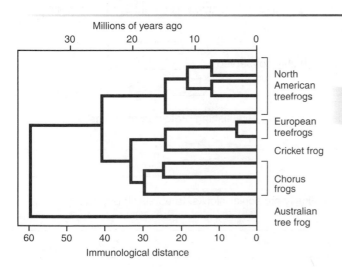

Millions of years ago

North American treefrogs

European treefrogs

Cricket frog

Chorus frogs

Australian tree frog

Immunological distance

The relationships among tree frogs have been established by immunological studies based on blood proteins such as immunoglobulins and albumins. The **immunological distance** is a measure of the number of amino acid substitutions between two groups. This, in turn, has been calibrated to provide a time scale showing when the various related groups diverged.

Related activities: Comparative Anatomy, DNA Homologies
Web links: Evidence for Evolution

Cytochrome *c* and the Molecular Clock Theory

Evolutionary change at the molecular level occurs primarily through fixation of neutral mutations by genetic drift. The rate at which one neutral mutation replaces another depends on the mutation rate, which is fairly constant for any particular gene.

If the rate at which a protein evolves is roughly constant over time, the amount of molecular change that a protein shows can be used as a molecular clock to date evolutionary events, such as the divergence of species.

The molecular clock for each species, and each protein, may run at different rates, so scientists calibrate the molecular clock data with other evidence (morphological, molecular) to confirm phylogenetic relationships.

For example, 20 amino acid substitutions in a protein since two organisms diverged from a known common ancestor 400 mya indicates an average substitution rate of 5 substitutions per 100 my.

	1					6				10				14		17	18		20		
Human	Gly	Asp	Val	Glu	Lys	Gly	Lys	Lys	Ile	Phe	Ile	Met	Lys	**Cys**	Ser	Gln	**Cys**	His	Thr	Val	Glu	Lys
Pig											Val	Gln			Ala							
Chicken			Ile								Val	Gln										
Dogfish										Val	Val	Gln			Ala							Asn
Drosophila	<<									Leu	Val	Gln	Arg		Ala							Ala
Wheat	<<	Asn	Pro	Asp	Ala		Ala				Lys	Thr			Ala						Asp	Ala
Yeast	<<	Ser	Ala	Lys			Ala	Thr	Leu		Lys	Thr	Arg		Glu	Leu						

This table shows the N-terminal 22 amino acid residues of human cytochrome *c*, with corresponding sequences from other organisms aligned beneath. Sequences are aligned to give the most position matches. A shaded square indicates no change. In every case, the cytochrome's heme group is attached to the Cys-14 and Cys-17. In *Drosophila*, wheat, and yeast, arrows indicate that several amino acids precede the sequence shown.

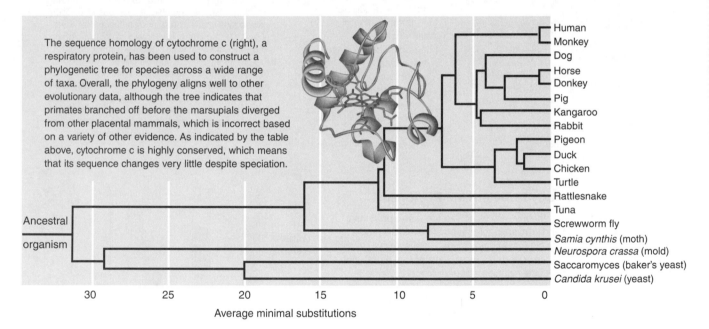

The sequence homology of cytochrome c (right), a respiratory protein, has been used to construct a phylogenetic tree for species across a wide range of taxa. Overall, the phylogeny aligns well to other evolutionary data, although the tree indicates that primates branched off before the marsupials diverged from other placental mammals, which is incorrect based on a variety of other evidence. As indicated by the table above, cytochrome c is highly conserved, which means that its sequence changes very little despite speciation.

Average minimal substitutions

1. Explain why chimpanzees and gorillas are considered most closely related to humans, while monkeys are less so:

_____

2. (a) Explain why a respiratory protein like cytochrome C would be highly conserved: _____

_____

   (b) Suggest why highly conserved proteins are good candidates for use in establishing protein homologies

_____

_____

3. Discuss some of the limitations of using protein homology, specifically molecular clocks, to establish phylogeny:

_____

_____

DNA Homologies

Establishing a phylogeny on the basis of homology in a protein, such as cytochrome c, is valuable, but it is also analogous to trying to see a complete picture through a small window. The technique of **DNA-DNA hybridization** provides a way to compare the total genomes of different species by measuring the degree of genetic similarity between pools of DNA sequences. It is usually used to determine the genetic distance between two species; the more closely two species are related, the fewer differences there will be between their genomes. This is because there has been less time for the point mutations that will bring about these differences to occur. This technique gives a measure of 'relatedness', and can be calibrated as a **molecular clock** against known fossil dates. It has been applied to primate DNA samples to help determine the approximate date of human divergence from the apes, which has been estimated to be between 10 and 5 million years ago.

DNA Hybridization

1. DNA from the two species to be compared is extracted, purified and cut into short fragments (e.g. 600-800 base pairs).

2. The DNA of one species is mixed with the DNAof another.

3. The mixture is incubated to allow DNA strands to dissociate and reanneal, forming hybrid double-stranded DNA.

4. The hybridized sequences that are highly similar will bind more firmly. A measure of the heat energy required to separate the hybrid strands provides a measure of DNA relatedness.

DNA Homologies Today

DNA-DNA hybridization has been criticized because duplicated sequences within a single genome make it unreliable for comparisons between closely related species.

Today, DNA sequencing and computed comparisons are more widely used to compare genomes, although DNA-DNA hybridization is still used to help identify bacteria.

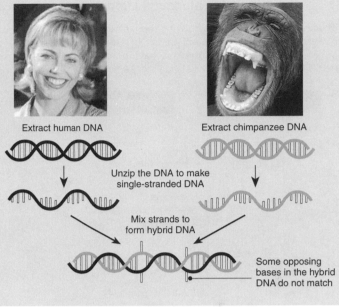

Extract human DNA   Extract chimpanzee DNA

Unzip the DNA to make single-stranded DNA

Mix strands to form hybrid DNA

Some opposing bases in the hybrid DNA do not match

The Origin and Evolution of Life

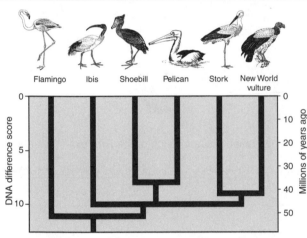

Flamingo   Ibis   Shoebill   Pelican   Stork   New World vulture

The relationships among the **New World vultures** and **storks** have been determined using DNA-DNA hybridization. It has been possible to estimate how long ago various members of the group shared a common ancestor.

Similarity of human DNA to that of other primates

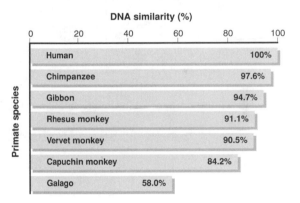

Primate species	DNA similarity (%)
Human	100%
Chimpanzee	97.6%
Gibbon	94.7%
Rhesus monkey	91.1%
Vervet monkey	90.5%
Capuchin monkey	84.2%
Galago	58.0%

The genetic relationships among the **primates** has been investigated using DNA-DNA hybridization. Human DNA was compared with that of the other primates. It largely confirmed what was suspected from anatomical evidence.

1. Explain how **DNA hybridization** can give a measure of genetic relatedness between species:

_____

_____

_____

2. Study the graph showing the results of a DNA hybridization between human DNA and that of other primates.

   (a) Identify which is the most closely related primate to humans: _____

   (b) Identify which is the most distantly related primate to humans: _____

3. State the DNA difference score for: (a) Shoebills and pelicans: _____   (b) Storks and flamingos: _____

4. On the basis of DNA hybridization, state how long ago the ibises and New World vultures shared a common ancestor:

_____

Periodicals:
Building a phylogenetic tree

Related activities: Protein Homologies

DA 2

Comparative Anatomy

The evolutionary relationships between groups of organisms is determined mainly by structural similarities called **homologous structures** (homologies), which suggest that they all descended from a common ancestor with that feature. The bones of the forelimb of air-breathing vertebrates are composed of similar bones arranged in a comparable pattern. This is indicative of a common ancestry. The early land vertebrates were amphibians and possessed a limb structure called the **pentadactyl limb**: a limb with five fingers or toes (below left). All vertebrates that descended from these early amphibians, including reptiles, birds and mammals, have limbs that have evolved from this same basic pentadactyl pattern. They also illustrate the phenomenon known as **adaptive radiation**, since the basic limb plan has been adapted to meet the requirements of different niches.

Generalized Pentadactyl Limb

The forelimbs and hind limbs have the same arrangement of bones but they have different names. In many cases bones in different parts of the limb have been highly modified to give it a specialized locomotory function.

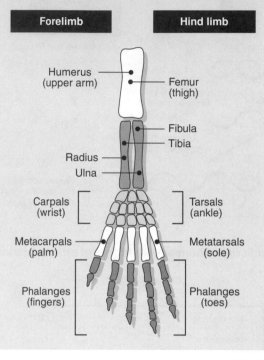

Forelimb | **Hind limb**

Humerus (upper arm) — Femur (thigh)

Fibula
Tibia

Radius
Ulna

Carpals (wrist) — Tarsals (ankle)

Metacarpals (palm) — Metatarsals (sole)

Phalanges (fingers) — Phalanges (toes)

Specializations of Pentadactyl Limbs

Bird's wing

Mole's forelimb

Bat's wing

Dog's front leg

Seal's flipper

Human arm

1. Briefly describe the purpose of the major anatomical change that has taken place in each of the limb examples above:

   (a) Bird wing: _Highly modified for flight. Forelimb is shaped for aerodynamic lift and feather attachment._

   (b) Human arm: _____

   (c) Seal flipper: _____

   (d) Dog foot: _____

   (e) Mole forelimb: _____

   (f) Bat wing: _____

2. Describe how homology in the pentadactyl limb is evidence for adaptive radiation: _____

   _____

   _____

   _____

3. Homology in the behavior of animals (for example, sharing similar courtship or nesting rituals) is sometimes used to indicate the degree of relatedness between groups. Suggest how behavior could be used in this way:

   _____

   _____

   _____

Related activities: Protein Homologies, The Evolution of Novel Forms
Web links: All in the family

Periodicals:
Using inquiry and
phylogeny to teach ...

Vestigial Organs

Some classes of characters are more valuable than others as reliable indicators of common ancestry. Often, the less any part of an animal is used for specialized purposes, the more important it becomes for classification. This is because common ancestry is easier to detect if a particular feature is unaffected by specific adaptations arising later during the evolution of the species. Vestigial organs are an example of this because, if they have no clear function and they are no longer subject to natural selection, they will remain unchanged through a lineage. It is sometimes argued that some vestigial organs are not truly vestigial, i.e. they may perform some small function. While this may be true in some cases, the features can still be considered vestigial if their new role is a minor one, unrelated to their original function.

Ancestors of Modern Whales

1.8 m long

2.5 m long

20-25 m long

Pakicetus (early Eocene) a carnivorous, four limbed, early Eocene whale ancestor, probably rather like a large otter. It was still partly terrestrial and not fully adapted for aquatic life.

Protocetus (mid Eocene). Much more whale-like than *Pakicetus*. The hind limbs were greatly reduced and although they still protruded from the body (arrowed), they were useless for swimming.

Basilosaurus (late Eocene). A very large ancestor of modern whales. The hind limbs contained all the leg bones, but were vestigial and located entirely within the main body, leaving a tissue flap on the surface (arrowed).

Vestigial organs are common in nature. The vestigial hind limbs of modern whales (right) provide anatomical evidence for their evolution from a carnivorous, four footed, terrestrial ancestor. The oldest known whale, *Pakicetus*, from the early Eocene (~54 mya) still had four limbs. By the late Eocene (~40 mya), whales were fully marine and had lost almost all traces of their former terrestrial life. For fossil evidence, see *Whale Origins* at: www.neoucom.edu/Depts/Anat/whaleorigins.htm

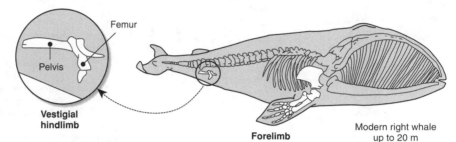

Femur

Pelvis

Vestigial hindlimb

Forelimb

Modern right whale up to 20 m

RM-DoC

Vestigial organs in birds and reptiles

In all snakes (far left), one lobe of the lung is vestigial (there is not sufficient room in the narrow body cavity for it). In some snakes there are also vestiges of the pelvic girdle and hind limbs of their walking ancestors. Like all ratites, kiwis (left) are flightless. However, more than in other ratites, the wings of kiwis are reduced to tiny vestiges. Kiwis evolved in the absence of predators to a totally ground dwelling existence.

1. In terms of natural selection explain how structures, that were once useful to an organism, could become vestigial:

_____

_____

_____

_____

2. Suggest why a vestigial structure, once it has been reduced to a certain size, may not disappear altogether:

_____

_____

3. Whale evolution shows the presence of **transitional forms** (fossils that are intermediate between modern forms and very early ancestors). Suggest how vestigial structures indicate the common ancestry of these forms:

_____

_____

_____

The Origin and Evolution of Life

Periodicals:
A waste of space

Related activities: Comparative Anatomy

A 2

The Evolution of Horses

The evolution of the horse from the ancestral *Hyracotherium* to modern *Equus* is well documented in the fossil record. For this reason it is often used to illustrate the process of evolution. The rich fossil record, which includes numerous **transitional fossils**, has enabled scientists to develop a robust model of horse phylogeny. Although the evolution of the line was once considered to be a gradual straight line process, it has been radically revised to a complex tree-like lineage with many divergences (below). It showed no inherent direction, and a diverse array of species coexisted for some time over the 55 million year evolutionary period. The environmental transition from forest to grasslands drove many of the changes observed in the equid fossil record. These include reduction in toe number, increased size of cheek teeth, lengthening of the face, and increasing body size.

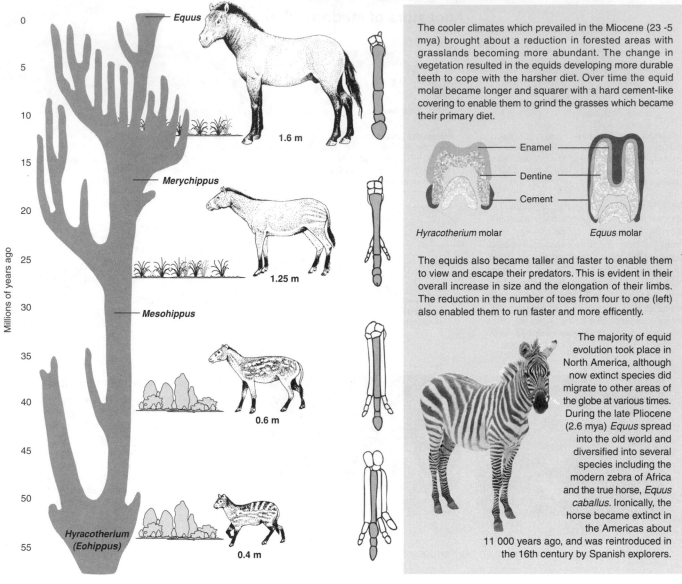

The cooler climates which prevailed in the Miocene (23 -5 mya) brought about a reduction in forested areas with grasslands becoming more abundant. The change in vegetation resulted in the equids developing more durable teeth to cope with the harsher diet. Over time the equid molar became longer and squarer with a hard cement-like covering to enable them to grind the grasses which became their primary diet.

Enamel — Dentine — Cement

Hyracotherium molar   *Equus* molar

The equids also became taller and faster to enable them to view and escape their predators. This is evident in their overall increase in size and the elongation of their limbs. The reduction in the number of toes from four to one (left) also enabled them to run faster and more efficently.

The majority of equid evolution took place in North America, although now extinct species did migrate to other areas of the globe at various times. During the late Pliocene (2.6 mya) *Equus* spread into the old world and diversified into several species including the modern zebra of Africa and the true horse, *Equus caballus*. Ironically, the horse became extinct in the Americas about 11 000 years ago, and was reintroduced in the 16th century by Spanish explorers.

1. Explain how the environmental change from forest to grassland influenced the following aspects of equid evolution:

    (a) Change in tooth structure: _____

    _____

    (b) Limb length: _____

    _____

    (c) Reduction in number of toes: _____

    _____

2. Explain why the equid fossil record provides a good example of the evolutionary process: _____

    _____

    _____

    _____

Related activities: *The History of Life on Earth, Fossil Formation*
Web links: *Horse Evolution, Fossil Horse Cybermuseum*

The Evolution of Novel Forms

The relatively new field of **evolutionary developmental biology** (or evo-devo) addresses the origin and evolution of embryonic development and looks at how modifications of developmental processes can lead to novel features. Scientists now know that specific genes in animals, including a subgroup of genes called *Hox* genes, are part of a basic **'tool kit'** of genes that control animal development. Genomic studies have shown that these genes are **highly conserved** (i.e. they show little change in different lineages). Very disparate organisms share the same **tool kit** of genes, but regulate them differently. The implication of this is that large changes in morphology or function are associated with changes in gene regulation, rather than the evolution of new genes, and natural selection associated with gene switches plays a major role in evolution.

The Role of *Hox* Genes

Hox genes control the development of back and front parts of the body. The same genes (or homologous ones) are present in essentially all animals, including humans.

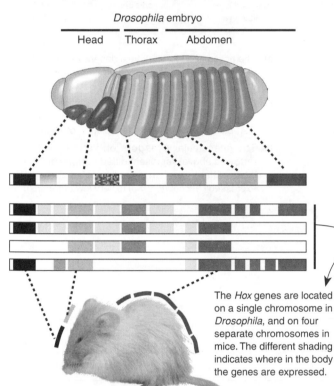

Drosophila embryo

Head | Thorax | Abdomen

The *Hox* genes are located on a single chromosome in *Drosophila*, and on four separate chromosomes in mice. The different shading indicates where in the body the genes are expressed.

The Evolution of Novel Forms

Even very small changes (mutations) in the *Hox* genes can have a profound effect on morphology. Such changes to the genes controlling development have almost certainly been important in the evolution of novel structures and body plans. Four principles underly the evo-devo thinking regarding the evolution of novel forms:

- **Evolution works with what is already present**: New structures are modifications of pre-existing structures.

- **Multifunctionality** and **redundancy**: Functional redundancy in any part of a multifunctional structure allows for specialisation and division of labour through the development of two separate structures.

  Example: the diversity of appendages (including mouthparts) in arthropods.

- **Modularity**: Modular architecture in animals (arthropods, vertebrates) allows for the modification and specialisation of individual body parts. Genetic switches allow changes in one part of a structure, independent of other parts.

Shifting *Hox* Expression

Huge diversity in morphology in organisms within and across phyla could have arisen through small changes in the genes controlling development.

Differences in neck length in vertebrates provides a good example of how changes in gene expression can bring about changes in morphology. Different vertebrates have different numbers of neck vertebrae. The boundary between neck and trunk vertebrae is marked by expression of the **Hox c6 gene** (c6 denotes the sixth cervical or neck vertebra) in all cases, but the position varies in each animal relative to the overall body. The forelimb (arrow) arises at this boundary in all four-legged vertebrates. In snakes, the boundary is shifted forward to the base of the skull and no limbs develop. As a result of these differences in expression, mice have a short neck, geese a long neck, and snakes, no neck at all.

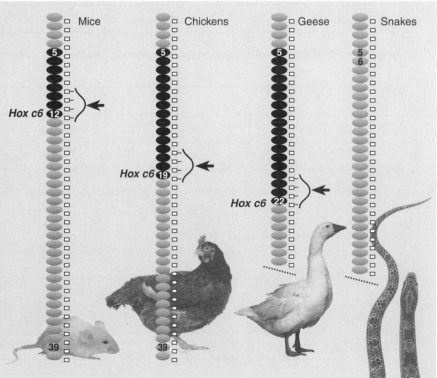

Mice | Chickens | Geese | Snakes

Hox c6 12

Hox c6 19

Hox c6 22

Periodicals:
Regulating evolution,
A fin is a limb is a wing…

Related activities: *The Modern Theory of Evolution*
Web links: *Genetic Tool Kit*

Genetic Switches in Evolution

The *Hox* genes are just part of the collection of genes that make up the genetic tool kit for animal development. The genes in the tool kit act as switches, shaping development by affecting how other genes are turned on or off. The distribution of genes in the tool kit indicates that it is ancient and was in place before the evolution of most types of animals. Differences in form arise through changes in genetic switches. One example is the evolution of eyespots in butterflies:

■ The ***Distal-less*** gene is one of the important **master body-building genes** in the genetic tool kit. Switches in the *Distal-less* gene control expression in the embryo (E), larval legs (L), and wing (W) in flies and butterflies, but butterflies have also evolved an extra switch (S) to control eyespot development.

■ Once *Distal-less* spots evolved, changes in *Distal-less* expression (through changes in the switch) produced more or fewer spots.

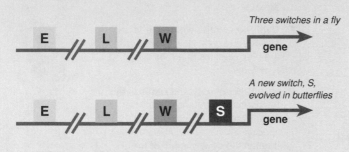

Three switches in a fly

E // L // W → **gene**

A new switch, S, evolved in butterflies

E // L // W // S → **gene**

Changes in *Distal-less* regulation were probably achieved by changing specific sequences of the *Distal-less* gene eyespot switch. The result? Changes in eyespot size and number.

Same Gene, New Tricks

Stichophthalma camadeva   *Junonia coenia (buckeye)*   *Taenaris macrops*

■ The action of a tool kit protein depends on context: where particular cells are located at the time when the gene is switched on.

■ Changes in the DNA sequence of a genetic switch can change the zone of gene expression without disrupting the function of the tool kit protein itself.

■ The spectacular **eyespots** on butterfly wings (arrowed above) represent different degrees of a basic pattern, from virtually all eyespot elements expressed (*Stichophthalma*) to very few (*Taenaris*).

1. Explain what is meant by "evo-devo" and state its aims: _____

_____

_____

2. Briefly describe the role of *Hox* genes in animal development: _____

_____

_____

_____

3. Outline the evidence that evo-devo provides for evolution and the mechanisms by which it occurs: _____

_____

_____

_____

_____

4. Using an example, discuss how changes in gene expression can bring about changes in morphology:

_____

_____

_____

_____

_____

_____

_____

Oceanic Island Colonizers

The distribution of organisms around the world lends powerful support to the idea that modern forms evolved from ancestral populations. **Biogeography** is the study of the geographical distribution of species, both present-day and extinct. It stresses the role of dispersal of species from a point of origin across pre-existing barriers. Studies from the island populations (below) indicate that flora and fauna of different islands are more closely related to adjacent continental species than to each other.

Galapagos and Cape Verde islands

Galapagos Is
South America
800 km
Pacific Ocean

Cape Verde Is
Western Africa
450 km
Atlantic Ocean

Biologists did not fully appreciate the uniqueness and diversity of tropical island biota until explorers began to bring back samples of flora and fauna from their expeditions in the 19th Century. The Galapagos Islands, the oldest of which arose 3-4 million years ago, had species similar to but distinct from those on the South American mainland. Similarly, in the Cape Verde Islands, species had close relatives on the West Africa mainland. This suggested to biologists that ancestral forms found their way from the mainland to the islands where they then underwent evolutionary changes.

South America
South Atlantic Ocean
Africa
4500 km
3000 km
Tristan da Cunha

Tristan da Cunha

The island of Tristan da Cunha in the South Atlantic Ocean is a great distance from any other land mass. Even though it is closer to Africa, there are more species closely related to South American species found there (see table on right). This is probably due to the predominant westerly trade winds from the direction of South America. The flowering plants of universal origin are found in both Africa and South America and could have been introduced from either land mass.

South American origin
7   Flowering plants
5   Ferns
30 Liverworts

African origin
2   Flowering plants
2   Ferns
5   Liverworts

Universal origin
19 Flowering plants

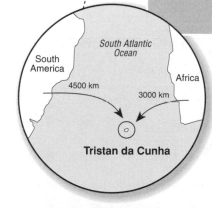

The flightless cormorant is one of a number of bird species that lost the power of flight after becoming resident on an island. Giant tortoises, such as the 11 subspecies remaining on the Galapagos today (center) were, until relatively recently, characteristic of many islands in the Indian Ocean including the Seychelles archipelago, Reunion,

Mauritius, Farquhar, and Diego Rodriguez. These were almost completely exterminated by early Western sailors, although a small population remains on the island of Aldabra. Another feature of oceanic islands is the adaptive radiation of colonizing species into different specialist forms. The three species of Galapagos iguana almost certainly arose,

through speciation, from a hardy traveler from the South American mainland. The marine iguana (above) feeds on shoreline seaweeds and is an adept swimmer. The two species of land iguana (not pictured) feed on cacti, which are numerous. The second of these (the pink iguana) was identified as a separate species only in 2009.

The Origin and Evolution of Life

Related activities: Continental Drift and Evolution

A 2

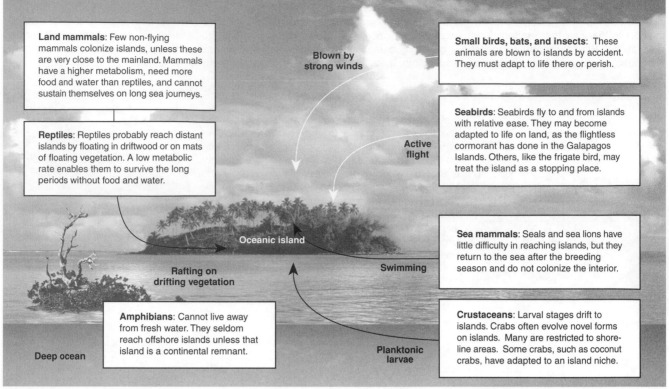

The diversity and uniqueness of biota of islands is determined by migration to and from the island and extinctions and diversifications following colonization. These events are themselves affected by a number of other factors (see table, below right). The animals that successfully colonize oceanic islands have to be marine in habit, or able to survive long periods at sea or in the air. This precludes large numbers from ever reaching distant islands.

Plants also have limited capacity to reach distant islands. Only some have fruits and seeds that are salt tolerant. Many plants are transferred to the islands by wind or migrating birds. The biota of the **Galapagos islands** provide a good example of the results of such a colonization process. For example, all the subspecies of giant tortoise evolved in Galápagos from a common ancestor that arrived from the mainland, floating with the ocean currents.

1. The Galapagos and the Cape Verde Islands are both tropical islands close to the equator, yet their biotas are quite different. Explain why this is the case:

_____

_____

_____

2. Explain why the majority of the plant species found on Tristan da Cunha originated from South America, despite its greater distance from the island:

_____

_____

3. The table (right) identifies some of the factors influencing the composition of island biota. Explain how each of the three following might affect the diversity and uniqueness of the biota found on an oceanic island:

(a) Large island area: _____

_____

(b) Long period of isolation from other land masses: _____

_____

(c) Relatively close to a continental land mass: _____

_____

4. Describe one feature typical of an oceanic island colonizer and explain its significance: _____

_____

Factors affecting final biota
Degree of isolation
Length of time of isolation
Size of island
Climate (tropical/Arctic, arid/humid)
Location relative to ocean currents
Initial plant and animal composition
The species composition of earliest arrivals (if always isolated)
Serendipity (chance arrivals)

Continental Drift and Evolution

Continental drift is a measurable phenomenon; it has happened in the past and continues today. Movements of up to 2-11 cm a year have been recorded between continents using laser technology. The movements of the Earth's 12 major crustal plates are driven by thermal convection currents in the mantle; a geological process known as **plate tectonics.** Some continents appear to be drifting apart while others are on a direct collision course. Various lines of evidence show that the modern continents were once joined together as 'supercontinents'. One supercontinent, called **Gondwana**, was made up of the southern continents some 200 million years ago. The diagram below shows some of the data collected that are used as evidence to indicate how the modern continents once fitted together.

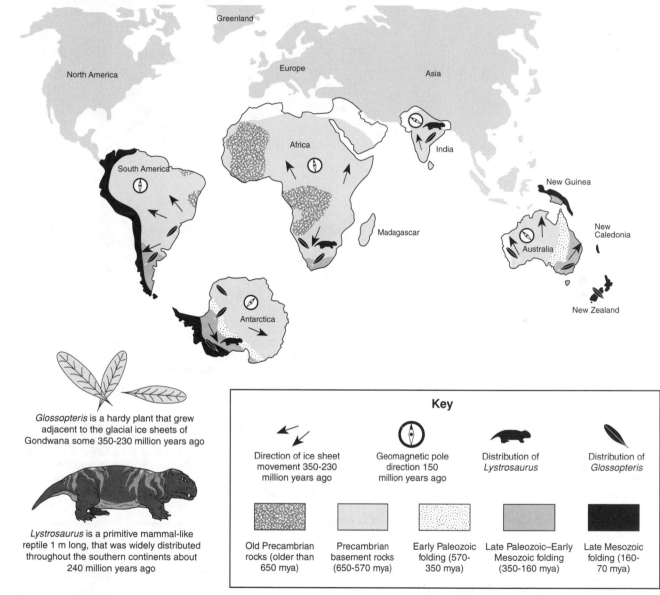

Glossopteris is a hardy plant that grew adjacent to the glacial ice sheets of Gondwana some 350-230 million years ago

Lystrosaurus is a primitive mammal-like reptile 1 m long, that was widely distributed throughout the southern continents about 240 million years ago

Key

| Direction of ice sheet movement 350-230 million years ago | Geomagnetic pole direction 150 million years ago | Distribution of *Lystrosaurus* | Distribution of *Glossopteris* |

| Old Precambrian rocks (older than 650 mya) | Precambrian basement rocks (650-570 mya) | Early Paleozoic folding (570-350 mya) | Late Paleozoic–Early Mesozoic folding (350-160 mya) | Late Mesozoic folding (160-70 mya) |

1. Name the modern landmasses (continents and large islands) that made up the supercontinent of Gondwana:

_____

2. Cut out the southern continents on page 315 and arrange them to recreate the supercontinent of Gondwana. Take care to cut the shapes out close to the coastlines. When arranging them into the space showing the outline of Gondwana on page 314, take into account the following information:
   (a) The location of ancient rocks and periods of mountain folding during different geological ages.
   (b) The direction of ancient ice sheet movements.
   (c) The geomagnetic orientation of old rocks (the way that magnetic crystals are lined up in ancient rock gives an indication of the direction the magnetic pole was at the time the rock was formed).
   (d) The distribution of fossils of ancient species such as *Lystrosaurus* and *Glossopteris*.

3. Once you have positioned the modern continents into the pattern of the supercontinent, mark on the diagram:
   (a) The likely position of the South Pole 350-230 million years ago (as indicated by the movement of the ice sheets).
   (b) The likely position of the geomagnetic South Pole 150 million years ago (as indicated by ancient geomagnetism).

4. State what general deduction you can make about the position of the polar regions with respect to land masses:

_____

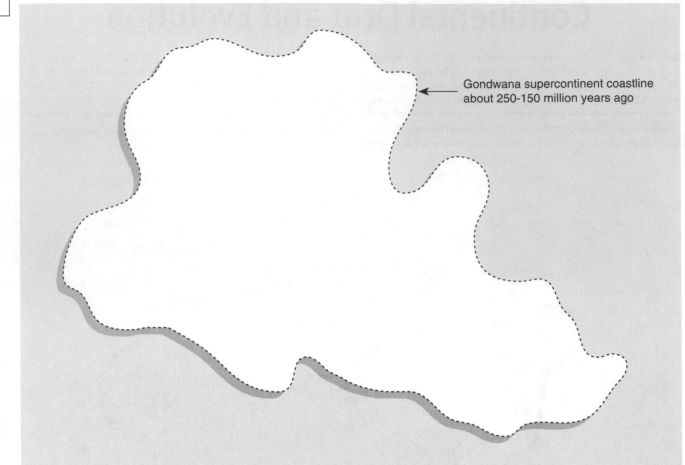

Gondwana supercontinent coastline about 250-150 million years ago

5. Fossils of *Lystrosaurus* are known from Antarctica, South Africa, India and Western China. With the modern continents in their present position, *Lystrosaurus* could have walked across dry land to get to China, Africa and India. It was not possible for it to walk to Antarctica, however. Explain the distribution of this ancient species in terms of continental drift:

_____

_____

6. The southern beech (*Nothofagus*) is found only in the southern hemisphere, in such places as New Caledonia, New Guinea, eastern Australia (including Tasmania), New Zealand, and southern South America. Fossils of southern beech trees have also been found in Antarctica. They have never been distributed in South Africa or India. The seeds of the southern beech trees are not readily dispersed by the wind and are rapidly killed by exposure to salt water.

(a) Suggest a reason why *Nothofagus* is not found in Africa or India: _____

_____

(b) Use a colored pen to indicate the distribution of *Nothofagus* on the current world map (on the previous page) and on your completed map of Gondwana above.

(c) State how the arrangement of the continents into Gondwana explains this distribution pattern:

_____

_____

7. The Atlantic Ocean is currently opening up at the rate of 2 cm per year. At this rate in the past, calculate how long it would have taken to reach its current extent, with the distance from Africa to South America being 2300 km (assume the rate of spreading has been constant):

_____

8. Explain how continental drift provides evidence to support evolutionary theory: _____

_____

_____

_____

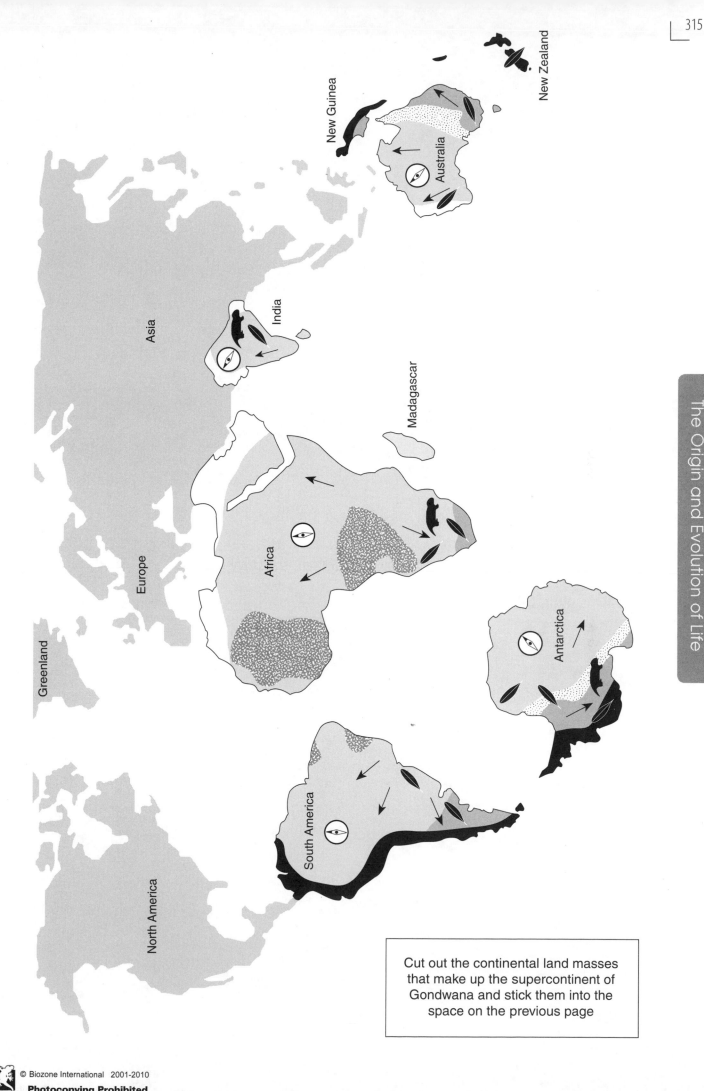

New Guinea

New Zealand

Australia

India

Asia

Madagascar

Africa

Antarctica

Europe

Greenland

South America

North America

Cut out the continental land masses that make up the supercontinent of Gondwana and stick them into the space on the previous page

This page has been deliberately left blank

KEY TERMS: Memory Card Game

The cards below have a keyword or term printed on one side and its definition printed on the opposite side. The aim is to win as many cards as possible from the table. To play the game.....

1) Cut out the cards and lay them definition side down on the desk. You will need one set of cards between two students.

2) Taking turns, choose a card and, BEFORE you pick it up, state your own best definition of the keyword to your opponent.

3) Check the definition on the opposite side of the card. If both you and your opponent agree that your stated definition matches, then keep the card. If your definition does not match then return the card to the desk.

4) Once your turn is over, your opponent may choose a card.

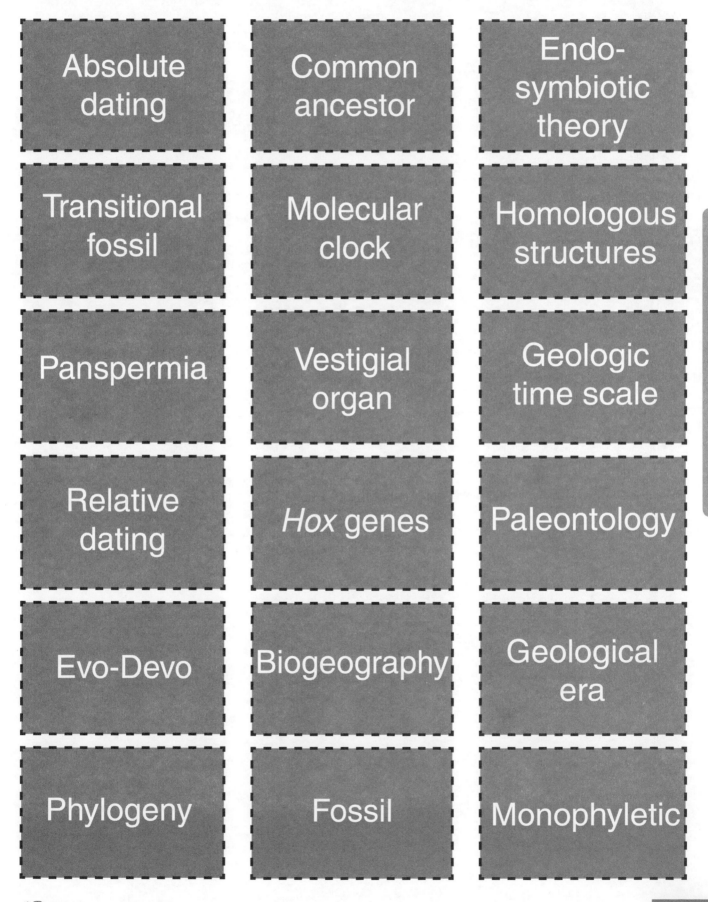

Absolute dating	Common ancestor	Endo-symbiotic theory
Transitional fossil	Molecular clock	Homologous structures
Panspermia	Vestigial organ	Geologic time scale
Relative dating	*Hox* genes	Paleontology
Evo-Devo	Biogeography	Geological era
Phylogeny	Fossil	Monophyletic

The Origin and Evolution of Life

When you've finished the game keep these cutouts and use them as flash cards!

A theory for the origins of eukaryotic mitochondria and plastids as bacterial endosymbionts.

The (usually most recent) individual from which all organisms a taxon are directly descended.

The process of determining a specific date for an archaeological or palaeontological site or artifact, usually based on its physical or chemical properties. Also called chronometric dating.

Structures in different but related species that are derived from the same ancestral structure but now serve different purposes, e.g. wings and fins.

A method, analogous to a timepiece, that uses molecular change to deduce the time in geologic history when two taxa diverged and so can be used to establish phylogenies.

Thee fossilized remains organisms that illustrate an evolutionary transition in that they possess both primitive and derived characteristics.

A system of chronological measurement relating stratigraphy to time. It is used by scientists to describe the timing and relationships between events in the history of the Earth.

Homologous structures that have apparently lost all or most of their original function in a species as a result of that species' evolution.

The hypothesis that life on Earth was 'seeded' from life already existing elsewhere in the Universe.

A historical science involving the study of prehistoric life, including the evolution of organisms.

Genes that play a role in development and are shared by almost all organisms.

A method that determines the sequential order in which past events occurred, without necessarily determining their absolute age. Also called chronostatic dating.

A clearly defined period of time that is of an arbitrary but well defined length, and is marked by a start event and an end event.

The study of how biodiversity is distributed over space and time.

The study relating the evolution of new characteristics to changes in the genes controlling development. Evolutionary developmental biology (acronym).

A taxon originating from (and including) a single common ancestor.

The preserved remains or traces (e.g. footprints) of past organisms.

The evolutionary history or genealogy of a group of organisms. Often represented as a 'tree' showing descent of new species from the ancestral one.

Speciation

KEY CONCEPTS

▶ Evolution (change in allele frequencies in a gene pool) is a consequence of populations rarely, if ever, being in genetic equilibrium.

▶ It is possible to calculate allele frequencies for a population.

▶ Reproductive isolation is essential for speciation. This is often preceded by allopatry.

▶ Natural selection sorts the variability in gene pools and establishes adaptive genotypes.

KEY TERMS

adaptation
allele
allele frequency
allopatric (allopatry)
balanced polymorphism
deme
directional selection
disruptive selection
evolution
fitness
founder effect
gene flow
gene pool
genetic bottleneck
genetic drift
genetic equilibrium
genotype
Hardy-Weinberg equation
hybrid
industrial melanism
microevolution
migration
mutation
natural selection
new synthesis
phenotype
population
postzygotic
prezygotic
reproductive isolation
ring species
selective breeding
sexual selection
speciation
species
stabilizing selection
sympatric (sympatry)

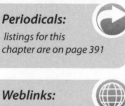

Periodicals:
listings for this chapter are on page 391

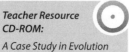

Weblinks:
www.thebiozone.com/
weblink/SB1-2597.html

Teacher Resource CD-ROM:
A Case Study in Evolution

OBJECTIVES

☐ 1. Use the **KEY TERMS** to help you understand and complete these objectives.

Evolution: A Review   pages 320-325

☐ 2. Recall the fundamental ideas in Darwin's *"Theory of evolution by natural selection"* and explain how his ideas have been since modified in the **new synthesis** to take into account our current understanding of genetics.

☐ 3. Explain what is meant by **fitness** and explain how evolution, through **adaptation**, equips species for survival.

☐ 4. Explain what is meant by a (biological) **species** and describe the limitations of its definition. Explain examples of **ring species** and their significance.

Processes in Gene Pools   pages 326-349, 357-361

☐ 5. Explain the concept of the **gene pool**. Show how **allele frequencies** are expressed for a population.

☐ 6. Describe the criteria for **genetic equilibrium** in a population. Use the **Hardy-Weinberg equation** to calculate allele frequencies, genotype frequencies, and phenotype frequencies for a gene in a population.

☐ 7. Recall **microevolution** as changes in the allele frequencies of gene pools. Explain how each of the following processes alters allele frequencies in a gene pool: **natural selection**, **genetic drift**, **gene flow**, and **mutation**.

☐ 8. Recall the role of **natural selection** in selectively changing genetic variation in a population. Explain **stabilizing**, **directional**, and **disruptive selection**. Describe examples of evolution by **natural selection**.

☐ 9. Describe the role of **sexual selection** in the evolution of sexual dimorphism.

☐ 10. Explain the genetic and evolutionary consequences of the **founder effect**.

☐ 11. Explain the genetic and evolutionary consequences of the **bottleneck effect**.

☐ 12. Explain what is meant by **selective breeding** (artificial selection) and explain its genetic basis. Describe examples of selective breeding, identifying the contribution of modern technology where appropriate.

Speciation   pages 350-356

☐ 13. Explain the significance of **demes** in the process of speciation. Distinguish between **allopatric** and **sympatric** populations.

☐ 14. Explain **allopatric speciation** in terms of **migration**, **isolation**, and **adaptation** leading to reproductive (genetic) isolation of gene pools.

☐ 15. Describe and explain mechanisms of **reproductive isolation**, distinguishing between **prezygotic** and **postzygotic** reproductive isolating mechanisms.

☐ 16. Describe and explain **sympatric speciation**. Discuss the role of **polyploidy** in **instant speciation** events and in the development of some crops.

Small Flies and Giant Buttercups

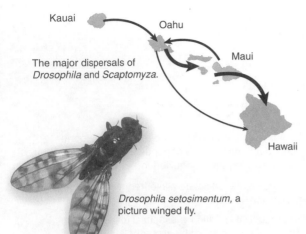

Kauai
Oahu
Maui
Hawaii

The major dispersals of *Drosophila* and *Scaptomyza*.

Drosophila setosimentum, a picture winged fly.

Photo: Karl Magnacca

Photo: Velela

Drosophilidae (commonly known as fruit flies) are a group of small flies found almost everywhere in the world. Two genera, *Drosophila* and *Scaptomyza* are found in the Hawaiian islands and between them there are more than 800 species present on a land area of just 16,500 km²; it is one of the densest concentrations of related species found anywhere. The flies range from 1.5 mm to 20 mm in length and display a startling range of wing forms and patterns, body shapes and colours, and head and leg shapes. This diverse array of species and characteristics has made these flies the subject of much evolutionary and genetics research. Genetic analyses show that they are all related to a single species that may have arrived on the islands around 8 million years ago and diversified to exploit a range of unoccupied niches. Older species appear on the older islands and more recent species appear as one moves from the oldest to the newest islands. Such evidence points to numerous colonisation events as new islands emerged from the sea. The volcanic nature of the islands means that newly isolated environments are a frequent occurrence. For example, forested areas may become divided by lava flows, so that flies in one region diverge rapidly from flies in another just tens of metres away. One such species is *D. silvestris*. Males have a series of hairs on their forelegs, which they brush against females during courtship. Males in the northeastern part of the island have many more of these hairs than the males on the southwestern side of the island. While still the same species, the two demes are already displaying structural and behavioural isolation. Behavioural isolation is clearly an important phenomenon in drosophilid speciation. A second species, *D. heteroneura*, is closely related to *D. silvestris* and the two species live sympatrically. Although hybrid offspring are fully viable, hybridisation rarely occurs because male courtship displays are very different.

New Zealand alpine buttercups (*Ranunculus*) are some of the largest in the world and are also the product of repeated speciation events. There are 14 species of *Ranunculus* in New Zealand; more than in the whole of North and South America combined. They occupy five distinct habitats ranging from snowfields and scree slopes to bogs. Genetic studies have shown that this diversity is the result of numerous isolation events following the growth and recession of glaciers. As the glaciers retreat, alpine habitat becomes restricted and populations are isolated at the tops of mountains. This restricts gene flow and provides the environment for species divergence. When the glaciers expand again, the extent of the alpine habitat increases, allowing isolated populations to come in contact and closely related species to hybridize.

1. Explain why so many drosophilidae are present in Hawaii: _____

_____

_____

_____

2. Explain why these flies are of interest: _____

_____

_____

3. Describe the relationship between the age of the islands and the age of the fly species: _____

_____

_____

_____

4. Explain why New Zealand has so many alpine buttercups: _____

_____

_____

_____

_____

_____

_____

Related activities: Gene Pools and Evolution, The Species Concept

The Species Concept

One of the best recognized definitions of a biological species is as "*a group of actually or potentially interbreeding natural populations that is reproductively isolated from other such groups*" (Ernst Mayr). However, the concept of a species is not as simple as it may first appear. The occurrence of cryptic species and closely related species that interbreed to produce fertile hybrids (e.g. species of *Canis*), indicate that the boundaries of a species gene pool can be unclear. Increasingly, biologists are turning to molecular analyses to help clarify relationships between the closely related populations that we regard as one species.

Geographical distribution of selected *Canis* species

The global distribution of most of the species of *Canis* (dogs and wolves) is shown on the map, right. The **gray wolf** inhabits the forests of North America, northern Europe, and Siberia. The **red wolf** and **Mexican wolf** (original distributions shown) were once distributed more widely, but are now extinct in the wild except for reintroduction efforts. In contrast, the **coyote** has expanded its original range and is now found throughout North and Central America. The range of the three **jackal** species overlap in the open savannah of Eastern Africa. The **dingo** is distributed throughout the Australian continent. Distribution of the domesticated **dog** is global as a result of the spread of human culture. The dog has been able to interbreed with all other members of the genus listed here to form fertile hybrids.

Interbreeding between *Canis* species

The *Canis* species illustrate problems with the traditional species concept. The domesticated dog is able to breed with other members of the same genus to produce fertile hybrids. Red wolves, gray wolves, Mexican wolves, and coyotes are all capable of interbreeding to produce fertile hybrids. Red wolves are very rare, and it is possible that hybridization with coyotes has been a factor in their decline. By contrast, the ranges of the three distinct species of jackal overlap in the Serengeti of Eastern Africa. These animals are highly territorial, but simply ignore members of the other jackal species and no interbreeding takes place.

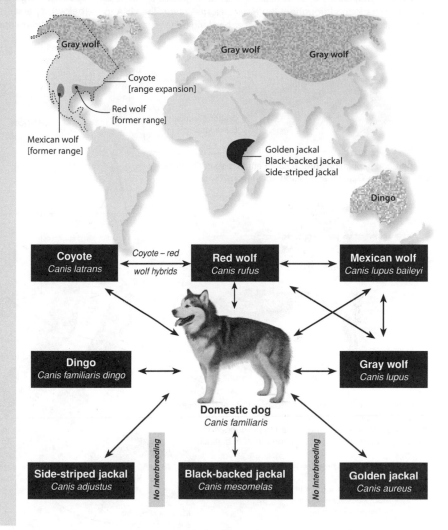

1. Describe the type of barrier that prevents the three species of jackal from interbreeding:

   _____

2. Describe the factor that has prevented the dingo from interbreeding with other *Canis* species (apart from the dog):

   _____

3. Describe a possible contributing factor to the occurrence of interbreeding between the coyote and red wolf:

   _____

4. The gray wolf is a widely distributed species. Explain why the North American population is considered to be part of the same species as the northern European and Siberian populations:

   _____

   _____

5. Explain what you understand by the term species, identifying examples where the definition is problematic:

   _____

   _____

   _____

Speciation

Periodicals:
Species and species formation

Related activities: Stages in Species Development, Geographical Distribution
Web links: Species and Speciation

A 2

Ensatina in the Pacific North-West: Ring Species or Cryptic Species?

A **ring species** is a connected series of closely related populations, distributed around a geographical barrier, in which the adjacent populations in the ring are able to interbreed, but those at the extremes of the ring are reproductively isolated. Such ring species are regarded as important because they are seen to illustrate what happens over time as populations diverge genetically. *Ensatina eschscholtzii* is a species of lungless salamanders found throughout the Pacific North-West of the USA to Baja California in Mexico. *E. eschscholtzii* has long been considered a

ring species, which probably expanded southwards from an ancestral population in Oregon along either side of California's Central Valley. However, molecular analyses are now indicating that the story of *Ensatina* is more complicated than first supposed. Geographically adjacent populations within the ring may be genetically isolated or comprise morphologically identical but genetically distinct **cryptic species**. Regardless of the conclusions drawn from the evidence (below), species such as *Ensatina* give us reason to reevaluate how we define species and quantify biodiversity.

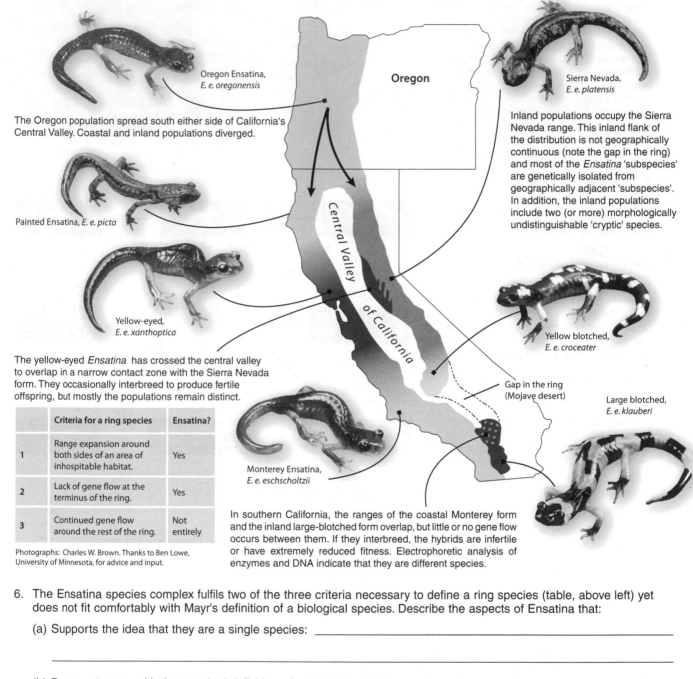

Oregon Ensatina, *E. e. oregonensis*

The Oregon population spread south either side of California's Central Valley. Coastal and inland populations diverged.

Painted Ensatina, *E. e. picta*

Yellow-eyed, *E. e. xanthoptica*

The yellow-eyed *Ensatina* has crossed the central valley to overlap in a narrow contact zone with the Sierra Nevada form. They occasionally interbreed to produce fertile offspring, but mostly the populations remain distinct.

Sierra Nevada, *E. e. platensis*

Inland populations occupy the Sierra Nevada range. This inland flank of the distribution is not geographically continuous (note the gap in the ring) and most of the *Ensatina* 'subspecies' are genetically isolated from geographically adjacent 'subspecies'. In addition, the inland populations include two (or more) morphologically undistinguishable 'cryptic' species.

Yellow blotched, *E. e. croceater*

Gap in the ring (Mojave desert)

Large blotched, *E. e. klauberi*

Monterey Ensatina, *E. e. eschscholtzii*

In southern California, the ranges of the coastal Monterey form and the inland large-blotched form overlap, but little or no gene flow occurs between them. If they interbreed, the hybrids are infertile or have extremely reduced fitness. Electrophoretic analysis of enzymes and DNA indicate that they are different species.

	Criteria for a ring species	Ensatina?
1	Range expansion around both sides of an area of inhospitable habitat.	Yes
2	Lack of gene flow at the terminus of the ring.	Yes
3	Continued gene flow around the rest of the ring.	Not entirely

Photographs: Charles W. Brown. Thanks to Ben Lowe, University of Minnesota, for advice and input.

6. The Ensatina species complex fulfils two of the three criteria necessary to define a ring species (table, above left) yet does not fit comfortably with Mayr's definition of a biological species. Describe the aspects of Ensatina that:

(a) Supports the idea that they are a single species: _____

_____

(b) Does not agree with the standard definition of a biological species: _____

_____

7. Yellow-eyed *Ensatina* is a mimic of the toxic California newt. What might this suggest about the selection pressures on this subspecies and their influence on the rate at which the population becomes genetically distinct?

_____

_____

_____

The Modern Theory of Evolution

Although **Charles Darwin** is credited with the development of the theory of evolution by natural selection, there were many people that contributed ideas upon which he built his own. Since Darwin first proposed his theory, aspects that were problematic (such as the mechanism of inheritance) have now been explained. The development of the modern theory of evolution has a history going back at least two centuries. The diagram below illustrates the way in which some of the major contributors helped to form the currently accepted model, or **new synthesis**. Understanding of evolutionary processes continued to grow through the 1980s and 1990s as comparative molecular sequence data were amassed and understanding of the molecular basis of developmental mechanisms improved. Most recently, in the exciting new area of evolutionary developmental biology (**evo-devo**), biologists have been exploring how developmental gene expression patterns explain how groups of organisms evolved.

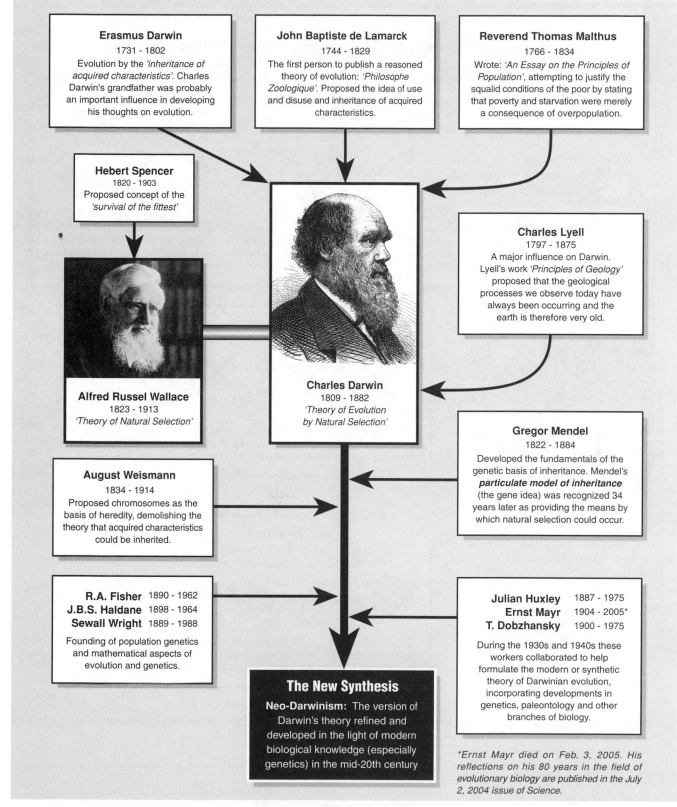

Erasmus Darwin
1731 - 1802
Evolution by the *'inheritance of acquired characteristics'*. Charles Darwin's grandfather was probably an important influence in developing his thoughts on evolution.

John Baptiste de Lamarck
1744 - 1829
The first person to publish a reasoned theory of evolution: *'Philosophe Zoologique'*. Proposed the idea of use and disuse and inheritance of acquired characteristics.

Reverend Thomas Malthus
1766 - 1834
Wrote: *'An Essay on the Principles of Population'*, attempting to justify the squalid conditions of the poor by stating that poverty and starvation were merely a consequence of overpopulation.

Hebert Spencer
1820 - 1903
Proposed concept of the *'survival of the fittest'*

Charles Lyell
1797 - 1875
A major influence on Darwin. Lyell's work *'Principles of Geology'* proposed that the geological processes we observe today have always been occurring and the earth is therefore very old.

Alfred Russel Wallace
1823 - 1913
'Theory of Natural Selection'

Charles Darwin
1809 - 1882
'Theory of Evolution by Natural Selection'

Gregor Mendel
1822 - 1884
Developed the fundamentals of the genetic basis of inheritance. Mendel's *particulate model of inheritance* (the gene idea) was recognized 34 years later as providing the means by which natural selection could occur.

August Weismann
1834 - 1914
Proposed chromosomes as the basis of heredity, demolishing the theory that acquired characteristics could be inherited.

R.A. Fisher 1890 - 1962
J.B.S. Haldane 1898 - 1964
Sewall Wright 1889 - 1988
Founding of population genetics and mathematical aspects of evolution and genetics.

Julian Huxley 1887 - 1975
Ernst Mayr 1904 - 2005*
T. Dobzhansky 1900 - 1975
During the 1930s and 1940s these workers collaborated to help formulate the modern or synthetic theory of Darwinian evolution, incorporating developments in genetics, paleontology and other branches of biology.

The New Synthesis
Neo-Darwinism: The version of Darwin's theory refined and developed in the light of modern biological knowledge (especially genetics) in the mid-20th century

Ernst Mayr died on Feb. 3, 2005. His reflections on his 80 years in the field of evolutionary biology are published in the July 2, 2004 issue of Science.

Speciation

1. From the diagram above, choose one of the contributors to the development of evolutionary theory (excluding Charles Darwin himself), and write a few paragraphs discussing their role in contributing to Darwin's ideas. You may need to consult an encyclopedia or other reference to assist you.

Darwin's Theory

In 1859, Darwin and Wallace jointly proposed that new species could develop by a process of natural selection. Natural selection is the term given to the mechanism by which better adapted organisms survive to produce a greater number of viable offspring. This has the effect of increasing their proportion in the population so that they become more common. It is Darwin who is best remembered for the theory of evolution by natural selection through his famous book: '**On the origin of species by means of natural selection**', written 23 years after returning from his voyage on the Beagle, from which much of the evidence for his theory was accumulated. Although Darwin could not explain the origin of variation nor the mechanism of its transmission (this was provided later by Mendel's work), his basic theory of evolution by natural selection (outlined below) is widely accepted today. The study of population genetics has greatly improved our understanding of evolutionary processes, which are now seen largely as a (frequently gradual) change in allele frequencies within a population. Students should be aware that scientific debate on the subject of evolution centers around the relative merits of various alternative hypotheses about the nature of evolutionary processes. The debate is not about the existence of the phenomenon of evolution itself.

Darwin's Theory of Evolution by Natural Selection

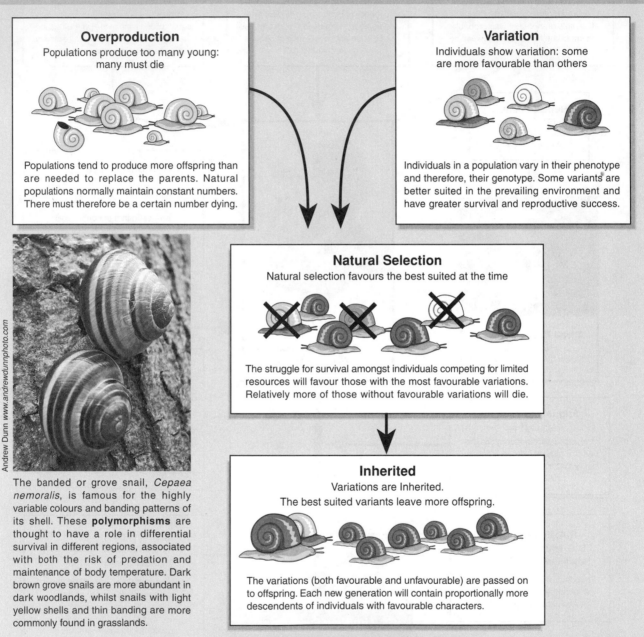

Overproduction
Populations produce too many young: many must die

Populations tend to produce more offspring than are needed to replace the parents. Natural populations normally maintain constant numbers. There must therefore be a certain number dying.

Variation
Individuals show variation: some are more favourable than others

Individuals in a population vary in their phenotype and therefore, their genotype. Some variants are better suited in the prevailing environment and have greater survival and reproductive success.

Natural Selection
Natural selection favours the best suited at the time

The struggle for survival amongst individuals competing for limited resources will favour those with the most favourable variations. Relatively more of those without favourable variations will die.

Inherited
Variations are Inherited.
The best suited variants leave more offspring.

The variations (both favourable and unfavourable) are passed on to offspring. Each new generation will contain proportionally more descendents of individuals with favourable characters.

Andrew Dunn www.andrewdunnphoto.com

The banded or grove snail, *Cepaea nemoralis*, is famous for the highly variable colours and banding patterns of its shell. These **polymorphisms** are thought to have a role in differential survival in different regions, associated with both the risk of predation and maintenance of body temperature. Dark brown grove snails are more abundant in dark woodlands, whilst snails with light yellow shells and thin banding are more commonly found in grasslands.

1. In your own words, describe how Darwin's theory of evolution by natural selection provides an explanation for the change in the appearance of a species over time:

_____

_____

_____

_____

Related activities: The Modern Theory of Evolution, Darwin's Finches
Web links: Variation : Snails

Periodicals:
Was Darwin wrong?

Adaptations and Fitness

An **adaptation**, is any heritable trait that suits an organism to its natural function in the environment (its niche). These traits may be structural, physiological, or behavioral. The idea is important for evolutionary theory because adaptive features promote fitness. **Fitness** is a measure of how well suited an organism is to survive in its habitat and its ability to maximize the numbers of offspring surviving to reproductive age. Adaptations are distinct from *properties* which, although they may be striking, cannot be described as adaptive unless they are shown to be functional in the organism's natural habitat. Genetic adaptation must not be confused with **physiological adjustment** (acclimatization), which refers to an organism's ability to adapt during its lifetime to changing environmental conditions. The physiological changes that occur when a person spends time at altitude provide a good example of acclimatization. Examples of adaptive features arising through evolution are illustrated below.

Ear Length in Rabbits and Hares

The external ears of many mammals are used as important organs to assist in thermoregulation (controlling loss and gain of body heat). The ears of rabbits and hares native to hot, dry climates, such as the jack rabbit of south-western USA and northern Mexico, are relatively very large. The Arctic hare lives in the tundra zone of Alaska, northern Canada and Greenland, and has ears that are relatively short. This reduction in the size of the extremities (ears, limbs, and noses) is typical of cold adapted species.

Body Size in Relation to Climate

Regulation of body temperature requires a large amount of energy and mammals exhibit a variety of structural and physiological adaptations to increase the effectiveness of this process. Heat production in any endotherm depends on body volume (heat generating metabolism), whereas the rate of heat loss depends on surface area. Increasing body size minimizes heat loss to the environment by reducing the surface area to volume ratio. Animals in colder regions therefore tend to be larger overall than those living in hot climates. This relationship is know as **Bergman's rule** and it is well documented in many mammalian species. Cold adapted species also tend to have more compact bodies and shorter extremities than related species in hot climates.

Arctic hare: *Lepus arcticus*

Black-tail jackrabbit: *Lepus californicus*

Fennec fox

Arctic fox

The **fennec fox** of the Sahara illustrates the adaptations typical of mammals living in hot climates: a small body size and lightweight fur, and long ears, legs, and nose. These features facilitate heat dissipation and reduce heat gain.

The **Arctic fox** shows the physical characteristics typical of cold-adapted mammals: a stocky, compact body shape with small ears, short legs and nose, and dense fur. These features reduce heat loss to the environment.

Number of Horns in Rhinoceroses

Not all differences between species can be convincingly interpreted as adaptations to particular environments. Rhinoceroses charge rival males and predators, and the horn(s), when combined with the head-down posture, add effectiveness to this behavior. Horns are obviously adaptive, but it is not clear that the possession of one (Indian rhino) or two (black rhino) horns is necessarily related directly to the environment in which those animals live.

Great Indian rhino

African black rhino

Speciation

1. Distinguish between adaptive features (genetic) and acclimatization: _____

_____

_____

2. Explain the nature of the relationship between the length of extremities (such as limbs and ears) and climate:

_____

_____

3. Explain the adaptive value of a larger body size at high latitude: _____

_____

_____

Periodicals:
Optimality

Related activities: Darwin's Theory

A 2

Natural Selection

Natural selection operates on the phenotypes of individuals, produced by their particular combinations of alleles. The differential survival of some genotypes over others is called **natural selection** and, as a result of it, organisms with phenotypes most suited to the prevailing environment are more likely to survive and breed than those with less suited phenotypes. Favorable phenotypes become more numerous while unfavorable phenotypes become less common or may disappear altogether. Natural selection is not a static phenomenon; it is always linked to phenotypic suitability in the prevailing environment. It may favor existing phenotypes or shift the phenotypic median one way or another, as is shown in the diagrams below. The top row of diagrams represents the population phenotypic spread before selection, and the bottom row the spread afterwards. Note that balancing selection is similar to disruptive selection, but the polymorphism that results is not associated with phenotypic extremes.

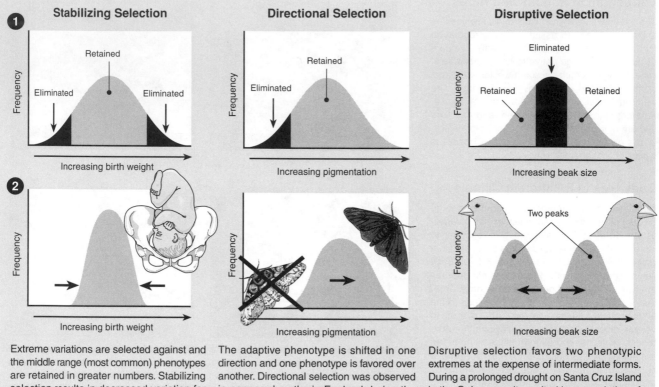

Extreme variations are selected against and the middle range (most common) phenotypes are retained in greater numbers. Stabilizing selection results in decreased variation for the phenotypic character involved. This type of selection operates most of the time in most populations and acts to prevent divergence of form and function, e.g. birth weight of human infants.

The adaptive phenotype is shifted in one direction and one phenotype is favored over another. Directional selection was observed in peppered moths in England during the Industrial Revolution. In England's current environment, the selection pressures on the moths are more typically balanced, and the proportions of each morph vary in different regions of the country.

Disruptive selection favors two phenotypic extremes at the expense of intermediate forms. During a prolonged drought on Santa Cruz Island in the Galapagos, it resulted in a population of ground finches that was bimodal for beak size. Competition for the usual seed sources was so intense that birds able to exploit either small or large seeds were favored, although intermediate phenotypes remained in low numbers.

1. Explain why fluctuating (as opposed to stable) environments favor disruptive (diversifying) selection:

2. Disruptive selection can be important in the formation of new species:

(a) Describe the evidence from the ground finches on Santa Cruz Island that lends support to this:

(b) The ground finches on Santa Cruz Island are one interbreeding population with a strongly bimodal distribution for the phenotypic character beak size. Suggest what conditions could lead to the two phenotypic extremes diverging further:

(c) Predict the consequences of the end of the drought and an increased abundance of medium size seeds as food:

Related activities: Industrial Melanism, Selection for Human Birth Weight, The Evolution of Darwin's Finches   *Web links:* Variation: Snails

Selection for Skin Color in Humans

Pigmented skin of varying tones is a feature of humans that evolved after early hominins lost the majority of their body hair. However, the distribution of skin color globally is not random; people native to equatorial regions have darker skin tones than people from higher latitudes. For many years, biologists postulated that this was because darker skins had evolved to protect against skin cancer. The problem with this explanation was that skin cancer is not tied to evolutionary fitness because it affects post-reproductive individuals and cannot therefore provide a mechanism for selection. More complex analyses of the physiological and epidemiological evidence has shown a more complex picture in which selection pressures on skin color are finely balanced to produce a skin tone that regulates the effects of the sun's ultraviolet radiation on the nutrients vitamin D and folate, both of which are crucial to successful human reproduction, and therefore evolutionary fitness.

Skin Color in Humans: A Product of Natural Selection

Alaska   France   The Netherlands   Iraq   China   Japan

80°   No data

Insufficient UV most of year

40°   Insufficient UV one month

0°   Sufficient UV all year

Sufficient UV all year

40°   Insufficient UV one month

Insufficient UV most of year

Adapted from Jablonski & Chaplin, Sci. Am. Oct. 2002

Peru   Liberia   Burundi   Botswana   Southern India   Malaysia

Speciation

Human skin color is the result of two opposing selection pressures. Skin pigmentation has evolved to protect against destruction of folate from ultraviolet light, but the skin must also be light enough to receive the light required to synthesize vitamin D. Vitamin D synthesis is a process that begins in the skin and is inhibited by dark pigment. Folate is needed for healthy neural development in humans and a deficiency is associated with fatal neural tube defects. Vitamin D is required for the absorption of calcium from the diet and therefore normal skeletal development.

Women also have a high requirement for calcium during pregnancy and lactation. Populations that live in the tropics receive enough ultraviolet (UV) radiation to synthesize vitamin D all year long. Those that live in northern or southern latitudes do not. In temperate zones, people lack sufficient UV light to make vitamin D for one month of the year. Those nearer the poles lack enough UV light for vitamin D synthesis most of the year (above). Their lighter skins reflect their need to maximize UV absorption (the photos illustrate skin color in people from different latitude).

Periodicals:
Skin deep, Fair enough

Related activities: Natural Selection

A 3

Long-term resident   Recent immigrant

① Southern Africa: ~20-30°S
Khoisan-Namibia   *Zulu: 1000 years ago*

② Australia: ~10-35°S
Aborigine   *European: 300 years ago*

③ Banks of the Red Sea: ~15-30°N
Nuba-Sudan   *Arab: 2000 years ago*

④ India: ~10-30°S
West Bengal   *Tamil: ~100 years ago*

The skin of people who have inhabited particular regions for millennia has adapted to allow sufficient vitamin D production while still protecting folate stores. In the photos above, some of these original inhabitants are illustrated to the left of each pair and compared with the skin tones of more recent immigrants (to the right of each pair, with the number of years since immigration). The numbered locations are on the map.

1. (a) Describe the role of folate in human physiology: _____

   (b) Describe the role of vitamin D in human physiology: _____

2. (a) Early hypotheses to explain skin color linked pigmentation level only to the degree of protection it gave from UV-induced skin cancer. Explain why this hypothesis was inadequate in accounting for how skin color evolved:

   _____

   _____

   (b) Explain how the new hypothesis for the evolution of skin color overcomes these deficiencies: _____

   _____

   _____

   _____

3. Explain why, in any given geographical region, women tend to have lighter skins (by 3-4% on average) than men:

   _____

   _____

4. The Inuit people of Alaska and northern Canada have a diet rich in vitamin D and their skin color is darker than predicted on the basis of UV intensity at their latitude. Explain this observation:

   _____

   _____

5. (a) Describe the health problems that could be expected for people of African origin now living in northern UK:

   _____

   _____

   (b) Explain how these people could avoid these problems in their new higher latitude environment: _____

   _____

   _____

Industrial Melanism

Natural selection may act on the frequencies of phenotypes (and hence genotypes) in populations in one of three different ways (through stabilizing, directional, or disruptive selection). Over time, natural selection may lead to a permanent change in the genetic makeup of a population. The increased prevalence of melanic forms of the peppered moth, *Biston betularia*, during the Industrial Revolution, is one of the best known examples of directional selection following a change in environmental conditions. Although the protocols used in the central experiments on *Biston*, and the conclusions drawn from them, have been queried, it remains one of the clearest documented examples of phenotypic change in a polymorphic population.

Industrial Melanism in Peppered Moths, *Biston betularia*

The **peppered moth**, *Biston betularia*, occurs in two forms (morphs): the gray mottled form, and a dark melanic form. Changes in the relative abundance of these two forms was hypothesized to be the result of selective predation by birds, with pale forms suffering higher mortality in industrial areas because they are more visible. The results of experiments by H.D. Kettlewell supported this hypothesis but did not confirm it, since selective predation by birds was observed but not quantified. Other research indicates that predation by birds is not the only factor determining the relative abundance of the different color morphs.

Gray or mottled morph: vulnerable to predation in industrial areas where the trees are dark.

Melanic or carbonaria morph: dark color makes it less vulnerable to predation in industrial areas.

Museum collections of the peppered moth made over the last 150 years show a marked change in the frequency of the melanic form. Moths collected in 1850 (above left), prior to the major onset of the industrial revolution in England. Fifty years later (above right) the frequency of the darker melanic forms had greatly increased. Even as late as the mid 20th century, coal-based industries predominated in some centers, and the melanic form occurred in greater frequency in these areas (see map, right).

Frequency of peppered moth forms in 1950

This map shows the relative frequencies of the two forms of peppered moth in the UK in 1950; a time when coal-based industries still predominated in some major centers.

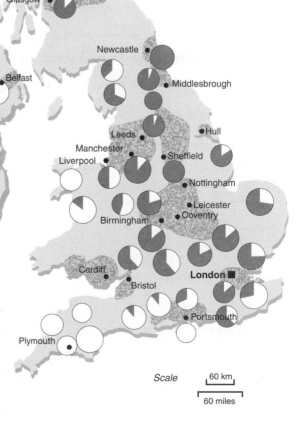

Scale 60 km / 60 miles

Key to frequency graphs

- Gray or speckled form
- Melanic or carbonaria form
- Industrial areas
- Non-industrial areas

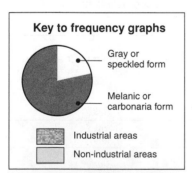

A gray (mottled) form of *Biston*, camouflaged against a lichen covered bark surface. In the absence of soot pollution, mottled forms appear to have the selective advantage.

A melanic form of *Biston*, resting on a dark branch, so that it appears as part of the branch. Note that the background has been faded out so that the moth can be seen.

Periodicals:
The moths of war, Black squirrels

Related activities: *Natural Selection*

RDA 2

Speciation

Changes in frequency of melanic peppered moths

In the 1940s and 1950s, coal burning was still at intense levels around the industrial centers of Manchester and Liverpool. During this time, the melanic form of the moth was still very dominant. In the rural areas further south and west of these industrial centers, the gray or speckled forms increased dramatically. With the decline of coal burning factories and the Clean Air Acts in cities, the air quality improved between 1960 and 1980. Sulfur dioxide and smoke levels dropped to a fraction of their previous levels. This coincided with a sharp fall in the relative numbers of melanic moths.

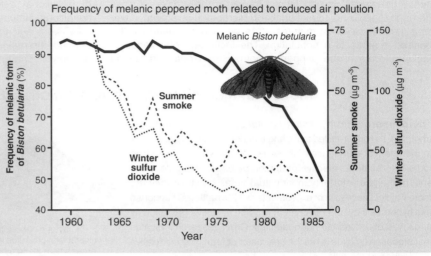

Frequency of melanic peppered moth related to reduced air pollution

1. The populations of peppered moth in England have undergone changes in the frequency of an obvious phenotypic character over the last 150 years. Describe the phenotypic character that changed in its frequency:

_____

2. (a) Identify the (proposed) selective agent for phenotypic change in *Biston*: _____

   (b) Describe how the selection pressure on the light colored morph has changed with changing environmental conditions over the last 150 years:

   _____

   _____

3. The industrial centers for England in 1950 were located around London, Birmingham, Liverpool, Manchester, and Leeds. Glasgow in Scotland also had a large industrial base. Comment on the relationship between the relative frequencies of the two forms of peppered moth and the geographic location of industrial regions:

   _____

   _____

4. The level of pollution dropped around Manchester and Liverpool between 1960 and 1985.

   (a) State how much the pollution dropped by: _____

   (b) Describe how the frequency of the darker melanic form changed during the period of reduced pollution:

   _____

5. In the *Biston* example, identify the selection pressure as disruptive, stabilizing, or directional, and explain your answer:

   _____

   _____

   _____

6. Outline the key difference between natural and artificial selection: _____

   _____

7. Discuss the statement "the environment directs natural selection": _____

   _____

   _____

   _____

   _____

Heterozygous Advantage

There are two mechanisms by which natural selection can affect allele frequencies. Firstly, there may be selection against one of the homozygotes. When one homozygous type (for example, aa), has a lower fitness than the other two genotypes (in this case, Aa or AA), the frequency of the deleterious allele will tend to decrease until it is completely eliminated. In some situations, both homozygous conditions (aa **and** AA) have lower fitness than the heterozygote; a situation that leads to **heterozygous advantage** and may result in the stable coexistence of both alleles in the population (**balanced polymorphism**). There are remarkably few well-documented examples in which the evidence for heterozygous advantage is conclusive. The maintenance of the sickle cell mutation in malaria-prone regions is one such example.

The Sickle Cell Allele (Hb^S)

Sickle cell disease is caused by a mutation to a gene that directs the production of the human blood protein called hemoglobin. The mutant allele is known as **Hb^S** and produces a form of hemoglobin that differs from the normal form by just one amino acid in the β-chain. This minute change however causes a cascade of physiological problems in people with the allele. Some of the red blood cells containing mutated hemoglobin alter their shape to become irregular and spiky; the so-called **sickle cells**.

Sickle cells have a tendency to clump together and work less efficiently. In people with just one sickle cell allele plus a normal allele (the heterozygote condition **Hb^{S}Hb**), there is a mixture of both red blood cell types and they are said to have the sickle cell trait. They are generally unaffected by the disease except in low oxygen environments (e.g. climbing at altitude). People with two Hb^S genes (**Hb^{S}Hb^S**) suffer severe illness and even death. For this reason Hb^S is considered **a lethal gene**.

Heterozygous Advantage in Malarial Regions

Falciparum malaria is widely distributed throughout central Africa, the Mediterranean, Middle East, and tropical and semi-tropical Asia (Fig. 1). It is transmitted by the *Anopheles* mosquito, which spreads the protozoan *Plasmodium falciparum* from person to person as it feeds on blood.

SYMPTOMS: These appear 1-2 weeks after being bitten, and include headache, shaking, chills, and fever. Falciparum malaria is more severe than other forms of malaria, with high fever, convulsions, and coma. It can be fatal within days of the first symptoms appearing.

THE PARADOX: The Hb^S allele offers considerable protection against malaria. Sickle cells have low potassium levels, which causes plasmodium parasites inside these cells to die. Those with a normal phenotype are very susceptible to malaria, but heterozygotes (**Hb^{S}Hb**) are much less so. This situation, called **heterozygous advantage**, has resulted in the Hb^S allele being present in moderately high frequencies in parts of Africa and Asia despite its harmful effects (Fig. 2). This is a special case of balanced polymorphism, called a **balanced lethal system** because neither of the homozygotes produces a phenotype that survives, but the heterozygote is viable.

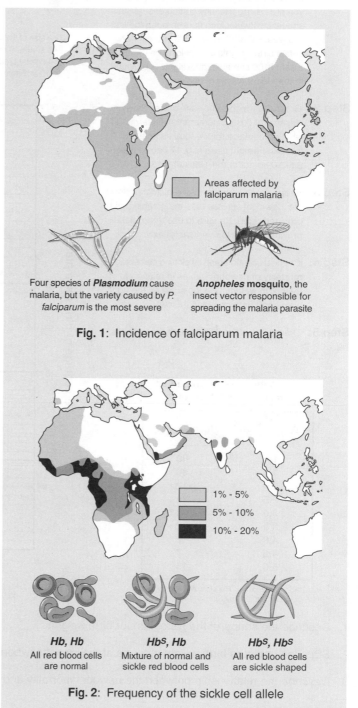

Four species of *Plasmodium* cause malaria, but the variety caused by *P. falciparum* is the most severe

Areas affected by falciparum malaria

Anopheles mosquito, the insect vector responsible for spreading the malaria parasite

Fig. 1: Incidence of falciparum malaria

1% - 5%
5% - 10%
10% - 20%

Hb, Hb
All red blood cells are normal

Hb^S, Hb
Mixture of normal and sickle red blood cells

Hb^S, Hb^S
All red blood cells are sickle shaped

Fig. 2: Frequency of the sickle cell allele

1. With respect to the sickle cell allele, explain how **heterozygous advantage** can lead to **balanced polymorphism**:

_____

_____

_____

_____

_____

Speciation

Selection for Human Birth Weight

Selection pressures operate on populations in such a way as to reduce mortality. For humans, giving birth is a special, but often traumatic, event. In a study of human birth weights it is possible to observe the effect of selection pressures operating to constrain human birth weight within certain limits. This is a good example of **stabilizing selection**. This activity explores the selection pressures acting on the birth weight of human babies. Carry out the steps below:

Step 1: Collect the birth weights from 100 birth notices from your local newspaper (or 50 if you are having difficulty getting enough; this should involve looking back through the last 2-3 weeks of birth notices). If you cannot obtain birth weights in your local newspaper, a set of 100 sample birth weights is provided in the Model Answers booklet.

Step 2: Group the weights into each of the 12 weight classes (of 0.5 kg increments). Determine what percentage (of the total sample) fall into each weight class (e.g. 17 babies weigh 2.5-3.0 kg out of the 100 sampled = 17%)

Step 3: Graph these in the form of a histogram for the 12 weight classes (use the graphing grid provided right). Be sure to use the scale provided on the left vertical (y) axis.

Step 4: Create a second graph by plotting percentage mortality of newborn babies in relation to their birth weight. Use the scale on the right y axis and data provided (below).

Step 5: Draw a line of 'best fit' through these points.

The size of the baby and the diameter and shape of the birth canal are the two crucial factors in determining whether a normal delivery is possible.

Mortality of newborn babies related to birth weight

Weight (kg)	Mortality (%)
1.0	80
1.5	30
2.0	12
2.5	4
3.0	3
3.5	2
4.0	3
4.5	7
5.0	15

Source: Biology: The Unity & Diversity of Life (4th ed) by Starr and Taggart

1. Describe the shape of the histogram for birth weights: _____

2. State the optimum birth weight in terms of the lowest newborn mortality: _____

3. Describe the relationship between the newborn mortality and the birth weights: _____

   _____

4. Describe the selection pressures that are operating to control the range of birth weight: _____

   _____

   _____

5. Explain how medical intervention methods during pregnancy and childbirth may have altered these selection pressures:

   _____

   _____

Related activities: Natural Selection, Drawing Histograms

The Evolution of Darwin's Finches

The Galápagos Islands, off the West coast of Ecuador, comprise 16 main islands and six smaller islands. They are home to a unique range of organisms, including 13 species of finches, each of which is thought to have evolved from a single species of grassquit. After colonising the islands, the grassquits underwent adaptive radiation in response to the availability of unexploited feeding niches on the islands. This adaptive radiation is most evident in the present beak shape of each species. The beaks are adapted for different purposes such as crushing seeds, pecking wood, or probing flowers for nectar. Current consensus groups the finches into ground finches, tree finches, warbler finches, and the Cocos Island finches. Between them, the 13 species of this endemic group fill the roles of seven different families of South American mainland birds. DNA analyses have confirmed Darwin's insight and have shown that all 13 species evolved from a flock of about 30 birds arriving a million years ago.

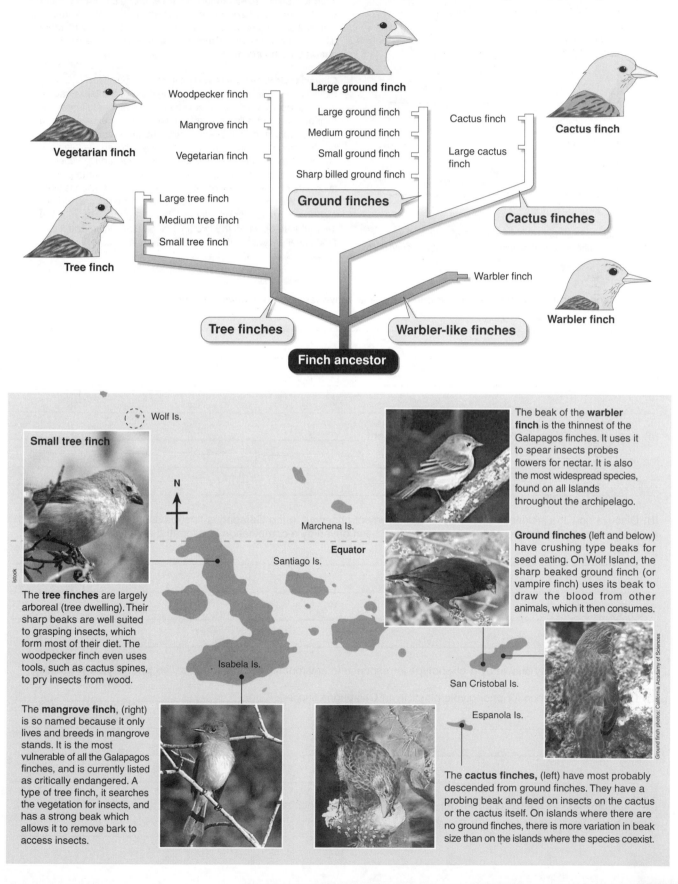

The beak of the **warbler finch** is the thinnest of the Galapagos finches. It uses it to spear insects probes flowers for nectar. It is also the most widespread species, found on all Islands throughout the archipelago.

Ground finches (left and below) have crushing type beaks for seed eating. On Wolf Island, the sharp beaked ground finch (or vampire finch) uses its beak to draw the blood from other animals, which it then consumes.

The **tree finches** are largely arboreal (tree dwelling). Their sharp beaks are well suited to grasping insects, which form most of their diet. The woodpecker finch even uses tools, such as cactus spines, to pry insects from wood.

The **mangrove finch**, (right) is so named because it only lives and breeds in mangrove stands. It is the most vulnerable of all the Galapagos finches, and is currently listed as critically endangered. A type of tree finch, it searches the vegetation for insects, and has a strong beak which allows it to remove bark to access insects.

The **cactus finches,** (left) have most probably descended from ground finches. They have a probing beak and feed on insects on the cactus or the cactus itself. On islands where there are no ground finches, there is more variation in beak size than on the islands where the species coexist.

Speciation

Related activities: *Darwin's Theory, Patterns of Evolution, Allopatric Speciation*
Web links: *Darwin's Finches*

A 3

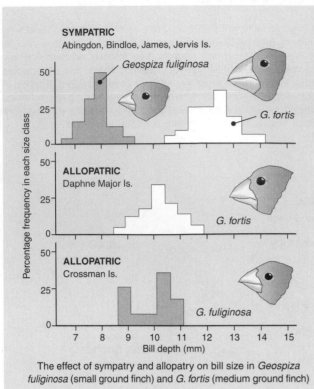

SYMPATRIC
Abingdon, Bindloe, James, Jervis Is.

Geospiza fuliginosa

G. fortis

ALLOPATRIC
Daphne Major Is.

G. fortis

ALLOPATRIC
Crossman Is.

G. fuliginosa

Bill depth (mm)

Percentage frequency in each size class

The effect of sympatry and allopatry on bill size in *Geospiza fuliginosa* (small ground finch) and *G. fortis* (medium ground finch)

Sympatry and Character Displacement

There is good evidence that finch evolution appears to be driven by a combination of allopatric and sympatric events. Coexisting species of ground finches on four islands (top graph) show large differences in bill sizes, enabling each species to feed on different sized seeds. However when either species exists in the absence of the other on different islands (lower graphs), it possesses intermediate bill sizes (about 10 mm) enabling it to feed without partitioning seed resources. This phenomenon, whereby competition causes two closely related species to become more different in regions where their ranges overlap, is referred to as **character displacement**.

Character displacement is evident in other populations of finches as well. There are well-studied populations of the large cactus finch (*G. conirostris*) on Genovesa and Espanola Islands, but their bill sizes are quite different. On Genovesa, the large ground finch (*G. magnirostris*) coexists with the large cactus finch. In these sympatric populations, the variability in bill size *within* each species is minimal but there is little overlap *between* the species with respect to this trait. On Espanola, where the large ground finch either never arrived, or became extinct, the situation is quite different. With no competition on Espanola, the large cactus finch displays a greater variability in bill size. Its bill is somewhat intermediate between the two finches on Genovesa, and it can feed equally well in both niches all year round.

Data based on an adaptation by Strickberger (2000)

1. Describe the main factors that have contributed to the adaptive radiation of Darwin's finches: _____

_____

_____

_____

_____

2. (a) Explain what is meant by **character displacement**: _____

_____

_____

(b) Discuss how the incidence of character displacement observed in the Galapagos finches supports the view that their adaptive radiation from a common ancestor has been driven by a combination of allopatric and sympatric events:

_____

_____

_____

_____

3. The range of variability shown by a phenotype in response to environmental variation is called **phenotypic plasticity**.

(a) Discuss the evidence for phenotypic plasticity in Galapagos finches: _____

_____

_____

_____

(b) Explain what this suggests about the biology of the original finch ancestor: _____

_____

_____

Insecticide Resistance

Insecticides are pesticides used to control insects considered harmful to humans, their livelihood, or environment. Insecticides have been used for hundreds of years, but their use has proliferated since the advent of synthetic insecticides (e.g. DDT) in the 1940s. **Insecticide resistance** develops when the target species becomes adapted to the effects of the control agent and it no longer controls the population effectively. Resistance can arise through a combination of behavioural, anatomical, biochemical, and physiological mechanisms, but the underlying process is a form of **natural selection**, in which the most resistant organisms survive to pass on their genes to their offspring. To combat increasing resistance, higher doses of more potent pesticides are sometimes used. This drives the selection process, so that increasingly higher dose rates are required to combat rising resistance. The increased application may also kill useful insects and birds, reducing biodiversity and leading to bioaccumulation in food chains. This cycle of increasing resistance in response to increased doses is made worse by the development of **multiple resistance** in some insect pest species. Insecticides are used in medical, agricultural, and environmental applications, so the development of resistance has serious environmental and economic consequences.

INSECTICIDE

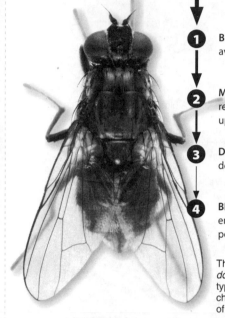

① **BEHAVIOR**
avoidance of contact

② **MECHANICAL RESISTANCE**
resistance to entry lowers uptake levels

③ **DESTRUCTION**
detoxification mechanisms

④ **BIOCHEMICAL RESISTANCE**
enzymatic changes at points of action

The housefly (*Musca domestica*) is a vector for typhoid, dysentery, and cholera, as well as for a range of less serious diseases.

Mechanisms of Resistance in Houseflies

Insecticide resistance in houseflies can arise through a combination of mechanisms. (1) Increased sensitivity to an insecticide will cause the pest to avoid a treated area. (2) The *Pen* gene confers stronger physical barriers, decreasing the rate at which the chemical penetrates the cuticle. (3) Chemical changes within the insect's body can render the pesticide harmless, and (4) structural changes to the target enzymes make the pesticide ineffective. No single mechanism provides total immunity, but together they transform the effect from potentially lethal to insignificant.

Susceptible
Resistant

A small proportion of the population will have the genetic makeup to survive the first application of a pesticide.

①

② *The genetic information for pesticide resistance is passed to the next generation.*

The proportion of resistant individuals increases following subsequent applications of insecticide. Eventually, almost all of the population is resistant.

The Development of Resistance

The application of an insecticide can act as a potent selection pressure for resistance in pest insects. The insecticide acts as a selective agent, and only individuals with greater natural resistance survive the application to pass on their genes to the next generation. These genes (or combination of genes) may spread through all subsequent populations.

1. Give two reasons why widespread insecticide resistance can develop very rapidly in insect populations:

   (a) _____

   (b) _____

2. Explain how repeated insecticide applications acts as a selective agent for evolutionary change in insect populations:

   _____

   _____

3. With reference to synthetic insecticides, discuss the implications of insecticide resistance to human populations:

   _____

   _____

   _____

   _____

Related activities: Resistance in Pathogens, Antibiotic Resistance

RA 2

Speciation

The Evolution of Antibiotic Resistance

Resistance to drugs results from an adaptive response that allows microbes to tolerate levels of antibiotic that would normally inhibit their growth. This resistance may arise spontaneously as the result of mutation, or by transfer of genetic material between microbes. Over the years, more and more bacteria have developed resistance to once-effective antibiotics. Methicillin resistant strains of the common bacterium *Staphylococcus aureus* (MRSA) have acquired genes that confer antibiotic resistance to all penicillins, including **methicillin** and other narrow-spectrum pencillin-type drugs. Such strains, called "superbugs", were discovered in the UK in 1961 and are now widespread, and the infections they cause are exceedingly difficult to treat.

The Evolution of Drug Resistance in Bacteria

Susceptible bacterium

Less susceptible bacterium

Mutations occur at a rate of one in every 10^8 replications.

Bacterium with greater resistance survives

Drug resistance genes can be transferred to non resistant strains.

Any population, including bacterial populations, includes variants with unusual traits, in this case reduced sensitivity to an antibiotic. These variants arise as a result of mutations in the bacterial chromosome. Such mutations are well documented.

When a person takes an antibiotic, only the most susceptible bacteria will die. The more resistant cells remain and continue dividing. Note that the antibiotic does not create the resistance; it provides the environment in which selection for resistance can take place.

If the amount of antibiotic delivered is too low, or the course of antibiotics is not completed, a population of resistant bacteria develops. Within this population too, there will be variation in susceptibility. Some will survive higher antibiotic levels.

A highly resistant population has evolved. The resistant cells can exchange genetic material with other bacteria (via horizontal gene transmission), passing on the genes for resistance. The antibiotic initially used against this bacterial strain will now be ineffective.

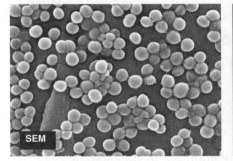

SEM

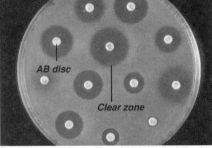

AB disc

Clear zone

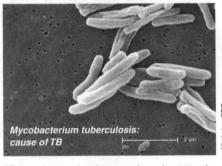

Mycobacterium tuberculosis: cause of TB

2 μm

All photos: CDC

Staphylococcus aureus is a common bacterium responsible various minor skin infections in humans. MRSA (above) is variant strain that has evolved resistance to penicillin and related antibiotics. MRSA is troublesome in hospital-associated infections where patients with open wounds, invasive devices (e.g. catheters), and weakened immune systems are at greater risk for infection than the general public.

The photo above shows an antibiogram plate culture of *Enterobacter sakazakii*, a rare cause of invasive infections in infants. An antibiogram measures the biological resistance of disease-causing organisms to antibiotic agents. The bacterial lawn (growth) on the agar plate is treated with antibiotic discs, and the sensitivity to various antibiotics is measured by the extent of the clearance zone in the bacterial lawn.

TB is a disease that has experienced spectacular ups and downs. Drugs were developed to treat it, but then people became complacent when they thought the disease was beaten. TB has since resurged because patients stop their medication too soon and infect others. Today, one in seven new TB cases is resistant to the two drugs most commonly used as treatments, and 5% of these patients die.

1. Explain what is meant by **antibiotic resistance**: _____

_____

2. (a) Explain how antibiotic resistance arises in a bacterial population: _____

_____

   (b) Describe two ways in which antibiotic resistance can become widespread:

_____

_____

3. With reference to tuberculosis, discuss the implications to humans of widespread antibiotic resistance:

_____

_____

_____

Related activities: Resistance in Pathogens, Insecticide Resistance
Web links: Why Evolution Matters Now, The Rise of Antibiotic Resistance

Periodicals:
MRSA: A hospital superbug

Gene Pool Exercise

Cut out each of the beetles on this page and use them to reenact different events within a gene pool as described in this topic (see

Gene Pools and Evolution, Changes in a Gene Pool, Founder Effect, Population Bottlenecks, Genetic Drift).

Speciation

Related activities: Gene Pools and Evolution, Changes in a Gene Pool, The Founder Effect, Population Bottlenecks, Genetic Drift

P

This page has deliberately been left blank

Gene Pools and Evolution

The diagram below illustrates the dynamic nature of **gene pools**. It portrays two imaginary populations of one beetle species. Each beetle is a 'carrier' of genetic information, represented here by the alleles (A and a) for a single **codominant gene** that controls the beetle's color. Normally, there are three versions of the phenotype: black, dark, and pale. Mutations may create other versions of the phenotype. Some of the **microevolutionary processes** that can affect the genetic composition (**allele frequencies**) of the gene pool are illustrated. See the activity *Gene Pool Exercise* for cut-out beetles to simulate this activity.

Immigration: Populations can gain alleles when they are introduced from other gene pools. Immigration is one aspect of gene flow.

Mutations: Spontaneous mutations can develop that alter the allele frequencies of the gene pool, and even create new alleles. Mutation is very important to evolution, because it is the original source of genetic variation that provides new material for natural selection.

Emigration: Genes may be lost to other gene pools.

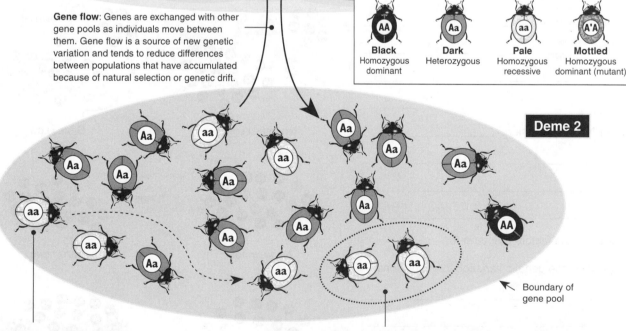

Natural selection: Selection pressure against certain allele combinations may reduce reproductive success or lead to death. Natural selection sorts genetic variability, and accumulates and maintains favorable genotypes in a population. It tends to reduce genetic diversity within the gene pool and increase differences between populations.

The term deme describes a local population that is genetically isolated from other populations in the species. Demes usually have some clearly definable genetic or other character that sets them apart from other populations.

Geographical barriers: Isolate the gene pool and prevent *regular* gene flow between populations.

Gene flow: Genes are exchanged with other gene pools as individuals move between them. Gene flow is a source of new genetic variation and tends to reduce differences between populations that have accumulated because of natural selection or genetic drift.

Key to genotypes and phenotypes

Black Homozygous dominant (AA)

Dark Heterozygous (Aa)

Pale Homozygous recessive (aa)

Mottled Homozygous dominant (mutant) (A'A)

Boundary of gene pool

Mate selection (non-random mating): Individuals may not select their mate randomly and may seek out particular phenotypes, increasing the frequency of these "favored" alleles in the population.

Genetic drift: Chance events can cause the allele frequencies of small populations to "drift" (change) randomly from generation to generation. Genetic drift can play a significant role in the microevolution of very small populations. The two situations most often leading to populations small enough for genetic drift to be significant are the **bottleneck effect** (where the population size is dramatically reduced by a catastrophic event) and the **founder effect** (where a small number of individuals colonize a new area).

Speciation

Related activities: *Gene Pool Exercise, Natural Selection*
Web links: *Mechanisms of Evolution, Processes in Gene Pools, Speciation*

PA 2

1. For each of the 2 demes shown on the previous page (treating the mutant in deme 1 as a AA):

   (a) Count up the numbers of **allele types** (**A** and **a**).

   (b) Count up the numbers of **allele combinations** (**AA, Aa, aa**).

2. Calculate the frequencies as percentages (%) for the allele types and combinations:

Deme 1		Number counted	%
Allele types	A		
	a		
Allele combinations	AA		
	Aa		
	aa		

Deme 2		Number counted	%
Allele types	A		
	a		
Allele combinations	AA		
	Aa		
	aa		

3. One of the fundamental concepts for population genetics is that of **genetic equilibrium**, stated as: *"For a very large, randomly mating population, the proportion of dominant to recessive alleles remains constant from one generation to the next"*. If a gene pool is to remain unchanged, it must satisfy all of the criteria below that favour gene pool stability. Few populations meet all (or any) of these criteria and their genetic makeup must therefore by continually changing. For each of the five factors (a-e) below, state briefly **how** and **why** each would affect the allele frequency in a gene pool:

   (a) Population size: _____

   _____

   _____

   _____

   (b) Mate selection: _____

   _____

   _____

   _____

   (c) Gene flow between populations:

   _____

   _____

   _____

   (d) Mutations: _____

   _____

   _____

   _____

   (e) Natural selection: _____

   _____

   _____

   _____

4. Identify the factors that tend to:

   (a) Increase genetic variation in populations:

   _____

   _____

   (b) Decrease genetic variation in populations:

   _____

   _____

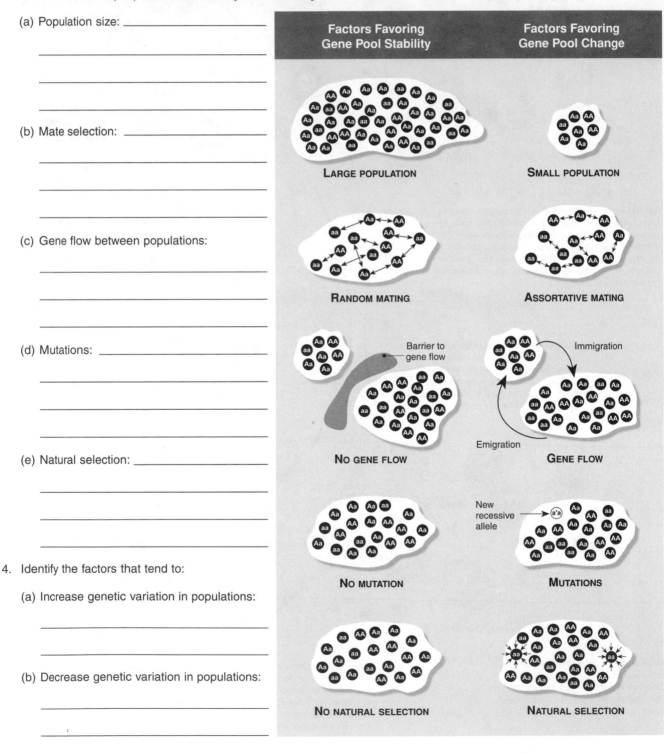

Factors Favoring Gene Pool Stability | Factors Favoring Gene Pool Change

LARGE POPULATION — SMALL POPULATION

RANDOM MATING — ASSORTATIVE MATING

Barrier to gene flow — Immigration / Emigration

NO GENE FLOW — GENE FLOW

NO MUTATION — MUTATIONS / New recessive allele

NO NATURAL SELECTION — NATURAL SELECTION

Changes in a Gene Pool

The diagram below shows an imaginary population of beetles undergoing changes as it is subjected to two 'events'. The three phases represent a progression in time, i.e. the same gene pool, undergoing change. The beetles have three phenotypes determined by the amount of pigment deposited in the cuticle. Three versions of this trait exist: black, dark, and pale. The gene controlling this character is represented by two alleles **A** and **a**. Your task is to analyze the gene pool as it undergoes changes.

Phase 1: Initial gene pool

Calculate the frequencies of the *allele types* and *allele combinations* by counting the actual numbers, then working them out as percentages.

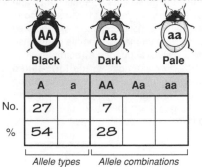

Black   Dark   Pale

	A	a	AA	Aa	aa
No.	27		7		
%	54		28		

Allele types   Allele combinations

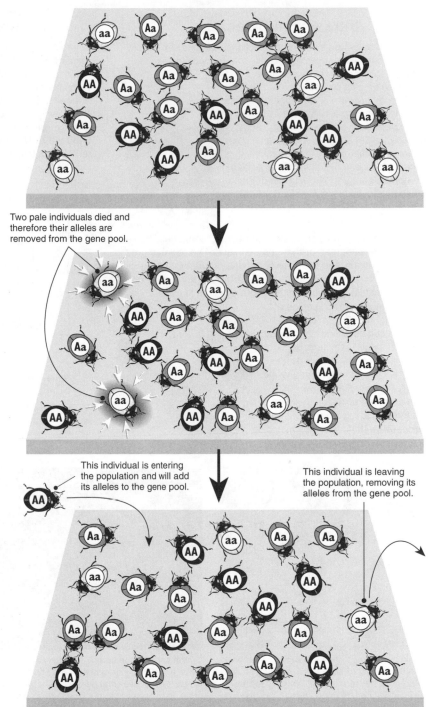

Two pale individuals died and therefore their alleles are removed from the gene pool.

This individual is entering the population and will add its alleles to the gene pool.

This individual is leaving the population, removing its alleles from the gene pool.

Phase 2: Natural selection

In the same gene pool at a later time there was a change in the allele frequencies. This was due to the loss of certain allele combinations due to natural selection. Some of those with a genotype of aa were eliminated (poor fitness).

Calculate as for above. Do not include the individuals surrounded by small white arrows in your calculations; they are dead!

	A	a	AA	Aa	aa
No.					
%					

Phase 3: Immigration and emigration

This particular kind of beetle exhibits wandering behavior. The allele frequencies change again due to the introduction and departure of individual beetles, each carrying certain allele combinations.

Calculate as above. In your calculations, include the individual coming into the gene pool (AA), but remove the one leaving (aa).

	A	a	AA	Aa	aa
No.					
%					

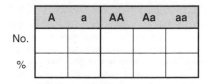

Speciation

1. Explain how the number of dominant alleles (A) in the genotype of a beetle affects its phenotype:

_____

_____

2. For each phase in the gene pool above (place your answers in the tables provided; some have been done for you):

   (a) Determine the relative frequencies of the two alleles: A and a. Simply total the **A** alleles and **a** alleles separately.

   (b) Determine the frequency of how the alleles come together as allele pair combinations in the gene pool (AA, Aa and aa). Count the number of each type of combination.

   (c) For each of the above, work out the frequencies as percentages:

         Allele frequency = Number of counted alleles ÷ Total number of alleles x 100

Sexual Selection

The success of an individual is measured not only by the number of offspring it leaves, but also by the quality or likely reproductive success of those offspring. This means that it becomes important who its mate will be. It was Darwin (1871) who first introduced the concept of sexual selection; a special type of natural selection that produces anatomical and behavioral traits that affect an individual's ability to acquire mates. Biologists today recognise two types: **intrasexual selection** (usually male-male competition) and **intersexual selection** or mate selection. One result of either type is the evolution of **sexual dimorphism**.

Intrasexual Selection

Intrasexual selection involves competition within one sex (usually males) with the winner gaining access to the opposite sex. Competition often takes place before mating, and males compete to establish dominance or secure a territory for breeding or mating. This occurs in many species of ungulates (**deer**, antelope, cattle) and in many birds. In deer and other ungulates, the males typically engage in highly ritualized battles with horns or antlers. The winners of these battles gain dominance over rival males and do most of the mating.

In other species, males compete vigorously for territories. These may contain resources or they may consist of an isolated area within a special arena used for communal courtship display (a **lek**). In lek species, males with the best territories on a lek (the dominant males) are known to get more chances to mate with females. In some species of grouse (right), this form of sexual selection can be difficult to distinguish from intersexual selection, because once males establish their positions on the lek the females then choose among them. In species where there is limited access to females and females are promiscuous, **sperm competition** (below, center) may also be a feature of male-male competition.

Intersexual Selection

In intersexual selection (or **mate choice**), individuals of one sex (usually the males) advertise themselves as potential mates and members of the other sex (usually the females), choose among them. Intersexual selection results in development of exaggerated ornamentation, such as elaborate plumages. Female preference for elaborate male ornaments is well supported by both anecdotal and experimental evidence. For example, in the **long-tailed widow bird** (*Euplectes progne*), females prefer males with long tails. When tails are artificially shortened or lengthened, females still prefer males with the longest tails; they therefore select for long tails, not another trait correlated with long tails.

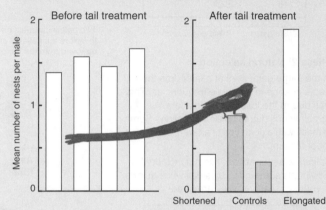

Mean number of nests per male

Before tail treatment / After tail treatment

Shortened   Controls   Elongated

As shown above, there was no significant difference in breeding success between the groups before the tails were altered. When the tails were cut and lengthened, breeding success went down and up respectively.

In male-male competition for mates, ornamentation is used primarily to advertise superiority to rival males, and not to mortally wound opponents. However, injuries do occur, most often between closely matched rivals, where dominance must be tested and established through the aggressive use of their weaponry rather than mere ritual duels.

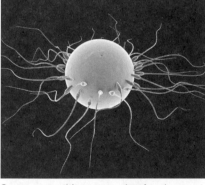

Sperm competition occurs when females remate within a relatively short space of time. The outcome of sperm competition may be determined by mating order. In some species, including those that guard their mates, the first male has the advantage, but in many the advantage accrues to the sperm of the second or subsequent males.

How do male features, such as the extravagant plumage of the peacock, persist when increasingly elaborate plumage must become detrimental to survival at some point? At first, preference for such traits must confer a survival advantage. Male adornment and female preference then advance together until a stable strategy is achieved.

1. Explain the difference between **intrasexual selection** and **mate selection**, identifying the features associated with each:

_____

_____

2. Suggest how sexual selection results in marked **sexual dimorphism**: _____

_____

_____

Related activities: Gene Pools and Evolution
Web links: Pheasant sexual selection

Periodicals:
Animal attraction

 © Biozone International 2001-2010
Photocopying Prohibited

Population Genetics Calculations

The **Hardy-Weinberg equation** provides a simple mathematical model of genetic equilibrium in a gene pool, but its main application in population genetics is in calculating allele and genotype frequencies in populations, particularly as a means of studying changes and measuring their rate. The use of the Hardy-Weinberg equation is described below.

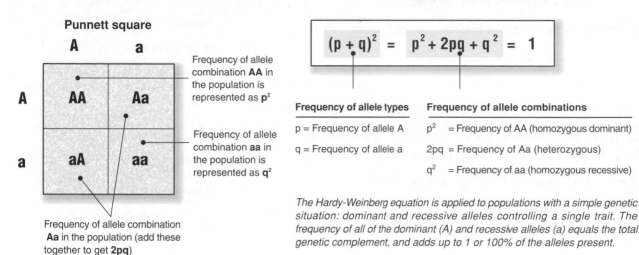

Punnett square

Frequency of allele combination **AA** in the population is represented as p^2

Frequency of allele combination **aa** in the population is represented as q^2

Frequency of allele combination **Aa** in the population (add these together to get **2pq**)

$$(p + q)^2 = p^2 + 2pq + q^2 = 1$$

Frequency of allele types

p = Frequency of allele A

q = Frequency of allele a

Frequency of allele combinations

p^2 = Frequency of AA (homozygous dominant)

2pq = Frequency of Aa (heterozygous)

q^2 = Frequency of aa (homozygous recessive)

The Hardy-Weinberg equation is applied to populations with a simple genetic situation: dominant and recessive alleles controlling a single trait. The frequency of all of the dominant (A) and recessive alleles (a) equals the total genetic complement, and adds up to 1 or 100% of the alleles present.

How To Solve Hardy-Weinberg Problems

In most populations, the frequency of two alleles of interest is calculated from the proportion of homozygous recessives (q^2), as this is the only genotype identifiable directly from its phenotype. If only the dominant phenotype is known, q^2 may be calculated (1 – the frequency of the dominant phenotype). The following steps outline the procedure for solving a Hardy-Weinberg problem:

Remember that all calculations must be carried out using proportions, NOT PERCENTAGES!

1. Examine the question to determine what piece of information you have been given about the population. In most cases, this is the percentage or frequency of the homozygous recessive phenotype q^2, or the dominant phenotype $p^2 + 2pq$ (see note above).

2. The first objective is to find out the value of p or q, If this is achieved, then every other value in the equation can be determined by simple calculation.

3. Take the square root of q^2 to find q.

4. Determine p by subtracting q from 1 (i.e. p = 1 – q).

5. Determine p^2 by multiplying p by itself (i.e. $p^2 = p \times p$).

6. Determine 2pq by multiplying p times q times 2.

7. Check that your calculations are correct by adding up the values for $p^2 + q^2 + 2pq$ (the sum should equal 1 or 100%).

Worked example

In the American white population approximately 70% of people can taste the chemical phenylthiocarbamide (PTC) (the dominant phenotype), while 30% are non-tasters (the recessive phenotype).

Determine the frequency of: **Answers**

(a) Homozygous recessive phenotype(**q^2**).   30% - provided

(b) The dominant allele (**p**).   45.2%

(c) Homozygous tasters (**p^2**).   20.5%

(d) Heterozygous tasters (**2pq**).   49.5%

Data: The frequency of the dominant phenotype (70% tasters) and recessive phenotype (30% non-tasters) are provided.

Working:

Recessive phenotype: **q^2**   =   30%
                       use 0.30 for calculation

therefore: **q**   =   0.5477
                       square root of 0.30

therefore: **p**   =   0.4523
                       1 – q = p
                       1 – 0.5477 = 0.4523

Use p and q in the equation (top) to solve any unknown:

Homozygous dominant   **p^2**   =   0.2046
                          (p × p = 0.4523 × 0.4523)

Heterozygous: **2pq** = 0.4953

1. A population of hamsters has a gene consisting of 90% M alleles (black) and 10% m alleles (gray). Mating is random.
   Data: Frequency of recessive allele (10% m) and dominant allele (90% M).

   Determine the proportion of offspring that will be black and the proportion that will be gray (show your working):

   _____

   _____

Recessive allele:	q =	
Dominant allele:	p =	
Recessive phenotype:	q^2 =	
Homozygous dominant:	p^2 =	
Heterozygous:	2pq =	

Speciation

Periodicals:
The Hardy-Weinberg principle

Related activities: *Analysis of a Squirrel Gene Pool*

RDA 2

2. You are working with pea plants and found 36 plants out of 400 were dwarf.
Data: Frequency of recessive phenotype (36 out of 400 = 9%)

(a) Calculate the frequency of the tall gene: _____

(b) Determine the number of heterozygous pea plants:

_____

_____

Recessive allele:	q =	
Dominant allele:	p =	
Recessive phenotype:	q^2 =	
Homozygous dominant:	p^2 =	
Heterozygous:	2pq =	

3. In humans, the ability to taste the chemical phenylthiocarbamide (PTC) is inherited as a simple dominant characteristic. Suppose you found out that 360 out of 1000 college students could not taste the chemical.
Data: Frequency of recessive phenotype (360 out of 1000).

(a) State the frequency of the gene for tasting PTC:

_____

(b) Determine the number of heterozygous students in this population:

_____

Recessive allele:	q =	
Dominant allele:	p =	
Recessive phenotype:	q^2 =	
Homozygous dominant:	p^2 =	
Heterozygous:	2pq =	

4. A type of deformity appears in 4% of a large herd of cattle. Assume the deformity was caused by a recessive gene.
Data: Frequency of recessive phenotype (4% deformity).

(a) Calculate the percentage of the herd that are carriers of the gene:

_____

(b) Determine the frequency of the dominant gene in this case:

_____

Recessive allele:	q =	
Dominant allele:	p =	
Recessive phenotype:	q^2 =	
Homozygous dominant:	p^2 =	
Heterozygous:	2pq =	

5. Assume you placed 50 pure bred black guinea pigs (dominant allele) with 50 albino guinea pigs (recessive allele) and allowed the population to attain genetic equilibrium (several generations have passed).
Data: Frequency of recessive allele (50%) and dominant allele (50%).

Determine the proportion (%) of the population that becomes white:

_____

Recessive allele:	q =	
Dominant allele:	p =	
Recessive phenotype:	q^2 =	
Homozygous dominant:	p^2 =	
Heterozygous:	2pq =	

6. It is known that 64% of a large population exhibit the recessive trait of a characteristic controlled by two alleles (one is dominant over the other).
Data: Frequency of recessive phenotype (64%). Determine the following:

(a) The frequency of the recessive allele: _____

(b) The percentage that are heterozygous for this trait: _____

(c) The percentage that exhibit the dominant trait: _____

(d) The percentage that are homozygous for the dominant trait: _____

(e) The percentage that has one or more recessive alleles: _____

7. Albinism is recessive to normal pigmentation in humans. The frequency of the albino allele was 10% in a population.
Data: Frequency of recessive allele (10% albino allele).

Determine the proportion of people that you would expect to be albino:

_____

_____

Recessive allele:	q =	
Dominant allele:	p =	
Recessive phenotype:	q^2 =	
Homozygous dominant:	p^2 =	
Heterozygous:	2pq =	

Analysis of a Squirrel Gene Pool

In Olney, Illinois, in the United States, there is a unique population of albino (white) and gray squirrels. Between 1977 and 1990, students at Olney Central College carried out a study of this population. They recorded the frequency of gray and albino squirrels. The albinos displayed a mutant allele expressed as an albino phenotype only in the homozygous recessive condition. The data they collected are provided in the table below. Using the **Hardy-Weinberg equation** for calculating genotype frequencies, it was possible to estimate the frequency of the normal 'wild' allele (G) providing gray fur coloring, and the frequency of the mutant albino allele (g) producing white squirrels. This study provided real, first hand, data that students could use to see how genotype frequencies can change in a real population.

Thanks to **Dr. John Stencel**, Olney Central College, Olney, Illinois, US, for providing the data for this exercise.

Gray squirrel, usual color form

Albino form of gray squirrel

Population of gray and white squirrels in Olney, Illinois (1977-1990)

Year	Gray	White	Total	GG	Gg	gg		Freq. of g	Freq. of G
1977	602	182	784	26.85	49.93	23.21		48.18	51.82
1978	511	172	683	24.82	50.00	25.18		50.18	49.82
1979	482	134	616	28.47	49.77	21.75		46.64	53.36
1980	489	133	622	28.90	49.72	21.38		46.24	53.76
1981	536	163	699	26.74	49.94	23.32		48.29	51.71
1982	618	151	769	31.01	49.35	19.64		44.31	55.69
1983	419	141	560	24.82	50.00	25.18		50.18	49.82
1984	378	106	484	28.30	49.79	21.90		46.80	53.20
1985	448	125	573	28.40	49.78	21.82		46.71	53.29
1986	536	155	691	27.71	49.86	22.43		47.36	52.64
1987	No data collected this year								
1988	652	122	774	36.36	47.88	15.76		39.70	60.30
1989	552	146	698	29.45	49.64	20.92		45.74	54.26
1990	603	111	714	36.69	47.76	15.55		39.43	60.57

1. **Graph population changes**: Use the data in the first 3 columns of the table above to plot a line graph. This will show changes in the phenotypes: numbers of gray and white (albino) squirrels, as well as changes in the total population. Plot: **gray**, **white**, and **total** for each year:

(a) Determine by how much (as a %) total population numbers have fluctuated over the sampling period:

_____

(b) Describe the overall trend in total population numbers and any pattern that may exist:

_____

_____

_____

_____

_____

_____

_____

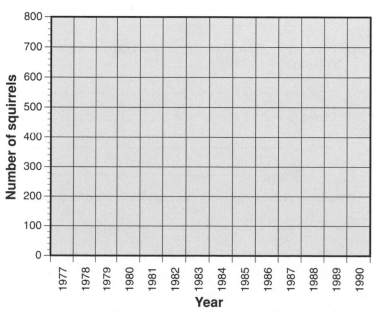

Number of squirrels (y-axis: 0, 100, 200, 300, 400, 500, 600, 700, 800)

Year (x-axis: 1977, 1978, 1979, 1980, 1981, 1982, 1983, 1984, 1985, 1986, 1987, 1988, 1989, 1990)

Speciation

2. **Graph genotype changes**: Use the data in the genotype columns of the table on the opposite page to plot a line graph. This will show changes in the allele combinations (**GG**, **Gg**, **gg**). Plot: **GG**, **Gg**, and **gg** for each year:

Describe the overall trend in the frequency of:

(a) Homozygous dominant (**GG**) genotype:

_____

_____

(b) Heterozygous (**Gg**) genotype:

_____

_____

(c) Homozygous recessive (gg) genotype:

_____

_____

_____

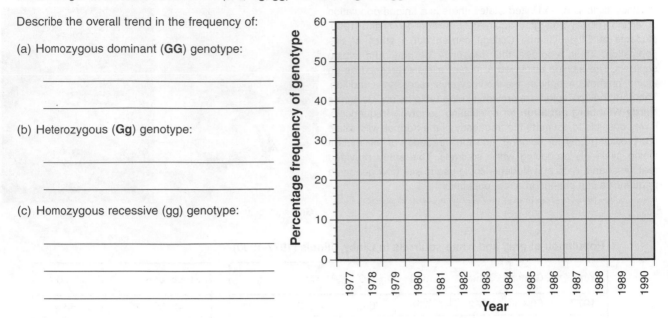

3. **Graph allele changes**: Use the data in the last two columns of the table on the previous page to plot a line graph. This will show changes in the *allele frequencies* for each of the dominant (**G**) and recessive (**g**) alleles.
Plot: the frequency of **G** and the frequency of **g**:

(a) Describe the overall trend in the frequency of the dominant allele (**G**):

_____

_____

_____

(b) Describe the overall trend in the frequency of the recessive allele (**g**):

_____

_____

_____

_____

_____

_____

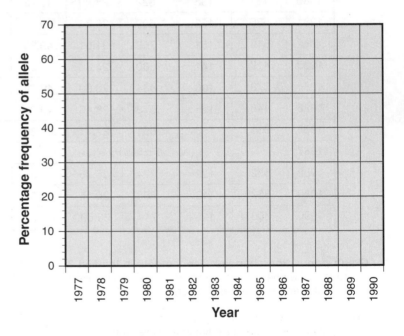

4. (a) State which of the three graphs best indicates that a significant change may be taking place in the gene pool of this population of squirrels:

_____

(b) Give a reason for your answer: _____

_____

_____

5. Describe a possible cause of the changes in allele frequencies over the sampling period: _____

_____

_____

_____

_____

The Founder Effect

Occasionally, a small number of individuals from a large population may migrate away, or become isolated from, their original population. If this colonizing or 'founder' population is made up of only a few individuals, it will probably have a *non-representative sample* of alleles from the parent population's gene pool. As a consequence of this **founder effect**, the

colonizing population may evolve differently from that of the parent population, particularly since the environmental conditions for the isolated population may be different. In some cases, it may be possible for certain alleles to be missing altogether from the individuals in the isolated population. Future generations of this population will not have this allele.

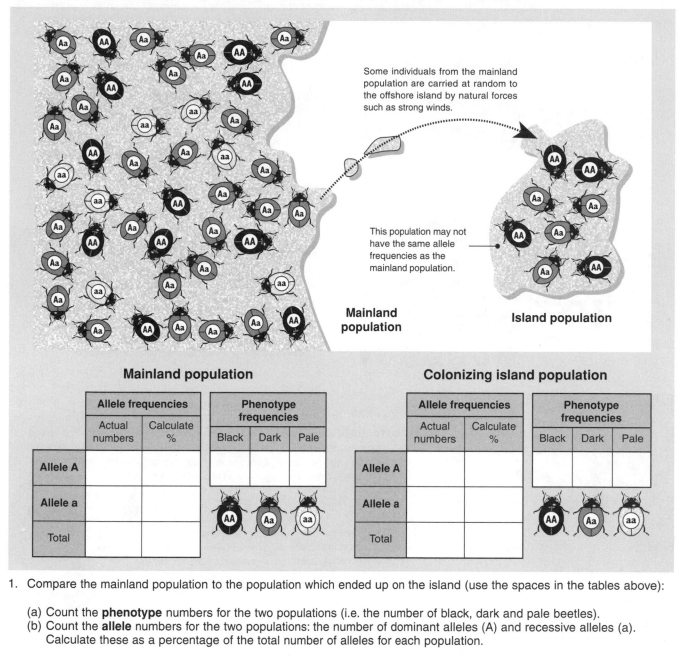

Some individuals from the mainland population are carried at random to the offshore island by natural forces such as strong winds.

This population may not have the same allele frequencies as the mainland population.

Mainland population

Island population

Mainland population

	Allele frequencies		Phenotype frequencies		
	Actual numbers	Calculate %	Black	Dark	Pale
Allele A					
Allele a					
Total					

Colonizing island population

	Allele frequencies		Phenotype frequencies		
	Actual numbers	Calculate %	Black	Dark	Pale
Allele A					
Allele a					
Total					

1. Compare the mainland population to the population which ended up on the island (use the spaces in the tables above):

   (a) Count the **phenotype** numbers for the two populations (i.e. the number of black, dark and pale beetles).
   (b) Count the **allele** numbers for the two populations: the number of dominant alleles (A) and recessive alleles (a). Calculate these as a percentage of the total number of alleles for each population.

2. Describe how the allele frequencies of the two populations are different: _____

   _____

3. Describe some possible ways in which various types of organism can be carried to an offshore island:

   (a) Plants: _____

   (b) Land animals: _____

   (c) Non-marine birds: _____

4. Since founder populations are often very small, describe another process that may further alter the allele frequencies:

   _____

   _____

Population Bottlenecks

Populations may sometimes be reduced to low numbers by predation, disease, or periods of climatic change. A population crash may not be 'selective': it may affect all phenotypes equally. Large scale catastrophic events (e.g. fire or volcanic eruption) are examples of such non-selective events. Humans may severely (and selectively) reduce the numbers of some species through hunting and/or habitat destruction. These populations may recover, having squeezed through a 'bottleneck' of low numbers.

The diagram below illustrates how population numbers may be reduced as a result of a catastrophic event. Following such an event, the small number of individuals contributing to the gene pool may not have a representative sample of the alleles in the pre-catastrophe population, i.e. the allele frequencies in the remnant population may be altered. Genetic drift may cause further changes to allele frequencies. The small population may return to previous levels but with a reduced genetic diversity.

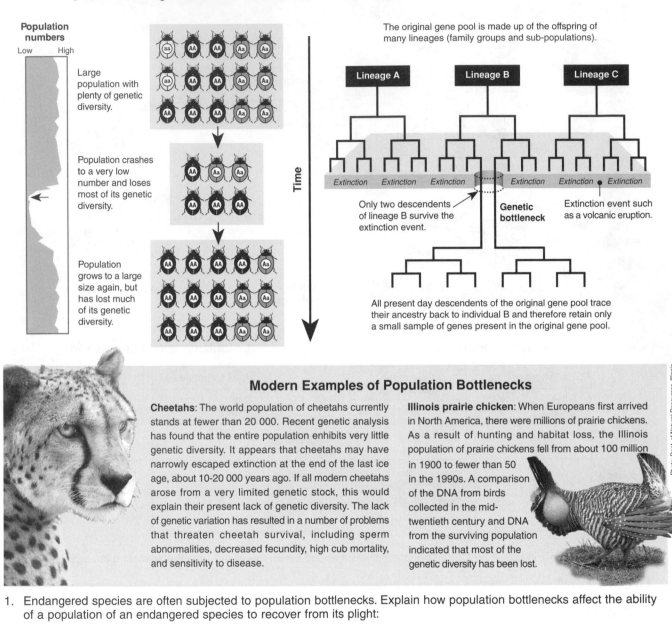

Population numbers

Low High

Large population with plenty of genetic diversity.

Population crashes to a very low number and loses most of its genetic diversity.

Population grows to a large size again, but has lost much of its genetic diversity.

Time

The original gene pool is made up of the offspring of many lineages (family groups and sub-populations).

Lineage A   Lineage B   Lineage C

Extinction   Extinction   Extinction   Extinction   Extinction   Extinction

Only two descendents of lineage B survive the extinction event.

Genetic bottleneck

Extinction event such as a volcanic eruption.

All present day descendents of the original gene pool trace their ancestry back to individual B and therefore retain only a small sample of genes present in the original gene pool.

Modern Examples of Population Bottlenecks

Cheetahs: The world population of cheetahs currently stands at fewer than 20 000. Recent genetic analysis has found that the entire population exhibits very little genetic diversity. It appears that cheetahs may have narrowly escaped extinction at the end of the last ice age, about 10-20 000 years ago. If all modern cheetahs arose from a very limited genetic stock, this would explain their present lack of genetic diversity. The lack of genetic variation has resulted in a number of problems that threaten cheetah survival, including sperm abnormalities, decreased fecundity, high cub mortality, and sensitivity to disease.

Illinois prairie chicken: When Europeans first arrived in North America, there were millions of prairie chickens. As a result of hunting and habitat loss, the Illinois population of prairie chickens fell from about 100 million in 1900 to fewer than 50 in the 1990s. A comparison of the DNA from birds collected in the mid-twentieth century and DNA from the surviving population indicated that most of the genetic diversity has been lost.

Photo: Dept. of Natural Resources, Illinois

1. Endangered species are often subjected to population bottlenecks. Explain how population bottlenecks affect the ability of a population of an endangered species to recover from its plight:

_____

_____

_____

2. Explain why the lack of genetic diversity in cheetahs has increased their sensitivity to disease:

_____

_____

3. Describe the effect of a population bottleneck on the potential of a species to adapt to changes (i.e. its ability to evolve):

_____

_____

Related activities: Gene Pool Exercise, Genetic Drift

Periodicals:
The Cheetah: Losing the Race?

Genetic Drift

Not all individuals, for various reasons, will be able to contribute their genes to the next generation. **Genetic drift** (also known as the Sewell-Wright Effect) refers to the *random changes in allele frequency* that occur in all populations, but are much more pronounced in small populations. In a small population, the effect of a few individuals not contributing their alleles to the next generation can have a great effect on allele frequencies. Alleles may even become **lost** from the gene pool altogether (frequency becomes 0%) or **fixed** as the only allele for the gene present (frequency becomes 100%).

The genetic makeup (allele frequencies) of the population changes randomly over a period of time

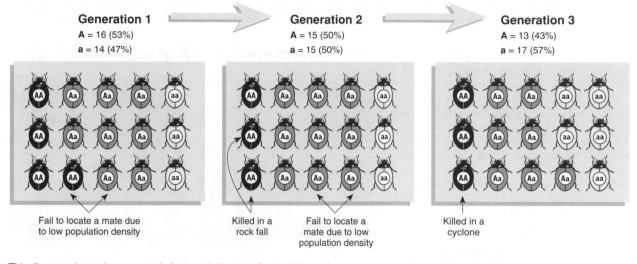

Generation 1
A = 16 (53%)
a = 14 (47%)

Fail to locate a mate due to low population density

Generation 2
A = 15 (50%)
a = 15 (50%)

Killed in a rock fall

Fail to locate a mate due to low population density

Generation 3
A = 13 (43%)
a = 17 (57%)

Killed in a cyclone

This diagram shows the gene pool of a hypothetical small population over three generations. For various reasons, not all individuals contribute alleles to the next generation. With the random loss of the alleles carried by these individuals, the allele frequency changes from one generation to the next. The change in frequency is directionless as there is no selecting force. The allele combinations for each successive generation are determined by how many alleles of each type are passed on from the preceding one.

Computer Simulation of Genetic Drift

Below are displayed the change in allele frequencies in a computer simulation showing random genetic drift. The breeding population progressively gets smaller from left to right. Each simulation was run for 140 generations.

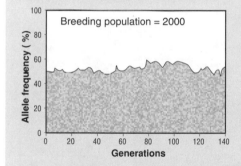

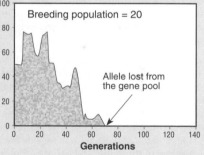

Large breeding population
Fluctuations are minimal in large breeding populations because the large numbers buffer the population against random loss of alleles. On average, losses for each allele type will be similar in frequency and little change occurs.

Small breeding population
Fluctuations are more severe in smaller breeding populations because random changes in a few alleles cause a greater percentage change in allele frequencies.

Very small breeding population
Fluctuations in very small breeding populations are so extreme that the allele can become fixed (frequency of 100%) or lost from the gene pool altogether (frequency of 0%).

Speciation

1. Explain what is meant by **genetic drift**: _____

_____

2. Describe how genetic drift affects the amount of genetic variation within very small populations: _____

_____

_____

_____

3. Identify a small breeding population of animals or plants in your country in which genetic drift could be occurring:

_____

Related activities: Gene Pool Exercise, The Founder Effect
Web links: Genetic Drift Simulation

PRA 3

Isolation and Species Formation

Isolating mechanisms are barriers to successful interbreeding between species. Reproductive isolation is fundamental to the **biological species concept**, which defines a species by its inability to breed with other species to produce fertile offspring. Prezygotic isolating mechanisms act before fertilisation occurs, preventing species ever mating, whereas postzygotic barriers take effect after fertilisation. **Geographical barriers** are not regarded as reproductive isolating mechanisms because they are not part of the species' biology, although they are often a necessary precursor to reproductive isolation in sexually reproducing populations. Ecological isolating mechanisms are those that isolate gene pools on the basis of ecological preferences, e.g habitat selection. Although ecological and geographical isolation are sometimes confused, they are quite distinct, as ecological isolation involves a component of the species biology. Similarly, the **temporal isolation** of species, through differences in the timing of important life cycle events, effectively prevents potentially interbreeding species from successfully reproducing.

Geographical Isolation

Geographical isolation describes the isolation of a species population (gene pool) by some kind of physical barrier, for example, mountain range, water body, isthmus, desert, or ice sheet. Geographical isolation is a frequent first step in the subsequent reproductive isolation of a species. For example, geological changes to the lake basins has been instrumental in the subsequent proliferation of cichlid fish species in the rift lakes of East Africa (right). Similarly, many Galapagos Island species (e.g. iguanas, finches) are now quite distinct from the Central and South American species from which they arose after isolation from the mainland.

Ecological (Habitat) Isolation

Ecological isolation describes the existence of a **prezygotic reproductive barrier** between two species (or sub-species) as a result of them occupying or breeding in different habitats within the same general geographical area. Ecological isolation includes small scale differences (e.g. ground or tree dwelling) and broad differences (e.g. desert vs grasslands). The red-browed and brown **treecreepers** (*Climacteris* spp.) are sympatric in south-eastern Australia and both species feed largely on ants. However the brown spends most of its time foraging on the ground or on fallen logs while the red-browed forages almost entirely in the trees.

Ecological isolation often follows geographical isolation, but in many cases the geographical barriers may remain in part. For example, five species of **antelope squirrels** occupy different habitat ranges throughout the southwestern United States and northern Mexico, a region divided in part by the Grand Canyon. The white tailed antelope squirrel is widely distributed in desert areas to the north and south of the canyon, while the smaller, more specialized Harris' antelope squirrel has a much more limited range only to the south in southern Arizona. The Grand Canyon still functions as a barrier to dispersal but the species are now ecologically isolated as well.

Geographical and Ecological Isolation of Species

Malawi cichlid species

L. Victoria

L. Tanganyika

L. Malawi

NASA Earth Observatory

Both photos: Aviceda

Red-browed treecreeper

Brown treecreeper

UtahCamera

White-tailed antelope squirrel

The Grand Canyon - a massive rift in the Colorado Plateau

Harris' antelope squirrel

Photo: Allan and Elaine Wilson

1. Describe the role of isolating mechanisms in maintaining the integrity of a species: _____

_____

_____

2. (a) Explain why geographical isolation is not regarded as a reproductive isolating mechanism: _____

_____

   (b) Explain why, despite this, it often precedes reproductive isolation: _____

_____

3. Distinguish between geographical and ecological isolation: _____

_____

Related activities: Reproductive Isolation

Periodicals:
Cichlids of the Rift lakes

Reproductive Isolation

Reproductive isolation prevents interbreeding (and therefore gene flow) between species. Any factor that impedes two species from producing viable, fertile hybrids contributes to reproductive isolation. Single barriers may not completely stop gene flow, so most species commonly have more than one type of barrier.

Single barriers to reproduction (including geographical barriers) often precede the development of a suite of reproductive isolating mechanisms (RIMs). Most operate before fertilisation (prezygotic RIMs) with postzyotic RIMs being important in preventing offspring between closely related species.

Temporal Isolation

Individuals from different species do not mate because they are active during different times of the day, or in different seasons. Plants flower at different times of the year or even at different times of the day to avoid hybridization (e.g. the orchid genus *Dendrobium*, which occupy the same location and flower on different days). Closely related animal species may have quite different breeding seasons or periods of emergence. **Periodical cicadas** of the genus *Magicicada* are so named because members of each species in a particular region are developmentally synchronized, despite very long life cycles. Once their underground period of development (13 or 17 years depending on the species) is over, the entire population emerges at much the same time to breed.

Gamete Isolation

The gametes from different species are often incompatible, so even if they meet they do not survive. For animals where fertilization is internal, the sperm may not survive in the reproductive tract of another species. If the sperm does survive and reach the ovum, chemical differences in the gametes prevent fertilization. Gamete isolation is particularly important in aquatic environments where the gametes are released into the water and fertilized externally, such as in reproduction in frogs. Chemical recognition is also used by flowering plants to recognize pollen from the same species.

Behavioral (ethological) Isolation

Behavioral isolation operates through differences in species courtship behaviors. Courtship is a necessary prelude to mating in many species and courtship behaviors are species specific. Mates of the same species are attracted with distinctive, usually ritualized, dances, vocalizations, and body language. Because they are not easily misinterpreted, the courtship behaviors of one species will be unrecognized and ignored by individuals of another species. Birds exhibit a remarkable range of courtship displays. The use of song is widespread but ritualized movements, including nest building, are also common. For example, the elaborate courtship bowers of bowerbirds are well known, and Galapagos frigatebirds have an elaborate display in which they inflate a bright red gular pouch (right). Amongst insects, empid flies have some of the most elaborate of courtship displays. They are aggressive hunters so ritualized behavior involving presentation of a prey item facilitates mating. The sexual organs of the flies are also like a lock-and-key, providing mechanical reproductive isolation as well (see below).

Mechanical (morphological) Isolation

Structural differences (incompatibility) in the anatomy of reproductive organs prevents sperm transfer between individuals of different species. This is an important isolating mechanism preventing breeding between closely related species of arthropods. Many flowering plants have coevolved with their animal pollinators and have flowers structures to allow only that insect access. Structural differences in the flowers and pollen of different plant species prevents cross breeding because pollen transfer is restricted to specific pollinators and the pollen itself must be species compatible.

Prezygotic Isolating Mechanisms

Amphibian ovary (*Rana*)

Mammalian sperm

Male frigatebird courtship display

Male tree frog calling

Wing beating in male sage grouse

Male

Female

Lock and key genitalia

Gift of prey keeps female occupied

Empid flies mating

Damselflies mating

Complex flowers in orchids

Speciation

Periodicals:
Listen, we're different

Related activities: Isolation and Species Formation

A 2

Postzygotic Isolating Mechanisms

Hybrid sterility

Even if two species mate and produce hybrid offspring that are vigorous, the species are still reproductively isolated if the hybrids are sterile (genes cannot flow from one species' gene pool to the other). Such cases are common among the horse family (such as the zebra and donkey shown on the right). One cause of this sterility is the failure of meiosis to produce normal gametes in the hybrid. This can occur if the chromosomes of the two parents are different in number or structure (see the **"zebronkey"** karyotype on the right). The **mule**, a cross between a donkey stallion and a horse mare, is also an example of **hybrid vigor** (they are robust) as well as **hybrid sterility**. Female mules sometimes produce viable eggs but males are infertile.

Zebra stallion (2N = 44) X Donkey jenny (2N = 62)

Karyotype of **'Zebronkey'** offspring (2N = 53)

Chromosomes contributed by zebra stallion — Y

Chromosomes contributed by donkey jenny — X

Hybrid inviability

Mating between individuals of two species may produce a zygote, but genetic incompatibility may stop development of the zygote. Fertilized eggs often fail to divide because of mis-matched chromosome numbers from each gamete. Very occasionally, the hybrid zygote will complete embryonic development but will not survive for long. For example, although sheep and goats seem similar and can be mated together, they belong to different genera. Any offspring of a sheep-goat pairing is generally stillborn.

Sheep (*Ovis*) 54 chromosomes

Goat (*Capra*) 60 chromosomes

Hybrid breakdown

Hybrid breakdown is common feature of some plant hybrids. The first generation (F₁) may be fertile, but the second generation (F₂) are infertile or inviable. Examples include hybrids between cotton species (near right), species within the genus *Populus*, and strains of the cultivated rice *Oryza* (far right)

---

1. In the following examples, classify the reproductive isolating mechanism as either **prezygotic** or **postzygotic** and describe the mechanisms by which the isolation is achieved (e.g. structrual isolation, hybrid sterility etc.):

   (a) Some different cotton species can produce fertile hybrids, but breakdown of the hybrid occurs in the next generation when the offspring of the hybrid die in their seeds or grow into defective plants:

   Prezygotic / postzygotic (delete one)   Mechanism of isolation: _____

   (b) Many plants have unique arrangements of their floral parts that stops transfer of pollen between plants:

   Prezygotic / postzygotic (delete one)   Mechanism of isolation: _____

   (c) Two skunk species do not mate despite having habitats that overlap because they mate at different times of the year:

   Prezygotic / postzygotic (delete one)   Mechanism of isolation: _____

   (d) Several species of the frog genus *Rana*, live in the same regions and habitats, where they may occasionally hybridize. The hybrids generally do not complete development, and those that do are weak and do not survive long:

   Prezygotic / postzygotic (delete one)   Mechanism of isolation: _____

2. Postzygotic isolating mechanisms are said to reinforce prezygotic ones. Explain why this is the case:

   _____

   _____

   _____

   _____

   _____

Allopatric Speciation

Allopatric speciation is a process thought to have been responsible for a great many instances of species formation. It has certainly been important in countries which have had a number of cycles of geographical fragmentation. Such cycles can occur as the result of glacial and interglacial periods, where ice expands and then retreats over a land mass. Such events are also accompanied by sea level changes which can isolate populations within relatively small geographical regions.

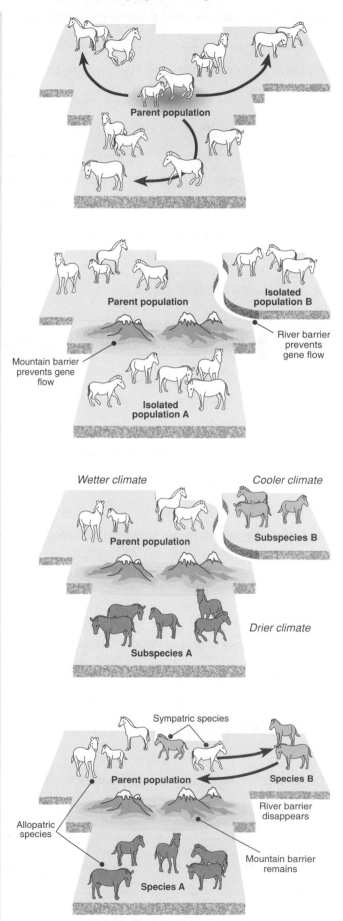

Stage 1: Moving into new environments

There are times when the range of a species expands for a variety of different reasons. A single population in a relatively homogeneous environment will move into new regions of their environment when they are subjected to intense competition (whether it is interspecific or intraspecific). The most severe form of competition is between members of the same species since they are competing for identical resources in the habitat. In the diagram on the right there is a 'parent population' of a single species with a common gene pool with regular 'gene flow' (theoretically any individual has access to all members of the opposite sex for mating purposes).

Stage 2: Geographical isolation

Isolation of parts of the population may occur due to the formation of **physical barriers**. These barriers may cut off those parts of the population that are at the extremes of the species range and gene flow is prevented or rare. The rise and fall of the sea level has been particularly important in functioning as an isolating mechanism. Climatic change can leave 'islands' of habitat separated by large inhospitable zones that the species cannot traverse.

Example: In mountainous regions, alpine species are free to range widely over extensive habitat during cool climatic periods. During warmer periods, however, they may become isolated because their habitat is reduced to 'islands' of high ground surrounded by inhospitable lowland habitat.

Stage 3: Different selection pressures

The isolated populations (A and B) may be subjected to quite different selection pressures. These will favor individuals with traits that suit each particular environment. For example, population A will be subjected to selection pressures that relate to drier conditions. This will favor those individuals with phenotypes (and therefore genotypes) that are better suited to dry conditions. They may for instance have a better ability to conserve water. This would result in improved health, allowing better disease resistance and greater reproductive performance (i.e. more of their offspring survive). Finally, as allele frequencies for certain genes change, the population takes on the status of a **subspecies**. Reproductive isolation is not yet established but the subspecies are significantly different genetically from other related populations.

Stage 4: Reproductive isolation

The separated populations (isolated subspecies) will often undergo changes in their genetic makeup as well as their behavior patterns. These ensure that the gene pool of each population remains isolated and 'undiluted' by genes from other populations, even if the two populations should be able to remix (due to the removal of the geographical barrier). Gene flow does not occur. The arrows (in the diagram to the right) indicate the zone of overlap between two species after the new Species B has moved back into the range inhabited by the parent population. Closely-related species whose distribution overlaps are said to be **sympatric species**. Those that remain geographically isolated are called **allopatric species**.

Related activities: Reproductive Isolation, Stages in Species Development
Web links: Species and Speciation, Mechanisms of Speciation, Allopatric Speciation

RA 2

Speciation

1. Describe why some animals, given the opportunity, move into new environments: _____
   _____
   _____

2. (a) Plants are unable to move. Explain how plants might disperse to new environments: _____
   _____

   (b) Describe the amount of **gene flow** within the parent population prior to and during this range expansion:
   _____
   _____
   _____

3. Identify the **process** that causes the formation of new **mountain ranges**: _____

4. Identify the event that can cause large changes in **sea level** (up to 200 metres): _____

5. Describe six **physical barriers** that could isolate different parts of the same population:
   _____
   _____
   _____
   _____

6. Describe the effect that physical barriers have on **gene flow**: _____
   _____
   _____
   _____

7. (a) Describe four different types of **selection pressure** that could have an effect on a gene pool: _____
   _____
   _____

   (b) Describe briefly how these selection pressures affect the isolated gene pool in terms of **allele frequencies**:
   _____
   _____
   _____

8. Describe two types of **prezygotic** and two types of **postzygotic** reproductive isolating mechanisms (see previous pages):

   (a) Prezygotic: _____
   _____

   (b) Postzygotic: _____
   _____

9. Distinguish between **allopatry** and **sympatry** in populations: _____
   _____
   _____
   _____
   _____
   _____

Sympatric Speciation

New species may be formed even where there is no separation of the gene pools by physical barriers. Called **sympatric speciation**, it is rarer than allopatric speciation, although not uncommon in plants which form **polyploids**. There are two situations where sympatric speciation is thought to occur. These are described below:

Speciation Through Niche Differentiation

Niche isolation
In a heterogeneous environment (one that is not the same everywhere), a population exists within a diverse collection of **microhabitats**. Some organisms prefer to occupy one particular type of 'microhabitat' most of the time, only rarely coming in contact with fellow organisms that prefer other microhabitats. Some organisms become so dependent on the resources offered by their particular microhabitat that they never meet up with their counterparts in different microhabitats.

Reproductive isolation
Finally, the individual groups have remained genetically isolated for so long because of their microhabitat preferences, that they have become reproductively isolated. They have become new species that have developed subtle differences in behavior, structure, and physiology. Gene flow (via sexual reproduction) is limited to organisms that share a similar microhabitat preference (as shown in the diagram on the right).

Example: When it is time for them to lay eggs, some beetles preferentially locate the same plant species as they grew up on. Individual beetles of the same species have different preferences.

Instant Speciation by Polyploidy

When polyploidy occurs, it is possible to form a completely new species without isolation from the parent species. This type of malfunction during the process of meiosis produces sudden reproductive isolation for the new group. Because the sex-determining mechanism is disturbed, animals are rarely able to achieve new species status this way (they are effectively sterile e.g. tetraploid XXXX). Many plants, on the other hand, are able to reproduce vegetatively, or carry out self pollination. This ability to reproduce on their own enables such polyploid plants to produce a breeding population.

Speciation by allopolyploidy
This type of polyploidy usually arises from the doubling of chromosomes in a hybrid between two different species. The doubling often makes the hybrid fertile.

Examples: Modern wheat. Swedes are polyploid species formed from a hybrid between a type of cabbage and a type of turnip.

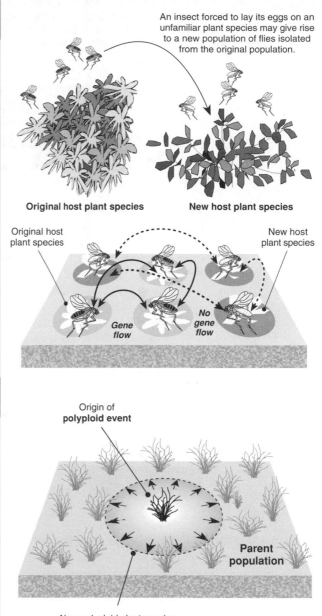

An insect forced to lay its eggs on an unfamiliar plant species may give rise to a new population of flies isolated from the original population.

Original host plant species   New host plant species

Original host plant species   New host plant species

Gene flow   No gene flow

Origin of **polyploid event**

Parent population

New polyploid plant species spreads outwards through the existing parent population

1. Explain what is meant by **sympatric speciation** and identify the mechanisms by which it can occur:

2. Explain briefly how polyploidy may cause the formation of a new species:

3. Identify an example of a species that has been formed by polyploidy:

4. Explain how niche differentiation may cause the formation of a new species:

Related activities: Allopatric Speciation, Breeding Modern Wheat
Web links: Sympatric Speciation

A 2

Speciation

Stages in Species Development

The diagram below represents a possible sequence of genetic events involved in the origin of two new species from an ancestral population. As time progresses (from top to bottom of the diagram) the amount of genetic variation increases and each group becomes increasingly isolated from the other. The mechanisms that operate to keep the two gene pools isolated from one another may begin with **geographical barriers**. This may be followed by **prezygotic** mechanisms which protect the gene pool from unwanted dilution by genes from other pools. A longer period of isolation may lead to **postzygotic** mechanisms (see the page on reproductive isolating mechanisms). As the two gene pools become increasingly isolated and different from each other, they are progressively labeled: population, race, and subspecies. Finally they attain the status of separate species.

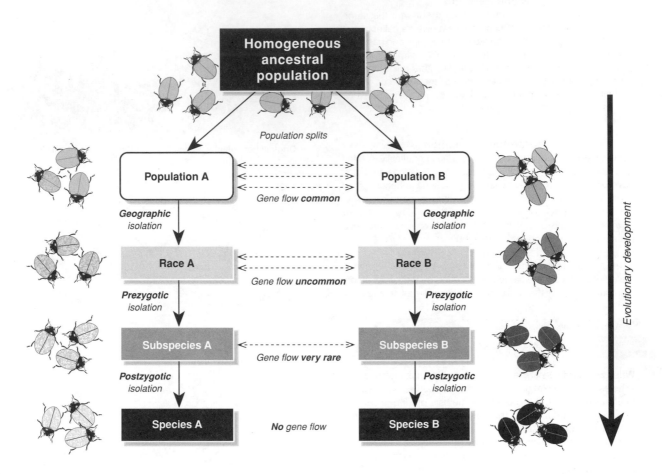

1. Explain what happens to the extent of gene flow between diverging populations as they gradually attain species status:

_____

_____

2. Early human populations about 500 000 ya were scattered across Africa, Europe, and Asia. This was a time of many regional variants, collectively called archaic *Homo sapiens*. The fossil skulls from different regions showed mixtures of characteristics, some modern and some 'primitive'. These regional populations are generally given subspecies status. Suggest reasons why gene flow between these populations may have been rare, but still occasionally occurred:

_____

_____

_____

3. In the USA, the species status of several duck species, including the black duck (*Anas rubripes*) and the mottled duck in Florida (*A. fulvigula*) is threatened by interbreeding with the now widespread and very adaptable mallard duck (*A. platyrhynchos*). Similar threatened extinction though hybridization has occurred in New Zealand, where the native gray duck has been virtually eliminated as a result of interbreeding with the introduced mallard.

(a) Suggest why these hybrids threaten the Species status of some native duck species: _____

_____

(b) Suggest what factor may deter mallards from hybridizing with other duck species: _____

_____

Selective Breeding in Animals

The domestication of livestock has a long history dating back at least 8000 years. Today's important stock breeds were all derived from wild ancestors that were domesticated by humans, who then used **selective breeding** to produce livestock to meet specific requirements. Selective breeding of domesticated animals involves identifying desirable qualities (e.g. high wool production or meat yield), and breeding together individuals with those qualities so the trait is reliably passed on. Practices such as **inbreeding**, **line-breeding**, and **outcrossing** are used to select and 'fix' desirable traits in varieties. Today, modern breeding techniques often employ reproductive technologies, such as artificial insemination, so that the desirable characteristics of one male can be passed on to many females. These new technologies refine the selection process and increase the rate at which stock improvements are made. Rates are predicted to accelerate further as new technologies, such as genomic selection, become more widely available and less costly. However, producing highly inbred lines of animals with specific traits can have disadvantages. **Homozygosity** for a number of desirable traits can cause physiological or physical problems to the animal itself. For example, animals bred specifically for rapid weight gain often grow so fast that they have skeletal and muscular difficulties.

The Origin of Domestic Animals

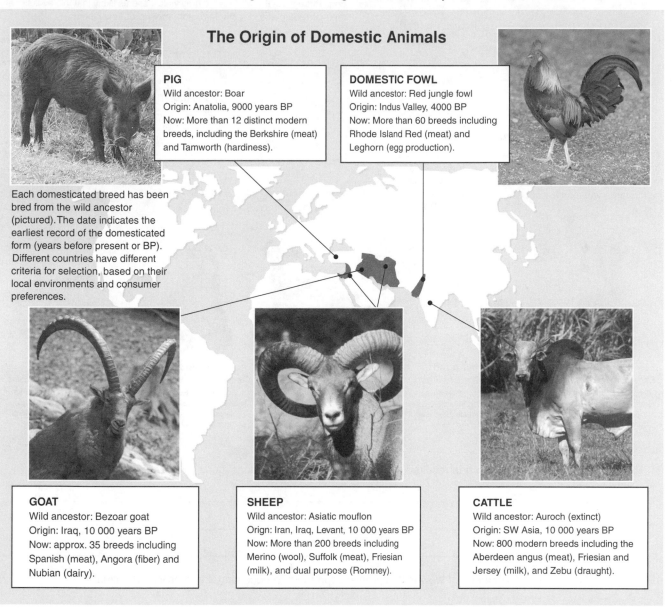

PIG
Wild ancestor: Boar
Origin: Anatolia, 9000 years BP
Now: More than 12 distinct modern breeds, including the Berkshire (meat) and Tamworth (hardiness).

DOMESTIC FOWL
Wild ancestor: Red jungle fowl
Origin: Indus Valley, 4000 BP
Now: More than 60 breeds including Rhode Island Red (meat) and Leghorn (egg production).

Each domesticated breed has been bred from the wild ancestor (pictured). The date indicates the earliest record of the domesticated form (years before present or BP). Different countries have different criteria for selection, based on their local environments and consumer preferences.

GOAT
Wild ancestor: Bezoar goat
Origin: Iraq, 10 000 years BP
Now: approx. 35 breeds including Spanish (meat), Angora (fiber) and Nubian (dairy).

SHEEP
Wild ancestor: Asiatic mouflon
Origin: Iran, Iraq, Levant, 10 000 years BP
Now: More than 200 breeds including Merino (wool), Suffolk (meat), Friesian (milk), and dual purpose (Romney).

CATTLE
Wild ancestor: Auroch (extinct)
Origin: SW Asia, 10 000 years BP
Now: 800 modern breeds including the Aberdeen angus (meat), Friesian and Jersey (milk), and Zebu (draught).

Speciation

1. Distinguish between inbreeding and out-crossing, explaining the significance of each technique in selective breeding:

_____

_____

_____

_____

2. Describe the contribution that new reproductive technologies are making to selective breeding: _____

_____

_____

Related activities: Selective Breeding in Crop Plants
Web links: Dogs and More Dogs

RA 2

Dogs provide a striking example of selective breeding, with more than 400 recognized breeds. Over centuries, humans have selected for desirable physical and behavioral traits. All breeds of dog are members of the same species, *Canis familiaris*. This species descended from a single wild species, the grey wolf *Canis lupus*, over 15,000 years ago. Five ancient dog breeds are recognized, from which all other breeds are thought to have descended by artificial selection.

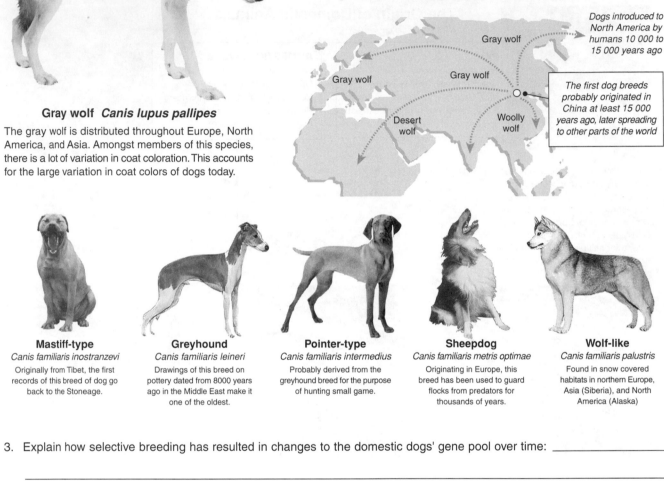

Gray wolf *Canis lupus pallipes*

The gray wolf is distributed throughout Europe, North America, and Asia. Amongst members of this species, there is a lot of variation in coat coloration. This accounts for the large variation in coat colors of dogs today.

Gray Wolf: Ancestor of Domestic Dogs

Until recently, it was unclear whether the ancestor to the modern domestic dogs was the desert wolf of the Middle East, the woolly wolf of central Asia, or the gray wolf of Northern Hemisphere. Recent genetic studies (mitochondrial DNA comparisons) now provide strong evidence that the ancestor of domestic dogs throughout the world is the gray wolf. It seems likely that this evolutionary change took place in a single region, most probably China.

Dogs introduced to North America by humans 10 000 to 15 000 years ago

The first dog breeds probably originated in China at least 15 000 years ago, later spreading to other parts of the world

Mastiff-type
Canis familiaris inostranzevi
Originally from Tibet, the first records of this breed of dog go back to the Stoneage.

Greyhound
Canis familiaris leineri
Drawings of this breed on pottery dated from 8000 years ago in the Middle East make it one of the oldest.

Pointer-type
Canis familiaris intermedius
Probably derived from the greyhound breed for the purpose of hunting small game.

Sheepdog
Canis familiaris metris optimae
Originating in Europe, this breed has been used to guard flocks from predators for thousands of years.

Wolf-like
Canis familiaris palustris
Found in snow covered habitats in northern Europe, Asia (Siberia), and North America (Alaska)

3. Explain how selective breeding has resulted in changes to the domestic dogs' gene pool over time: _____

_____

_____

_____

4. Describe the behavioral tendency of wolves that predisposed them to becoming a domesticated animal:

_____

_____

_____

5. List the physical and behavioral traits that would be desirable (selected for) in the following uses of a dog:

   (a) Hunting large game (e.g. boar and deer): _____

   (b) Game fowl dog: _____

   (c) Stock control (sheep/cattle dog): _____

   (d) Family pet (house dog): _____

   (e) Guard dog: _____

Selective Breeding in Crop Plants

For thousands of years, farmers have used the variation in wild and cultivated plants to develop crops. Genetic diversity gives species the ability to adapt to new environmental challenges, such as new pests, diseases, or growing conditions. The genetic diversity within different crop varieties provides options to develop, through selection, new and more productive crop plants.

Brassica oleracea is a good example of the variety that can be produced by selectively growing plants with desirable traits. Not only are there six varieties of *Brassica oleracea*, but each of those has a number of sub varieties as well. Although brassicas have been cultivated for several thousand years, cauliflower, broccoli and brussels sprouts appeared only in the last 500 years.

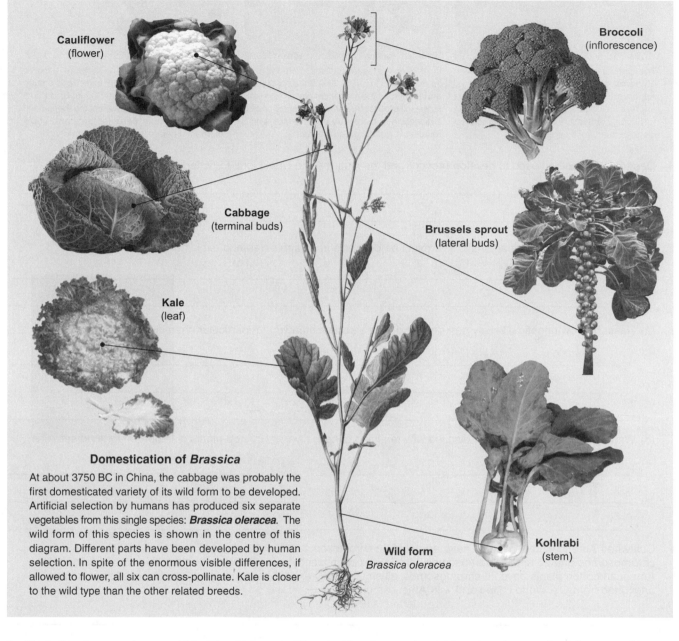

Cauliflower (flower)

Broccoli (inflorescence)

Cabbage (terminal buds)

Brussels sprout (lateral buds)

Kale (leaf)

Kohlrabi (stem)

Wild form *Brassica oleracea*

Domestication of *Brassica*

At about 3750 BC in China, the cabbage was probably the first domesticated variety of its wild form to be developed. Artificial selection by humans has produced six separate vegetables from this single species: ***Brassica oleracea***. The wild form of this species is shown in the centre of this diagram. Different parts have been developed by human selection. In spite of the enormous visible differences, if allowed to flower, all six can cross-pollinate. Kale is closer to the wild type than the other related breeds.

Speciation

1. Study the diagram above and identify which part of the plant has been selected for to produce each of the vegetables:

    (a) Cauliflower: _____   (d) Brussels sprout: _____

    (b) Kale: _____   (e) Cabbage: _____

    (c) Broccoli: _____   (f) Kohlrabi: _____

2. Describe the feature of these vegetables that suggests they are members of the same species: _____

    _____

3. Human artificial selection pressures can also influence the development of characteristics in 'unwanted' species. Suggest how human weed control measures may inadvertently select for weed plants that have a resistance to the measures:

    _____

    _____

The number of apple varieties is now a fraction of the many hundreds grown a century ago. Apples are native to Kazakhstan and breeders are now looking back to this centre of diversity to develop apples resistant to the bacterial disease that causes fireblight.

In 18th-century Ireland, potatoes were the main source of food for about 30% of the population, and farmers relied almost entirely on one very fertile and productive variety. That variety proved susceptible to the potato blight fungus which resulted in a widespread famine.

Hybrid corn varieties have been bred to minimize harm inflicted by insect pests such as corn rootworm (above). Hybrids are important because they recombine the genetic characteristics of parental lines and show increased heterozygosity and hybrid vigor.

4. Describe the method used to develop broccoli and the features one would look for when doing so:

5. Describe a phenotypic characteristic that might be desirable in an apple tree and explain your choice:

6. (a) Describe how genetic diversity can decline during selective breeding for particular characteristics:

(b) With reference to an example, discuss why retaining genetic diversity in crop plants is important for food security:

7. Cultivated American cotton plants have a total of 52 chromosomes (2N = 52). In each cell there are 26 large chromosomes and 26 small chromosomes. Old World cotton plants have 26 chromosomes (2N = 26), all large. Wild American cotton plants have 26 chromosomes, all small. Briefly explain how cultivated American cotton may have originated from Old World cotton and wild American cotton:

8. The Cavendish is the variety of banana most commonly sold in world supermarkets. It is seedless, sterile, and under threat of extinction by Panama disease Race 4. Explain why Cavendish banana crops are so endangered by this fungus:

9. Discuss the need to maintain the biodiversity of wild plants and ancient farm breeds:

Breeding Modern Wheat

Wheat has been cultivated for more than 9000 years and has undergone many genetic changes during its domestication (below). Increasingly, researchers are focused on enhancing the genetic diversity of this important crop to provide for future crop development. Several research centers and **seed** (gene) **banks** play a key role in this for wheat and other crop plants. They store the seeds of the species that collectively provide most of the food consumed by humans, and also keep a bank of seeds from less common or non-commercial varieties that may be threatened with loss.

The Evolution of Wheat

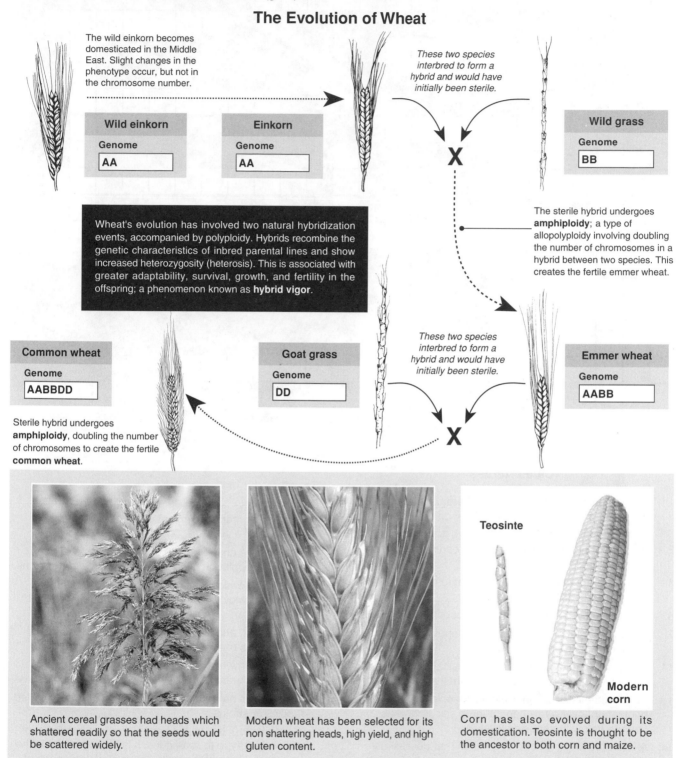

The wild einkorn becomes domesticated in the Middle East. Slight changes in the phenotype occur, but not in the chromosome number.

Wild einkorn
Genome
AA

Einkorn
Genome
AA

These two species interbred to form a hybrid and would have initially been sterile.

Wild grass
Genome
BB

Wheat's evolution has involved two natural hybridization events, accompanied by polyploidy. Hybrids recombine the genetic characteristics of inbred parental lines and show increased heterozygosity (heterosis). This is associated with greater adaptability, survival, growth, and fertility in the offspring; a phenomenon known as **hybrid vigor**.

The sterile hybrid undergoes **amphiploidy**; a type of allopolyploidy involving doubling the number of chromosomes in a hybrid between two species. This creates the fertile emmer wheat.

Common wheat
Genome
AABBDD

Goat grass
Genome
DD

These two species interbred to form a hybrid and would have initially been sterile.

Emmer wheat
Genome
AABB

Sterile hybrid undergoes **amphiploidy**, doubling the number of chromosomes to create the fertile **common wheat**.

Ancient cereal grasses had heads which shattered readily so that the seeds would be scattered widely.

Modern wheat has been selected for its non shattering heads, high yield, and high gluten content.

Teosinte

Modern corn

Corn has also evolved during its domestication. Teosinte is thought to be the ancestor to both corn and maize.

1. Describe why producing a hybrid from two inbred lines often has desirable effects in terms of crop characteristics:

_____

_____

2. Describe two phenotypic characteristics that would be desirable in a cereal plant:

(a) _____

(b) _____

Related activities: Selective Breeding in Crop Plants

A 1

Speciation

KEY TERMS Crossword

Complete the crossword below, which will test your understanding of key terms in this chapter and their meanings

Clues Across

1. The occurrence of two or more different forms within the same species usually caused by different allele combinations of the same gene.

2. A group of organisms capable of breeding together to produce viable offspring.

4. The number of times an allele appears in a population is termed the allele __ __ __ __ __ __ __ __ __.

5. The collective group of genes in a population (2 words).

7. The movement of genes between populations as a result of mating or migration.

11. Prezygotic isolation as result of physical separation of populations by environmental features is called this.

13. Small-scale changes in allele frequencies in a population, over a few generations, also known as change below the species level.

14. A interbreeding part of a population that possesses clearly definable characteristics.

15. A population sharing the same gene pool where groups next to each other are able to interbreed but groups at the extreme ends of the population can not (2 words).

16. Speciation as a result of reproductive isolation without any physical separation of the populations, i.e. populations remain within the same range. A term for populations sharing the same range.

17. An isolating mechanism that occurs after the formation of the zygote.

Clues Down

1. An isolating mechanism that occurs before the formation of a zygote.

3. A cumulative change in the characteristics and gene pool of a population over many successive generations.

6. Selection in which a single phenotype and therefore allele frequency is favoured and continuously shifts in one direction.

8. The merging and refinement of Darwin and Mendel's explanations of evolution and genetics (2 words).

9. Speciation in which the populations are physically separated. A term for physically separated populations.

10. An organism produced by the crossing of two unrelated species or strains.

12. In evolutionary terms, the capability of an individual of certain genotype to reproduce. It is usually equal to the proportion of the individual's genes in all the genes of the next generation.

Patterns of Evolution

KEY CONCEPTS

▶ Larger scale patterns of evolution involve the diversification and extinction of species.

▶ Divergent evolution is often associated with the diversification of species into new niches.

▶ Adaptive radiation and subsequent sympatry are associated with character displacement.

▶ Evolutionary developmental biology provides a explanation for the diversity of organisms and the rapid evolution of novel forms.

KEY TERMS

adaptive radiation

anagenesis

analogous structures (analogies)

background extinction rate

cladogenesis

coevolution

convergent evolution

divergent evolution

evo-devo

extinction

homologous structures (homologies)

mass extinction

parallel evolution

phyletic gradualism

phylogeny

punctuated equilibrium

sequential evolution

OBJECTIVES

☐ 1. Use the **KEY TERMS** to help you understand and complete these objectives.

Patterns of Evolution   pages 309-310, 364-377

☐ 2. Describe patterns of species formation: **sequential evolution** (also known as phyletic gradualism or anagenesis), **coevolution**, **divergent evolution** (also known as cladogenesis), **adaptive radiation**.

☐ 3. Explain how evolutionary change over time has resulted in a great diversity among living organisms.

☐ 4. Describe and explain **convergent evolution**.

☐ 5. Explain how **analogous structures** (analogies) may arise as a result of convergence. Distinguish between **analogies** and **homologies** and explain the role of homology in identifying evolutionary relationships.

☐ 6. Describe examples of **coevolution**, including in flowering plants and their pollinators, parasites and their hosts, and predators and their prey (including herbivory). Discuss the evidence for coevolution in species with close ecological relationships.

☐ 7. Understand that some biologists also recognize **parallel evolution** to indicate evolution along similar lines in related groups.

☐ 8. Distinguish between the **punctuated equilibrium** and **gradualism** models for the pace of evolutionary change. Discuss the evidence for each model.

☐ 9. Describe the role of **extinction** in evolution. Distinguish clearly between **background extinction rates** and **mass extinction**. Identify the major **mass extinctions** and discuss the theories for their causes.

☐ 10. Describe examples of evolution (including speciation). Include important features of the species divergence:
   • Geographical barriers between populations.
   • Habitat range and niche differentiation.
   • Any zones of overlap in distribution (sympatry).
   • Recent range expansions.

☐ 11. Explain how evolutionary developmental biology (**evo-devo**) has provided some of the strongest evidence for the mechanisms of evolution, particularly for the evolution of novel forms.

Classification and Phylogeny   Senior Biology 2

☐ 12. Appreciate how classification systems (should) reflect the evolutionary relationships and history (**phylogeny**) of organisms. Describe the evolution and classification of a taxonomic group.

Periodicals:
listings for this chapter are on page 391

Weblinks:
www.thebiozone.com/
weblink/SB1-2597.html

Teacher Resource CD-ROM:
The Evolution of Humans

Patterns of Evolution

The diversification of an ancestral group into two or more species in different habitats is called **divergent evolution**. This process is shown below, where two species have diverged from a **common ancestor**. Note that another species budded off, only to become extinct. Divergence is common in evolution. When divergent evolution involves the formation of a large number of species to occupy different niches, this is called an **adaptive radiation**. The example below (right) describes the radiation of the mammals that occurred after the extinction of the dinosaurs; an event that made niches available. Note that the evolution of species may not necessarily involve branching: a species may accumulate genetic changes that, over time, result in the emergence of what can be recognized as a different species. This is known as **sequential evolution** (also called anagensis or phyletic gradualism).

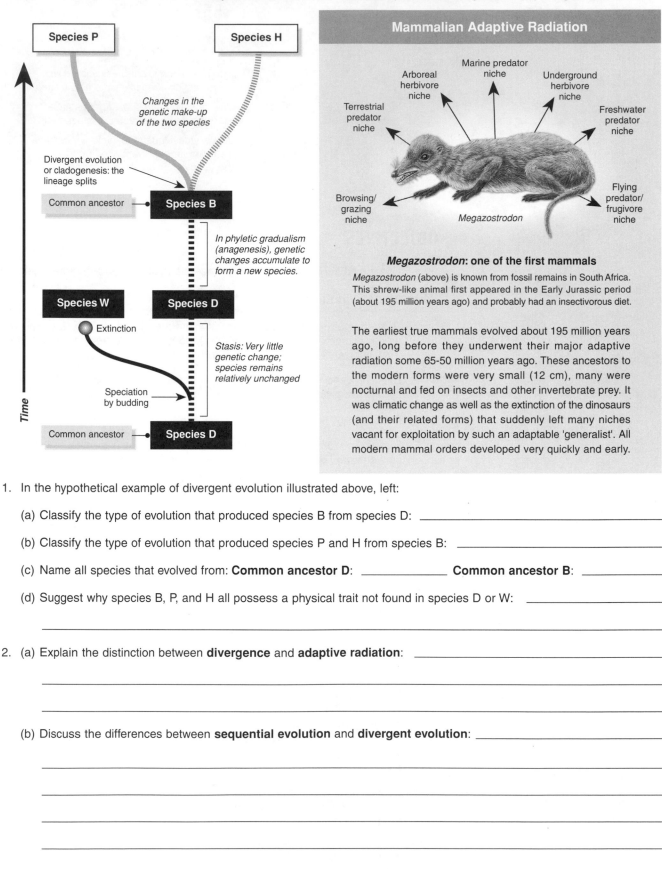

Mammalian Adaptive Radiation

Megazostrodon

Megazostrodon: one of the first mammals

Megazostrodon (above) is known from fossil remains in South Africa. This shrew-like animal first appeared in the Early Jurassic period (about 195 million years ago) and probably had an insectivorous diet.

The earliest true mammals evolved about 195 million years ago, long before they underwent their major adaptive radiation some 65-50 million years ago. These ancestors to the modern forms were very small (12 cm), many were nocturnal and fed on insects and other invertebrate prey. It was climatic change as well as the extinction of the dinosaurs (and their related forms) that suddenly left many niches vacant for exploitation by such an adaptable 'generalist'. All modern mammal orders developed very quickly and early.

1. In the hypothetical example of divergent evolution illustrated above, left:

   (a) Classify the type of evolution that produced species B from species D: _____

   (b) Classify the type of evolution that produced species P and H from species B: _____

   (c) Name all species that evolved from: **Common ancestor D**: _____ **Common ancestor B**: _____

   (d) Suggest why species B, P, and H all possess a physical trait not found in species D or W: _____

   _____

2. (a) Explain the distinction between **divergence** and **adaptive radiation**: _____

   _____

   _____

   (b) Discuss the differences between **sequential evolution** and **divergent evolution**: _____

   _____

   _____

   _____

   _____

   _____

A 2

Related activities: The Species Concept, Adaptive Radiation in Mammals
Web links: Macroevolution

Periodicals:
Evolution: five big
questions

© Biozone International 2001-2010
Photocopying Prohibited

Convergent Evolution

Not all similarities between species are a result of common ancestry. Species from different evolutionary lines may come to resemble each other if they have similar ecological roles and natural selection has shaped similar adaptations. This is called **convergent evolution** (**convergence**). Similarity of form due to convergence is called **analogy**.

Convergence in Swimming Form

Although similarities in body form and function can arise because of common ancestry, it may also be a result of **convergent evolution**. Selection pressures in a particular environment may bring about similar adaptations in unrelated species. These selection pressures require the solving of problems in particular ways, leading to the similarity of body form or function. The development of succulent forms in unrelated plant groups (*Euphorbia* and the cactus family) is an example of convergence in plants. In the example (right), the selection pressures of the aquatic environment have produced a similar **streamlined** body shape in unrelated vertebrate groups. Icthyosaurs, penguins, and dolphins each evolved from terrestrial species that took up an aquatic lifestyle. Their general body form has evolved to become similar to that of the shark, which has always been aquatic. Note that flipper shape in mammals, birds, and reptiles is a result of convergence, but its origin from the pentadactyl limb is an example of **homology**.

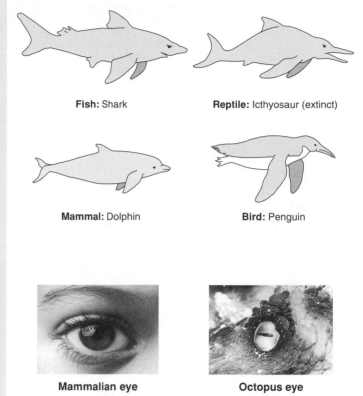

Fish: Shark

Reptile: Icthyosaur (extinct)

Mammal: Dolphin

Bird: Penguin

Analogous Structures

Analogous structures are those that have the same function and often the same basic external appearance, but **quite different origins**. The example on the right illustrates how a complex eye structure has developed independently in two unrelated groups. The appearance of the **eye** is similar, but there is no genetic relatedness between the two groups (mammals and cephalopod molluscs). The **wings** of birds and insects are also an example of analogy. The wings perform the same function, but the two groups share no common ancestor. *Longisquama*, a lizard-like creature that lived about 220 million years ago, also had 'wings' that probably allowed gliding between trees. These 'wings' were not a modification of the forearm (as in birds), but highly modified long scales or feathers extending from its back.

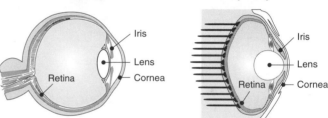

Mammalian eye

Octopus eye

Iris
Lens
Retina
Cornea

Iris
Lens
Retina
Cornea

1. In the example above illustrating convergence in swimming form, describe two ways in which the body form has evolved in response to the particular selection pressures of the aquatic environment:

(a) _____

_____

(b) _____

_____

2. Describe two of the selection pressures that have influenced the body form of the swimming animals above:

(a) _____

(b) _____

3. Early taxonomists, when encountering new species in the Pacific region and the Americas, were keen to assign them to existing taxonomic families based on their apparent similarity to European species. In recent times, many of the new species have been found to be quite unrelated to the European families they were assigned to. Explain why the traditional approach did not reveal the true evolutionary relationships of the new species:

_____

_____

_____

_____

Periodicals:
Which came first?

Related activities: Comparative Anatomy, Convergent Evolution

RA 2

4. For each of the paired examples (b)-(f), briefly describe the adaptations of body shape, diet and locomotion that appear similar in both forms, and the likely selection pressures that are acting on these mammals to produce similar body forms:

Convergence Between Marsupials and Placentals

Marsupials and **placental** mammals were separated from each other very early in mammalian evolution (about 120 mya). Marsupials were initially widely distributed throughout the ancient supercontinent of Gondwana, and there are some modern species still living in the American continent. Gondwana split up about 100 million years ago. As the placentals developed, they displaced the marsupials in most habitats around the world. The island continent of Australia, because of its early isolation by the sea, escaped this competition and placentals did not reach the continent until the arrival of humans 35 000 to 50 000 years ago. The Australian marsupials evolved into a wide variety of forms (below left) that bear a remarkable resemblance to ecologically equivalent species of North American placentals (below right).

Australia   **North America**

Marsupial mammals   **Placental mammals**

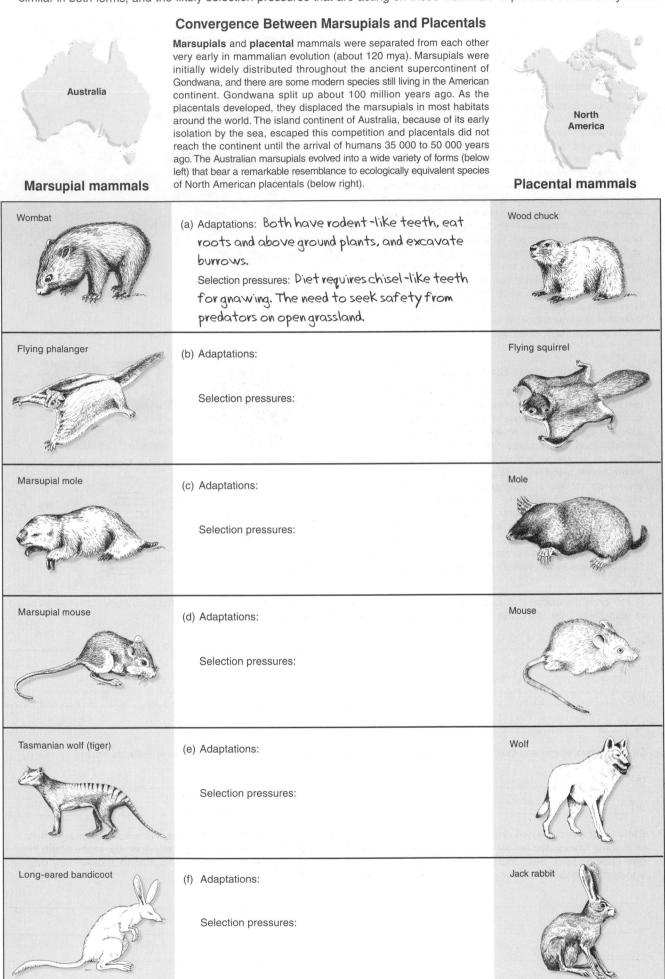

Marsupial	Answer	Placental
Wombat	(a) Adaptations: Both have rodent-like teeth, eat roots and above ground plants, and excavate burrows. Selection pressures: Diet requires chisel-like teeth for gnawing. The need to seek safety from predators on open grassland.	**Wood chuck**
Flying phalanger	(b) Adaptations: Selection pressures:	**Flying squirrel**
Marsupial mole	(c) Adaptations: Selection pressures:	**Mole**
Marsupial mouse	(d) Adaptations: Selection pressures:	**Mouse**
Tasmanian wolf (tiger)	(e) Adaptations: Selection pressures:	**Wolf**
Long-eared bandicoot	(f) Adaptations: Selection pressures:	**Jack rabbit**

Coevolution

The term **coevolution** is used to describe cases where two (or more) species reciprocally affect each other's evolution. Each party in a coevolutionary relationship exerts selective pressures on the other and, over time, the species develop a relationship that may involve mutual dependency. Coevolution is a likely consequence when different species have close ecological interactions with one another. These ecological relationships include predator-prey and parasite-host relationships and mutualistic relationships such as those between plants and their pollinators (see *Pollination Syndromes*). There are many examples of coevolution amongst parasites or pathogens and their hosts, and between predators and their prey, as shown on the following page.

Swollen-thorn *Acacia* lack the cyanogenic glycosides found in related *Acacia* spp. and the thorns are large and hollow, providing living space for the aggressive, stinging *Pseudomyrmex* ants which patrol the plant and protect it from browsing herbivores. The *Acacia* also provides the ants with protein rich food.

Hummingbirds (above) are important pollinators in the tropics. Their needle-like bills and long tongues can take nectar from flowers with deep tubes. Their ability to hover enables them to feed quickly from dangling flowers. As they feed, their heads are dusted with pollen, which is efficiently transferred between flowers.

Butterflies find flowers by vision and smell them after landing to judge their nectar source. Like bees, they can remember characteristics of desirable flowers and so exhibit constancy, which benefits both pollinator and plant. Butterfly flowers are very fragrant and are blue, purple, deep pink, red, or orange.

Bees are excellent pollinators; they are strong enough to enter intricate flowers and have medium length tongues which can collect nectar from many flower types. They have good colour vision, which extends into the UV, but they are red-blind, so bee pollinated flowers are typically blue, purplish, or white and they may have nectar guides that are visible as spots.

Beetles represent a very ancient group of insects with thousands of modern species. Their high diversity has been attributed to extensive coevolution with flowering plants. Beetles consume the ovules as well as pollen and nectar and there is evidence that ovule herbivory by beetles might have driven the evolution of protective carpels in angiosperms.

NZ's short tailed bat pollinates *Dactylanthus* flowers on the forest floor

Bats are nocturnal and colour-blind but have an excellent sense of smell and are capable of long flights. Flowers that have coevolved with bat pollinators are open at night and have light or drab colours that do not attract other pollinators. Bat pollinated flowers also produce strong fragrances that mimic the smell of bats and have a wide bell shape for easy access.

1. Using examples, explain what you understand by the term coevolution: _____

_____

_____

_____

_____

2. Describe some of the strategies that have evolved in plants to attract pollinators: _____

_____

_____

_____

_____

_____

Patterns of Evolution

Related activities: Pollination Syndromes
Web links: The Evolutionary Arms Race, The Coevolutionary Arms Race

RA 3

368

Predators, Parasites, and Coevolution

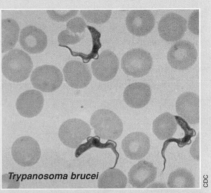

Trypanosoma brucei

Predators have obviously evolved to exploit their prey, with effective offensive weapons and hunting ability being paramount. Prey have evolved numerous strategies to protect themselves from predators, including large size and strength, protective coverings, defensive weapons, and toxicity. Lions have evolved the ability to hunt cooperatively to increase their chance of securing a kill from swift herding species such as zebra and gazelles.

Female *Helicornius* butterflies will avoid laying their eggs on plants already occupied by eggs, because their larvae are highly cannibalistic. Passionfruit plants (*Passiflora*) have exploited this by creating fake, yellow eggs on leaves and buds. *Passiflora* has many chemical defences against herbivory, but these have been breached by *Heliconius*. It has thus counter-evolved new defences against this herbivory by this genus.

Trypanosomes provide a good example of **host-parasite coevolution**. Trypanosomes must evolve strategies to evade their host's defences, but their virulence is constrained by needing to keep their host alive. Molecular studies show that *Trypanosoma brucei* coevolved in Africa with the first hominids around 5 mya, but *T. cruzi* contact with human hosts occurred in South America only after settlements were made by nomadic cultures.

3. Explain how coevolution could lead to an increase in biodiversity: _____

_____

_____

_____

_____

_____

4. Discuss some of the possible consequences of species competition: _____

_____

_____

_____

_____

_____

_____

5. The analogy of an "arms race" is often used to explain the coevolution of exploitative relationships such as those of a parasite and its host. Form a small group to discuss this idea and then suggest how the analogy is flawed:

_____

_____

_____

_____

_____

_____

_____

Pollination Syndromes

The mutualistic relationship between plants and their pollinators represents a classic case of coevolution. Flower structure has evolved in many different ways in response to the many types of animal pollinators. Flowers and pollinators have coordinated traits known as **pollination syndromes**. This makes it relatively easy to deduce pollinators type from the appearance of flowers (and vice versa). Plants and animals involved in such pollination associations often become highly specialized in ways that improve pollination efficiency: innovation by one party leads to some response from the other.

Controlling Pollinator Access

Flowers control pollinator access by flower shape and position.

Dandelion

Rigid inflorescences offer a stable landing platform to small or heavy insects, such as bumblebees.

Fuschia

Only animals that can hover can collect rewards from and pollinate flowers that hang upside down.

Attracting Pollinators

Flowers advertise the presence of nectar and pollen, with colour, scent, shape, and arrangement.

Rose

Daisy

Nectar guides help the pollinator to locate nectar and pollen. In this flower, the inner petals reflect UV.

While many flowers, like roses, are fragrant, flowers pollinated by flies (right) can give off dung or rotten meat smells.

Common Pollination Syndromes: Insects

Beetles

Ancient insect group
Good sense of smell
Hard, smooth bodies

Beetle-pollinated flowers

Ancient plant groups
Strong, fruity odours
Large, often flat, with easy access

Nectar-feeding flies

Sense nectar with feet
Tubular mouthparts

Nectar-feeding fly-pollinated flowers

Simple flowers with easy access
red or light colour, little odour

Moths

Many active at night
Good sense of smell
Feed with long, narrow tongues
Some need landing platforms

Moth-pollinated flowers

Flowers may be open at night
Fragrant; with heavy, musky scent
Nectar in narrow, deep tubes
landing platforms often provided

Carrion flies

Attracted by heat, odours, or
or colour of carrion or dung.
Food in the form of nectar or
pollen not required.

Carrion fly-pollinated flowers

Coloured to resemble dung or carrion
Produce heat or foul odours
No nectar or pollen reward offered

Common Pollination Syndromes: Vertebrates

Birds

Most require a perching site
Good colour vision, including red
Poor sense of smell
Feed during daylight
High energy requirements

Bird-pollinated flowers

Large and damage resistant
Often red or other bright colours
Not particularly fragrant
Open during the day
Copious nectar produced

Bats

Active at night
High food requirements
Colour blind
Good sense of smell
Cannot fly in foliage
High blossom intelligence

Bat-pollinated flowers

Open at night
Plentiful nectar and pollen offered
Light or dingy colours
Strong, often bat-like odours
Open shape, easy access
Pendulous or on the trunks of trees

Non-flying mammals

Relatively large size
High energy requirements
Colour vision may be lacking
Good sense of smell

Non-bat mammal-pollinated flowers

Robust, damage resistant
Copious, sugar-rich nectar
Dull coloured
Odorous, but not necessarily fragrant

1. (a) Describe a common pollination syndrome of an insect: _____

_____

(b) Describe a common pollination syndrome of a vertebrate: _____

_____

2. Suggest how knowledge of pollination syndromes might be used to develop testable predictions about plant and animal pollination relationships:

_____

_____

_____

_____

_____

Related activities: Coevolution

A 2

Patterns of Evolution

The Rate of Evolutionary Change

The pace of evolution has been much debated, with two models being proposed: **gradualism** and **punctuated equilibrium**. Some scientists believe that both mechanisms may operate at different times and in different circumstances. Interpretations of the fossil record will vary depending on the time scales involved. During its formative millennia, a species may have accumulated its changes gradually (e.g. over 50 000 years). If that species survives for 5 million years, the evolution of its defining characteristics would have been compressed into just 1% of its (species) lifetime. In the fossil record, the species would appear quite suddenly.

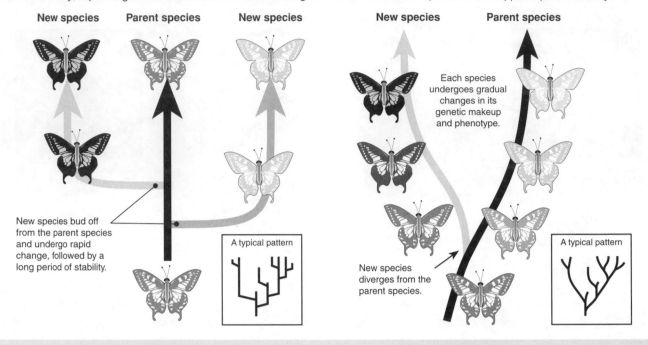

New species bud off from the parent species and undergo rapid change, followed by a long period of stability.

A typical pattern

Each species undergoes gradual changes in its genetic makeup and phenotype.

New species diverges from the parent species.

A typical pattern

Punctuated Equilibrium

There is abundant evidence in the fossil record that, instead of gradual change, species stayed much the same for long periods of time (called stasis). These periods were punctuated by short bursts of evolution which produce new species quite rapidly. According to the punctuated equilibrium theory, most of a species' existence is spent in stasis and little time is spent in active evolutionary change. The stimulus for evolution occurs when some crucial factor in the environment changes.

Phyletic Gradualism

Phyletic gradualism assumes that populations slowly diverge by accumulating adaptive characteristics in response to different selective pressures. If species evolve by gradualism, there should be transitional forms seen in the fossil record, as is seen with the evolution of the horse. Trilobites, an extinct marine arthropod, are another group of animals that have exhibited gradualism. In a study in 1987 a researcher found that they changed gradually over a three million year period.

1. Suggest the kinds of environments that would support the following paces of evolutionary change:

   (a) Punctuated equilibrium: _____

   _____

   _____

   (b) Gradualism: _____

   _____

   _____

2. In the fossil record of early human evolution, species tend to appear suddenly, linger for often very extended periods before disappearing suddenly. There are few examples of smooth inter-gradations from one species to the next. Explain which of the above models best describes the rate of human evolution:

   _____

3. Some species apparently show little evolutionary change over long periods of time (hundreds of millions of years).

   (a) Name two examples of such species: _____

   (b) State the term given to this lack of evolutionary change: _____

   (c) Suggest why such species have changed little over evolutionary time: _____

   _____

Related activities: Patterns of Evolution, The Evolution of Novel Forms
Web links: The Big Issues in Evolution

Periodicals:
What missing link?

Adaptive Radiation in Ratites

The **ratites** evolved from a single common ancestor; they are a monophyletic group of birds that lost the power of flight very early on in their evolutionary development. Ratites possess two features distinguishing them from other birds: a flat breastbone (instead of the more usual keeled shape) and a primitive palate (roof to the mouth). Flightlessness in itself is not unique to this group. There are other examples of birds that have lost the power of flight, particularly on remote, predator-free islands. Fossil evidence indicates that the ancestors of ratites were flying birds living about 80 million years ago. These ancestors also had a primitive palate, but they possessed a keeled breastbone.

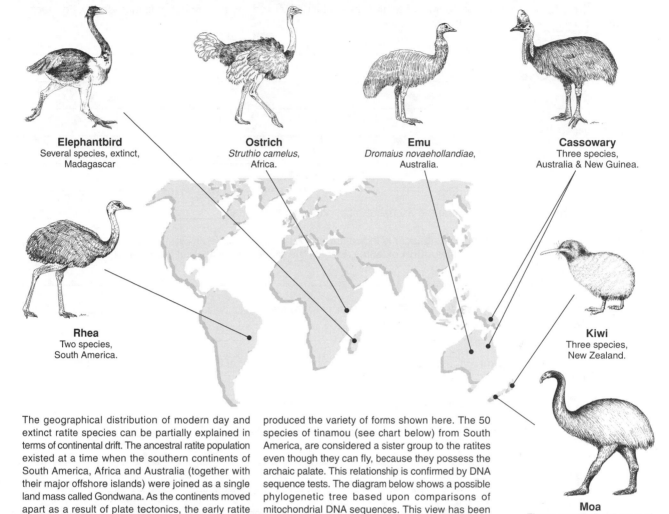

Elephantbird
Several species, extinct,
Madagascar

Ostrich
Struthio camelus,
Africa.

Emu
Dromaius novaehollandiae,
Australia.

Cassowary
Three species,
Australia & New Guinea.

Rhea
Two species,
South America.

Kiwi
Three species,
New Zealand.

Moa
Eleven species (Lambert *et al*.
2004*), all extinct, New Zealand.

The geographical distribution of modern day and extinct ratite species can be partially explained in terms of continental drift. The ancestral ratite population existed at a time when the southern continents of South America, Africa and Australia (together with their major offshore islands) were joined as a single land mass called Gondwana. As the continents moved apart as a result of plate tectonics, the early ratite populations were carried with them. Subsequent speciation on each continent and some of the islands produced the variety of forms shown here. The 50 species of tinamou (see chart below) from South America, are considered a sister group to the ratites even though they can fly, because they possess the archaic palate. This relationship is confirmed by DNA sequence tests. The diagram below shows a possible phylogenetic tree based upon comparisons of mitochondrial DNA sequences. This view has been supported by the extensive comparison of skeletons from the different ratite species.

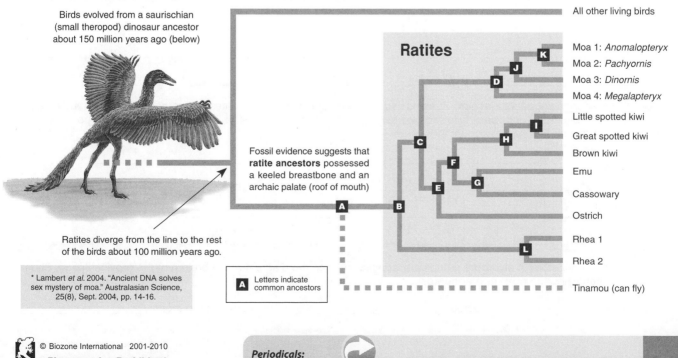

Mesozoic Era

Birds evolved from a saurischian (small theropod) dinosaur ancestor about 150 million years ago (below)

Ratites diverge from the line to the rest of the birds about 100 million years ago.

* Lambert *et al*. 2004. "Ancient DNA solves sex mystery of moa." Australasian Science, 25(8), Sept. 2004, pp. 14-16.

Cenozoic Era

Fossil evidence suggests that **ratite ancestors** possessed a keeled breastbone and an archaic palate (roof of mouth)

A Letters indicate common ancestors

Ratites

All other living birds
Moa 1: *Anomalopteryx*
Moa 2: *Pachyornis*
Moa 3: *Dinornis*
Moa 4: *Megalapteryx*
Little spotted kiwi
Great spotted kiwi
Brown kiwi
Emu
Cassowary
Ostrich
Rhea 1
Rhea 2
Tinamou (can fly)

Periodicals:
Ancient DNA solves sex
mystery of the moa

1. (a) Describe three physical features distinguishing all ratites from most other birds: _____

_____

(b) Identify the primitive feature shared by ratites and tinamou: _____

_____

2. Describe two anatomical changes, common to all ratites, which have evolved as a result of flightlessness. For each, describe the selection pressures for the anatomical change:

(a) Anatomical change: _____

Selection pressure: _____

_____

(b) Anatomical change: _____

Selection pressure: _____

_____

3. Name the ancient supercontinent that the ancestral ratite population inhabited: _____

4. (a) The extinct elephantbird from Madagascar is thought to be very closely related to another modern ratite. Based purely on the **geographical distribution** of ratites, identify the modern species that is the most likely relative:

_____

(b) Explain why you chose the modern ratite in your answer to (a) above: _____

_____

(c) Draw lines on the diagram at the bottom of the previous page to represent the divergence of the elephantbird from the modern ratite you have selected above.

5. (a) Name two other flightless birds that are not ratites: _____

(b) Explain why these other flightless species are not considered part of the ratite group: _____

_____

_____

6. Eleven species of moa is an unusually large number compared to the species diversity of the kiwis, the other ratite group found in New Zealand. The moas are classified into at least four genera, whereas kiwis have only one genus. The diets of the moas and the kiwis are thought to have had a major influence on each group's capacity to diverge into separate species and genera. The moas were herbivorous, whereas kiwis are nocturnal feeders, feeding on invertebrates in the leaf litter. Explain why, on the basis of their diet, moas diverged into many species, whereas kiwis diverged little:

_____

_____

_____

7. The DNA evidence suggests that New Zealand had two separate invasions of ratites, an early invasion from the moas (before the breakup of Gondwana) followed by a second invasion of the ancestors of the kiwis. Describe a possible sequence of events that could account for this:

_____

_____

_____

_____

8. The common ancestors of divergent groups are labelled (A-L) on the diagram at the bottom of the previous page. State the **letter** identifying the **common ancestor** for:

(a) The kiwis and the Australian ratites: _____   (b) The kiwis and the moas: _____

Adaptive Radiation in Mammals

Adaptive radiation is diversification (both structural and ecological) among the descendants of a single ancestral group to occupy different niches. Immediately following the sudden extinction of the dinosaurs, the mammals underwent an adaptive radiation. Most of the modern mammal groups became established very early. The diagram below shows the divergence of the mammals into major orders; many occupying niches left vacant by the dinosaurs. The vertical extent of each gray shape

shows the time span for which that particular mammal order has existed (note that the scale for the geological time scale in the diagram is not linear). Those that reach the top of the chart have survived to the present day. The width of a gray shape indicates how many species were in existence at any given time (narrow means there were few, wide means there were many). The dotted lines indicate possible links between the various mammal orders for which there is no direct fossil evidence.

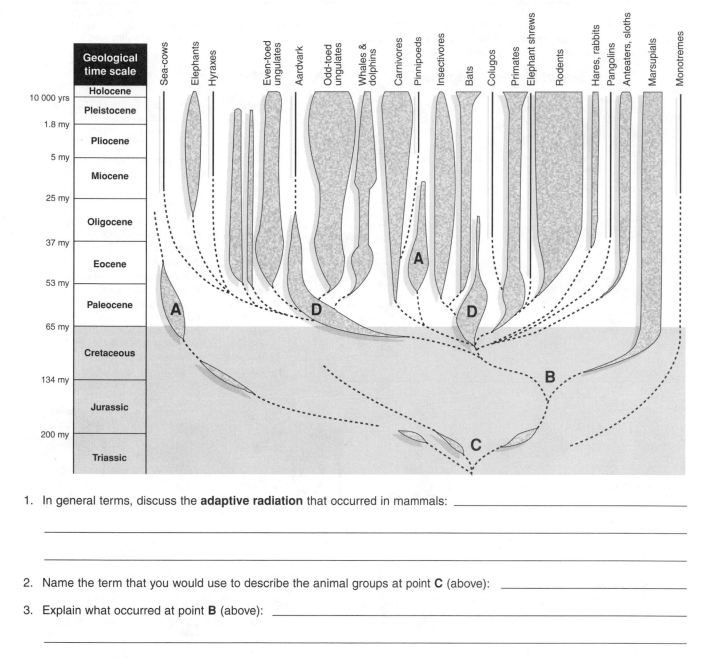

1. In general terms, discuss the **adaptive radiation** that occurred in mammals: _____
   _____
   _____

2. Name the term that you would use to describe the animal groups at point **C** (above): _____

3. Explain what occurred at point **B** (above): _____
   _____

4. Describe two things that the animal orders labeled **D** (above) have in common:

   (a) _____

   (b) _____

5. Identify the two orders that appear to have been most successful in terms of the number of species produced:
   _____

6. Explain what has happened to the mammal orders labeled **A** in the diagram above: _____
   _____

7. Identify the **epoch** during which there was the most adaptive radiation: _____

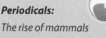

Periodicals:
The rise of mammals

Related activities: Patterns of Evolution

RDA 3

Patterns of Evolution

8. Describe two key features that distinguish mammals from other vertebrates:

(a) _____ (b) _____

9. Describe the principal reproductive features distinguishing each of the major mammalian lines (sub-classes):

(a) Monotremes: _____

(b) Marsupials: _____

(c) Placentals: _____

10. There are 18 orders of placental mammals (or 17 in schemes that include the pinnipeds within the Carnivora). Their names and a brief description of the type of mammal belonging to each group is provided below. Identify and label each of the diagrams with the correct name of their Order:

Orders of Placental Mammals

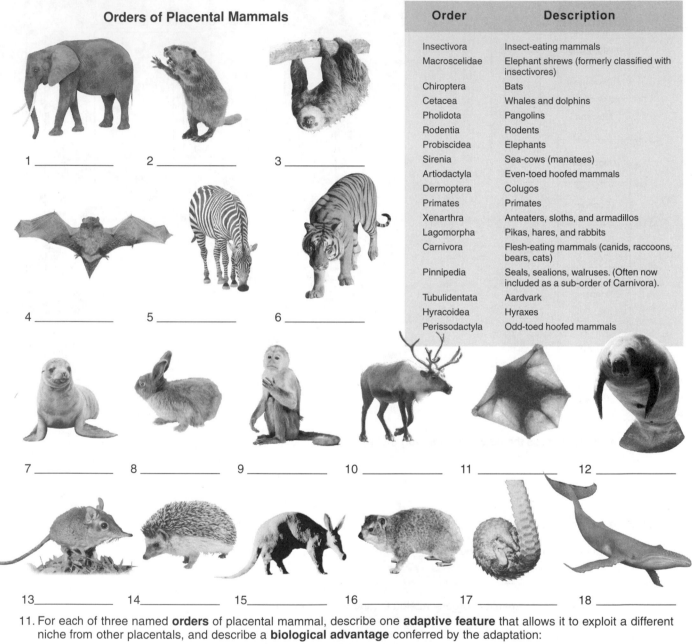

Order	Description
Insectivora	Insect-eating mammals
Macroscelidae	Elephant shrews (formerly classified with insectivores)
Chiroptera	Bats
Cetacea	Whales and dolphins
Pholidota	Pangolins
Rodentia	Rodents
Probiscidea	Elephants
Sirenia	Sea-cows (manatees)
Artiodactyla	Even-toed hoofed mammals
Dermoptera	Colugos
Primates	Primates
Xenarthra	Anteaters, sloths, and armadillos
Lagomorpha	Pikas, hares, and rabbits
Carnivora	Flesh-eating mammals (canids, raccoons, bears, cats)
Pinnipedia	Seals, sealions, walruses. (Often now included as a sub-order of Carnivora).
Tubulidentata	Aardvark
Hyracoidea	Hyraxes
Perissodactyla	Odd-toed hoofed mammals

1 _____ 2 _____ 3 _____

4 _____ 5 _____ 6 _____

7 _____ 8 _____ 9 _____ 10 _____ 11 _____ 12 _____

13 _____ 14 _____ 15 _____ 16 _____ 17 _____ 18 _____

11. For each of three named **orders** of placental mammal, describe one **adaptive feature** that allows it to exploit a different niche from other placentals, and describe a **biological advantage** conferred by the adaptation:

(a) Order: _____ Adaptive feature: _____

Biological advantage: _____

(b) Order: _____ Adaptive feature: _____

Biological advantage: _____

(c) Order: _____ Adaptive feature: _____

Biological advantage: _____

Geographical Distribution

The camel family, Camelidae, consists of six modern-day species that have survived on three continents: Asia, Africa and South America. They are characterized by having only two functional toes, supported by expanded pads for walking on sand or snow. The slender snout bears a cleft upper lip. The recent distribution of the camel family is fragmented. Geophysical forces such as plate tectonics and the ice age cycles have controlled the extent of their distribution. South America, for example, was separated from North America until the end of the Pliocene, about 2 million years ago. Three general principles about the dispersal and distribution of land animals are:

- When very closely related animals (as shown by their anatomy) were present at the same time in widely separated parts of the world, it is highly probable that there was no barrier to their movement in one or both directions between the localities in the past.
- The most effective barrier to the movement of land animals (particularly mammals) was a sea between continents (as was caused by changing sea levels during the ice ages).
- A scattered distribution of modern species may be explained by the movement out of the area they originally occupied, or by extinction in those regions between modern species.

Origin and Dispersal of the Camel Family

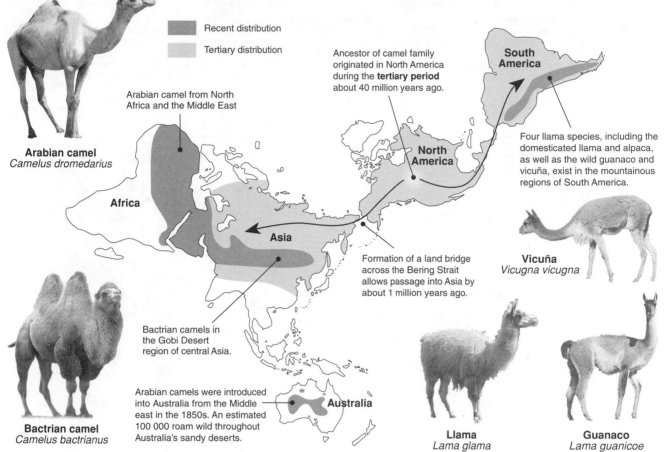

Recent distribution

Tertiary distribution

Ancestor of camel family originated in North America during the **tertiary period** about 40 million years ago.

Arabian camel from North Africa and the Middle East

Four llama species, including the domesticated llama and alpaca, as well as the wild guanaco and vicuña, exist in the mountainous regions of South America.

Formation of a land bridge across the Bering Strait allows passage into Asia by about 1 million years ago.

Bactrian camels in the Gobi Desert region of central Asia.

Arabian camels were introduced into Australia from the Middle east in the 1850s. An estimated 100 000 roam wild throughout Australia's sandy deserts.

Arabian camel
Camelus dromedarius

Bactrian camel
Camelus bactrianus

Vicuña
Vicugna vicugna

Llama
Lama glama

Guanaco
Lama guanicoe

1. The early camel ancestors were able to move into the tropical regions of Central and South America. Explain why this did not happen in southern Asia and southern Africa:

   _____

2. Arabian camels are found wild in the Australian Outback. Explain how they got there and why they were absent during prehistoric times:

   _____

3. The camel family originated in North America. Explain why there are no camels in North America now:

   _____

4. Suggest how early camels managed to get to Asia from North America: _____

   _____

5. Describe the present distribution of the camel family and explain why it is scattered (discontinuous):

   _____

   _____

Related activities: The Founder Effect, Allopatric Speciation

RA 2

Patterns of Evolution

Extinction

Extinction is an important process in evolution as it provides opportunities, in the form of vacant niches, for the development of new species. Most species that have ever lived are now extinct. The species alive today make up only a fraction of the total list of species that have lived on Earth throughout its history. Extinction is a natural process in the life cycle of a species. Background extinction is the steady rate of species turnover in a taxonomic group (a group of related species). The duration of a species is thought to range from as little as 1 million years for complex larger organisms, to as long as 10-20 million years for simpler organisms. Superimposed on this constant background extinction are catastrophic events that wipe out vast numbers of species in relatively brief periods of time in geological terms. The diagram below shows how the number of species has varied over the history of life on Earth. The number of species is indicated on the graph by families; a taxonomic group comprising many genera and species. There have been five major extinction events and two of these have been intensively studied by palaeontologists.

Major Mass Extinctions

The Permian extinction
(225 million years ago)

This was the most devastating mass extinction of all. Nearly all life on Earth perished, with 90% of marine species and probably many terrestrial ones also, disappearing from the fossil record. This extinction event marks the **Paleozoic-Mesozoic** boundary.

The Cretaceous extinction
(65 million years ago)

This extinction event marks the boundary between the Mesozoic and Cenozoic eras. More than half the marine species and many families of terrestrial plants and animals became extinct, including nearly all the dinosaur species (the birds are now known to be direct descendants of the dinosaurs).

Megafaunal extinction
(10 000 years ago)

This mass extinction occurred when many giant species of mammal died out. This is known as the Pleistocene overkill because their disappearance was probably hastened by the hunting activities of prehistoric humans. Many large marsupials in Australia and placental species elsewhere became extinct.

The sixth extinction
(now)

The current mass extinction is largely due to human destruction of habitats (e.g. coral reefs, tropical forests) and pollution. It is considered far more serious and damaging than some earlier mass extinctions because of the speed at which it is occurring. The increasing human impact is making biotic recovery difficult.

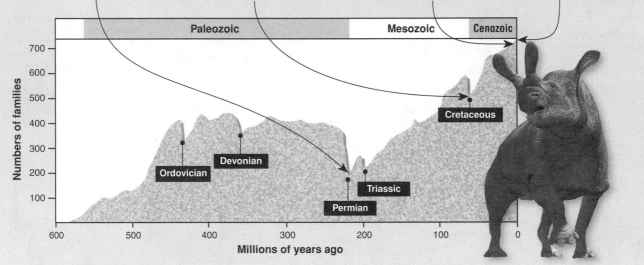

1. Describe the main features (scale and type of organisms killed off) of each of the following major extinction events:

   (a) Permian extinction: _____

   _____

   _____

   (b) Cretaceous extinction: _____

   _____

   _____

   (c) Megafaunal extinction: _____

   _____

   _____

2. Explain how human activity has contributed to the most recent mass extinction: _____

   _____

3. In general terms, describe the effect that past mass extinctions had on the way the surviving species **further evolved**:

   _____

   _____

Related activities: Causes of Mass Extinctions
Web links:: What Killed the Dinosaurs?

Periodicals:
The sixth extinction

© Biozone International 2001-2010
Photocopying Prohibited

Causes of Mass Extinctions

Over the last 540 million years, marine life has experienced about 24 bouts of mass extinction: five of these were major and some 19 were minor (judged by the percentage of genera that became extinct). Many of these extinction peaks coincided with known comet/asteroid impact events, strongly implying that they may have been the cause. A general deterioration of the environment, caused by climatic change or some major cosmic, geological, or biological event, is deemed the principal cause of extinctions. The ability of the biosphere to recover from such crises is evident by the fact that life continues to exist today.

Large asteroid/comet impacts | Large solar flares | Volcanism

Possible causes of mass extinction	Effects on the extinction rate	Examples of extinction events and their likely causes
Impacts by large asteroid/comet or 'showers': Shock waves, heat-waves, wildfires, impact 'winters' caused by global dust clouds, super-acid rain, toxic oceans, superwaves and superfloods from an oceanic impact.	Global extinction of much of the planet's biodiversity. Smaller comet showers could cause stepwise, regional extinctions.	**Late Pleistocene** (15 000-10 000 ya) *Extinction of:* Many large mammals and large flightless birds. *Probable cause:* Warming of the global climate after the last ice age plus predation (hunting) by humans.
Supernovae radiation: Direct exposure to X-rays and cosmic rays. Ozone depletion and subsequent exposure to excessive UV radiation from the sun.	Causes mutations and kills organisms. Selective mass extinctions, particularly of animals (but not plants) exposed to the atmosphere, as well as shallow-water aquatic forms.	**Late Cretaceous** (65 Mya) *Extinction of:* Dinosaurs, plesiosaurs, icthyosaurs, mosasaurs, pterosaurs, ammonites and belemnites (squid-like animals), and many other groups.
Large solar flares: Exposure to large doses of X-rays, and UV radiation. Ozone depletion.	Mass extinctions.	*Probable cause:* An asteroid impact (probably Yucatan peninsula) produced catastrophic environmental disturbance.
Geomagnetic reversals: Increased flux of cosmic rays.	Mass extinctions.	**Late Permian** (250 Mya) *Extinction of:* 90% of marine species and 70% of land species. Coral reefs, trilobites, some amphibians, and mammal-like reptiles, were eliminated (the Great Dying).
Continental drift: Climatic changes, such as glaciations and droughts, occur when continents move towards or away from the poles.	Global cooling due to changes in the pattern of oceans currents caused by shifting land masses. Extinctions as species find themselves in inhospitable climates.	*Probable cause:* An asteroid impact, followed by furious volcanic activity, a rapid heating of the atmosphere, and depletion of life-giving oxygen from the oceans.
Intense volcanism: Cold conditions caused by volcanic dust reducing solar input. Volcanic gases causing acid rain and reduced alkalinity of oceans. Toxic trace elements.	Stepwise mass extinctions.	**Late Devonian** (360 Mya) *Extinction of:* Many corals, bivalves, fish, sponges (21% of all marine families). Collapse of tropical reef communities.
Sea level change: Loss of habitat.	Mass extinctions of susceptible species (e.g. marine reptiles, coral reefs, coastal species).	*Probable cause:* Global cooling associated with (or causing?) widespread oxygen deficiency of shallow seas.
Arctic spill over: Release of cold fresh water or brackish water from an isolated Arctic Ocean. Ocean temperature falls 10°C, resulting in atmospheric cooling, drought.	Mass extinctions in marine ecosystems. Change of vegetation with drastic effect on large reptiles.	
Anoxia: Shortage of oxygen.	Mass extinctions in the oceans.	
Spread of disease/predators: Direct effects due to changing geographic distribution.	Mass extinctions.	

Source: *Environmental Change – The Evolving Ecosphere* (1997), by R.J. Huggett

Source: *Evolution: A Biological and Palaeontological Approach*, Skelton, P. (ed.), Addison-Wesley (1993)

1. Describe how each of the following events might have caused mass extinctions in the past:

   (a) Large asteroid/comet impact: _____

   _____

   (b) Continental drift: _____

   _____

   (c) Volcanism: _____

   _____

2. Explain how the arrival of a new plant or animal species onto a continent may cause the demise of other species there:

   _____

Periodicals:
Mass extinctions

Related activities: Extinction

A 3

Patterns of Evolution

KEY TERMS Word Find

Use the clues below to find the relevant key terms in the WORD FIND grid

```
S N H S I X T H E X T I N C T I O N R T A S M T
D I V E R G E N T E V O L U T I O N B X E O A K
E S G E Y J E X T I N C T I O N L T T I J X Q Q
B V P V D D I Z C S S F N V T U W E C C X G Y M
O K H O R G R A D U A L I S M I E E H R N G Y M
Z P Y D F V Y H C O B T C G D M P I R H P J K O
K R L E A Z X C U F U E X E O S S G U R N L P D
Z I O V R S O L O L G D E V N I L U I Y I H N T
F A G O Y M R C O E V O L U T I O N Q S Y Q F Y
K E E H O M O L O G O U S S T R U C T U R E S S
O K N X Y C L H O X G E N E S O B M N Y K G G T
P O Y M O N O P H Y L E T I C M Y C I N O P U J
N V S Q Z U I H D M W E H T Z X P E N Z D H O A
A T I X C O N V E R G E N T E V O L U T I O N X
S I E P A R A L L E L E V O L U T I O N N X B K
P U N C T U A T E D E Q U I L I B R I U M C O S
L V A D A P T I V E R A D I A T I O N O H N R A
S E Q U E N T I A L E V O L U T I O N A X A Q I
W Q I J S O H Z Y U O X S W G V E V Z V Z G O H
W P O A N A L O G O U S S T R U C T U R E V Y P
```

Evolution in unrelated species occupying similar niches that causes them to arrive at similar structural, physiological and behavioral solutions.

Structures in different but related species that are derived from the same ancestral structure but now serve different purposes, e.g. wings and fins.

A model for the evolution of different forms over a long period of time but with only slight changes occurring between many successive generations.

Mass extinction that many believe to be currently underway. Different from other mass extinctions because of the speed at which species are becoming extinct (2 words).

Group or population of individuals that are able to interbreed to produce viable offspring.

Evolutionary process in which a species or related species follow different evolutionary pathways to eventually become less related (2 words).

A structure present in unrelated species that also has an unrelated mode of origin, e.g. vertebrate and squid eyes (2 words).

A model for the evolution of lineages in which long periods of stasis are interrupted by brief periods of rapid speciation (2 words).

Genes that play a role in development and are shared by almost all organisms (2 words).

Gradual evolution within a lineage (2 words).

Reciprocal evolution in two species as a result of the selection pressure each imposes on the other.

The independent evolution of similar traits, starting from a similar ancestral condition (2 words).

The complete dying out of a species so that there are no representatives of the species remaining anywhere.

The study relating the evolution of new characteristics to changes in the genes controlling development. Evolutionary developmental biology (acronym).

The evolutionary history or genealogy of a group of organisms. Often represented as a 'tree' showing descent of new species from the ancestral one.

A taxon originating from (and including) a single common ancestor.

A form of divergent evolution in which there is rapid speciation of one ancestral species to fill many different ecological niches (2 words).

Appendix

PERIODICAL REFERENCES

SKILLS IN BIOLOGY

▶ **The Truth Is Out There**

New Scientist, 26 February 2000 (Inside Science). *The philosophy of scientific method: starting with an idea, formulating a hypothesis, and following the process to theory.*

▶ **Experiments**

Biol. Sci. Rev., 14(3) February 2002, pp. 11-13. *The basics of experimental design and execution: determining variables, measuring them, and establishing a control.*

▶ **Descriptive Statistics**

Biol. Sci. Rev., 13(5) May 2001, pp. 36-37. *An account of descriptive statistics using text, tables and graphs.*

▶ **Percentages**

Biol. Sci. Rev., 17(2) Nov. 2004, pp. 28-29. *The calculation of percentage and the appropriate uses of this important transformation.*

▶ **It's a Plot!**

Biol. Sci. Rev., 22 (2) Nov. 2009, pp. 16-19. *Using graphs to evaluate and explain data.*

▶ **Dealing with Data**

Biol. Sci. Rev., 12 (4) March 2000, pp. 6-8. *A short account of the best ways in which to deal with the interpretation of graphically presented data in examinations.*

▶ **Drawing Graphs**

Biol. Sci. Rev., 19(3) Feb. 2007, pp. 10-13. *A guide to creating graphs. The use of different graphs for different tasks is explained and there are a number of pertinent examples described to illustrate points.*

▶ **Size Does Matter**

Biol. Sci. Rev., 17 (3) February 2005, pp. 10-13. *Measuring the size of organisms and calculating magnification and scale.*

▶ **Describing the Normal Distribution**

Biol. Sci. Rev., 13(2) Nov. 2000, pp. 40-41. *The normal distribution: data spread, mean, median, variance, and standard deviation.*

▶ **Estimating the Mean and Standard Deviation**

Biol. Sci. Rev., 13(3) January 2001, pp. 40-41. *Simple statistical analysis. Includes formulae for calculating sample mean and standard deviation.*

▶ **The Variability of Samples**

Biol. Sci. Rev., 13(4) March 2001, pp. 34-35. *The variability of sample data and the use of sample statistics as estimators for population parameters.*

THE CHEMISTRY OF LIFE

▶ **Water, Life, and Hydrogen Bonding**

Biol. Sci. Rev., 21(2) Nov. 2008, pp. 18-20. *The molecules of life and the important role of hydrogen bonding.*

▶ **Glucose & Glucose-Containing Carbohydrates**

Biol. Sci. Rev., 19(1) Sept. 2006, pp. 12-15. *The structure of glucose and its polymers.*

▶ **Designer Starches**

Biol. Sci. Rev., 19(3) Feb. 2007, pp. 18-20. *The composition of starch, and an excellent account of its properties and functions.*

▶ **What is Tertiary Structure?**

Biol. Sci. Rev., 21(1) Sept. 2008, pp. 10-13. *How amino acid chains fold into the functional shape of a protein.*

▶ **Modeling Protein Folding**

The American Biology Teacher 66(4) Apr. 2004, pp. 287-289. *How protein folding produces physical structures (teacher's reference).*

▶ **Enzymes: Nature's Catalytic Machines**

Biol. Sci. Rev., 22(2) Nov. 2009, pp. 22-25. *Enzymes as catalysts: a very up-to-date description of enzyme specificity and binding, how enzymes work, and how they overcome the energy barriers for a reaction. Some well known enzymes are described.*

▶ **Enzymes: Fast and Flexible**

Biol. Sci. Rev., 19(1) Sept. 2006, pp. 2-5. *The structure of enzymes and how they work so efficiently at relatively low temperatures.*

▶ **Enzyme Technology**

Biol. Sci. Rev., 12 (5) May 2000, pp. 26-27. *The range and importance of industrial enzymes in modern biotechnology.*

CELL STRUCTURE

▶ **The Living Dead**

New Scientist, 13 October 2001, (Inside Science). *The non-cellular nature of viruses and how they operate.*

▶ **Are Viruses Alive?**

Scientific American, December 2004, pp. 77-81. *Although viruses challenge our concept of what "living" means,* they are vital members of the web of life. This excellent account covers the nature of viruses, including an account of viral replication and a critical evaluation of the status of viruses in the natural world.

▶ **Bacteria**

National Geographic, 184(2) August 1993, pp. 36-61. *The structure and diversity of bacteria; our most abundant and useful organisms.*

▶ **Size Does Matter**

Biol. Sci. Rev., 17 (3) Feb. 2005, pp. 10-13. *Measuring the size of organisms and calculating magnification and scale.*

▶ **Cellular Factories**

New Scientist, 23 November 1996 (Inside Science). *The role of different organelles in plant and animal cells*

▶ **Chloroplasts: Biosynthetic Powerhouses**

Biol. Sci. Rev., 21(4) April 2009, pp. 25-27. *The features of chloroplasts and their origin from proplastids.*

▶ **Light Microscopy**

Biol. Sci. Rev., 13(1) Sept. 2000, pp. 36-38. *An excellent account of the basis and various techniques of light microscopy.*

▶ **Transmission Electron Microscopy**

Biol. Sci. Rev., 19(4) April 2007, pp. 6-9. *An excellent account of the techniques and applications of TEM. Includes an excellent diagram comparing features of TEM and light microscopy.*

▶ **Scanning Electron Microscopy**

Biol. Sci. Rev., 13(3) January 2001, pp. 6-9. *An excellent account of the techniques and applications of SEM. Includes details of specimen preparation and recent advancements in the technology.*

▶ **The Power Behind an Electron Microscopist**

Biol. Sci. Rev., 18(1) Sept. 2005, pp. 16-20. *The use of TEMs to obtain greater resolution of finer details than is possible from optical microscopes.*

PROCESSES IN CELLS

▶ **Cellular Factories**

New Scientist, 23 November 1996 (Inside Science). *The role of different organelles in plant and animal cells*

▶ **Border Control**

New Scientist, 15 July 2000 (Inside

Appendix

Science). *The role of the plasma membrane in cell function: membrane structure, transport processes, and the role of receptors on the cell membrane.*

▶ **The Fluid-Mosaic Model for Membranes**

Biol. Sci. Rev., 22(2), Nov. 2009, pp. 20-21. *Diagrammatic revision of membrane structure and function.*

▶ **What Happens When Proteins Lose their Shape?**

Biol. Sci. Rev., 21(1) Sept. 2008, pp. 21-24. *The tertiary structure of proteins, their production as functional entities and their secretion from cells.*

▶ **Getting in and Out**

Biol. Sci. Rev., 20(3), Feb. 2008, pp. 14-16. *Diffusion: some adaptations and some common misunderstandings.*

▶ **How Biological Membranes Achieve Selective Transport**

Biol. Sci. Rev., 21(4), April 2009, pp. 32-36. *The structure of the plasma membrane and the proteins that enable the selective transport of molecules.*

▶ **What is Endocytosis?**

Biol. Sci. Rev., 22(3), Feb. 2010, pp. 38-41. *The mechanisms of endocytosis and the role of membrane receptors in concentrating important molecules before ingestion.*

▶ **The Cell Cycle and Mitosis**

Biol. Sci. Rev., 14(4) April 2002, pp. 37-41. *Cell growth and division, stages in the cell cycle, and the complex control over different stages of mitosis.*

▶ **Rebels Without a Cause**

New Scientist, 13 July 2002, (Inside Science). *The causes of cancer: the uncontrolled division of cells that results in tumor formation. Breast cancer is a case example.*

▶ **What is Cell Suicide?**

Biol. Sci. Rev., 20(1) Sept. 2007, pp. 17-20. *An account of the mechanisms behind cell suicide and its role in normal growth and development.*

▶ **Cell Differentiation**

Biol. Sci. Rev., 20(4), April 2008, pp. 10-13. *How tissues arise through the control of cellular differentiation during development. The example provided is the differentiation of blood cells.*

▶ **What is a Stem Cell?**

Biol. Sci. Rev., 16(2) Nov. 2003, pp. 22-23. *The nature of stem cells and their therapeutic applications.*

▶ **Grown to Order**

New Scientist, 3 May, 2008, pp. 40-43. *The breakthrough in creating stem cells could be a step towards generating new tissues.*

▶ **Embryonic Stem Cells**

Biol. Sci. Rev., 22(1) Sept. 2009, pp. 28-31. *The future of embryonic stem cell research. Problems and solutions.*

CELLULAR ENERGETICS

▶ **The Role of ATP in Cells**

Biol. Sci. Rev., 19(2) Nov. 2006, pp. 30-33. *Synthesis and uses of ATP.*

▶ **The Double Life of ATP**

Scientific American, Dec. 2009, pp. 60-67. *ATP the fuel inside living cells, also serves as a molecular messenger that affects cell behavior.*

▶ **Fuelled for Life**

New Scientist, 13 January 1996 (Inside Science). *Energy and metabolism: ATP, glycolysis, electron transport, Krebs cycle, and enzymes and cofactors.*

▶ **AcetylCoA: A Central Metabolite**

Biol. Sci. Rev., 20(4) April 2008, pp.38-40. *The role of acetyl coenzyme A in metabolising fat and carbohydrate.*

▶ **Lactic Acid: Who Needs It?**

Biol. Sci. Rev., 18(2) Nov. 2005, pp. 6-90. *An account of the biological roles of lactic acid, including its production in anaerobic metabolism in muscle.*

▶ **Photosynthesis....Most Hated Topic?**

Biol. Sci. Rev., 20(1) Sept. 2007, pp. 13-16. *A useful account documenting key points when learning about processes in photosynthesis.*

CHROMOSOMES AND MEIOSIS

▶ **Mechanisms of Meiosis**

Biol. Sci. Rev., 15(4), April 2003, pp. 20-24. *A clear and thorough account of the events and mechanisms of meiosis.*

HEREDITY

▶ **What is Variation?**

Biol. Sci. Rev., 13(1) Sept. 2000, pp. 30-31. *The nature of continuous and discontinuous variation. The distribution pattern of traits that show continuous variation is discussed.*

▶ **Mendel's Legacy**

Biol. Sci. Rev., 18(4), April 2006,

pp. 34-37. *Explores the accuracy of Mendel's laws in light of today's knowledge.*

▶ **The Y Chromosome: It's a Man Thing**

Biol. Sci. Rev., 20(4) April 2008, pp. 2-6. *The Y chromosome is at the root of sex determination. This account discusses the nature of the Y chromosome, non-disjunction and Y chromosome disorders and the inheritance of Y linked diseases*

▶ **Strange Inheritance**

New Scientist, 12 July 2008, pp. 28-33. *Studies of epigenetic inheritance is forcing us to rethink accepted knowledge about genetics and evolution.*

▶ **Secrets of The Gene**

National Geographic, 196(4) Oct. 1999, pp. 42-75. *Thorough coverage of the nature of genes and the inheritance of particular genetic traits through certain populations.*

▶ **Genetic Screening - Controlling the Future**

Biol. Sci. Rev., 12 (4) March 2000, pp. 36-38. *The techniques, applications, and ethical questions posed by genetic screening.*

▶ **The Colour Code**

New Scientist, 10 March 2002, pp. 34-37. *Researchers are uncovering the five to ten genes responsible for skin pigmentation.*

MOLECULAR GENETICS

▶ **DNA: 50 Years of the Double Helix**

New Scientist, 15 March 2003, pp. 35-51. *A special issue on DNA: structure and function, repair, the new-found role of histones, and the functional significance of chromosome position in the nucleus.*

▶ **What is a Gene?**

Biol. Sci. Rev., 21(4) April 2009, pp. 10-12. *The molecular basis of genes, gene transcription, and production of a functional mRNA by removal of introns.*

▶ **The Alternative Genome**

Scientific American, April, 2005, pp. 40-47. *The old axiom "one gene, one protein" no longer holds true. The more complex an organism, the more likely it became that way by extracting multiple protein meanings from individual genes.*

▶ **Gene Structure and Expression**

Biol. Sci. Rev., 12 (5) May 2000, pp. 22-25. *An account of the structure*

Appendix

PERIODICAL REFERENCES

and function of genes, and the basis of gene regulation in eukaryotes and prokaryotes.

▶ **Transfer RNA**

Biol. Sci. Rev., 15(3) Feb. 2003, pp. 26-29. *A good account of the structure and role of tRNA in protein synthesis.*

▶ **Tyrosine**

Biol. Sci. Rev., 12 (4) March 2000, pp. 29-30. *Tyrosine and its role. This article includes a brief discussion of errors in tyrosine metabolism.*

▶ **Are Viruses Alive?**

Scientific American, December 2004, pp. 77-81. *Although viruses challenge our concept of what "living" means, they are vital members of the web of life. This excellent account covers the nature of viruses, including an account of viral replication and a critical evaluation of the status of viruses in the natural world.*

MUTATION

▶ **What is a Mutation?**

Biol. Sci. Rev., 20(3) Feb. 2008, pp. 6-9. *The nature of mutations: causes, timing, and effects. Sickle cell disease is described.*

▶ **Radiation and Risk**

New Scientist, 18 March 2000 (Inside Science). *In large doses radiation can kill you in hours. In low doses, it can lead to slow death by cancer. How do we quantify the effects?.*

▶ **How do Mutations Lead to Evolution?**

New Scientist, 14 June 2003, pp. 32-39, 48-51. *An account of the five most common points of discussion regarding evolution and the mechanisms by which it occurs.*

▶ **Genetics of Sickle Cell Anaemia**

Biol. Sci. Rev., 20(4) April 2008, pp. 14-17. *The molecular and physiological basis of sickle cell disease.*

▶ **The Biological Aspects of Down Syndrome**

Biol. Sci. Rev., 10(5) May 1998, pp. 11-15. *Chromosome trisomy: how it arises and its phenotypic effects. Includes methods of diagnosis.*

NUCLEIC ACID TECHNOLOGY

▶ **The Polymerase Chain Reaction**

Biol. Sci. Rev., 16(3) Feb. 2004, pp. 10-13. *This well illustrated*

account explains the techniques and applications of PCR in easy-to-understand language.

▶ **The Magic of Microarrays**

Scientific American, Feb. 2002, pp. 34-41. *DNA chips and their use in identifying health and disease, along with implications for drug treatment.*

▶ **Tailor-Made Proteins**

Biol. Sci. Rev., 13(4) March 2001, pp. 2-6. *Recombinant proteins and their uses in industry and medicine.*

▶ **Rice, Risk and Regulations**

Biol. Sci. Rev., 20(2) Nov. 2007, pp. 17-20. *The genetic engineering of one of the world's most important cereal crops is an ethical concern for many.*

▶ **The Engineering of Crop Plants**

Biol. Sci. Rev., 20(4) April 2008, pp. 30-36. *Crop plants can be engineered to increase the nutritional value of foods and to improve non-food crops as sources of raw materials for industry.*

▶ **Genes Can Come True**

New Scientist, 30 Nov. 2002, pp. 30-33. *An overview of gene therapy, and a note about future directions.*

▶ **Genes, the Genome, & Disease**

New Scientist, 17 Feb. 2001, (Inside Science). *Producing gene maps, the role of introns in gene regulation, and the future of genomic research.*

▶ **Understanding the HGP**

The Am. Biology Teacher, 67(8), Oct. 2005, pp. 475-484. *Lessons designed to introduce some of the techniques used to study and manipulate DNA.*

▶ **Genomes for All**

Scientific American, Jan. 2006, pp. 32-40. *New approaches such as genome-reading technology, and ethical issues involved in using personal genetic information.*

▶ **What is…Genomics?**

Biol. Sci. Rev., 20(2) Nov. 2007, pp. 38-41. *The nature of genome projects and how genomics are being used to study the pathogenicity of bacteria.*

▶ **Birds, Bees, and Superweeds**

Biol. Sci. Rev., 17(2) Nov. 2004, pp. 24-27. *GM crops: their advantages and applications, as well as the risks and concerns associated with their use.*

▶ **Food / How Altered?**

National Geographic, May 2002, pp. 32-50. *Biotech foods: what are they, how are they made, and are they safe?*

THE ORIGIN AND EVOLUTION OF LIFE

▶ **The Ice of Life**

Scientific American, August 2001, pp. 37-41. *Space ice may promote organic molecules and may have seeded life on Earth (teacher's reference).*

▶ **A Simpler Origin of Life**

Scientific American, June 2007, pp. 24-31. *Two conflicting theories on how a complicated molecule such as RNA formed (teacher's reference).*

▶ **Primeval Pools**

New Scientist, 2 July 2005, pp. 40-43. *An ecosystem where microbes still dominate as they did millions of years in the past.*

▶ **Earth in the Beginning**

National Geographic, 210(6) Dec. 2006, pp. 58-67. *Modern landscapes offer glimpses of the way Earth may have looked billions of years ago.*

▶ **The Rise of Life on Earth**

National Geographic, 193(3) March 1998, pp. 54-81. *A series of excellent, readable articles covering the theories for the origins of life on Earth, the evolution of life's diversity, and the origin of eukaryotic cells.*

▶ **Life's Long Fuse**

New Scientist, 14 April 2007, pp. 34-38. *A look at recently discovered fossils of a peculiar form, a type of Ediacaran or fern-like rangeomorph, that existed in the Precambian era.*

▶ **A Cool Early Life**

Scientific American, Oct. 2005, pp. 40-47. *Evidence suggests that the Earth cooled sooner than once thought; as early as 4.4 bya. These cooler, wet surroundings were necessary for life to evolve.*

▶ **Meet your Ancestor**

New Scientist, 9 Sept. 2006, pp. 35-39. *The significance of a recent fossil find: the missing link between fish and tetrapods.*

▶ **The Quick and the Dead**

New Scientist, 5 June 1999, pp. 44-48. *The formation of fossils: fossil types and preservation in different environments.*

▶ **The Accidental Discovery of a Feathered Giant Dinosaur**

Biol. Sci. Rev., 20(4), April 2008, pp. 18-20. *How scientists piece together and interpret sometimes confusing fossil evidence.*

Appendix

▶ **How Old is...**

Nat. Geographic, 200(3) Sept. 2001, pp. 79-101. *A comprehensive discussion of dating methods and their application.*

▶ **Building a Phylogenetic Tree of the Human & Ape Superfamily Using DNA Hybridization Data**

The American Biology Teacher, 66(8), Oct. 2004, pp. 560-566. *A how-to-do-it activity determining genetic differences between species.*

▶ **Using Inquiry and Phylogeny to Teach Comparative Morphology**

The Am. Biology Teacher, 67(6), Aug. 2005, pp. 412-417. *A hands-on, inquiry based approach to teaching comparative vertebrate skeletal morphology.*

▶ **A Waste of Space**

New Scientist, 25 April 1998, pp. 38-39. *Vestigial organs: how they arise in an evolutionary sense and what role they may play.*

▶ **A Fin is a Limb is a Wing-How Evolution Fashioned its Masterworks**

National Geographic, 210(5) Nov. 2006, pp. 110-135. *An excellent account of the role of developmental genes in the evolution of complex organs and structures in animals. Beautifully illustrated, compelling evidence for the mechanisms of evolutionary change.*

▶ **Regulating Evolution**

Sci. American, May 2008, pp. 34-45. *Mutations in the DNA switches controlling body-shaping genes, rather than the genes themselves, have been significant in the evolution of morphological differences.*

SPECIATION

▶ **Species and Species Formation**

Biol. Sci. Rev., 20(3), Feb. 2008, pp. 36-39. *A feature covering the definition of species and how new species come into being through speciation.*

▶ **Evolution: Five Big Questions**

New Scientist, 14 June 2003, pp. 32-39, 48-51. *A synopsis of the five most common points of discussion regarding evolution and the mechanisms by which it occurs.*

▶ **Was Darwin Wrong?**

National Geographic, 206(5) Nov. 2004, pp. 2-35. *An account of the scientific evidence for evolution. A good way to remind students that the scientific debate around evolutionary theory is associated with the mechanisms by which evolution occurs, not the fact of evolution itself.*

▶ **Optimality**

Biol. Sci. Rev., 17(4), April 2005, pp. 2-5. *Environmental stability and optimality of structure and function can explain evolutionary stasis in animals. Examples are described.*

▶ **Skin Deep**

Scientific American, October 2002, pp. 50-57. *This article examines the evolution of skin colour in humans and presents powerful evidence for skin colour ("race") being the end result of opposing selection forces. Of high interest, this is a must for student discussion and a perfect vehicle for examining natural selection.*

▶ **Fair Enough**

New Scientist, 12 Oct. 2002, pp. 34-37. *Skin color in humans: this article examines the argument for there being a selective benefit to being dark or pale in different environments.*

▶ **The Moths of War**

New Scientist, 8 Dec. 2007, pp 46-49. *New research into the melanism of the peppered moth reaffirms it as an example of evolution.*

▶ **Black Squirrels**

Biol. Sci. Rev., 21(2) Nov. 2008, pp. 39-41. *A recently recognized black morph of the gray squirrel is becoming regionally more common in Britain.*

▶ **Polymorphism**

Biol. Sci. Rev., 14(1) Sept. 2001, pp. 19-21. *A good account of genetic polymorphism. Examples include the carbonaria gene (Biston), the sickle cell gene, and aphids.*

▶ **MRSA: A Hospital Superbug**

Biol. Sci. Rev., 19(4) April 2007, pp. 30-33. *An excellent account of how the evolution of MRSA has been driven by the misuse of antibiotics.*

▶ **Animal Attraction**

National Geographic, July 2003, pp. 28-55. *An engaging and expansive account of mating in the animal world.*

▶ **The Hardy-Weinberg Principle**

Biol. Sci. Rev., 15(4), April 2003, pp. 7-9. *A succinct explanation of the basis of the Hardy-Weinberg principle, and its uses in estimating genotype frequencies and predicting change in populations.*

▶ **Evolution in New Zealand**

Biol. Sci. Rev., 21(3) Feb. 2009, pp. 33-37. *NZ offers a unique suite of case studies in evolution*

▶ **The Cheetah: Losing the Race?**

Biol. Sci. Rev., 14(2) Nov. 2001, pp.

7-10. *The inbred status of cheetahs and its evolutionary consequences.*

▶ **Cichlids of the Rift Lakes**

Biol. Sci. Rev., 14(2) Nov. 2001, pp. 7-10. *The inbred status of cheetahs and its evolutionary consequences.*

▶ **Listen, We're Different**

New Scientist, 17 July 1999, pp. 32-35. *An account of speciation in periodic cicadas as a result of behavioural and temporal isolating mechanisms.*

▶ **Speciation**

Biol. Sci. Rev., 16(2) Nov. 2003, pp. 24-28. *An excellent account of speciation. It covers the nature of species, reproductive isolation, how separated populations diverge, and sympatric speciation. Case examples include the cichlids of Lake Victoria and the founder effect in mynahs.*

PATTERNS OF EVOLUTION

▶ **Evolution: Five Big Questions**

New Scientist, 14 June 2003, pp. 32-39, 48-51. *A synopsis of the five most common points of discussion regarding evolutionary mechanisms.*

▶ **Which Came First?**

Scientific American, Feb. 1997, pp. 12-14. *Shared features among fossils; convergence or common ancestry?*

▶ **What Missing Link?**

New Scientist, 1 March 2008, pp. 35-41. *An account of how the fossil record reveals large scale evolutionary patterns.*

▶ **Ancient DNA Solves Sex Mystery of Moa**

Australasian Science, 25(8), Sept. 2004, pp. 14-16. *Recent mtDNA analyses of moa indicate reverse sexual dimorphism and have resulted in a reassessment of the number of moa species represented in New Zealand.*

▶ **The Rise of Mammals**

National Geographic, 203(4), pp. April 2003, p. 2-37. *An account of the adaptive radiation of mammals and the significance of the placenta in mammalian evolution.*

▶ **Mass Extinctions**

New Scientist, 11 Dec. 1999 (Inside Science). *The nature and likely causes of the mass extinctions of the past.*

▶ **The Sixth Extinction**

National Geographic, 195(2) Feb. 1999, pp. 42-59. *High extinction rates have occurred five times in the past. Human impact is driving the sixth extinction.*

Appendix

INDEX OF LATIN & GREEK ROOTS

INDEX OF LATIN & GREEK ROOTS, PREFIXES, & SUFFIXES

Many biological terms have a Latin or Greek origin. Understanding the meaning of these components in a word will help you to understand and remember its meaning and predict the probable meaning of new words. Recognizing some common roots, suffixes, and prefixes will make learning and understanding biological vocabulary easier.

The following terms are identified, together with an example illustrating their use in biology.

a(n)- without .... anoxic
ab- away from .... abductor
ad- towards .... adductor
affer- carrying to .... afferent
amphi- both .... amphibian
amyl- starch .... amylase
anemo- wind .... anemometer
ante- before .... antenatal
anthro- human .... anthropology
anti- against, opposite .... antibiotic
apo- separate, from .... apoenzyme
aqua- water .... aquatic
arach- spider .... arachnoid
arbor- tree .... arboreal
arch(ae/i)- ancient .... Archaea
arthro- joint .... arthropod
artic- jointed .... articulation
artio- even-numbered .... artiodactyl
auto- self .... autologous
avi- bird .... avian
axi- axis .... axillary

blast- germ .... blastopore
brachy- short .... brachycardia
brady- slow .... bradycardia
branch- gill .... branchial
bronch- windpipe .... bronchial
bucca- mouth cavity .... buccal

caec- blind .... caecum
card- heart .... cardiac
cauda- tail .... caudal
centi- hundred .... centimorgan
ceph(al)- head .... cephalothorax
cera(s)(t)- horn .... ceratopsian
cerebro- brain .... cerebrospinal

cerv- neck .... cervix
chrom- color .... chromoplast
chym- juice .... chyme
cili- eyelash .... cilia
cloaca- sewer .... cloacal
coel- hollow .... coelomate
contra- opposite .... contraception
cotyl- cup .... hypocotyl
crani- skull .... cranium
crypt- hidden .... crptic
cyan- blue .... cyanobacteria
cyt- cell .... cytoplasm

dactyl- finger .... polydactylic
deci-(a) ten .... decibel, decapod
dendr- tree .... dendrogram
dent- tooth .... edentate
derm- skin .... pachyderm
di- two .... dihybrid
dors- back .... dorsal
dur- hard .... dura mater

echino- spiny .... echinoderm
ecto- outside .... ectoderm
effer- carrying away .... efferent
endo- inside .... endoparasite
equi- horse, equal .... equilibrium
erect- upright .... Homo erectus
erythr- red .... erthyrocyte
eu- well, very .... eukaryote
eury- wide .... eurythermal
ex- out of .... explant
exo- outside .... exoskeleton
extra- beyond .... extraperitoneal

foramen- opening ..... foramen magnum

gast(er)- stomach, pouch ..... gastropod
gymn- naked .... gymnosperm

hal- salty .... halophyte
haplo- single, simple .... hapolid
holo- complete, whole .... holozoic
hydr- water .... hydrophyte
hyper- above .... hypertoic
hypo- beneath .... hypotonic

infra- under .... infrared
inter- between .... interspecific
intra- within .... intraspecific
iso- equal .... isotonic

kilo- thousand .... kilogram

labi- lips .... labial palps
lacuna- space .... lacunae
lamella- leaf, layer .... lamellar bone
leuc- white .... leu(k)cocyte
lip- fat .... lipoprotein
lith- stone .... Palaeolithic
lumen- cavity .... lumen
lute- yellow .... corpus luteum
lymph- clear water .... lymphatic

magni- large .... magnification
mamma- breast .... mammal
mat(e)ri- mother .... maternal
mega- large .... megakaryocyte
melan- black .... melanocyte
meso- middle .... Mesolithic
meta- after .... metamorphosis
micro- small .... microorganism
milli- thousand .... millimetre
mirabile- wonderful .... rete mirabile
mono- one .... monohybrid
morph- form .... morphology
motor- mover .... motor nerve
multi- many .... multicellular
myo- muscle .... myofibril

necro- dead .... necrosis
neo- new .... Neolithic
nephr- kidney .... nephro
neur- nerve .... neural
notho- southern .... Nothofagus
noto- back, south .... notochord

oecious- house of .... monoecious
oed- swollen .... oedema
olfact- smelling .... olfactory
opistho- behind .... opisthosoma
os(s/t)- bone .... osteocyte
ovo- egg .... ovoviviparous

pachy- thick .... pachyderm
palae- old .... Palaeocene
pect(or)- chest .... pectoral fin
ped- foot .... quadraped
pent- five .... pentose sugar
per(i)- through, beyond .... peristalsis
peri- around .... periosteum
phaeo- dark .... phaeomelanin
phag- eat .... phagosome
phyll- leaf .... sclerophyll

Appendix

MULTIPLES AND SI UNITS

physio- nature .... physiology
phyto- plant ....phytohormone
pisc- fish ....piscivorous
plagio- oblique .... plagioclimax
pneu(mo/st)- air, lung ....pneumonia
pod- foot ....sauropod
poly- many .... polydactyly
pre- before ....premolar
pro- in front of ....Prokaryote
prot- first .... protandry
pseud- false .... pseudopodia
pter- wing, fern ....Pterophyta
pulmo- lung ....pulmonary

radi- root ....radicle
ren- kidney ....renal
retic- network ....reticulated
rhin- nose, snout ....rhinoceros
rostr- beak, prow ....rostrum

sacchar- sugar .... polysaccharide

schizo- split ....schizocoelomate
scler- hard ....sclerophyll
seba- tallow, wax .... sebaceous
semi- half ....semi-conservative
sept- seven, wall ....septum
soma- body ....opisthosoma, somatic
sperm- seed ....spermatophyte
sphinct- closing ....sphincter
stereo- solid ....stereocilia
stom- mouth .... stoma
strat- layer .... stratification
sub- below .... subtidal
sucr- sugar .... sucrase
sulc- furrow .... sulci
super- beyond ....superior
supra- above ....supracoracoideus
sym- with ....symbiosis
syn- with ....synapsis

tact- touch .... tactile
tachy- fast ....tachycardia

taenia- ribbon ....Taenia (tapeworm)
trans- across .... transmembrane
tri- three .... triploblastic
trich- hair ....trichome

ultra- above .... ultraviolet
un- one .... unicellular
uro- tail ....urodele

vas- vessel .... vascular
ven- vein .... venous
ventr- belly .... ventral
vern- spring .... vernal
visc- organs of body cavity ....viscera
vitr- glass ....in vitro

xanth- yellow ....xanthophyll
xen- stranger ....xenotransplant
xer- dry .... xerophyte
xyl- wood ....xylem
zo- animal ....zoological

TEST YOURSELF!
Use the index of Latin and Greek terms to deduce the meaning of the following terms:

1. *Sclerophyll*: _____

2. *Osteocyte*: _____

3. *Polydactyly*: _____

4. *Gymnosperm*: _____

5. *Tachycardia*: _____

MULTIPLES

MULTIPLE	PREFIX	SYMBOL	EXAMPLE
10^9	giga	G	gigawatt (GW)
10^6	mega	M	megawatt (MW)
10^3	kilo	k	kilogram (kg)
10^2	hecto	h	hectare (ha)
10^{-1}	deci	d	decimetre (dm)
10^{-2}	centi	c	centimetre (cm)
10^{-3}	milli	m	milliimetre (mm)
10^{-6}	micro	μ	microsecond (μs)
10^{-9}	nano	n	nanometre (nm)
10^{-12}	pico	p	picosecond (ps)

INTERNATIONAL SYSTEM OF UNITS (SI)

Examples of SI derived units

DERIVED QUANTITY	NAME	SYMBOL
area	square metre	m^2
volume	cubic metre	m^3
speed, velocity	metre per second	ms^{-1}
acceleration	metre per second squared	ms^{-2}
mass density	kilogram per cubic metre	kgm^{-3}
specific volume	cubic meter per kilogram	m^3kg^{-1}
amount-of-substance concentration	mole per cubic meter	$molm^{-3}$
luminance	candela per square meter	cdm^{-2}

Index

Index

    © Biozone International 2001-2010